Handbook of Fibrous Materials

Handbook of Fibrous Materials

Edited by
Jinlian Hu
Bipin Kumar
Jing Lu

Volume 2

WILEY-VCH

Editors
Jinlian Hu
The Hong Kong Polytechnic University
Institute of Textiles and Clothing
Room QT715, Q Core, 7/F
Hung Hom
Kowloon 999077
Hong Kong S.A.R.
P.R. China

Bipin Kumar
Indian Institute of Technology Delhi
Department of Textile and Fibre Engineering
TX135, Hauz Khas
New Delhi 110016
India

Jing Lu
The Hong Kong Polytechnic University
Institute of Textiles and Clothing
Room QT807
Hung Hom
Kowloon 999077
Hong Kong S.A.R.
P.R. China

Library of Congress Card No.:
applied for

British Library Cataloguing-in-Publication Data
A catalogue record for this book is available from the British Library.

Bibliographic information published by the Deutsche Nationalbibliothek
The Deutsche Nationalbibliothek lists this publication in the Deutsche Nationalbibliografie; detailed bibliographic data are available on the Internet at <http://dnb.d-nb.de>.

Print ISBN: 978-3-527-34220-4
ePDF ISBN: 978-3-527-34256-3
ePub ISBN: 978-3-527-34259-4
oBook ISBN: 978-3-527-34258-7

Cover Design Adam-Design, Weinheim, Germany
Typesetting SPi Global, Chennai, India
Printing and Binding Markono

Printed on acid-free paper

10 9 8 7 6 5 4 3 2 1

Contents

Preface

The art of using a fiber material is as old as human civilization. Initially, they were primarily used to make textiles such as yarn or fabric, serving the basic requirement of garment, storage, protection, building, ropes, fishing nets, etc. In textile industry, fibers are characterized by having a length at least 100 times its diameter (ASTM D123-15) and offer several unique features due to its unique configuration – high flexibility, high specific surface area, easy transformability into textile structures, and high load-bearing ability along axial direction – which makes them perfect material for apparel use. Further with the development of advanced materials and their technologies, it is now possible to generate novel fibers, which may not possess the same properties like textile fibers (cotton, PET, Nylon, wool) but have unique functions such as shape memory, superhydrophobicity, phase change, optics, and conductivity that could solve new scientific and technological challenges of different advanced fields. This book focuses on the research and development in fibrous materials and innovative application potentials. Each of the chapters is exclusive and selectively chosen, edited by the internationally recognized experts in the field to cover latest developments and future trends in their fields.

Chapter 1 introduces the different types of fibrous materials including natural and synthetic fibers and their fundamental characteristics. Chapters 2–6 include topics from animal fibers (wool, silk, collagen, chitosan) and plant fibers (cellulose). Chapters 7–16 focus on research and application of different synthetic fibers including electrospun fibers, aramid fibers, conductive fibers, phase change fibers, bicomponent fibers, superabsorbent fibers, elastic fibers, smart fibers, optical fibers, and memory fibers. Chapters 17–19 describes the scientific principles and technologies for converting fibrous materials into textile yarns and fabrics. Chapters 20 and 21 list some characterization methods for analyzing fibrous materials and their structures. More emphasis is given on the wide application potential and scope of advanced fibrous materials in electronics (Chapter 22), optics (Chapter 23), energy (Chapter 24), sensors and actuators (Chapter 25), wearable (Chapter 26), medical (Chapter 27), environmental protection (Chapter 28), filtration (Chapter 29), protection (Chapters 30 and 32), health (Chapter 33), and auxetic (Chapter 34).

The content has been designed to cover different types of advanced fibers, their materials, and devices as well as different properties, diversified functions, and applications.

Finally, we would like to acknowledge the time and efforts of our contributors, who are experts in the respective areas described in this book.

Jinlian Hu
The Hong Kong Polytechnic University

20

Chemical Characterization of Fibrous Materials

Chi-wai Kan and Ka-po Maggie Tang

The Hong Kong Polytechnic University, Institute of Textiles and Clothing, Hung Hom, Kowloon, Hong Kong, China

20.1 Introduction

New research trends focus on enhancing the functionality of fibrous materials. In the past, fibrous materials were mainly used for clothing to give warmth to the wearer, filter paper for household or industrial purposes, or hygiene products. Nowadays, the development of fibrous materials is rapid, and many advanced applications are found in the market, such as electronic textiles, medical textiles, and self-cleaning textiles.

Electronic textiles refer to a textile substrate with electrical functioning similar to electronics, which physically functions as a textile product [1]. Functional electronic textiles should be lightweight, should be comfortable to wear, and should have easy care properties [2, 3]. Conducting polymers, because of their lightweight, processability, relatively high conductivity, stability, and flexibility, are well suited for fabrication of conductive textiles [4].

Medical textiles usually have antimicrobial properties that inhibit the growth of microorganisms and can minimize a range of undesirable effects not only on the textile itself but also on the wearer. These effects include unpleasant odors, color degradation, allergic responses, deterioration of the textiles, and even potential health risks [5, 6]. A number of antimicrobial agents have been developed to impart antimicrobial properties to textile goods. These include silver, inorganic salts, organometallics, iodophors (substance that slowly releases iodine), phenols and thiophenols, onium salts, antibiotics, heterocyclics with anionic groups, nitro compounds, urea and related compounds, formaldehyde derivatives, and amines [5]. Chitosan [7], natural dyes [8], and other natural herbal products such as aloe vera, tea tree oil, eucalyptus oil, and tulsi leaf extracts can also provide antimicrobial effects.

Self-cleaning material is water, oil, and soil repellent. The superhydrophobicity is achieved by chemical and geometrical modification of the surface. By creating micro- and nanoscale roughness accompanied by a hydrophobic coating, water, and dirt on the surface can be removed easily, thus keeping the surface clean [9]. On the other hand, photocatalytic self-cleaning is based on chemical breakdown

Handbook of Fibrous Materials, First Edition. Edited by Jinlian Hu, Bipin Kumar, and Jing Lu.

of dirt through photooxidation and photoreduction reactions in the presence of light. Titanium dioxide or titania is the most frequently used substance in this field [10]. Surfaces exhibit superhydrophobicity when the contact angle is greater than 150°, and contact angle hysteresis and sliding angle are low [11]. Self-cleaning materials have attracted tremendous attention in academic research and industrial areas [12].

The three advanced applications mentioned above suggest that nowadays more sophisticated treatment are being used to achieve these advanced features. The following sections describe the chemical finishing applied for these advanced applications, followed by discussion of the chemical characterization methods including Fourier transform infrared (FTIR) spectroscopy, X-ray diffraction (XRD), X-ray photoelectron spectroscopy (XPS), X-ray fluorescence, chromatographic methods, and energy-dispersive X-ray analysis (EDX). Apart from that, the performance of the treated samples and their applications are also described.

20.2 Chemical Finishing of Fibrous Materials for Advanced Applications

20.2.1 Electronic Textiles

Electronic textiles can be produced with conductive fibers, conductive yarns, or conductive inks. Properties of electronic textiles depend on the type of conductive materials used and the manufacturing method. Metallic coatings, sputtering, vacuum deposition, and filling fibers are used commonly. Applying carbon black, carbon nanotubes (CNTs), and intrinsically conducting polymers (ICPs) to a fiber surface through solution casting, inkjet printing, chemical vapor deposition, or vapor-phase polymerization methods can improve the quality of conductive fibers [13].

20.2.1.1 Conductive Fibers

Depending on the manufacturing processes adopted, conductive fibers can be divided into *intrinsic* or *extrinsic* [14], as shown in Table 20.1.

20.2.1.2 Conductive Fabrics

Apart from conductive fibers, another method of achieving conductivity is coating yarns or fabrics with a metal layer. A fine metal layer such as copper (Cu), silver (Ag), or gold (Au) can be deposited on the surface of the fibers by an electrochemical process [21]. The electrical conductivity of Cu, 99% Ag, and bronze is 58.5, 62.5, and 7.5 S·m/mm^2, respectively [22].

20.2.1.3 Conductive Inks

Conductive ink is usually applied to fibrous materials through a coating process. It must contain a highly conductive metal precursor such as Ag, Cu, or Au nanoparticles, specific metal alloys, core–shell systems, and a carrier vehicle. Most of them are water based and can be printed onto various materials to

Table 20.1 List of conductive fibers.

Intrinsic	– Metallic and metallic alloy fibers are very thin metal filaments. They are brittle, and it is difficult to incorporate them into clothing, and they are not comfortable to wear. Also, they can damage the spinning machinery easily [15]. – Carbon fibers have a graphite-like structure [16]. – Intrinsically conducting polymers (ICPs) are organic conductive materials based on polyaniline (PANI), poly(3,4-ethylenedioxithiophene):polystyrene sulfonate (PEDOT:PSS), or polypyrrole (PPy) [17, 18]. PEDOT is a well-known and well-studied intrinsically conductive polymer (ICP). When coupled with acidic PSS, it forms a conductive, solution processable ionomer.
Extrinsic	– Conductive filled fibers, these are made by adding conductive fillers (e.g. metallic powder, metallic nanowires, carbon nanotubes (CNTs), or ICPs into nonconductive polymers such as polypropylene, polystyrene, or polyethylene [19]. These fibers are commonly made by melt spinning and solution spinning. – Conductive coated fibers are produced by coating insulating materials with conductive ones (e.g. carbon black, metals, CNTs, or ICPs) [20].

create electrically active patterns [22]. However, printing on the rough, uneven, and porous surface of the fabric can be an issue [23].

The chemical structure of conductive polymer is shown in Figures 20.1 and 20.2. Table 20.2 summarizes literature about chemicals used and treatment methods for electronic textiles.

20.2.2 Medical Textiles

For medical textiles, an active agent can be added to the fabric during different processing stages.

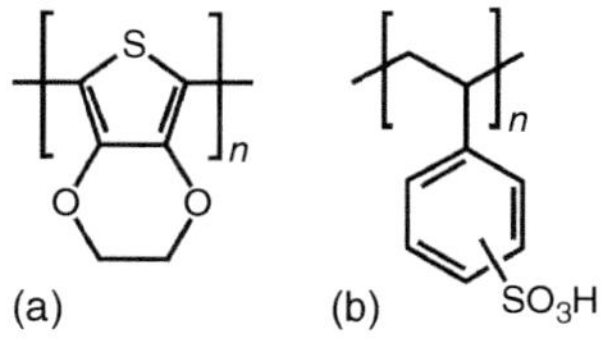

Figure 20.1 Chemical structure of (a) PEDOT and (b) PSS. Source: Irwin et al. 2011 [24]. Reproduced with permission of Springer Nature.

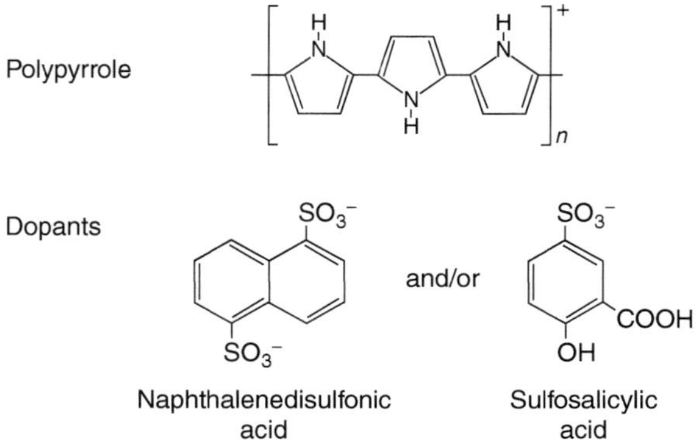

Figure 20.2 Chemical composition of electrically conductive polymer coating on Contex textiles. Source: Heisey et al. 1993 [25]. Reproduced with permission of SAGE.

Table 20.2 Literatures about chemicals used and treatment method for electronic textiles.

	Authors	Chemical used	Treatment
1	Irwin et al. [24]	The ionomer mixture of poly(3,4-ethylenedioxythiophene) and poly(styrenesulfonate) (PEDOT:PSS; 1 : 2.5, w:w) was cast onto silk fibers from a 50 : 50 (v:v) ethylene glycol solution	Dip coating
2	Yamashita et al. [26]	(1) Coating polyethylene terephthalate (PET) ribbon cable with hydrophilic poly(3,4-ethylenedioxythiophene): poly(4-styrenesulfonate) (PEDOT:PSS) followed by hydrophobic PEDOT:PSS using die-coating method (2) The PEDOT:PSS is thermally cured (3) A 10 μl drop of silicone emulsion is dropped on the hydrophobic PEDOT:PSS-coated cable, and it is then annealed	Coating
3	Rehnby and coworkers [27]	Conductive polymers (polyaniline, polythiophene, and polypyrrole) were mixed with an acrylic binder polymer and coated on a polyester fabric	Coating
4	Wu et al. [28]	Nylon Lycra fabric was soaked in an aqueous solution containing both monomer and dopant. Then an oxidant solution was added into the container to trigger polymerization on the fabric surface, followed by washing	—
5	Ding et al. [17]	(1) Soaking Spandex fabric in a conductive polymer aqueous dispersion, PEDOT–PSS (2) Secondary doping with D-sorbitol enhances the conductivity of the fabric	—
6	Kazani et al. [29]	Electrodag PF 410 conductive ink (consist of PES resin with silver particles) and 5025 conductive ink by Du-Pont (Epoxy resin with silver particles)	Screen printing
7	Karaguzel et al. [30]	(1) Different conductive silver inks of different viscosities and percentages of silver (2) The printed nonwoven fabrics were laminated using thermoplastic urethane (TPU) meltblown layer in order to secure the printed ink traces in place	Screen printing
8	Heisey et al. [25]	Polypyrrole-coated nylon	Coating
9	Lee et al. [31]	Nanosized silver colloids, synthesized by chemical reduction with formaldehyde, were dispersed by a cosolvent system made of diethylene glycol and water to get a 25 wt% silver ink	Ink-jet printing
10	Lin et al. [32]	– The polyester textile was initially immersed in either pyrrole/anthraquinone-2-sulfonic acid sodium salt (AQSA) (monomer/dopant) or a ferric chloride (oxidant) aqueous solution for a specified period of time before the polymerization reaction took place – Mixing the reactant solutions initialized the polymerization reaction – Two competitive processes occurred simultaneously in the reaction: polymerization on the surface of the substrate and in the solution.	—
11	Garg et al. [33]	Polymerization was carried out in an aqueous solution that contained pyrrole, AQSA monohydrate 97% (Sigma–Aldrich), ferric chloride hexahydrate, and Albegal FFA (Ciba Specialty Chemicals), resulting in a black PPy coating on the fabric surface	Coating

20.2.2.1 Add into the Initial Polymer Solution or Chemical Chain Prior to Fiber Extrusion

Numerous natural polymers, such as collagen, chitosan (Figure 20.3), and alginate (Figure 20.4), can be used for medical applications, including wound dressings, nonwoven felts, and meshes and sutures. These are bioabsorbable materials composed of cross-linked polymer chains and are degraded through hydrolysis of bonds, for example, peptide, hemiacetal, ester, and phosphate in the main chain into the water-soluble low molecular weight compounds. These are then absorbed by the surrounding body fluids. Similarly, anhydride, carbonate, ester, and orthoester hydrolysable units containing synthetic polymers are also bioabsorbable into the body tissues [35]. Chitosan (Figure 20.3), a major component of crustacean shells, is an effective natural antimicrobial agent derived from chitin. Burnett-Boothroyd and McCarthy [34] reported that combining these with kelp (seaweed) can increase the likelihood that the wound will heal more rapidly (Figures 20.5–20.7).

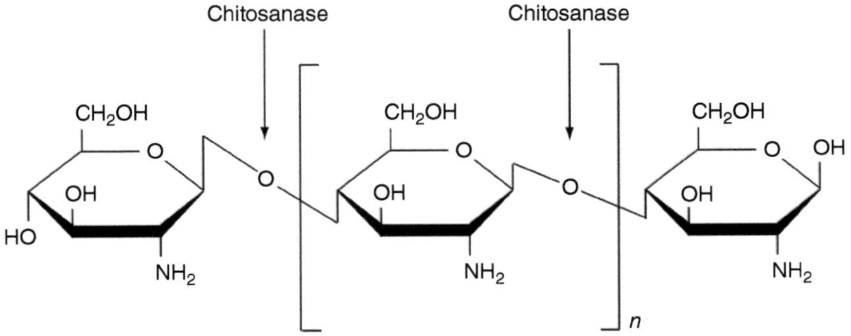

Figure 20.3 Chemical structure of chitosan polymer of β-(1-4)-D-glucosamine units. Source: Burnett-Boothroyd and McCarthy 2011 [34]. Reproduced with permission of Elsevier.

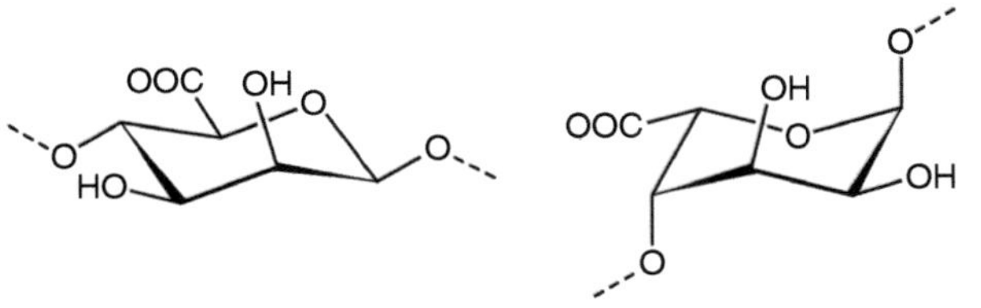

Figure 20.4 Chemical structure of alginate. Source: Rajendran et al. 2016 [35]. Reproduced with permission of Taylor & Francis.

Figure 20.5 Triclosan. Source: Burnett-Boothroyd and McCarthy 2011 [34]. Reproduced with permission of Elsevier.

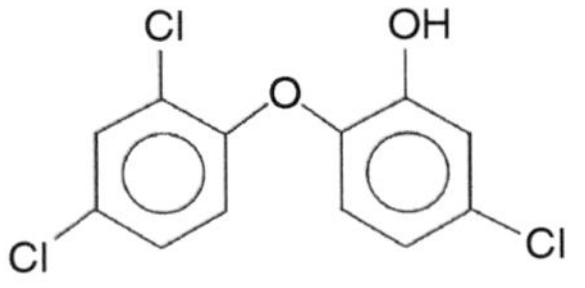

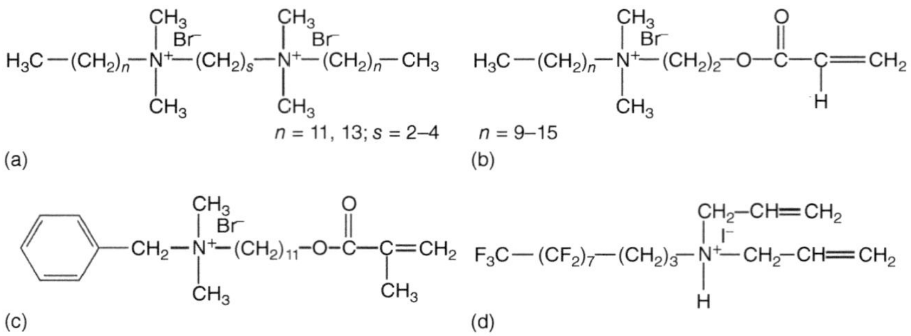

Figure 20.6 Chemical structure of quaternary ammonium compounds (QACs) used for textile applications: (a) diquaternary ammonium salt (alkanediyl-α, ω-bis(dimethylalkyl ammonium bromide)), (b) alkyl(2-(acryloyloxy)ethyl)dimethyl ammonium bromide, (c) benzyl(11-(acryloyloxy)undecyl) dimethyl ammonium bromide, and (d) *N*-(4,4,5,5,6,6,7,7,8,8,9,9,10,10,11,11-heptadecafluoroundecyl)-*N*,*N*-diallylmethylammonium iodide. Source: Nayak and Padhye 2015 [36]. Reproduced with permission of Elsevier.

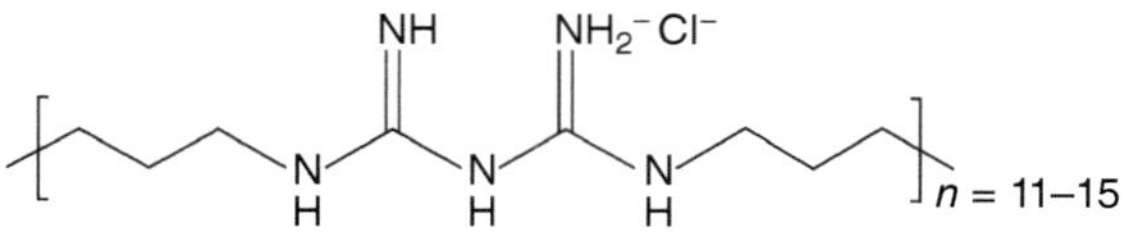

Figure 20.7 Structure of polyhexamethylene biguanide (PHMB). Source: Nayak and Padhye 2015 [36]. Reproduced with permission of Elsevier.

20.2.2.2 During Textile Finishing

The finishing agent can be applied to the textile substrate by pad-dry-cure, coating, spraying, and foam techniques and by exhaust application via a dyebath. Pad-dry-cure technique is the most commonly used. These techniques in most cases result in nondurable finishes. There are generally three mechanisms to treat fibers and fabrics for antimicrobial properties: (i) a controlled release mechanism, (ii) a regeneration principle, or (iii) a barrier or blocking action [37].

With a view to develop antimicrobial textile materials, considerable research has been carried out by making use of organic and inorganic compounds [38]. The chemicals used to achieve antibacterial function are summarized in Table 20.3. A variety of metals and metal oxides have been explored for antimicrobial finishing of textiles, including silver, titanium dioxide, zinc, copper, and cobalt. Research has also showed that metal nanoparticles are more effective than the corresponding bulk materials due to the smaller particle size, which provides larger surface area, leading to greater interaction with microorganisms [39]. A proposed scheme for silver nanoparticles attachment with 3-mercaptopropyltrimethoxysilane (3-MPTMS) on the glycidyltrimethylammonium chloride (GTAC)-treated cotton fiber is shown in Figure 20.8 [40]. Literatures about chemicals used and treatment methods for medical textiles are shown in Table 20.4.

Common polymers are used as hydrogels as scaffolds in tissue engineering, biosensors, disposable diapers, sanitary towels, and medical electrodes (Figure 20.9), including chitosan, poly(ethylene oxide) (PEO), poly(vinyl alcohol) (PVA), and poly(*N*-vinyl pyrrolidone) (PVP).

Table 20.3 Antimicrobial agents used for finishing of textiles.

Antimicrobial agents	Details
Chitosan (extracted from shrimps and crustacean shells)	Bactericidal mechanism: electrostatic interaction
Silk sericin (protein from silk worm)	Natural compounds based on animals
Triclosan (see Figure 20.5)	Inhibit growth of microorganisms using an electrochemical mode of action to penetrate and disrupt their cell walls
Quaternary ammonium compounds (see Figure 20.6)	Bind microorganisms to their cell membrane and disrupt the structure resulting in breakdown of the cell
Polybiguanides (see Figure 20.7)	– Polymeric polycationic amines that include cationic biguanide repeat units separated by hydrocarbon chain linkers of identical or dissimilar lengths [36] – One of the most used polybiguanides is a heterodisperse mixture of polyhexamethylene biguanides (PHMBs) (Figure 20.7)
Metallic compounds	– Metal-based, including cadmium, silver, copper, and mercury – Several heavy metals in the free state or in compounds are toxic to microbes at very low concentrations [36] – Cause inhibition of the active enzyme centers (inhibition of metabolism)
Silver	– The most used antibacterial agent [5] – A high degree of biocompatibility – An excellent resistance to sterilization conditions (usually temperatures in excess of 82 °C) – Antibacterial properties with respect to different bacteria with a long-term efficiency – Nanosilver can destruct the disease-causing microorganisms without any adverse effect on the human skin [5] – Low toxicity to humans
Dyes	– Mainly metal-based dyes. Chromium (Cr)- and copper (Cu)-based dyes on silk have shown effective antimicrobial efficacy [37] – Based on their respective molecular structures

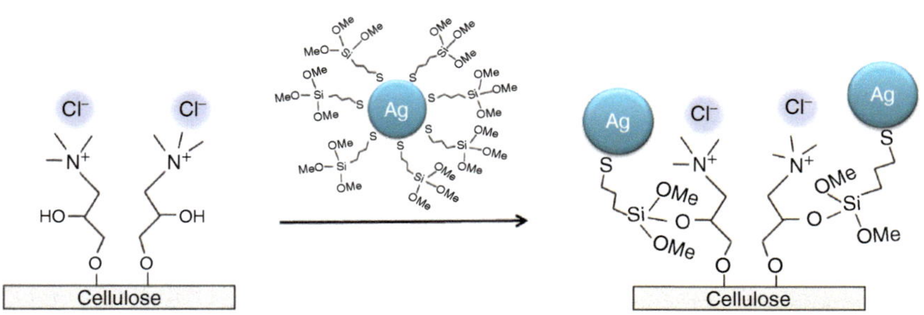

Figure 20.8 Proposed reaction scheme between silver nanoparticles/3-MPTMS and GTAC-treated cotton. Source: Kang et al. 2016 [40]. Reproduced with permission of Elsevier.

Table 20.4 Literatures about chemicals used and treatment methods for medical textiles.

	Author	Chemical used	Treatment
1	Kang et al. [40]	Cotton fibers were treated chemically with glycidyltrimethylammonium chloride (GTAC), a quaternary ammonium salt, and coated with silver nanoparticles/3-mercaptopropyltrimethoxysilane (3-MPTMS)	Coating
2	Shafei and Abou-Okeil [41]	ZnO/carboxymethyl chitosan bionano-composite	Pad-dry-cure method
3	Petkova et al. [42]	Cotton with zinc oxide nanoparticles (ZnO NPs)	Coating

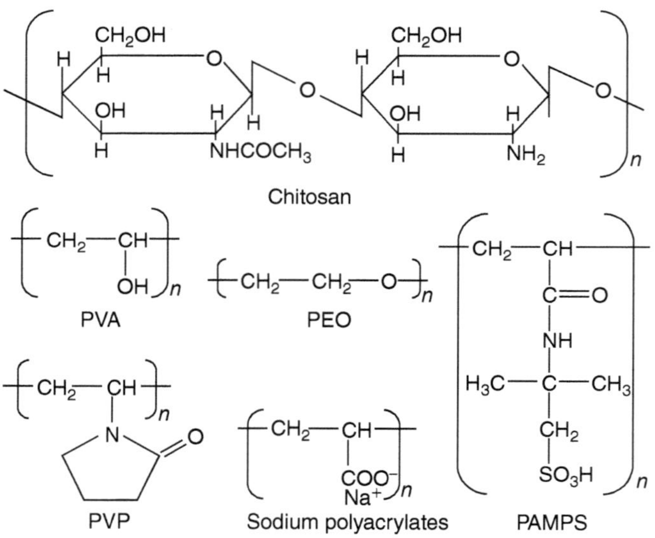

Figure 20.9 Chemical structure of polymers used as hydrogel. Source: Adapted from Gupta et al. 2010 [43].

20.2.3 Self-Cleaning Textiles

Superhydrophobic surface can be achieved by two techniques: making a rough surface from a low surface energy material or modifying a rough surface using a material of low surface energy.

The conventional padding, coating, and layer by layer (LbL) assembly technique have been used to deposit nano-coatings on textiles. Researchers have also tried to apply different materials onto the fibrous materials, including particles, rod arrays or pores, CNTs, silica particles, ZnO nanorods, and silver nanoparticles. Some examples are shown in Table 20.5.

Schematic illustration of the preparation of superhydrophobic cotton fabric in one-step treatment using EDTA and HDTMS (Figure 20.10).

Table 20.5 Literatures about chemicals used and treatment methods for self-cleaning textiles.

	Author	Chemical used	Treatment
1	Xue et al. [44]	– Coating silica nanoparticles with functional groups onto cotton textiles to generate a dual-size surface roughness – Followed by hydrophobization with stearic acid, 1H, 1H, 2H, 2H-perfluorodecyltrichlorosilane or their combination	Coating
2	Lee and Michielsen [45]	– Coating of nylon 6,6 fibers on polyester fabric – Grafted poly(acrylic acid) chains onto nylon-6,6 surfaces followed by grafting of 1H,1H-perfluorooctylamine to create a SH surface	Flock coating
3	Xu and Cai [46]	– Apply ZnO nanocrystals to cotton – Followed by fabrication of oriented ZnO nanorod arrays to develop nanoscale roughness – Fabrics were coated with dodecyltrimethoxysilane to impart surface hydrophobicity	Hydrothermal method
4	Hoefnagels et al. [47]	– Silica particles coated with amine groups were generated in situ and covalently bonded to cotton by a one or two step reaction – The amine groups were reacted with mono-epoxy functionalized polydimethylsiloxane to hydrophobize the surface	Silane-based sol–gel treatment
5	Zhao et al. [48]	Assembly of polyelectrolyte/silica nanoparticle multilayers on cotton fibers, followed by a fluoroalkylsilane treatment	Electrostatic layer by layer (LbL) assembly

A new class of self-cleaning textiles, which is superhydrophobic and photocatalytic simultaneously under visible light irradiation, is shown in Figure 20.11. Self-assembled monolayers of meso-tetra(4-carboxyphenyl)porphyrin (TCPP) were formed on TiO_2-coated cotton by a simple post-adsorption method, followed by hydrophobization with trimethoxy-(octadecyl)silane (OTMS) [50].

20.3 Principles and Methods in Chemical Characterization of Fibrous Materials

20.3.1 Fourier Transform Infrared (FTIR) Spectroscopy

Infrared spectroscopy can be used to identify compounds or investigate sample compositions. The infrared spectrum of a sample is collected by passing a beam of infrared light, in the 600–4000 cm^{-1} range, through the sample. The chemical bonds in the sample absorb specific wavelengths of the incident radiation. The amount of the transmitted light reveals how much energy is absorbed at each wavelength. A transmittance or absorbance spectrum can be produced, showing at which infrared wavelengths the sample absorbs. It depends on the nature of each molecule's vibrations or rotations. The molecular structure of the sample

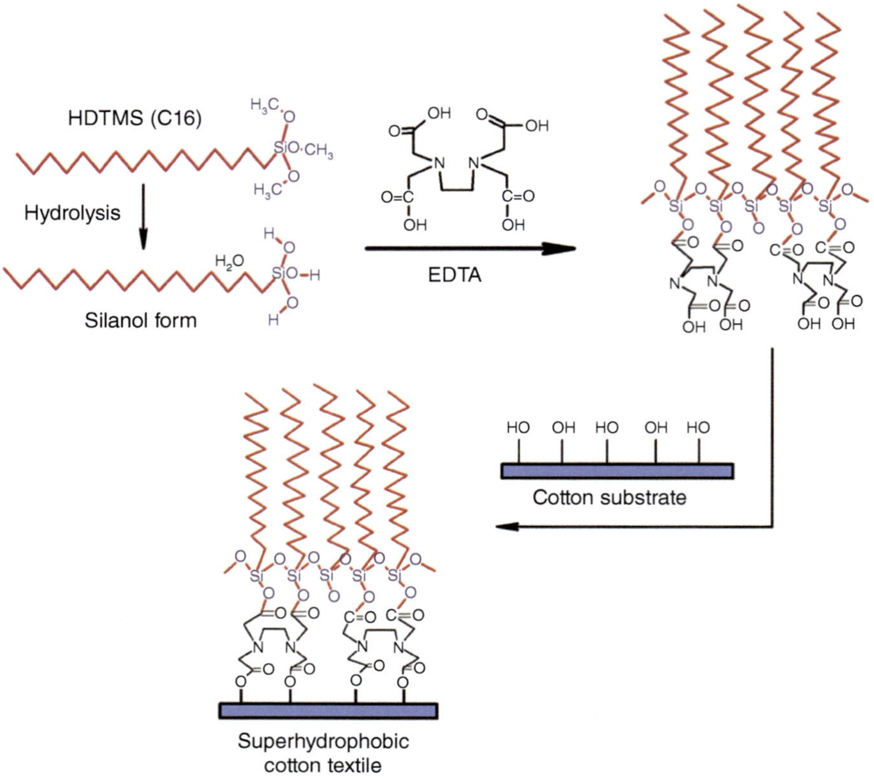

Figure 20.10 Schematic illustration of one step process for the preparation of superhydrophobic cotton textile. Source: Abbas et al. 2015 [49]. Reproduced with permission of Springer Nature.

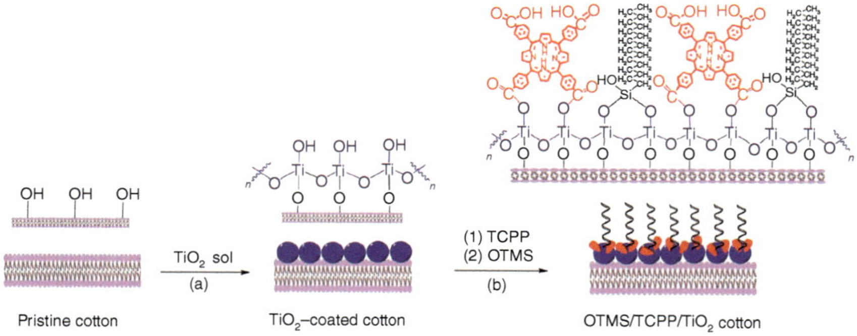

Figure 20.11 Formation of hydrophobized OTMS/TCPP/TiO_2-coated cotton. Source: Afzal et al. 2014 [50]. Reproduced with permission of Royal Society of Chemistry.

Table 20.6 Infrared frequencies associated with chemical finishes.

Absorption frequency (cm^{-1})	Chemical bond
3700–3300	O—H
3500–3200	N—H
2967–2857	C—H
2252–2062	C≡N
1750–1735	C=O (esters)
1725–1700	C=O (saturated aliphatic acids)
1725–1705	C=O (saturated aliphatic ketones)
1680–1630	C=O (amides)
1600–1500	C=C (aromatic)
1570–1515	N—H (secondary amides)
1250–1150	C—O (esters)
~1250	P=O
720–730	C—H

Source: Hauser 2005 [52]. Reproduced with permission of Elsevier.

and the presence of specific functional groups can be inferred in the transmittance or absorbance spectrum [51].

Table 20.6 lists some of the frequencies associated with chemical finishes used with textiles [52].

Krishnaveni and Thambidurai [53] confirmed the presence of chitosan by FTIR spectroscopic analysis (Figure 20.12a and b). FTIR spectra of cross-linked cotton fabric treated with chitosan (AN–CS) and chitosan–ZnO composite (AN–CZO) are shown in Figure 20.12c and d, respectively. Compared with chitosan and cotton cellulose, the new peak at 2372 cm^{-1} (Figure 20.12c) can be attributed to the C–N asymmetric band stretching [54]. The peak that appeared at 2252 cm^{-1} for AN–CZO (Figure 20.12d) is due to the stretching vibration of C≡N of the beta-cyanoethyl group resulting from the addition of acrylonitrile. The presence of ZnO in the cross-linked fabric was also confirmed from the peak that appeared at 470 cm^{-1} in Figure 20.12c [55]. This example demonstrates the use of FTIR to identify compounds or investigate sample compositions.

20.3.2 X-Ray Diffraction (XRD)

XRD is a powerful nondestructive technique for characterizing (identification/quantification) crystalline materials. It provides information on crystal structure, phase, preferred crystal orientation (texture), and other structural parameters, such as average grain size, crystallinity, strain, and crystal defects. Figure 20.13 shows the XRD pattern of bacterial cellulose (BC), polypyrrole (PPy), and a composite. The XRD spectrum of PPy exhibits no diffraction peaks, indicating its amorphous structure. Three main peaks can be identified in spectra of BC and BC/PPy, which can be assigned to the $(1\bar{1}0)$, (110), and (200) diffraction planes, which is typical for cellulose I [57, 58]. The estimated degree

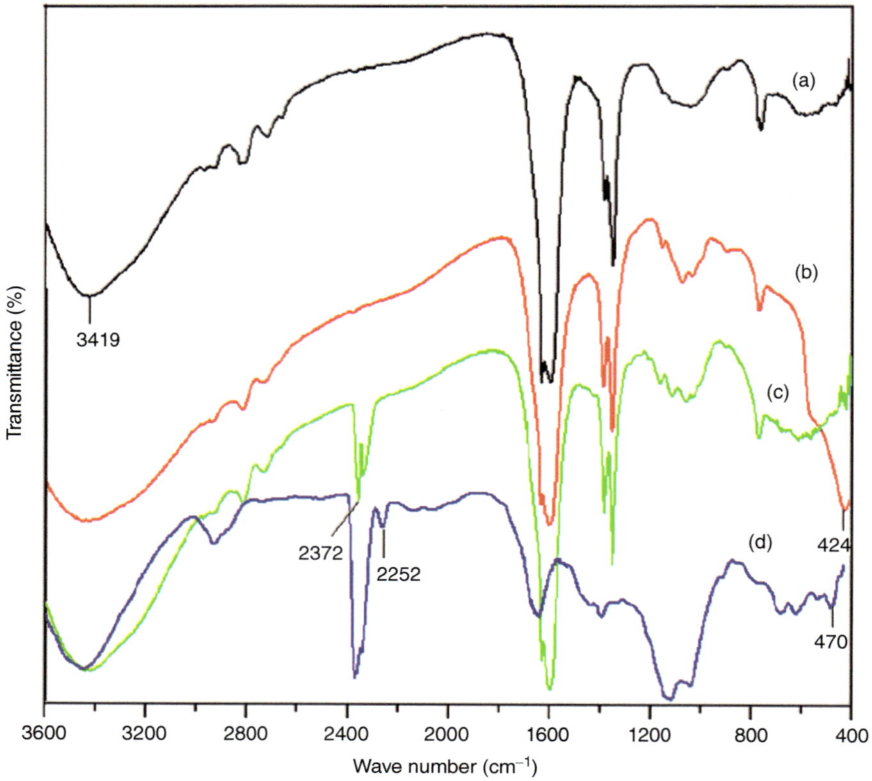

Figure 20.12 The FTIR spectrum of (a) chitosan, (b) CZO, (c) AN–CS, and (d) AN–CZO. Source: Krishnaveni and Thambidurai [53]. Reproduced with permission of Elsevier.

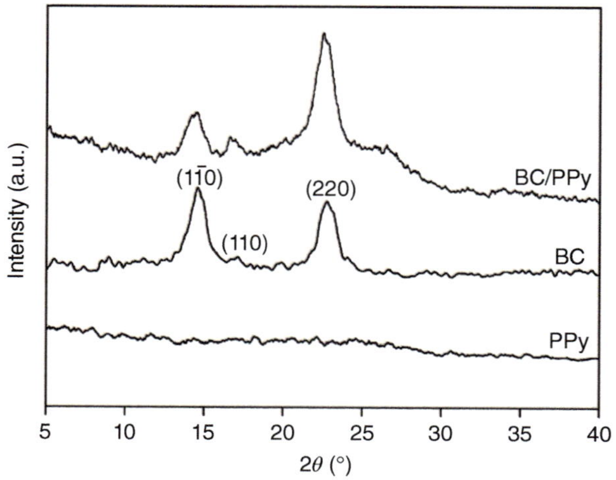

Figure 20.13 XRD spectra of pure BC, pure PPy, and a BC/PPy composite. Source: Müller et al. 2011 [56]. Reproduced with permission of Elsevier.

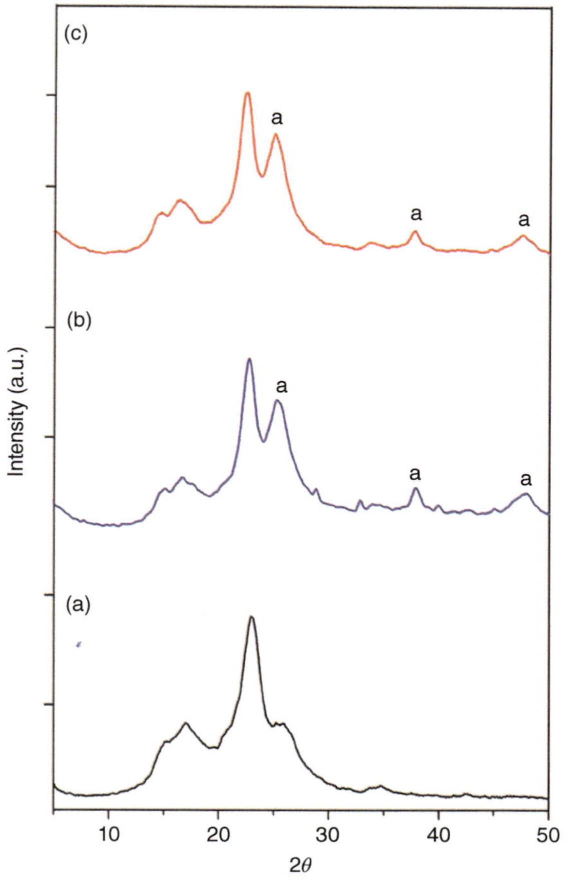

Figure 20.14 XRD spectra of cotton: (a) pristine, (b) TiO_2-coated cotton, (c) OTMS/TCPP/TiO_2-coated cotton (a = peaks associated with anatase). Source: Afzal et al. 2014 [50]. Reproduced with permission of Royal Society of Chemistry.

of crystallinity was 56 ± 5% for BC membrane and 51 ± 5% for the BC/PPy composite. This demonstrates that the crystalline structure of cellulose matrix was not significantly modified in the BC/PPy composites.

Another research work used XRD to determine crystallinity of TiO_2 films on cotton. Figure 20.14a shows that there is no anatase-associated peak for the pristine cotton while cotton samples coated with TiO_2 show the characteristic diffraction peaks for anatase at $2\theta = 25.4°$, 38.0°, and 48.0° (Figure 20.14b). After deposition of TCPP and OTMS, the same anatase peaks can still be observed (Figure 20.14c). Hence, TiO_2 retains its crystallinity even after modification with TCPP and OTMS [50].

20.3.3 X-Ray Photoelectron Spectroscopy (XPS)

XPS is also known as electron spectroscopy for chemical analysis (ESCA). Photoelectron spectroscopy uses photo-ionization and energy-dispersive analysis of the emitted photoelectrons to study the composition and electronic state of the surface region of a sample. The ejected electron energies are characteristic of

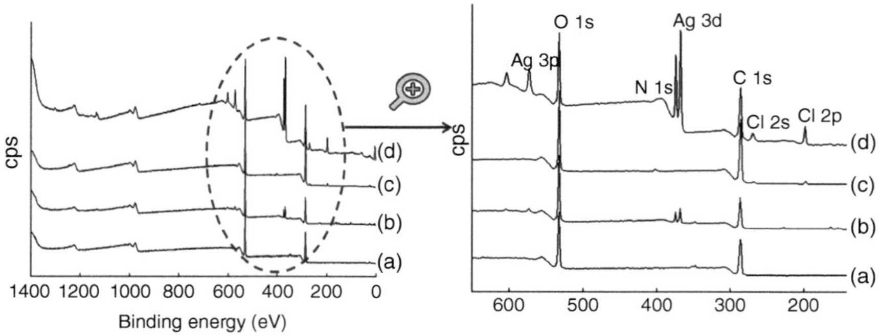

Figure 20.15 XPS spectra of (a) naïve cotton, (b) cotton treated with silver nanoparticles/0.1% (v/v) of 3-MPTMS, (c) cotton treated with 50% (v/v) of GTAC, (d) treated cotton with 50% (v/v) of GTAC and silver nanoparticles/0.1% (v/v) of 3-MPTMS. Source: Kang et al. 2016 [40]. Reproduced with permission of Elsevier.

binding energies of the atoms on the material's surface. The shapes of the peaks in the electron energy spectrum are influenced by chemical bonding, so XPS can provide both elemental and chemical bond information [51]. This can be used to characterize some treatments that are restricted to the surface of the material, such as plasma treatment and conductive polymer coating [25].

Kang et al. [40] used XPS to confirm the presence of silver in cotton fibers treated with 3-MPTMS/silver nanoparticles (Figure 20.15); the binding energy of the silver nanoparticles was 368 and 374 eV, which were assigned to Ag [59].

Figure 20.16 shows XPS results of various cotton samples. Ordinary cotton exhibits distinctive signals at 285.0, 532.0, and 400 eV indicating the presence of carbon, oxygen, and nitrogen, respectively. With Ar and H_2 plasma treatment, XPS results show an increase in carbon (C) and a decrease in oxygen (O) content, whereas the signal for oxygen plasma-pretreated cotton shows values comparable with pristine cotton [60].

20.3.4 X-Ray Fluorescence

When a material is exposed to a beam of high-energy X-ray radiation, the electrons emit characteristic fluorescence. The intensity of each individual fluorescence is proportional to the amount of that element present. The elements with atomic number 12 (magnesium) and higher are most suitable to be studied by X-ray fluorescence [52].

A study has shown that the amount of TiO_2 loaded on the cotton samples can be determined by X-ray fluorescence measurements. TiO_2 loaded on the cotton was found to be 1.5–2 wt% approximately, while pristine cotton was used as a reference [50].

20.3.5 Chromatographic Methods

The finishing must be separated from the fibrous materials and solubilized prior to chromatographic analysis. For pyrolysis chromatography, the samples should

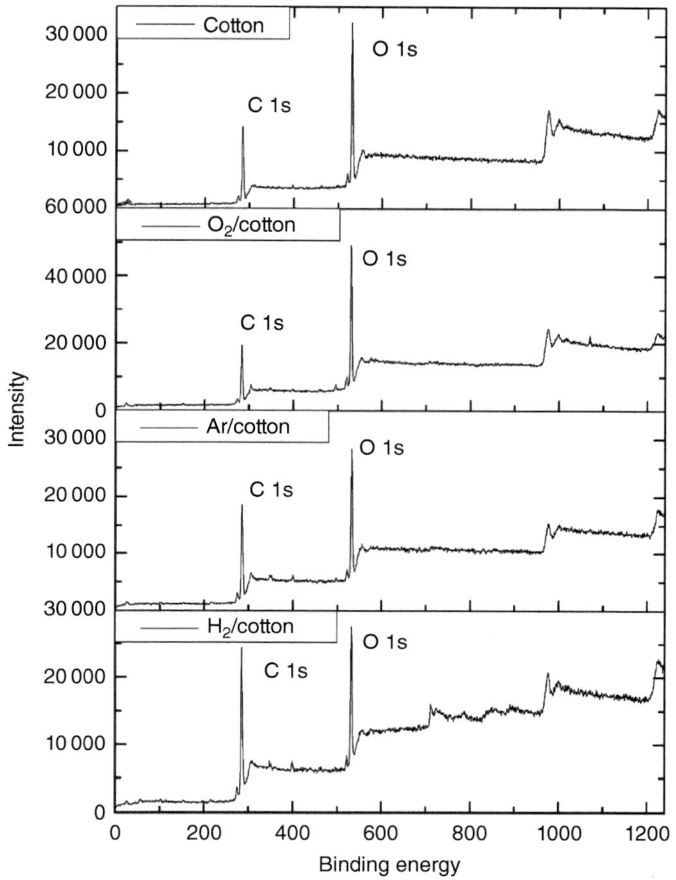

Figure 20.16 XPS survey scan analysis of pristine cotton compared with the cotton that has undergone various plasma pretreatments. Source: Caschera et al. 2013 [60]. Reproduced with permission of American Chemical Society.

be heated to very high temperature (600–1000 °C) for a very short period of time. The volatile products of this pyrolysis can then be analyzed by gas chromatography (PY–GC), mass spectrometry (PY–MS), or a combination of the two (PY–GC–MS) [52].

20.3.6 Energy-Dispersive X-Ray Analysis (EDX)

EDX analysis is an analytical technique commonly used for analysis of chemical compositions. It analyzes X-rays emitted by a material when it is hit with electromagnetic radiation [51]. Figure 20.17 shows the EDX spectrum of a treated sample confirming the existence of silver element on the surface of the fabric treated with 100 ppm of nanosilver particles. As a tiny peak of silver was observed, it suggests that there was only a small amount of nanosilver particles on the surface of the treated fabrics. Also the presence of sodium element in the EDX

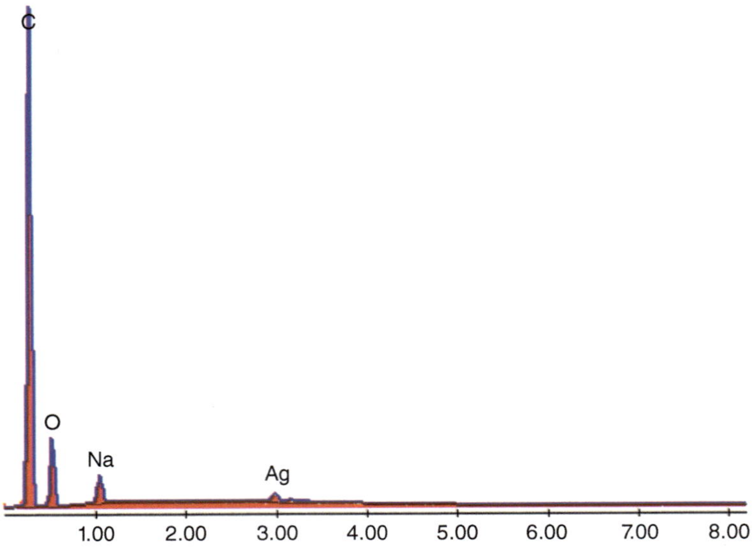

Figure 20.17 EDX image for cotton-coated fabric with 100 ppm AgNPs. Source: El-Rafie et al. 2014 [61]. Reproduced with permission of American Chemical Society.

spectrum is related to the alkaline solution of nanosilver particles that contains sodium hydroxide [61].

In another study, energy-dispersive spectroscopy analysis was conducted to examine chemical composition of the surface of the fabric. As shown in Figure 20.18a, only peaks of C and O were detected on the pristine fabric. After the polyvinylidene fluoride (PVDF)/polydimethylsiloxane (PDMS) treatment, new peaks for F and Si were observed (Figure 20.18b). F is mainly from PVDF, while Si is from PDMS. It suggests that the fabric was successfully coated with PVDF and PDMS [12].

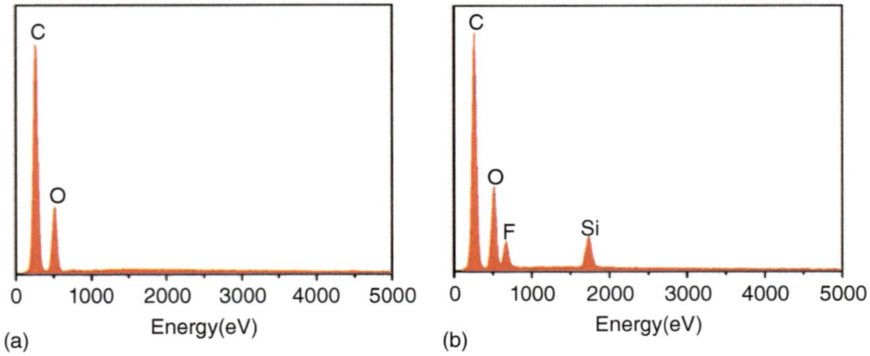

Figure 20.18 EDX of the samples (a) pristine fabric and (b) PVDF/PDMS-coated fabric. Source: Xue et al. 2016 [12]. Reproduced with permission of Royal Society of Chemistry.

20.3.7 Scanning Electron Microscope (SEM)

The scanning electron microscope (SEM) has the benefit of providing a great depth of field and extremely high resolution images at magnification levels in excess of 10 000× of a surface [62]. With the help of SEM, the fibrous surface characteristics can be analyzed extensively. The information obtained from SEM is normally qualitative, but with the use of software, surface properties of average roughness (R_a), rms roughness (R_{rms}), and fractal parameters can be obtained [62].

20.3.8 Atomic Force Microscopy (AFM)

The atomic force microscopy (AFM) comprises a scanning probe, with a tip at the end of a cantilever. The probe is made to scan in the very near surface region (in the van der Waals' forces) of a surface, and an intricate control system enables the tip to follow the contours of the surface in great details [62]. Imaging of individual atoms has been possible with the use of AFM for analyzing the fibrous surface [62]. The surface properties of fibrous materials for advanced application would play a fundamental role in affecting some key technical properties, which may closely be related to the processing condition [62]. With the understanding of the atomic changes in the surface of fibrous materials after different treatments, the processing parameters can be changed and optimized for achieving better final properties. The AFM can serve this purpose to obtain important topographical information of a fibrous material [62].

20.4 Performance, Evaluations, and Applications of Chemical Treatment on Fibrous Materials

20.4.1 Electronic Textiles

Electrically conducting fibers can be used for antistatic, antimicrobial, anti-odor, shielding, and other applications. In electronic textiles, the conducting elements can provide power, deliver input and output signals, or act as a transducer. E-textiles, depending on the purpose of usage, can be classified into two basic categories [62]: (i) fashion and (ii) utility. Table 20.7 summarizes some helpful literature on evaluations, performances, and applications of electronic textiles.

20.4.1.1 Fashion Statements

Electronic textiles can be used for costumes, catwalks, and consumer entertainment.

20.4.1.2 Utility Functions

Military This is highly desirable in applications such as uniform for warfighters where they integrate solar energy, optics, communication, and global positioning systems into clothing or an accessory [24].

Table 20.7 Literatures on evaluations, performance, and applications of electronic textiles.

Author	Evaluations	Performance	Applications
Irwin et al. [24]	– Dynamic mechanical analysis (DMA) – Electrical conductivity – SEM imaging	– Young's modulus and mechanical strength are maintained – The electrical conductivity of PEDOT:PSS-coated samples is 8.5 S/cm, which is only 10× less than sample made by Ag-coated thread	Electrical interconnects used in flexible, fully functional 555 timer circuits stitched into fabric substrates
Yamashita et al. [26]	– SEM imaging – Bending radius – Electrical resistance – Durability (cycle load test)	– Functions as the electrical contact between weft and warp (interlaced) fiber ribbons – Enhances the durability, flexibility, and stability of electrical contact in the woven e-textile better than those of the ribbons without it	Electrical contact structure in flexible device technology
Rehnby and coworkers [27]	– SEM imaging – Surface resistivity – Fastness to rubbing and shear	– The concentration of the conductive polymer and the number of coated layers should be further investigated	Conductive surfaces for smart textile applications
Wu et al. [28]	– Scanning electron microscopy – Thermogravimetric analysis (TGA) – Surface resistivity versus aging time – Force versus stretching	– Conductive polypyrrole-coated nylon Lycra fabric was found less stable in air due to its reactivity with a variety of atmospheric chemicals, especially oxygen – PPy-coated nylon Lycra is suitable to apply as a strain gauge when stretched up to 50–60%	Functional wearable textile sensing systems can monitor human motion and provide immediate, individual, and objective biofeedback. This innovative technique can be widely used for injury prevention, rehabilitation, sport techniques modification, and medical treatment
Ding et al. [17]	– Hydrophilicity – SEM imaging	– Fabrics with higher water uptake resulted in higher conductivity – Allowing for perception of the electrochromic color change on the surface of the fabric	Wearable displays and truly integrated fabric electronics

Kazani et al. [29]	Square resistance at different stages (i.e. after printing, abrading, and washing)	– Exhibit good electrical properties after printing and abrading	Produce flexible, conductive, lightweight, practical and comfortable "smart textiles"
Karaguzel et al. [30]	– Surface tension – Viscosity – Fastness to washing – Time domain reflectometry (TDR) measurement	Coating of the printed lines with a meltblown layer could improve the durability of the finishing	Fabric biomedical sensors
Heisey et al. [25]	– X-ray photoelectron spectroscopy – SEM imaging – Dynamic contact angle measurements – TGA	– Using TGA–MS, the 4.5% weight loss from the polypyrrole-coated nylon fiber by 298 °C was due to evolution of water and carbon dioxide – The adhesion of epoxy to the control nylon fiber was better than to the polypyrrole-coated nylon fiber	—
Lee et al. [31]	– Electrical resistivity	The electrical resistivity of the printed line is close to the bulk resistivity of silver	Serve as conducting lines for electronic applications
Lin et al. [32]	– FTIR – SEM imaging – Optical fibre diameter analyser (OFDA) – coating thickness – Transmission electron microscopy (TEM) – identifying the conducting polymer layer – Surface resistance – AATCC 76 – Abrasion resistance	Surface resistance changed as the thickness of the coating layer varied from 0.1–0.6 μm	Electromagnetic shielding, chemical sensors, and heating fabrics
Garg et al. [33]	– Surface contact angle – Wettability – Surface energy change – Abrasion resistance – Surface resistivity – Reflectance	– Treated fabrics exhibited better hydrophilicity and increased surface energy – Surface treatment by an APGD gas mixture of 95% helium/5% nitrogen yielded the best results with respect to coating uniformity, abrasion resistance, and conductivity	Sensors, actuators, electromagnetic shields, and absorbers and heating

Medical The electrocardiogram (ECG) is a skin surface measurement of the electrical activity of the heart muscle. The electrical potential on the body surface at different locations varies, generating different ECG vectors. These vectors are known as leads. The ECG leads are formed from three different electrodes placed on the body. Silver-coated yarns are commonly used to make the electrodes [63]. An electromyogram (EMG) records measurements of a muscle's electrical activity that occurs during muscle contraction and relaxation cycles [64]. Finni et al. [65] use textile electron embedded into shorts to measure EMG activity. An electroencephalogram (EEG) records voltage differences between points on the scalp generated by brain structures. The electrodes are often fixed onto a cap, made of elastic fabric, to guarantee a proper configuration [66].

Electrically conducting fibers can be used directly to measure strain. A strain sensor can measure a variation in length of an object. Most commonly the sensing mechanism is resistive, meaning that stretching the sensor causes a measurable change in resistance [67]. Cochrane et al. [68] made a flexible textile compatible strain sensor, which is based on conductive polymer composites. Xue et al. [69] studied a PPy-coated XLA elastic fiber as a textile strain sensor. These textile-based strain sensors can be used to measure breathing rate, posture, or body motion. It is reported that PPy-coated Lycra compares well with sensitive strain gauge materials and inorganic thermistors [70].

Electrically conducting fibers can also be used as pressure sensors. A pressure sensor can measure the change in resistance or in capacitance with the change in pressure. This can be used to evaluate the comfort when wearing some tight-fitting garment such as intimate apparel. Pressure sensors can be made by coating a composite of PDMS [71]. Apart from that, electronic fibers can also be used as temperature sensor, heat flux sensor, and sweat/humidity sensor.

Safety Gas sensors are used to detect volatile compounds, such as odor of urine, armpit sweat, or exhaled breath. The fibers used are able to change their electrical properties when a particular chemical agent is presented near the sensor [72]. Using a polymer coating such as PPy or polyaniline (PANI) onto polyethylene terephthalate (PET) can fabricate toxic gas sensors [73]. Poly(3,4-ethylenedioxithiophene) (PEDOT) nanotubes are suitable for monitoring concentration levels of nitric oxide (NO) [74].

Pragmatic Applications These include charging and controlling digital devices, caller ID and other smart phone applications, and visual feedback transferred to wearable displays.

Figure 20.19 depicts an example showing the application of electronic textiles. An intelligent knee sleeve is composed of a strip of Lycra coated with a thin layer of conducting PPy. When the coated fabric is stretched, resistance of the textile changes, resulting in changes of the output of an electronic circuit, so different sounds are emitted according to the strain from the coated fabrics.

20.4.2 Medical Textiles

An antimicrobial agent is defined as a natural or synthetic substance that kills or inhibits the growth of microorganisms such as bacteria, fungi, and algae.

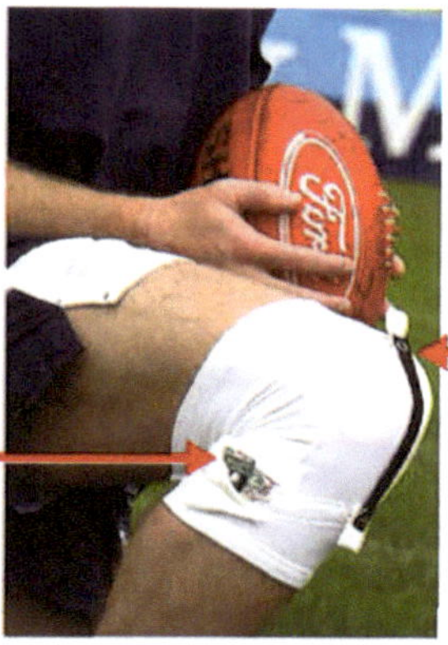

Figure 20.19 The knee sleeve . Source: Wu et al. 2005 [28]. Reproduced with permission of Elsevier. Photo: Courtesy of CSIRO TFT.

The application of this agent not only helps to protect the user of a textile material against microbes related to aesthetic, hygienic, or medical problems but also to protect the textile material itself. Some of the specific applications in medical and hygiene textiles include sutures, surgical gowns, face masks, bandages, drapes, bedding or blankets, surgical hosiery, and incontinence diapers [37].

Antimicrobial treatments for textiles materials are necessary [34]: (i) to avoid cross-contamination by pathogenic microorganisms, (ii) to control infestation by microorganisms, (iii) to reduce the formation of odor caused by metabolism in bacteria, and (iv) to prevent deterioration of the quality of a textile product. Therefore, the antimicrobial property, anti-odor property, and mechanical property of the medical textiles should be evaluated.

Antibacterial tests are used to determine the survival, growth, or inhibition of the isolate. Generally, testing involves using bacteria such as *Staphylococcus aureus* or *Klebsiella pneumoniae* (common isolates that can be replicated easily). The usual test methods for testing the antimicrobial efficacy of textiles are JIS 1907, AATCC 147, and AATCC 100. Table 20.8 summarizes some helpful literature on evaluations, performance, and applications of medical textiles.

20.4.3 Self-Cleaning Textiles

The most important method for evaluation of self-cleaning textiles is water and soil repellency, which can be measured using the contact angle method. Both static and sliding contact angles should be measured. Tablecloths, men's suits, awnings, tents, and other architectural structures can stay spotlessly clean without requiring any washing or cleaning [9] if they are self-cleaned. Apart from being water and soil repellent, self-cleaning textiles should not absorb any body tobacco odors. Table 20.9 summarizes some helpful literatures on evaluations, performance, and applications of self-cleaning textiles.

Table 20.8 Literatures on evaluations, performance, and applications of medical textiles.

Author	Evaluations	Performance	Applications
Kang et al. [40]	– Scanning electron microscopy – X-ray photoelectron spectroscopy (XPS) – Thermogravimetric analysis – Antibacterial property	The cotton fibers treated with both GTAC and silver nanoparticles showed synergistic antibacterial properties against *Pseudomonas aeruginosa*	A new approach to prepare hybrid antibacterial cotton fibers with a synergistic antibacterial efficacy in the medical and biomaterial industry
Shafei and Abou-Okeil [41]	– UV spectroscopy – FTIR – Transmission electron microscope (TEM) – X-ray diffraction (XRD) – UPF rating – Antibacterial property	Finished cotton fabric exhibits very good antibacterial properties against Gram-positive and Gram-negative bacteria, which increased with increasing the composite concentration, and also has a good UV protection, which increased with increasing the temperature of curing	Protection of the body against solar radiation, bacterial action, and for other technological applications
Cho (2009) [70]	– SEM imaging – Antibacterial property	The NP-coated cotton fabrics inhibited the growth of the medically relevant *Staphylococcus aureus* and *Escherichia coli*, respectively, by 67 and 100%	Diminish the risk of hospital-acquired infections

Table 20.9 Literatures on evaluations, performance, and applications of self-cleaning textiles.

Author	Evaluations	Performance	Applications
Xue et al. [44]	– Contact angle measurement – SEM imaging – Transmission electron microscopy – Fourier transformation infrared spectroscopy – Thermal gravimetric analysis	The incorporation of functionalized SiO_2 particles not only generates a dual-size surface roughness but also facilitates further hydrophobization of the surfaces to achieve a superhydrophobic property	Simple fabrication, easy availability of raw materials, and production of large superhydrophobic surfaces
Lai et al. [75]	– SEM imaging – TEM – XPS – XRD – UV–vis transmittance – AFM – Water contact angle	Upon heat treatment, samples with three very different wetting states can be obtained, viz. highly hydrophobic with strong adhesion, superhydrophobic with weak adhesion, and superhydrophilic with rapid water spreading	A wide range of applications such as microdroplet transportation, self-cleaning, antifogging, ultrafast spreading for biomolecules analysis, ultrafast absorption in printing

20.5 Performance Tests

In this chapter, we reviewed the development of fibrous materials with advanced applications such as electronic textiles, medical textiles, and self-cleaning textiles and discussed their specific performances. Although the development of each type of fibrous materials has its specific use, if they are applied for wearable use, they should meet certain wearable performance like general apparel. Despite their own applications, the functionality of fibrous materials with advanced application should be improved to meet the increasing expectations of the customers.

Table 20.10 lists various fabric performance tests, including easy care, flame retardancy, water and soil repellency, UV protection property, and anti-insect

Table 20.10 Summary of various performance tests.

Finishing	Test method
Durable press testing	AATCC 66 Wrinkle recovery of woven fabrics: Recovery Angle
	AATCC 88C Retention of creases in fabrics after repeated home laundering
	AATCC 124 Appearance of fabric after repeated home launderings
	AATCC 128 Wrinkle recovery of fabrics: Appearance method
	AATCC 135 Dimensional changes in automatic home laundering
Flame retardancy testing	Test 16 CFR 1610 Standard for the Flammability of Clothing Textiles
	ASTM D2863 Standard Test Method for Measuring the Minimum Oxygen Concentration to Support Candle-Like Combustion of Plastics (Oxygen Index)
Soil release testing	AATCC 130 Soil release: Oily stain release method
Repellency testing	AATCC 22 Water repellency: Spray test
	AATCC 35 Water resistance: Rain test AATCC 118 Oil repellency: Hydrocarbon resistance test
	ISO 9865 Textiles: Determination of water repellency of fabrics by the Bundesmann rain-shower test
UV Protective testing	AATCC 183 Transmittance or blocking of erythemally weighted ultraviolet radiation through fabrics
Anti-insect and mite testing	AATCC 24 Insects, resistance of textiles to AATCC 28 Insect pest deterrents on textiles
Strength	ASTM D6797 Standard Test Method for Bursting Strength of Fabrics Constant-Rate-of-Extension (CRE) Ball Burst Test
	ASTM D3787 Standard Test Method for Bursting Strength of Textiles—Constant-Rate-of-Traverse (CRT) Ball Burst Test
	ASTM D3786 Standard Test Method for Bursting Strength of Textile Fabrics—Diaphragm Bursting Strength Tester Method
	ASTM D5034 Standard Test Method for Breaking Strength and Elongation of Textile Fabrics (Grab Test)
	ASTM D1424 Standard Test Method for Tearing Strength of Fabrics by Falling-Pendulum (Elmendorf-Type) Apparatus

and mite testing, which should be considered when developing fibrous materials with advanced applications.

20.6 Conclusion

In this chapter, we have reviewed the performance and applications of electronic textiles, medical textiles, and self-cleaning textiles briefly. These textile applications are growing rapidly in textile and fashion product markets. Being comfortable, functional, and safe textile and fashion products, they should meet different criteria from the customers, and therefore, different existing evaluation methods are discussed. Due to the huge research on these products, it is not surprising that new technology and international standards will develop in the future for characterizing these products effectively.

Acknowledgment

The authors would like to thank the financial support from the Hong Kong Polytechnic University for this work.

References

1 Stoppa, M. and Chiolerio, A. (2016). 4 – Testing and evaluation of wearable electronic textiles and assessment thereof. In: *Performance Testing of Textiles* (ed. Lijing Wang), 65–101. Woodhead Publishing.
2 Mattila, H.R. (2006). *Intelligent Textiles and Clothing*. Cambridge, Boca Raton, FL: Woodhead Publishing.
3 Tao, X. and Textile, I. (2005). *Wearable Electronics and Photonics*. Boca Raton, FL; Cambridge: Woodhead Publishing.
4 Malinauskas, A. (2001). Chemical deposition of conducting polymers. *Polymer* 42: 3957–3972.
5 Murugesh Babu, K. and Ravindra, K. (2015). Bioactive antimicrobial agents for finishing of textiles for health care products. *Journal of the Textile Institute* 106: 706–717.
6 Zhang, Y., Xu, Q., Fu, F., and Liu, X. (2016). Durable antimicrobial cotton textiles modified with inorganic nanoparticles. *Cellulose* 23: 2791–2808.
7 Lim, S.-H. and Hudson, S.M. (2003). Review of chitosan and its derivatives as antimicrobial agents and their uses as textile chemicals. *Journal of Macromolecular Science, Part C: Polymer Reviews* 43: 223–269.
8 Gupta, D., Khare, S.K., and Laha, A. (2004). Antimicrobial properties of natural dyes against Gram-negative bacteria. *Coloration Technology* 120: 167–171.
9 Gupta, D. and Gulrajani, M.L. (2015). 8 – Self cleaning finishes for textiles. In: *Functional Finishes for Textiles* (ed. Roshan Paul), 257–281. Woodhead Publishing.

10 Fujishima, A. and Zhang, X. (2006). Titanium dioxide photocatalysis: present situation and future approaches. *Comptes Rendus Chimie* 9: 750–760.
11 Zhang, X., Shi, F., Niu, J. et al. (2008). Superhydrophobic surfaces: from structural control to functional application. *Journal of Materials Chemistry* 18: 621–633.
12 Xue, C.-H., Li, X., Jia, S.-T. et al. (2016). Fabrication of robust superhydrophobic fabrics based on coating with PVDF/PDMS. *RSC Advances* 6: 84887–84892.
13 Im, S.G. and Gleason, K.K. (2007). Systematic control of the electrical conductivity of poly (3,4-ethylenedioxythiophene) via oxidative chemical vapor deposition. *Macromolecules* 40: 6552–6556.
14 Bashir, T. (2013). Conjugated polymer-based conductive fibers for smart textile applications. Chalmers University of Technology. Ph.D.
15 Araki, T., Makikawa, M., and Hirai, S. (2012). Experimental investigation of surface identification ability of a low-profile fabric tactile sensor. In: *2012 IEEE/RSJ International Conference on Intelligent Robots and Systems (IROS)*, 4497–4504. IEEE.
16 Dalmas, F., Dendievel, R., Chazeau, L. et al. (2006). Carbon nanotube-filled polymer composites. Numerical simulation of electrical conductivity in three-dimensional entangled fibrous networks. *Acta Materialia* 54: 2923–2931.
17 Ding, Y., Invernale, M.A., and Sotzing, G.A. (2010). Conductivity trends of PEDOT-PSS impregnated fabric and the effect of conductivity on electrochromic textile. *ACS Applied Materials & Interfaces* 2: 1588–1593.
18 Bocchini, S., Chiolerio, A., Porro, S. et al. (2013). Synthesis of polyaniline-based inks, doping thereof and test device printing towards electronic applications. *Journal of Materials Chemistry C* 1: 5101–5109.
19 Thongruang, W., Spontak, R.J., and Balik, C.M. (2002). Correlated electrical conductivity and mechanical property analysis of high-density polyethylene filled with graphite and carbon fiber. *Polymer* 43: 2279–2286.
20 Xue, P., Tao, X., Kwok, K.W. et al. (2004). Electromechanical behavior of fibers coated with an electrically conductive polymer. *Textile Research Journal* 74: 929–936.
21 Schwarz, A., Hakuzimana, J., Westbroek, P., and Van Langenhove, L. (2009). How to equip para-aramide yarns with electro-conductive properties. In: *2009 6th International Workshop on Wearable and Implantable Body Sensor Networks. BSN 2009*, 278–281. IEEE.
22 Stoppa, M. and Chiolerio, A. (2014). Wearable electronics and smart textiles: a critical review. *Sensors* 14: 11957–11992.
23 Merilampi, S., Björninen, T., Haukka, V. et al. (2010). Analysis of electrically conductive silver ink on stretchable substrates under tensile load. *Microelectronics Reliability* 50: 2001–2011.
24 Irwin, M.D., Roberson, D.A., Olivas, R.I. et al. (2011). Conductive polymer-coated threads as electrical interconnects in e-textiles. *Fibers and Polymers* 12: 904.
25 Heisey, C., Wightman, J., Pittman, E., and Kuhn, H. (1993). Surface and adhesion properties of polypyrrole-coated textiles. *Textile Research Journal* 63: 247–256.

26 Yamashita, T., Miyake, K., and Itoh, T. (2012). Conductive polymer coated elastomer contact structure for woven electronic textile. In: *2012 IEEE 25th International Conference on Micro Electro Mechanical Systems (MEMS)*, 408–411. IEEE.
27 Skrifvars, M., Rehnby, W., and Gustafsson, M. (2008). Coating of textile fabrics with conductive polymers for smart textile applications. Ambience'08, Borås, Sweden.
28 Wu, J., Zhou, D., Too, C.O., and Wallace, G.G. (2005). Conducting polymer coated lycra. *Synthetic Metals* 155: 698–701.
29 Kazani, I., Hertleer, C., De Mey, G. et al. (2012). Electrical conductive textiles obtained by screen printing. *Fibres & Textiles in Eastern Europe* 20: 57–63.
30 Karaguzel, B., Merritt, C., Kang, T. et al. (2009). Flexible, durable printed electrical circuits. *Journal of the Textile Institute* 100: 1–9.
31 Lee, H.-H., Chou, K.-S., and Huang, K.-C. (2005). Inkjet printing of nanosized silver colloids. *Nanotechnology* 16: 2436.
32 Lin, T., Wang, L., Wang, X., and Kaynak, A. (2005). Polymerising pyrrole on polyester textiles and controlling the conductivity through coating thickness. *Thin Solid Films* 479: 77–82.
33 Garg, S., Hurren, C., and Kaynak, A. (2007). Improvement of adhesion of conductive polypyrrole coating on wool and polyester fabrics using atmospheric plasma treatment. *Synthetic Metals* 157: 41–47.
34 Burnett-Boothroyd, S.C. and McCarthy, B.J. (2011). 13 – Antimicrobial treatments of textiles for hygiene and infection control applications: an industrial perspective. In: *Textiles for Hygiene and Infection Control* (ed. McCarthy, B.J.), 196–209. Woodhead Publishing.
35 Rajendran, S., Anand, S.C., and Rigby, A.J. (2016). 5- Textiles for healthcare and medical applications. In: *Handbook of Technical Textiles* (eds. A. Richard Horrocks and Subhash C. Anand), 2e, 135–168. Woodhead Publishing.
36 Nayak, R. and Padhye, R. (2015). 12 – Antimicrobial finishes for textiles. In: *Functional Finishes for Textiles* (ed. Roshan Paul), 361–385. Woodhead Publishing.
37 Dhende, V.P., Hardin, I.R., and Locklin, J. (2012). 8 – Durable antimicrobial textiles: types, finishes and applications. In: *Understanding and Improving the Durability of Textiles* (ed. Patricia Annis), 145–173. Woodhead Publishing.
38 Montazer, M. and Afjeh, M.G. (2007). Simultaneous X-linking and antimicrobial finishing of cotton fabric. *Journal of Applied Polymer Science* 103: 178–185.
39 Morones, J.R., Elechiguerra, J.L., Camacho, A. et al. (2005). The bactericidal effect of silver nanoparticles. *Nanotechnology* 16: 2346.
40 Kang, C.K., Kim, S.S., Kim, S. et al. (2016). Antibacterial cotton fibers treated with silver nanoparticles and quaternary ammonium salts. *Carbohydrate Polymers* 151: 1012–1018.
41 Shafei, A.E. and Abou-Okeil, A. (2011). ZnO/carboxymethyl chitosan bionano-composite to impart antibacterial and UV protection for cotton fabric. *Carbohydrate Polymers* 83: 920–925.

42 Petkova, P., Francesko, A., Perelshtein, I. et al. (2016). Simultaneous sonochemical-enzymatic coating of medical textiles with antibacterial ZnO nanoparticles. *Ultrasonics Sonochemistry* 29: 244–250.
43 Gupta, B., Agarwal, R., and Alam, M. (2010). Textile-based smart wound dressings. *Indian Journal of Fibre and Textile Research* 35: 174–187.
44 Xue, C.-H., Jia, S.-T., Zhang, J. et al. (2008). Preparation of superhydrophobic surfaces on cotton textiles. *Science and Technology of Advanced Materials* 9: 035008.
45 Lee, H.J. and Michielsen, S. (2007). Preparation of a superhydrophobic rough surface. *Journal of Polymer Science Part B: Polymer Physics* 45: 253–261.
46 Xu, B. and Cai, Z. (2008). Fabrication of a superhydrophobic ZnO nanorod array film on cotton fabrics via a wet chemical route and hydrophobic modification. *Applied Surface Science* 254: 5899–5904.
47 Hoefnagels, H., Wu, D., De With, G., and Ming, W. (2007). Biomimetic superhydrophobic and highly oleophobic cotton textiles. *Langmuir* 23: 13158–13163.
48 Zhao, Y., Tang, Y., Wang, X., and Lin, T. (2010). Superhydrophobic cotton fabric fabricated by electrostatic assembly of silica nanoparticles and its remarkable buoyancy. *Applied Surface Science* 256: 6736–6742.
49 Abbas, R., Khereby, M.A., Sadik, W.A., and El Demerdash, A.G.M. (2015). Fabrication of durable and cost effective superhydrophobic cotton textiles via simple one step process. *Cellulose* 22: 887–896.
50 Afzal, S., Daoud, W.A., and Langford, S.J. (2014). Superhydrophobic and photocatalytic self-cleaning cotton. *Journal of Materials Chemistry A* 2: 18005–18011.
51 Wei, Q., Huang, F., and Cai, Y. (2009). 2 – Textile surface characterization methods. In: *Surface Modification of Textiles* (ed. Wei, Q.), 26–57. Woodhead Publishing.
52 Hauser, P.J. (2005). 6 – Chemical analysis of fabric finishes and performance-related tests. In: *Chemical Testing of Textiles* (ed. Fan, Q.), 107–125. Woodhead Publishing.
53 Krishnaveni, R. and Thambidurai, S. (2013). Industrial method of cotton fabric finishing with chitosan–ZnO composite for anti-bacterial and thermal stability. *Industrial Crops and Products* 47: 160–167.
54 Singh, J. and Dutta, P. (2010). Preparation, antibacterial and physicochemical behavior of chitosan/ofloxacin complexes. *International Journal of Polymeric Materials* 59: 793–807.
55 Al-Gaashani, R., Radiman, S., Tabet, N., and Daud, A.R. (2011). Effect of microwave power on the morphology and optical property of zinc oxide nano-structures prepared via a microwave-assisted aqueous solution method. *Materials Chemistry and Physics* 125: 846–852.
56 Müller, D., Rambo, C., Recouvreux, D. et al. (2011). Chemical in situ polymerization of polypyrrole on bacterial cellulose nanofibers. *Synthetic Metals* 161: 106–111.
57 Watanabe, K., Tabuchi, M., Morinaga, Y., and Yoshinaga, F. (1998). Structural features and properties of bacterial cellulose produced in agitated culture. *Cellulose* 5: 187–200.

58 Czaja, W., Krystynowicz, A., Bielecki, S., and Brown, R.M. (2006). Microbial cellulose—the natural power to heal wounds. *Biomaterials* 27: 145–151.
59 Scaini, M.J., Bancroft, G.M., Lorimer, J.W., and Maddox, L.M. (1995). The interaction of aqueous silver species with sulphur-containing minerals as studied by XPS, AES, SEM, and electrochemistry. *Geochimica et Cosmochimica Acta* 59: 2733–2747.
60 Caschera, D., Cortese, B., Mezzi, A. et al. (2013). Ultra hydrophobic/superhydrophilic modified cotton textiles through functionalized diamond-like carbon coatings for self-cleaning applications. *Langmuir* 29: 2775–2783.
61 El-Rafie, M., Ahmed, H.B., and Zahran, M. (2014). Characterization of nanosilver coated cotton fabrics and evaluation of its antibacterial efficacy. *Carbohydrate Polymers* 107: 174–181.
62 Wainwright, H.L. (2016). 9 – Design, evaluation, and applications of electronic textiles. In: *Performance Testing of Textiles* (ed. Lijing Wang), 193–213. Woodhead Publishing.
63 Alzaidi, A., Zhang, L., and Bajwa, H. (2012). Smart textiles based wireless ECG system, systems, applications and technology conference (LISAT). In: *2012 IEEE Long Island*, 1–5. IEEE.
64 Benatti, S., Farella, E., and Benini, L. (2014). Towards EMG control interface for smart garments. In: *Proceedings of the 2014 ACM International Symposium on Wearable Computers: Adjunct Program*, 163–170. Seattle, WA: ACM.
65 Finni, T., Hu, M., Kettunen, P. et al. (2007). Measurement of EMG activity with textile electrodes embedded into clothing. *Physiological Measurement* 28: 1405.
66 Löfhede, J., Seoane, F., and Thordstein, M. (2012). Textile electrodes for EEG recording—a pilot study. *Sensors* 12: 16907.
67 Cochrane, C., Hertleer, C., and Schwarz-Pfeiffer, A. (2016). 2 – Smart textiles in health: an overview. In: *Smart Textiles and their Applications* (ed. Vladan Koncar), 9–32. Oxford: Woodhead Publishing.
68 Cochrane, C., Koncar, V., Lewandowski, M., and Dufour, C. (2007). Design and development of a flexible strain sensor for textile structures based on a conductive polymer composite. *Sensors* 7: 473–492.
69 Xue, P., Wang, J., and Tao, X. (2014). Flexible textile strain sensors from polypyrrole-coated XLA™ elastic fibers. *High Performance Polymers* 26: 364–370.
70 Cho, G. (2009). *Smart Clothing: Technology and Applications*. CRC Press.
71 Wang, F., Zhu, B., Shu, L., and Tao, X. (2013). Flexible pressure sensors for smart protective clothing against impact loading. *Smart Materials and Structures* 23: 015001.
72 Windmiller, J.R. and Wang, J. (2013). Wearable electrochemical sensors and biosensors: a review. *Electroanalysis* 25: 29–46.
73 Hong, K.H., Oh, K.W., and Kang, T.J. (2004). Polyaniline–nylon 6 composite fabric for ammonia gas sensor. *Journal of Applied Polymer Science* 92: 37–42.

74 Lu, H.-H., Lin, C.-Y., Fang, Y.-Y. et al. (2008). NO gas sensor of PEDOT: PSS nanowires by using direct patterning DPN. In: *2008 30th Annual International Conference of the IEEE Engineering in Medicine and Biology Society. EMBS 2008*, 3208–3211. IEEE.

75 Lai, Y., Tang, Y., Gong, J. et al. (2012). Transparent superhydrophobic/superhydrophilic TiO_2-based coatings for self-cleaning and anti-fogging. *Journal of Materials Chemistry* 22: 7420–7426.

21

Soft Computing in Fibrous Materials

Abhijit Majumdar[1], Piyali Hatua[2], and Mirela Blaga[3]

[1] *Indian Institute of Technology, Department of Textile and Fibre Engineering, Hauz Khas, New Delhi 110016, Delhi, India*
[2] *Veermata Jijabai Technological Institute, Department of Textile Technology, Mumbai 400019, India*
[3] *University of Iaşi, Faculty of Textiles-Leather and Industrial Management, Iaşi 700050, Romania*

21.1 Introduction

Soft computing (SC) is a new breed of computing systems. Prof. Zadeh introduced the concept of soft computing in 1990. According to him "soft computing is a collection of methodologies that aim to exploit the tolerance for imprecision and uncertainty to achieve tractability, robustness, and low solution cost" [1]. Its principal constituents are fuzzy logic (FL), artificial neural networks (ANN), and genetic algorithms (GA). Hard computing is based on precise and accurate mathematical models. In contrast, the major advantage of soft computing lies in its tolerance to imprecision, uncertainty, and partial truth. It can produce effective solutions to difficult problems, which cannot be solved with the conventional analytical approach or other computational tools. Moreover, the potential of fusing soft computing with conventional hard computing techniques increases its acceptance in scientific community. Soft computing techniques have also been successfully applied in engineering design area, which increased their popularity in industrial applications. Soft computing techniques mimic the functioning of biological systems and get advantage over other hard computing techniques in imprecision handling and decision-making. ANN mimics the working of biological neurons, FL gets motivation from the highly imprecise nature of human speech, and GA follows nothing but the Darwinian evolution theory. Furthermore, soft computing systems are complementary to each other and therefore present more potent solutions when combined together.

For centuries, the main use of the textile materials has remained confined to clothing and interior applications. However, last few decades have witnessed a sharp increase in technical applications of textile materials and structures in various sectors like in agriculture, horticulture, construction, automobiles, protection, aeronautics, medicine, sports, etc. Technical textiles are mainly manufactured keeping focus on their functional and technical performances rather than on their aesthetic or decorative characteristics. Technical woven

Handbook of Fibrous Materials, First Edition. Edited by Jinlian Hu, Bipin Kumar, and Jing Lu.

fabrics found various diversified application such as sports, protective clothing, filtration, belting, packaging, and in many other areas. Technical fabrics are also used as reinforcement material in composites [2]. Now these technical applications need a paradigm shift in designing principles of textile materials: from aesthetic design to engineering design so that a textile structure can meet certain functional properties according to its end use requirement. Textile structures are highly complex, flexible, and porous in nature. Structure-property modeling, process modeling, and process optimization should be carried out before making any attempt for engineering design of textile structures. However, these are highly complex tasks. In general, a woven textile material consists of yarns, and yarns consist of fibers. Thus the mechanical properties of fabrics are dependent on the structural complexity and nonlinearity. Furthermore, complexity amplifies when the inherent nonlinearities of the materials are added to the problem. This double nonlinear behavior of the textile fabric increases the difficulty in the fabric design and engineering processes. Conventional and precise analytical models often fail or perform poorly due to the intricacy induced by the complex structure and the nonlinearity in the behavior of raw materials. In many areas, models based on soft computing have been adapted to overcome these problems. Till date, soft computing techniques have been applied in almost every domain of textile such as fiber, yarn, woven and knitted fabrics, nonwovens, clothing and garment, textile quality control, etc. [3, 4]. In the subsequent sections of this chapter, soft computing techniques, namely, ANN, FL, GA, etc., have been described followed by their applications in fibrous materials.

21.2 Soft Computing Techniques

The following sections describe some of the soft computing techniques, namely, ANN, FL, GA, and adaptive network-based fuzzy inference system (FIS). These techniques have been widely used in textile research for modeling, prediction, and optimization.

21.2.1 Artificial Neural Network (ANN)

ANN is a powerful data modeling tool that can effectively capture and represent any kind of input–output relationship using historical or experimental data. It uses simple mathematical functions to represent highly complex functional relationships. Each mathematical function used in ANN mimics a specific operation of biological neurons. It can exhibit characteristics such as mapping or pattern recognition, generalization, robustness, fault tolerance, parallel and high-speed information processing, etc. [5].

ANN, in general, is a multilayered network consisting of a large number of structurally interconnected artificial neurons or nodes, which are the computing elements (Figure 21.1). The input parameters and output parameters are positioned at the input and output layer, respectively. One or more hidden layers are placed between the input and output layers.

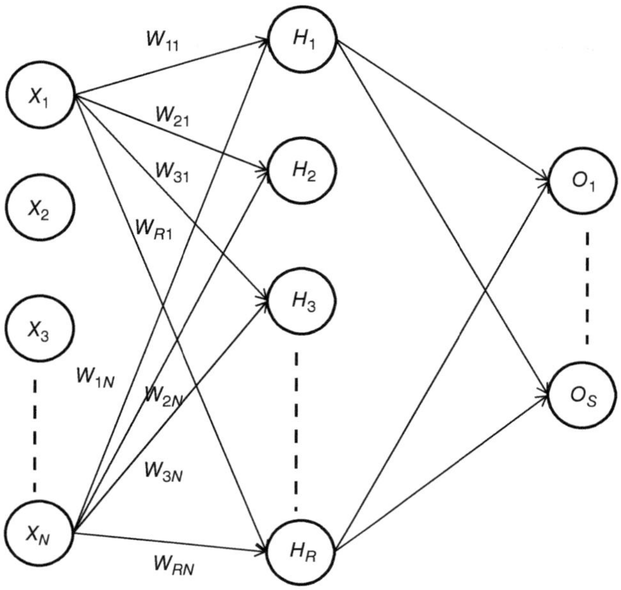

Figure 21.1 Multilayered artificial neural network structure.

Figure 21.2 Simple model of an artificial neuron. Source: Majumdar 2011 [4].

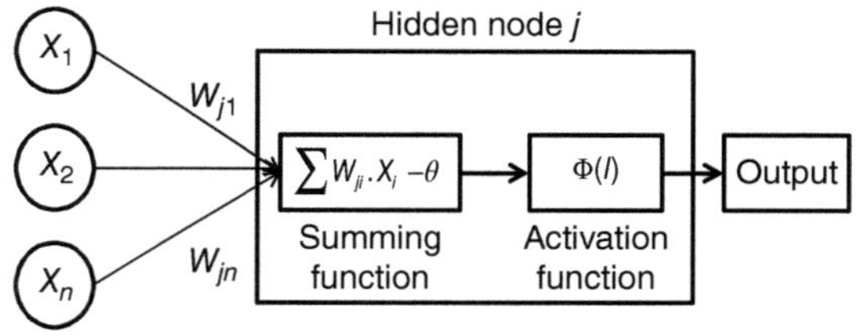

The functioning of an ANN is explained by the simple model of an artificial neuron as shown in Figure 21.2. Here X_1, X_2, X_3, …, X_n are the n input parameters to the artificial neuron, and W_{j1}, W_{j2}, W_{j3}, …, W_{jn} are n weights by which inputs are connected to the hidden layer. W_{ji} is the synaptic weight connecting the hidden neuron j of hidden layer to the input neuron i. It acts as a multiplying factor for the signals coming out from the neurons of the previous layer and thus accelerates or retards the input signal. The weighted inputs are summed up to get the weighted sum of inputs received by an artificial neuron. The weighted sum of inputs is then compared with a threshold value called bias weight, which represents the membrane potential of a biological neuron. For neuron j, the net input (I) can be expressed as follows:

$$I = (W_{j1}X_1 + W_{j2}X_2 + \cdots + W_{jn}X_n) - \theta = \sum_{i=1}^{n} W_{ji}X_i - \theta \quad (21.1)$$

where θ is the bias weight.

The net input is then passed through a transfer function, which converts the output to a fixed range of values (either between 0 and 1 or between −1 and +1) as shown below:

$$Z = \phi(I) \tag{21.2}$$

Several kinds of transfer functions such as threshold or hard limit, saturating linear, log-sigmoid, tan-sigmoid, etc. can be used depending on the nature of problem [4]. The output of the transfer function is transmitted to the nodes of the next layer, and similar kind of computations is done again. Finally, the output is produced and is compared with the expected output, and an error signal is generated. The prediction accuracy of ANN depends on finding the optimum set of weights connecting various nodes. The weights are optimized by a process termed as "training." There are three types of training algorithm for ANN, namely, supervised learning, unsupervised learning, and reinforced learning [6]. Back-propagation and Levenberg–Marquardt training algorithms are popularly used for training of ANN.

21.2.1.1 Back-Propagation Algorithm

Back-propagation is a supervised training algorithm developed by Rumelhart et al. [7]. The details of this algorithm can be found in many standard textbooks of ANN authored by Haykin [8], Zurada [9], and Bose and Liang [10]. According to this algorithm, training occurs in two phases, namely, forward pass and backward pass. In the forward pass, a set of data is presented to ANN as input, and finally, a set of outputs is produced. The error vector is calculated according to the following equation:

$$E = \sum_{j=1}^{p} E_j \tag{21.3}$$

where E is the error vector, E_j is the error associated with the jth pattern, and p is the total number of training patterns.

The expression of E_j is given in the following equation:

$$E_j = \frac{1}{2} \sum_{k=1}^{s} (T_k - \text{out}_k)^2 \tag{21.4}$$

where T_k and out_k are the target output and predicted output, respectively, at output neuron k and s is the total number of output neurons.

In the backward pass, this error signal is propagated backward, and the synaptic weights are adjusted so that the error signal reduces in each iteration. The corrections necessary in synaptic weights between hidden and output layer are carried out by a delta rule. At the start of the training, ANN uses random combinations of synaptic weights, which cause very high error signal. Therefore, the gradient (slope) of the error function (surface) is determined with respect to the weights. Once the training converges to the point of minimum error, the gradient of the error surface becomes zero, and no further update of the weights takes place. Mathematically this process is represented as shown below.

$$\Delta W_{jk} = W_{jk(\text{new})} - W_{jk(\text{old})} = -\eta[\partial E/\partial W_{jk}] = -\eta\left[\frac{\partial E}{\partial \text{out}_k} \times \frac{\partial \text{out}_k}{\partial \text{net}_k} \times \frac{\partial \text{net}_k}{\partial W_{jk}}\right] \tag{21.5}$$

$$\text{Now, } \frac{\partial E}{\partial \text{out}_k} = \frac{\partial}{\partial \text{out}_k}\left[\frac{1}{2}\sum_k (T_k - \text{out}_k)^2\right] = -(T_k - \text{out}_k) \tag{21.6}$$

$$\frac{\partial \text{out}_k}{\partial \text{net}_k} = f_0'(\text{net}_k) = \text{out}_k\,(1 - \text{out}_k) \text{ for log sigmoid transfer function} \tag{21.7}$$

$$\frac{\partial \text{net}_k}{\partial W_{jk}} = \frac{\partial}{\partial W_{jk}}\sum_k w_{jk}\text{out}_j = \text{out}_j \tag{21.8}$$

$$\text{Therefore, } \Delta W_{jk} = \eta[(T_k - \text{out}_k)\text{out}_k(1 - \text{out}_k)]\text{out}_j = \eta\delta_k\text{out}_j \tag{21.9}$$

where W_{jk} is the weight connecting the neurons j of hidden layer and neuron k of the output layer, ΔW_{jk} is the correction applied to W_{jk} at a particular iteration, η is a constant known as learning rate, and out_j is the output of neuron j.

The weight change between hidden layer and input layer is calculated as shown below.

$$\Delta W_{ij} = \eta f_H'(\text{net}_j)X_i\sum_k \delta_k W_{jk} \tag{21.10}$$

where ΔW_{ij} is the correction applied to the weight connecting input neuron i and hidden neuron j and X_i is the input received by the neuron i.

21.2.1.2 Levenberg–Marquardt Algorithm

Back-propagation training algorithm has poor convergence rate. Levenberg–Marquardt is another training algorithm that updates weight and bias values according to Levenberg–Marquardt optimization [11, 12]. Levenberg–Marquardt algorithm can be considered as a combination of steepest descent and the Gauss–Newton method. It can be expressed by Eq. (21.11):

$$(J^tJ + \lambda I)\delta = J^tE \tag{21.11}$$

where J is the Jacobian matrix for the system, λ is the Levenberg's damping factor, I is the identity matrix, δ is the weight update vector, and E is the error vector containing the output errors for each input vector used on training the network. The Jacobian is a matrix containing all first-order partial derivatives of a vector-valued function. It is a $N \times W$ matrix, where N is the number of entries in training set and W is the total number of parameters (weights and biases) of the network. It can be created by taking the partial derivatives of each output with respect to each weight as given below:

$$J = \begin{bmatrix} \frac{\partial F(x_1, w)}{\partial w_1} & \cdots & \frac{\partial F(x_1, w)}{\partial w_w} \\ \vdots & \ddots & \vdots \\ \frac{\partial F(x_N, w)}{\partial w_1} & \cdots & \frac{\partial F(x_N, w)}{\partial w_w} \end{bmatrix} \tag{21.12}$$

where $F(x_i, w)$ is the network function evaluated for the ith input vector of the training set using the weight vector w and w_j is the jth element of the weight vector w of the network.

The δ quantifies the change in network weights to achieve an improved solution. The J^tJ matrix is also known as the approximated Hessian. The damping factor (λ) is adjusted at each iteration, and it guides the weight optimization process. If reduction of error E is rapid, a smaller value of λ can be used, making the algorithm closer to the Gauss–Newton algorithm. On the other hand, if an iteration gives insufficient reduction in the residual, λ can be increased, making the algorithm closer to the gradient descent direction.

21.2.1.3 Important Parameters of Artificial Neural Network

The structure of ANN and training parameters play important roles in determining the performance of ANN model. The important parameters are as follows.

Number of Hidden Layers and Nodes in Each Hidden Layer The number of hidden layers and number of nodes in each hidden layer depends on the complexity of the function that is being modeled by ANN. In most of the cases, one hidden layer can capture the nonlinear relationship between the input and output variables. However, for very complex functions, more than one hidden layer may be needed.

The number of nodes in the input layer and output layer is equal to the number of input and output variables, respectively. There is no exact rule to determine the optimum number of nodes in the hidden layers as it is dependent on the type of problem. Therefore, the number of nodes in the hidden layers is often determined by trial-and-error approach. If the number of nodes is too less, then ANN will fail to capture the input–output relationship. On the other hand, if the number of nodes is too high, then ANN may capture even the noise present in data. This will also cause poor prediction performance in the unseen testing data.

Learning Rate and Momentum Learning rate (η) determines the magnitude of weight adjustment during ANN training, and hence it influences the rate of convergence. If the learning rate is too small, then the training advances in small steps and training time may become very large. On the other hand, if the learning rate is too large, then the search path may oscillate, and "global minima" may not be attained as depicted in Figure 21.3. The learning rate has to be chosen as large as possible without inducing oscillations during ANN training. The value of learning rate ranges from 0 to 1.

The error function, in most of the practical problems, is not smooth. There can be several "local minima" in the error function, as depicted in

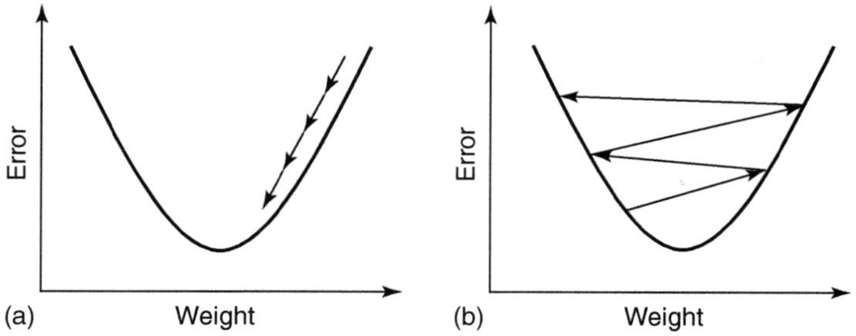

Figure 21.3 Effect of (a) very small and (b) very large learning rate on convergence.

Figure 21.4 Global and local minima of error surface.

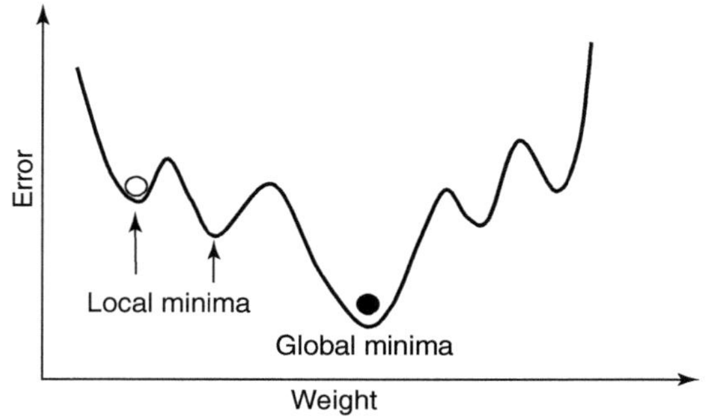

Figure 21.4. The back-propagation training algorithm may stuck as the slope of the error surface is zero at these local minima. To overcome this problem, back-propagation algorithm is often modified by adding a momentum term. The momentum is a constant that determines the effect of past weight changes on current direction of movement in weight space. If a pith ball is allowed to fall from the left side of error surface shown in Figure 21.4, then it will be caught in the local minima. However, if a marble is allowed to fall from the same position, it will be able to overcome curvatures of local minima due to its momentum, and finally it will rest at the global minima. This concept is used while adding the momentum term in the back-propagation algorithm.

The modified rule of weight adjustment with momentum term is shown below:

$$\Delta W_{jk}(t+1) = \eta \delta_k \text{out}_j + \mu \cdot \Delta W_{jk}(t) \tag{21.13}$$

where $\Delta W_{jk}\ (t+1)$ and $\Delta W_{jk}\ (t)$ are the weight changes in $(t+1)$ th and tth iteration, respectively, and μ is the momentum term and it ranges from 0 to 1.

The addition of momentum helps to smooth out the descent path by preventing extreme changes in the gradient due to local minima or potholes. If the momentum is 0, the smoothing is the least, and the weight adjustments are done solely by the newly calculated gradient. If the momentum is between 0 and 1, the weight adjustment is smoothed by an amount proportional to the momentum factor.

21.2.2 Fuzzy Logic

FL was propounded by Prof. Zadeh at the University of California, Berkeley, USA, in 1965 [1]. The concept of FL originated from the way human brain represents and reasons with real-world knowledge in spite of uncertainty, vagueness, and incomplete knowledge. Fuzzy set theory and approximate reasoning have been used to deal with imprecision and ambiguity in decision-making [5].

FL has its foundation built on fuzzy sets. Fuzzy sets, an extension of their crisp counterpart, do not have clearly defined boundaries. A crisp set is defined by the characteristic function that can assume only two values, for example, 0 or 1, true or false, black or white. If A is a crisp set and x is an element in A, then testing of the element x using the characteristic function χ can be expressed as follows [13]:

$$\chi_A(x) = \begin{cases} 1, & \text{if } x \in A \\ 0, & \text{if } x \in A \end{cases} \tag{21.14}$$

On the other hand, in fuzzy set theory, the concept of membership function is introduced to handle uncertainty of classes, which is having unclear or overlapping boundaries [14, 15]. Thus a membership function is associated with a fuzzy set A, and this function maps every element of the universe of discourse X to the interval [0, 1]. If X is the universe of discourse and its elements are denoted by x, then a fuzzy set A in X is defined as a set of ordered pairs as shown below:

$$\chi_A(x) = \{x, \mu_A(x) | x \in X\} \tag{21.15}$$

where $\mu_A(x)$ is the membership function of x in fuzzy set A.

Membership function always takes values between 0 and 1. It may be in discrete form as well as in continuous function. Membership function can have various forms, such as triangle, trapezoid, sigmoid, and Gaussian as depicted in Figure 21.5. Dubois and Prade [16] defined the triangular membership function as shown below:

$$\mu_A(x) = \begin{cases} \dfrac{x-L}{m-L}, & \text{for } L < x < m \\ \dfrac{R-x}{R-m}, & \text{for } m < x < R \\ 0, & \text{otherwise} \end{cases} \tag{21.16}$$

where m is the most promising or modal value and L and R are the left and right spread (the smallest and largest value, respectively, that m can take).

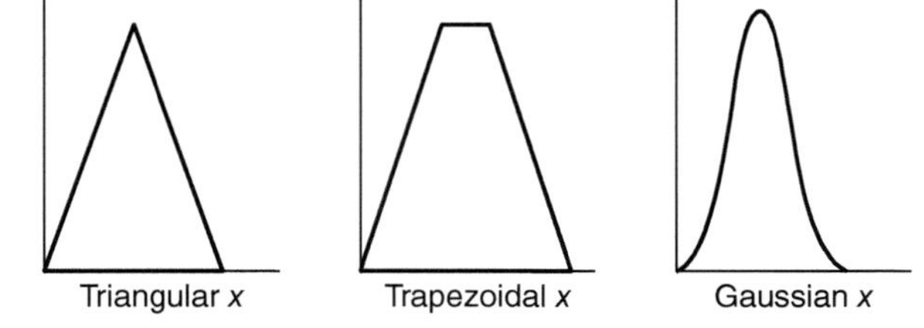

Figure 21.5 Different forms of membership function. Source: Majumdar 2011 [4].

The trapezoidal membership function is defined by four scalar parameters a, b, c, and d, as shown below:

$$f(x;a,b,c,d) = \begin{cases} 0, & \text{for } x \le a \text{ or } x \ge d \\ \dfrac{x-a}{b-a}, & \text{for } a \le x \le b \\ 1, & \text{for } b \le x \le c \\ \dfrac{d-x}{d-c}, & \text{for } c \le x \le d \end{cases} \tag{21.17}$$

The trapezoidal membership curve has a flat top. It is a truncated triangle producing $\mu_A(x) = 1$ over large regions of universe of discourse. The Gaussian membership function depends on two parameters, namely, standard deviation (σ) and mean (μ), and it is represented as shown below:

$$\mu_A(x) = e^{\frac{-(x-\mu)^2}{2\sigma^2}} \tag{21.18}$$

The general bell-shaped membership function is defined by three parameters (a, b, and c) as shown below:

$$\mu_A(x) = \frac{1}{1 + \left|\dfrac{x-c}{a}\right|^{2b}} \tag{21.19}$$

Membership function converts the crisp inputs to fuzzy membership values. This step is known as "fuzzification." Different linguistic terms such as high, medium, low, etc. are used to describe fuzzy sets. Then fuzzy if–then rules are established to relate input fuzzy sets with output fuzzy sets using linguistic terms. Fuzzy if–then rules are developed based on the experience and knowledge of experts. Fuzzy rules have two parts, namely, antecedent or if part and consequent or then part. The fuzzy sets in antecedent part are joined by "fuzzy AND" or "fuzzy OR" operators. For example, rules for predicting the yarn strength from cotton fiber properties can be as follows:

Rule 1: If fiber tenacity is *high* AND fiber length is *high*, then yarn strength is *very high*.
Rule 2: If fiber tenacity is *high* AND fiber length is *moderate*, then yarn strength is *high*.
Rule 3: If fiber tenacity is *moderate* AND fiber length is *high*, then yarn strength is *moderate*.
Rule 4: If fiber tenacity is *moderate* AND fiber length is *moderate*, then yarn strength is *low*.
Rule 5: If fiber tenacity is *low* AND fiber length is *low*, then yarn strength is *very low*.

The rule base shown above is nonexhaustive. It should be noted that there are two input variables, namely, fiber tenacity and fiber length, and only one output variable, i.e. yarn tenacity. Each of the input variables has three fuzzy sets, which are linguistically termed as high, moderate, and low. The output variable

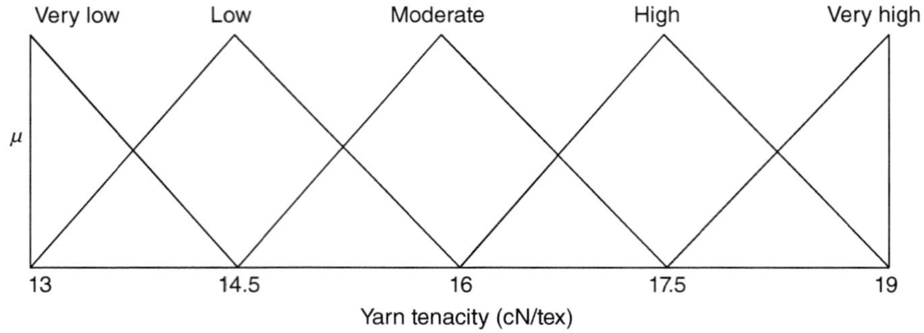

Figure 21.6 Membership functions of yarn tenacity.

has five fuzzy sets, which are linguistically termed as very high, high, moderate, low, and very low. As fuzzy sets do not have sharp boundaries, there will be overlaps between two neighboring fuzzy sets. Hypothetical fuzzy sets for yarn tenacity are shown in Figure 21.6.

The output of each rule is also a fuzzy set. Output fuzzy sets are aggregated into a single fuzzy set generally by the "MAX" operator. This step is known as "aggregation." Finally, the resulting fuzzy set is converted to a crisp output by the "defuzzification" method. There are many methods available for defuzzification as mentioned below.

- Centroid (center of gravity or center of area)
- Center of sums (COS)
- Mean of maxima (MOM)
- Smallest of maxima (SOM)
- Largest of maxima (LOM)

Centriod method is one of the most popular methods of defuzzification. In this method, the overlapped areas of output fuzzy sets are considered only once. According to this method, the defuzzified crisp value is calculated as follows:

$$x^* = \frac{\int \mu_A(x)x \ dx}{\int \mu_A(x) \ dx} \tag{21.20}$$

where x^* is the defuzzified output and $\mu_A(x)$ is the membership value in output fuzzy set after aggregation of individual implication results.

The steps involved in fuzzy modeling can be summarized as shown in Figure 21.7.

21.2.2.1 Types of Fuzzy Inference System

There are two types of FIS, namely, Mamdani type and Sugeno (Takagi, Sugeno, and Kang or TSK) type. In case of Mamdani FIS, the consequent membership functions are also fuzzy in nature. Mamdani FIS is more popularly used as it provides reasonably good results with a relatively simple structure. Besides, the intuitive and interpretable nature of the rule base makes it more appealing. Mamdani FIS can be used directly for multiple input single output (MISO) systems and multiple input multiple output (MIMO) systems. On the contrary, the consequent

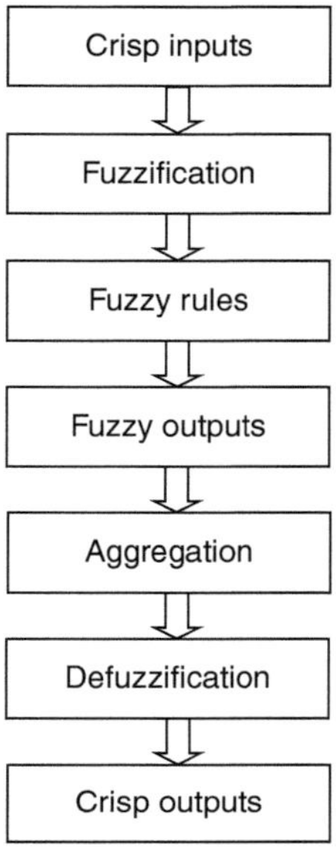

Figure 21.7 Steps involved in fuzzy modeling.

membership functions in a Sugeno FIS are non-fuzzy (either linear or constant), and thus they lack interpretability. However, in Sugeno FIS, the consequent membership functions can have as many parameters, per rule, as the number of input variables. This gives more degrees of freedom and flexibility in the design of Sugeno FIS as compared with Mamdani FIS. Sugeno FIS can only be used in the case of the MISO systems. Some advantages of Sugeno FIS over Mamdani FIS are listed below:

- Sugeno FIS is more apt for functional analysis than the Mamdani FIS.
- In computational terms, the Sugeno FIS is more effective because the complex defuzzification process of the Mamdani FIS is supplanted with weighted average method.
- Sugeno FIS is more flexible than the Mamdani FIS because the former allows more parameters in the output. Moreover, as the output is a function of the inputs Sugeno FIS expresses a more explicit relation among them.

21.2.3 Genetic Algorithm

GA was invented by John Holland in the early 1970s [17]. It is an unorthodox search and nondeterministic optimization algorithm inspired by Darwinian theory for the survival of the fittest. GA is very different from most of the other optimization methods. It can efficiently find optimal solutions to complicated optimization problem, which cannot be solved by conventional optimization techniques [18]. GA randomly searches for the best solution with respect to the given criterion of goodness expressed in terms of an objective function or fitness function. The fitness function is either to be maximized or minimized. Unlike traditional optimization methods, it works on a population of solution points at a time instead of one solution point approach.

In GA, first a population of solutions is randomly generated. GA requires design space to be converted into genetic space. So each probable solution point, called "individuals," are converted to form a string using binary (0, 1) or real numbers (0, 9) coding. These strings are called "chromosomes," and each variable within a solution point (chromosome) is called "gene." Different genetic inheritance operators are used to create new and improved solution points called "offspring," which will in turn form the next generation population. The genetic inheritance operators are reproduction, crossover, mutation, inversion, deletion, duplication, etc. However, reproduction, crossover, and mutation are the most common operators used in optimization problems. Reproduction operator is generally the first operator applied to a population. It selects good individuals or chromosomes from the present population to form mating pool. It utilizes the fitness score of the individuals for selection. "Roulette wheel selection" and "tournament selection" are the two popular methods of selection in GA. The next operator is the crossover operator. It recombines the genes of two parents hoping to create better offspring, i.e. solution point. First, two parents are randomly chosen from mating pool, then a cross-site is selected along the length of the strings, and then their position values are swapped between each other to generate two new offspring as depicted in Figure 21.8. "Single-point crossover," "two-point crossover," and "matrix crossover" are some of the commonly used crossover operators. Mutation operator diversifies the population. It changes from 1 to 0 and vice versa in a point along the length of the chromosome. The probability of mutation should be very low; otherwise it can spoil the good individuals. The typical range of crossover probability is 0.6–0.8, whereas mutation probability is in the range of 0.001–0.01. Finally, the GA program stops searching when one of the termination criteria is achieved.

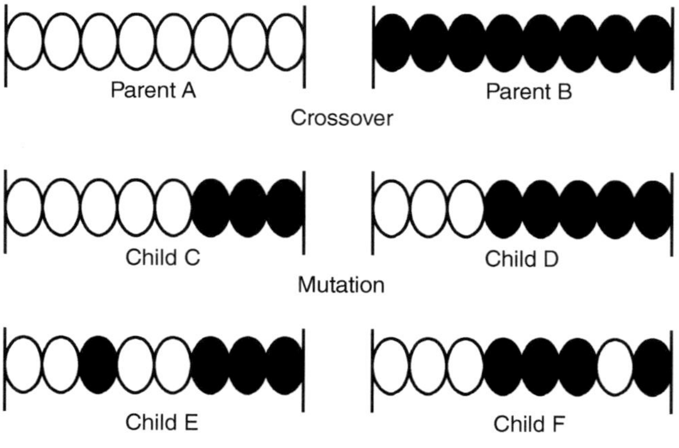

Figure 21.8 Schematic representation of single-point crossover and mutation. Source: Majumdar 2011 [4].

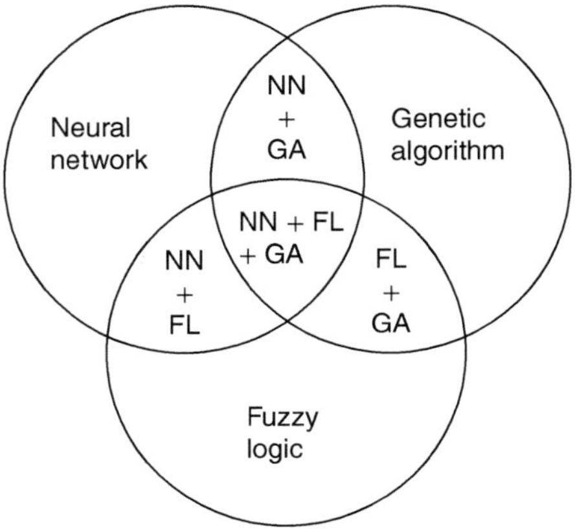

Figure 21.9 Possible hybrid combinations of three soft computing techniques.

21.2.4 Hybrid Systems

Hybrid systems employ more than one technology or system to solve a given problem. Hybridization is generally performed when an individual technology or system fails to obtain an acceptable solution. It overcomes the drawbacks of individual technologies while synergizing the strengths of each other [5]. Soft computing methods give efficient solutions to a wide range of problems. However, hybrid systems combining two or three of these methods have a tremendous potential to solve complex problems. The possible combinations are neuro-fuzzy, neuro-genetic, fuzzy-genetic, and neuro-fuzzy-genetic hybrids as shown in Figure 21.9. GA-based back propagation, ANN, and adaptive network-based fuzzy inference system (ANFIS) are some of the popular hybrid soft computing systems, which have been used to solve textile problems.

21.2.4.1 Adaptive Network-Based Fuzzy Inference System (ANFIS)

ANFIS is a multilayered adaptive network in which each node performs a particular function, often called node function, on incoming signals. The formula for the node functions varies from node to node. Mainly two types of nodes, namely, circle and square, are used in a network to reflect different adaptive capabilities (Figure 21.10). A square node is an adaptive node having parameters, while a circle node is a fixed node having no parameter. These parameters are updated according to given training data to achieve a desired input–output mapping [19]. The learning capability of ANN is used to tune the fuzzy membership function parameters using given input–output data sets. Generally two learning methods, namely, back-propagation learning and hybrid learning, are used for training of ANFIS model. In hybrid learning, ANFIS uses back-propagation learning (gradient descent method) to determine premise parameters (related to the input fuzzy sets) and least mean squares estimation to determine the consequent parameters (related to the output fuzzy sets). A step in the hybrid learning procedure is composed of two passes. In the forward pass, functional signals go forward, and the consequent parameters are estimated by an iterative least mean square procedure, while the premise parameters remain fixed. In the backward pass, the error is propagated backward to update the premise parameters by gradient descent

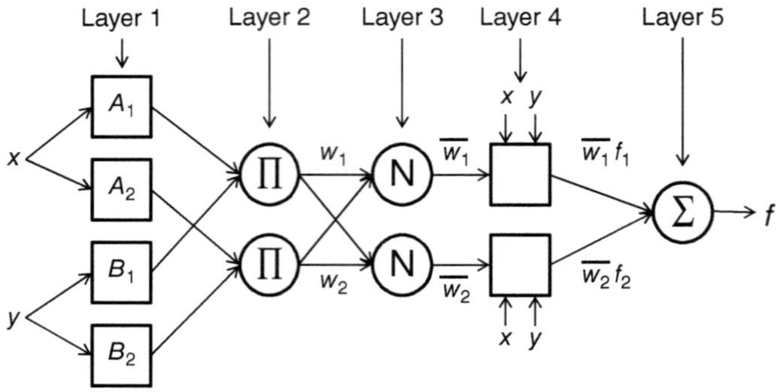

Figure 21.10 ANFIS architecture.

method, while the consequent parameters remain fixed. This procedure is then iterated until the error criterion is satisfied or preset number of cycles is reached.

ANFIS Architecture Figure 21.10 depicts the ANFIS architecture having five layers assuming two inputs x and y and one output z. A common first-order Sugeno fuzzy model with only two rules can be represented as follows:

Rule 1: If x is A_1 and y is B_1, then $f_1 = p_1x + q_1y + r_1$
Rule 2: If x is A_2 and y is B_2, then $f_2 = p_2x + q_2y + r_2$
Layer 1: Every node in this layer is an adaptive node with a node function as shown below:

$$O_{1,i} = \mu_{A_i}(x) \text{ for } i = 1, 2 \text{ or} \tag{21.21}$$

$$O_{1,i} = \mu_{B_{i-2}}(y) \text{ for } i = 3, 4 \tag{21.22}$$

where x and y are the input to node i; A_i and B_{i-2} are fuzzy sets associated with inputs x and y, respectively; and $\mu_{A_i}(x)$ is the membership of x in fuzzy set $A_i(A_1, A_2)$. $O_{1,i}$ is the output at layer 1 for the ith node. Here, the membership function for A can be any appropriate parameterized membership function, such as the generalized Gaussian function as shown in Eq. (21.24):

$$\mu_A(x) = e^{\frac{-(x-\mu)^2}{2\sigma^2}} \tag{21.23}$$

where $\{\mu, \sigma\}$ is the parameter set containing mean and standard deviation of the distribution, respectively. As the values of these parameters change, the Gaussian function varies accordingly, thus exhibiting various shapes of membership functions for fuzzy set A. Parameters in this layer are referred to as premise parameters.

Layer 2: Every node in this layer is a fixed node labeled Π, whose output is the product of all the incoming signals as shown in Eq. (21.24):

$$O_{2,i} = \mu_{A_i}(x)\mu_{B_i}(y) = w_i, \text{ for } i = 1, 2 \tag{21.24}$$

Each node output represents the firing strengths of a fuzzy rule. In general, any T-norm operator that performs fuzzy AND operation can be used as the node function in this layer.

Layer 3: Every node in this layer is a fixed node labeled N. The ith node calculates the ratio of the ith rule's firing strength to the sum of all rules' firing strengths as shown in Eq. (21.25). Outputs of this layer represent normalized firing strength of rules:

$$O_{3,i} = \frac{w_i}{w_1 + w_2} = \overline{w}_i \tag{21.25}$$

Layer 4: Every node i in this layer is an adaptive node with a node function as shown in Eq. (21.26):

$$O_{4,i} = \overline{w_i} f_i = \overline{w_i}(p_i x + q_i y + r_i) \tag{21.26}$$

where $\overline{w_i}$ is a normalized firing strength from layer 3 and $\{p_i, q_i, r_i\}$ is the parameter set for this node. Parameters in this layer are referred to as consequent parameters.

Layer 5: The single node in this layer is a fixed node labeled Σ, which calculates the overall output as the summation of all incoming signals.

$$O_{5,i} = \sum \overline{w_i} f_i = \frac{\sum w_i f_i}{\sum w_i} \tag{21.27}$$

ANFIS Parameters Selection of input variables and determining the number of fuzzy sets for each of the inputs is very important for ANFIS modeling since they determine the number of rules to be trained. If p is the number of membership function for each input and q is the number of inputs, then there are p^q rules to be trained. Table 21.1 shows the number of nodes and number of parameters in various layers of ANFIS considering two parameter membership functions for the input variables.

The learning scheme for ANFIS is represented in Table 21.2. In the forward pass, the consequent parameters (p, q, and r) are optimized by least square method. In the backward pass, the premise parameters are optimized by gradient descent algorithm.

Table 21.1 Number of ANFIS parameters.

Layer number	Layer type	Number of nodes	Number of parameters
0	Inputs	q	0
1	Fuzzification	$p \times q$	$2 \times p \times q$
2	Rules	p^q	0
3	Normalization	p^q	0
4	Linear functions	p^q	$(q+1)\,p^q$
5	Summation	1	0

Table 21.2 Hybrid learning for ANFIS.

Parameters and signals	Forward pass	Backward pass
Premise parameters (nonlinear)	Fixed	Gradient descent
Consequent parameters (linear)	Least square	Fixed
Signals	Node outputs	Error signals

21.3 Applications of Soft Computing in Fibrous Materials

21.3.1 Applications in Yarn Manufacturing

Prediction of yarn properties from fiber properties and process parameters are one of the most favorite topics in yarn manufacturing research. Majority of these researches focuses on the prediction of yarn strength, elongation, unevenness, and hairiness. The advent of soft computing techniques has provided a new impetus in yarn modeling research. A glimpse of soft computing aided researches in yarn manufacturing is chronologically presented in Table 21.3.

21.3.1.1 Yarn Engineering and Process Optimization

Soft computing models have performed extremely well for the prediction of yarn properties from the cotton fiber properties or process parameters. However, reverse modeling that can predict the fiber properties or process parameters from the given yarn properties has been seldom attempted. Majumdar et al. [32] attempted to engineer ring spun yarn properties using ANN model and linear programming. Four yarn properties, namely, tenacity, elongation, unevenness, and hairiness, were considered for the yarn engineering and were used as input parameters to the ANN model. The outputs of the ANN model were spinning consistency index (SCI) and micronaire. The ANN model was able to predict the values of SCI and micronaire with mean absolute percentage error (MAPE) of 4.38% and 1.92%, respectively. From the five testing samples, three were finally chosen for the engineering. Linear programming problems were formulated with the objective of minimization of cotton mixing cost. Constraint equations were formed to ensure that the ANN predicted combinations of SCI and micronaire are fulfilled. A typical linear programming problem for the 40^s yarn is shown below:

$$\begin{aligned}
&\text{Minimize } Z = P_A \cdot 65 + P_B \cdot 52\\
&P_A \cdot 160 + P_B \cdot 126 \geq 139\\
&P_A \cdot \frac{1}{4.01} + P_B \cdot \frac{1}{4.35} \geq \frac{1}{4.19}\\
&P_A + P_B = 1, P_A \geq 0, P_B \geq 0
\end{aligned} \tag{21.28}$$

where Z is the overall cost of the mixing; P_A and P_B are the proportions of two cottons A and B, respectively; cotton cost per kilogram, SCI, and micronaire values of cotton A and B are 65 and 52, 160 and 126, and 4.01 and 4.35, respectively.

Table 21.3 Glimpse of soft computing applications in yarn technology.

Name of researchers	Predicted properties/application	System used	Accuracy of prediction
Pynckels et al. [20]	Spinnability	ANN	Accuracy 90–95%.
Cheng and Adams [21]	CSP	ANN	$R = 0.85$
Ethridge and Zhu [22]	CSP, tenacity, elongation, and unevenness	ANN	$R^2 = 0.80$, 0.78, 0.87, and 0.95, respectively
Zhu and Ethridge [23]	Unevenness	ANN	$R^2 = 0.88$
Sette et al. [24]	Tenacity and elongation	ANN and GA	Normalized error of 0.01 and 0.03, respectively
Rajamanickam et al. [25]	Tenacity	ANN	Error up to 8.5%
Zhu and Ethridge [26]	Hairiness	ANN	$R^2 = 0.84$ and 0.77, respectively
Pynckels et al. [27]	Tenacity and elongation	ANN	Error less than 5% in 90% data
Guha et al. [28]	Tenacity	ANN	Error = 6.9%
Chattopadhyay et al. [29]	Tenacity, CSP, U%, and imperfections	ANN	Error = 7.1% and 14.7%, respectively
Majumdar and Majumdar [30]	Elongation	ANN	Error = 4.536%
Majumdar et al. [31]	Tenacity	ANFIS	Error = 4.72% (ring) and 2.03% (rotor)
Majumdar et al. [32]	Yarn engineering	ANN	Error = 1.617–6.242%
Majumdar et al. [33]	Unevenness	ANFIS	Error = 2.367% in testing data
Majumdar and Ghosh [34]	Tenacity	FL	Error = 4.04%
Ghosh and Chatterjee [35]	Tenacity, elongation, unevenness, and hairiness	SVM	Accuracy > 95%
Ghosh et al. [36]	Stress–strain behavior	GA	$R^2 = 0.99$
Mishra [37]	Yarn strength utilization	ANN	$R^2 = 0.97$, error = 2.6%

Source: Majumdar 2011 [4]. Reproduced with permission of Taylor & Francis.

The linear programming problems were solved, and the solutions obtained in terms of proportion of fiber mix (P_A and P_B) were actually used for spinning of engineered yarns. Finally, the targeted and the experimentally achieved yarn properties were compared to evaluate the efficacy of yarn engineering. The results are shown in Table 21.4. It is observed that the yarn properties can be engineered with overall average error of less than 7%.

21.3.2 Applications in Fabric Property Prediction

In the field of fabrics, researchers have successfully applied various soft computing techniques for modeling and prediction of different properties such as tensile,

Table 21.4 The accuracy of yarn engineering system.

Test sample number	Yarn count (Ne)	Mean absolute error (%)				Overall % error
		Tenacity	Elongation	Unevenness	Hairiness	
1	40	2.727	1.515	3.924	11.777	4.986
3	50	3.510	0.549	0.197	2.210	1.617
5	60	2.982	13.871	2.661	5.455	6.242

shear, bending, comfort, drape, air permeability, thermal conductivity, etc. Apart from this, soft computing techniques have also been attempted for fabric defect detection both in static and dynamic conditions, fabric classification, and fabric engineering.

21.3.2.1 Prediction of Mechanical Properties of Fabrics

A plentiful of research works have been published on soft computing based modeling and prediction of fabric mechanical properties. Tensile properties of fabrics have been predicted using ANN model by Majumdar et al. [38] and Hadizadeh et al. [39]. Fan and Hunter [40, 41] and Behera and Guruprasad [42] have used ANN model to predict bending and shear behavior of fabrics. FIS and ANFIS have also been used for predicting fabric properties and knitting machine parameters [42, 43]. Bilisik and Demiryurek [44], Zeydan [45], and Hadizadeh et al. [46] have compared the efficacy of mathematical, statistical, and ANN models for the prediction of tensile and bending properties of fabrics. All of them reported lowest prediction error in case of ANN model.

21.3.2.2 Prediction of Transmission Properties of Fabrics

Comfort is a very important aspect of apparel fabric. A number of research studies have been reported in this area using soft computing techniques. Bhattacharjee and Kothari [47] developed two different types of ANN model to predict thermal resistance and instantaneous heat flow (Q_{max}) of woven fabrics. One model was having two parallel ANNs working in tandem, while another had a single network with two outputs. They used weave, yarn count and spacing of warp and weft, fabric thickness, and areal density as inputs to the ANN model. The first model with two networks working in tandem gave better result.

Fayala et al. [48] used ANN model to predict the thermal conductivity of knitted fabrics. The input parameters were yarn thermal conductivity, fabric areal density, porosity, and air permeability. They reported good prediction ($R^2 = 0.913$) in the training data set. In a similar study, Majumdar [49] predicted thermal conductivity of cotton–bamboo viscose blended knitted fabrics. Knitted fabric structure along with yarn count (tex), bamboo viscose fiber %, thickness, and areal density were considered as ANN inputs. Very good prediction accuracy was obtained in training and testing data set. The trend analysis given by ANN model revealed that thermal conductivity increases with increasing yarn count (tex), areal density and decreasing fabric thickness, and bamboo viscose %.

ANN has been utilized in predicting air permeability of woven fabrics. Tokarska [50] predicted impact air permeability using a simple ANN model. The author focused on the real pressure impulse $p(t)$ generated on fabrics. Therefore, an integral of actual pressure impulse $p(t)$ was considered as output, whereas fabric density (kg/m^3), warp twist, and weft twist were input parameters. A very low average error of prediction (0.00241 Pa) was obtained in this study. Brasquet and Le Cloirec [51] studied pressure drops across the fabric structure. They tried different classical pressure drop models and ANN model for predicting pressure drop. Apart from the fabric parameters (type, thickness, number of openings, and opening specific surface area), fluid parameters (viscosity, density, and Reynolds number) were also included as input parameters to the ANN model. It was revealed from the input variable analysis that characteristics of macroscopic interstices of fabric have significant influence on pressure drop. In another study, Brasquet and Le Cloirec [52] modeled pressure drop of activated carbon cloth with the help of ANN. Ten different weaves were used for this investigation. Six input parameters related to fabric and fluid characteristics were selected on the basis of knowledge of the experimental flow behavior of activated carbon cloth and cross-correlation analysis. The high R^2 value (0.992) confirmed high generalization ability of the ANN model. Analysis of importance of input variables was performed to figure out significant fabric and fluid characteristics influencing pressure drop.

21.3.2.3 Modeling of Fabric UV Protection

Hatua et al. [53, 54] have modeled the UV protection factor (UPF) of plain woven fabrics using nonlinear regression, ANN, and ANFIS models. Ring spun yarns made of 100% cotton, 100% polyester, and polyester–cotton blends (50 : 50 and 65 : 35) were used in this research. The yarn counts were 20, 30, and 40 Ne for each of the four blends making a total of 12 yarn samples. Fabrics were produced by varying the pick density at three levels (16, 20, and 24 cm^{-1}) for each of these 12 yarns, which were used in weft. So a total of 36 (3 × 12) fabric samples, differing in proportion of polyester (%), weft count (Ne), and pick density (cm^{-1}), were produced. Cotton warp (100%) with 40 Ne count and end density of 40 cm^{-1} was used for all the fabrics. Out of 36 data set, 27 data set were used for the model (nonlinear regression, ANN, and ANFIS) development. These are termed as training data set. Prediction accuracy of the models was evaluated by using the remaining nine data set (65:35 polyester–cotton blend), which are termed as testing data set.

Model Development The nonlinear regression model, as shown in Eq. (21.29), was developed using the experimental data. Here X_1, X_2, and X_3 are proportion of polyester (%), weft count (Ne), and pick density (cm^{-1}), respectively. Coded values were used to develop the regression model.

$$\begin{aligned}\text{UPF} = {} & 8.26 + 2.56X_1 - 1.88X_2 + 1.91X_3 - 1.04X_1X_2 + 1.28X_1X_3 \\ & - 0.91X_2X_3 - 1.58X_1^2 + 0.76X_2^2 \end{aligned} \tag{21.29}$$

$[R^2 = 0.973]$

Table 21.5 ANFIS parameters.

Layer number	Layer type	Number of nodes	Number of parameters
0	Inputs	Number of inputs = 3	0
1	Fuzzification	No. of fuzzy sets per input × No. of inputs = 2 × 3 = 6	No. of parameters per fuzzy set × No. of fuzzy sets = 2 × 6 = 12
2	Rules	(No. of fuzzy sets per input)$^{\text{No. of inputs}}$ = 2^3	0
3	Normalization	(No. of fuzzy sets per input)$^{\text{No. of inputs}}$ = 2^3	0
4	Constant function	(No. of fuzzy sets per input)$^{\text{No. of inputs}}$ = 2^3	No. of fuzzy rules × No. of constants per rule = 8 × 1 = 8
5	Summation	1	0

An ANN model comprising three nodes in input layer, one hidden layer with four nodes, and one node in the output layer was also developed. Tan-sigmoid transfer function was used for both the hidden and output layers. Levenberg–Marquardt algorithm was used for training of ANN. The learning rate was kept at 0.3. Maximum iteration was set at 500, and mean squared error (MSE) was set at 0.01 as stopping criteria.

In ANFIS model, only two fuzzy sets (low and high) were used for each of the three input parameters so that the number of fuzzy rules becomes 8 (2^3). Gaussian membership function, which is defined by mean and standard deviation, was selected as the input membership function, and constant function was selected as output membership function. Hybrid optimization technique was employed for the training of ANFIS model. The number of nodes and parameters in different layers of ANFIS has been presented in Table 21.5.

Table 21.6 summarizes the UPF prediction accuracy of nonlinear, ANN, and ANFIS models. The nonlinear regression model can explain 97.3% variability of UPF for training data set. In the training data set, MAPE is 6.725, and MSE is 0.319 implying good prediction accuracy of the regression model. It is observed that R^2 value of nonlinear regression model has reduced marginally from 0.973 to 0.96 in testing data sets. The MAPE and MSE values are 4.748 and 0.704, respectively, which indicates quite good efficacy of the nonlinear regression model in unseen testing data set.

However, the prediction performance of both the soft computing models is found to be much better than that of nonlinear regression model. Coefficient of determination (R^2) of ANN model is very high (0.999), and prediction error is low (MAPE = 1.12 and MSE = 0.007) in training data set. The R^2, MAPE, and MSE values in the testing data set are 0.994, 2.590, and 0.007, respectively. The prediction accuracy is relatively poor in the case of testing data set as compared with that of training data set. For ANFIS model, coefficient of determination (R^2), in training and testing data set, is not only very high (0.988 and 0.992, respectively)

Table 21.6 Summary of UPF prediction accuracy of regression, ANN, and ANFIS models.

Model	Performance parameter	Data set	
		Training	Testing
Nonlinear regression	R^2	0.973	0.960
	MAPE	6.725	4.748
	MSE	0.319	0.704
	No. of samples with >5% error	11	2
ANN	R^2	0.999	0.994
	MAPE	1.120	2.590
	MSE	0.007	0.098
	No. of samples with >5% error	1	1
ANFIS	R^2	0.988	0.992
	MAPE	2.276	2.506
	MSE	0.047	0.173
	No. of samples with >5% error	3	1

but also very close to each other. The MAPE is 2.276 and 2.506 for training and testing data set, respectively. These results imply very good generalizing ability of ANFIS model.

Trend Analysis by ANN model Trend analysis was performed with the developed ANN model to check whether the model is able to decipher the effect of various input parameters on UPF of fabrics. Trend analysis results are depicted in Figures 21.11 and 21.12. Figure 21.11 shows the effect of proportion of polyester and weft yarn count on UPF keeping pick density constant at 20 cm^{-1}. It is observed that UPF improves with the increase in the proportion of polyester. However, the effect is very prominent when proportion of polyester is changed from 20% to 60%. Coarser weft yarn gives higher UPF, due to increased fabric cover, at same proportion of polyester. Figure 21.12 presents the effect of proportion of polyester and pick density keeping weft count constant at 30 Ne. It is observed that at low pick density (16 cm^{-1}), UPF increases marginally with the increase in the proportion of polyester as the fabric is very open. However, at high pick density (24 cm^{-1}), UPF continues to improve with the increase in the proportion of polyester. It is also observed that improvement in UPF with increasing pick density is negligible at low proportion of polyester. However, the difference amplifies continuously as the proportion of polyester increases. This implies the synergistic role of fiber material and fabric construction parameters on UPF.

Linguistic Rules of ANFIS Model Figure 21.13 shows eight fuzzy rules that are relating three input parameters (proportion of polyester, weft count, and pick density) with the output parameter (UPF). The fuzzy sets for each of the three input parameters have two levels, namely, high and low. The output variable (UPF) has

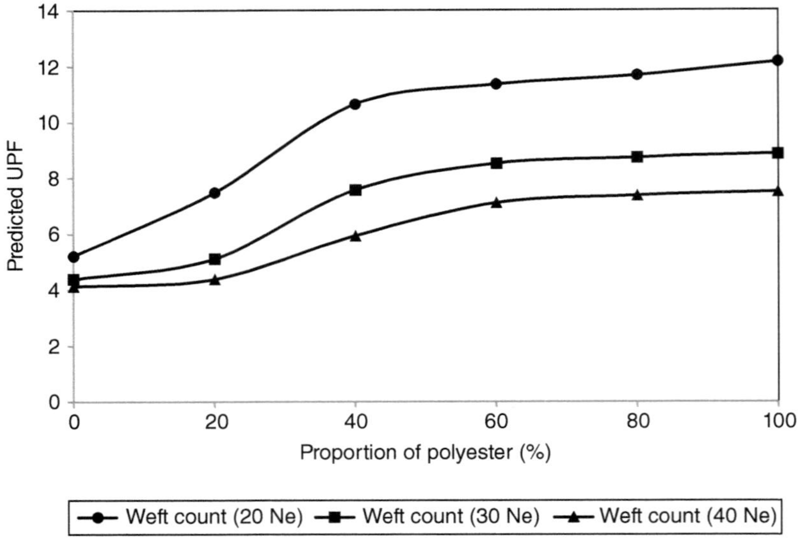

Figure 21.11 Effect of proportion of polyester and weft count on UPF. Source: Hatua et al. 2013 [53].

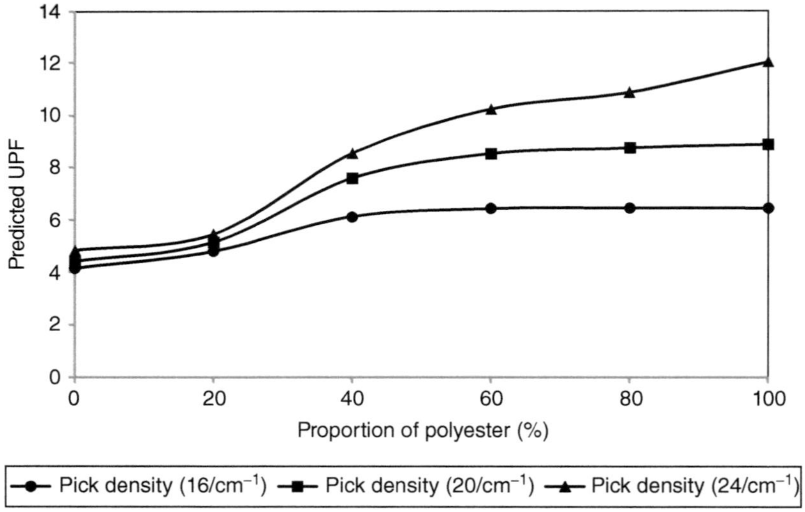

Figure 21.12 Effect of proportion of polyester and pick density on UPF. Source: Hatua et al. 2013 [53].

eight levels. From Figure 21.13, it is observed that when proportion of polyester, weft count, and pick density are 95%, 30 Ne, and 20 cm^{-1} respectively, four fuzzy rules (rules 5–8) are firing with different strength indicated by the heights of the blue pillars. Final output (UPF) of 8.87 is obtained by defuzzyfying, i.e. calculating the weighted average of these four fuzzy sets. It can be understood from rule 3 that the lowest UPF is obtained with the combination of the lowest proportion of

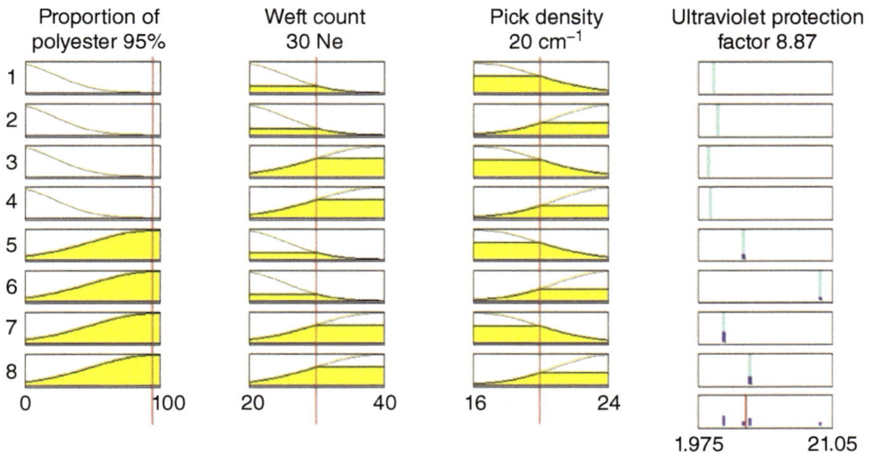

Figure 21.13 Fuzzy linguistic rules. Source: Hatua et al. 2014 [54].

polyester, the finest weft yarn, and the lowest pick density. Similarly, rule 6 implies that the highest UPF is obtained with the combination of the highest proportion of polyester, the coarsest weft yarn, and the highest pick density. Comparing rules 1 and 2, it can be inferred that UPF increases with increasing pick density keeping the proportion of polyester and weft count constant. Similarly, it can be said from rules 1 and 3 that UPF increases with coarser weft yarns if the other two input parameters are kept constant. Comparing rule 1 and 5, it can be inferred that UPF increases with the increasing proportion of polyester keeping the remaining two input parameters constant. These results are in agreement with the outcome of trend analysis of ANN model.

21.3.2.4 Applications in Nonwoven Fabrics

Rawal et al. [55] have used ANN modeling to predict the tensile properties of needlepunched nonwoven fabrics by relating them with the process parameters, namely, web area density, punch density, and depth of needle penetration. Out of 27 samples, 21 data sets were randomly chosen for training of ANN model, and remaining six data sets were for the testing purpose. Table 21.7 shows the summary of prediction performance of ANN models in training and testing data sets. Correlation coefficients between the experimental and predicted values were found to be excellent. The prediction error was also at acceptable level even in the testing or unseen data sets. Furthermore, the experimental and predicted values for individual test samples were reasonably close, as shown in Table 21.8.

21.4 Conclusions

Soft computing has opened a new horizon for fibrous material modeling and optimization research. These techniques have complementary strengths as ANN is good in modeling, FL is potent in imprecision handling, and GA is very powerful

Table 21.7 Prediction performance of ANN model.

Statistical parameter	Bulk density		TMD		TXMD	
	Training	Testing	Training	Testing	Training	Testing
Coefficient of correlation (R)	0.986	0.907	0.997	0.986	0.997	0.982
Mean absolute % error	5.40	6.70	3.17	9.21	3.69	6.71

Table 21.8 Comparison between experimental and predicted properties of needlepunched nonwoven fabrics.

Sample	TXMD (kN/m)			TMD (kN/m)		
	Measured	Predicted	Absolute error (%)	Measured	Predicted	Absolute error (%)
1	50.90	53.61	5.32	29.61	30.94	4.49
2	54.10	50.82	6.06	40.04	34.07	14.91
3	97.22	90.07	7.35	65.83	57.3	12.96
4	75.39	82.98	10.07	58.10	55.64	4.23
5	126.53	133.5	5.51	88.22	76.3	13.51
6	92.16	97.66	5.97	67.11	63.64	5.17

in solving optimization problems. These techniques in virgin or hybrid forms have been used to model gamut of yarn, fabric, and clothing properties with very good accuracy. However, soft computing tools are not panacea for modeling problems, and therefore they should be used with caution. These techniques should be used only in complex situations where simple modeling techniques like regression is not able to yield satisfactory prediction accuracy. It is also important to analyze whether soft computing tools are able to understand the underlying interactions among the input variables when the former is used as a modeling technique. It is envisaged that soft computing techniques will be used to solve complicated yarn, fabric, and clothing engineering related problems in future.

References

1 Zadeh, L.A. (1965). Fuzzy sets. *Information and Control* 8: 338–353.
2 Vassiliadis, S., Rangoussi, M., Cay, A., and Provatidis, C. (2010). Artificial neural networks and their applications in the engineering of fabrics. In: *Woven Fabric Engineering* (ed. P.D. Dubrovski), 111–134. Rijeka: Sciyo.
3 Chattopadhyay, R. and Guha, A. (2004). *Artificial Neural Networks: Applications to Textiles*. Manchester: The Textile Institute.
4 Majumdar, A. (2011). *Soft Computing in Fibrous Materials Engineering*. Manchester: The Textile Institute.
5 Rajasekaran, S. and Pai, G.A.V. (2003). *Neural Networks, Fuzzy Logic and Genetic Algorithms: Synthesis and Applications*. New Delhi: Prentice-Hall of India Pvt. Ltd.

6 Kartalopoulos, S.V. (2000). *Understanding Neural Networks and Fuzzy Logic: Basic Concepts and Applications*. New Delhi: Prentice-Hall of India Pvt. Ltd.
7 Rumelhart, D.E., Hinton, G.E., and Williams, R.J. (1986). Learning internal representations by error propagation. In: *Parallel Distributed Processing*, vol. 1, 318–362. Cambridge: MIT Press.
8 Haykin, S. (2004). *Neural Networks: A Comprehensive Foundation*, 2e. Singapore: Pearson Education.
9 Jurada, J.M. (1992). *Introduction to Artificial Neural Networks*. New York: West Publishing Company.
10 Bose, N.K. and Liang, P. (1998). *Neural Network Fundamentals with Graphs, Algorithms and Applications*. New Delhi: Tata McGraw Hill Publishing Company.
11 Hagan, M.T. and Menhaj, M.B. (1994). Training feed forward networks with the Marquardt algorithm. *IEEE Transactions on Neural Networks* 5 (6): 989–993.
12 Yu, H. and Wilamowski, B.M. (2010). Levenberg–Marquardt training. In: *Industrial Electronics Handbook* (eds. B.M. Wilamowski and J.D. Irwin), 12-1–12-16. Boca Raton, FL: CRC Press.
13 Zimmerman, H.J. (1996). *Fuzzy Set Theory and Its Applications*, 2e. New Delhi: Allied Publishers Limited.
14 Klir, G.J. and Yuan, B. (2000). *Fuzzy Sets and Fuzzy Logic: Theory and Applications*. New Delhi: Prentice-Hall of India Pvt. Ltd.
15 Berkan, R.C. and Trubatch, S.L. (2000). *Fuzzy Systems Design Principles*. New Delhi: Standard Publishers Distributors.
16 Dubois, D. and Prade, H. (1979). Fuzzy real algebra, some results. *Fuzzy Sets and Systems* 2: 327–348.
17 Holland, J.L. (1973). Genetic algorithms and their optimal allocation of trials. *SIAM Journal of Computing* 2 (2): 88–105.
18 Deb, K. (2001). *Multiobjective Optimization Using Evolutionary Algorithms*. Chichester: Wiley.
19 Jang, J.S.R. (1993). ANFIS: adaptive network based fuzzy inference system. *IEEE Transactions on Systems, Man, and Cybernetics* 23: 665–685.
20 Pynckels, F., Kiekens, P., Sette, S. et al. (1995). Use of neural nets for determining the spinnability of fibres. *Journal of the Textile Institute* 86: 425–437.
21 Cheng, L. and Adams, D.L. (1995). Yarn strength prediction using neural networks, Part I: Fibre properties and yarn strength relationship. *Textile Research Journal* 65: 495–500.
22 Ethridge, D. and Zhu, R. (1996). Prediction of rotor spun cotton yarn quality: a comparison of neural network and regression algorithms. In: *Proceedings of the Beltwide Cotton Conference*, Memphis TN, vol. 2, 1314–1317.
23 Zhu, R. and Ethridge, M.D. (1996). The prediction yarn irregularity based on the AFIS measurement. *Journal of the Textile Institute* 87: 509–512.
24 Sette, S., Boullart, L., and Van Langenhove, L. (1996). Optimizing a production process by a neural network/genetic algorithm approach. *Engineering Applications of Artificial Intelligence* 9: 681–689.

25 Rajamanickam, R., Hansen, S.M., and Jayaraman, S. (1997). Analysis of the modelling methodologies for predicting the strength of air-jet spun yarns. *Textile Research Journal* 67: 39–44.

26 Zhu, R. and Ethridge, M.D. (1997). Predicting hairiness for ring and rotor spun yarns and analyzing the impact of fibre properties. *Textile Research Journal* 67: 694–698.

27 Pynckels, F., Kiekens, P., Sette, S. et al. (1997). The use of neural nets to simulate the spinning process. *Journal of the Textile Institute* 88: 440–447.

28 Guha, A., Chattopadhyay, R., and Jayadeva (2001). Predicting yarn tenacity: a comparison of mechanistic, statistical and neural network models. *Journal of the Textile Institute* 92: 139–145.

29 Chattopadhyay, R., Guha, A., and Jayadeva (2004). Performance of neural networks for predicting yarn properties using principal component analysis. *Journal of Applied Polymer Science* 91: 1746–1751.

30 Majumdar, P.K. and Majumdar, A. (2004). Prediction of ring spun cotton yarn elongation using mathematical, statistical and artificial neural network models. *Textile Research Journal* 74: 652–655.

31 Majumdar, A., Majumdar, P.K., and Sarkar, B. (2005). Application of adaptive neuro-fuzzy system for the prediction of cotton yarn strength from HVI fibre properties. *Journal of the Textile Institute* 96: 55–60.

32 Majumdar, A., Sarkar, B., and Majumdar, P.K. (2006). An investigation on yarn engineering using artificial neural networks. *Journal of the Textile Institute* 97: 429–434.

33 Majumdar, A., Ciocoiu, M., and Blaga, M. (2008). Modelling of ring yarn unevenness by soft computing approach. *Fibers and Polymers* 9: 210–216.

34 Majumdar, A. and Ghosh, A. (2008). Yarn strength modelling by fuzzy expert systems. *Journal of Engineered Fibres and Fabrics* 3 (4): 61–68.

35 Ghosh, A. and Chatterjee, P. (2010). Prediction of cotton yarn properties using support vector machine. *Fibers and Polymers* 11: 84–88.

36 Ghosh, A., Das, S., and Saha, B. (2016). Simulation of stress–strain curves of polyester and viscose filaments. *Journal of Institution of Engineers India Series E* 96 (2): 139–143.

37 Mishra, S. (2016). Prediction of yarn strength utilization in cotton woven fabrics using artificial neural network. *Journal of Institution of Engineers India Series E* 96 (2): 151–157.

38 Majumdar, A., Ghosh, A., Saha, S.S. et al. (2008). Empirical modelling of tensile strength of woven fabrics. *Fibers and Polymers* 9 (2): 240–245.

39 Hadizadeh, M., Jeddi, A.A.A., and Tehran, M.A. (2009). The prediction of initial load extension behaviour of woven fabrics using artificial neural network. *Textile Research Journal* 79: 1599–1609.

40 Fan, J. and Hunter, L. (1998). A worsted fabric expert system: Part I: System development. *Textile Research Journal* 68 (9): 680–686.

41 Fan, J. and Hunter, L. (1998). A worsted fabric expert system, Part II: An artificial neural network model for predicting the properties of worsted fabrics. *Textile Research Journal* 68 (10): 763–771.

42 Behera, B.K. and Guruprasad, R. (2012). Predicting bending rigidity of woven fabrics using adaptive neuro-fuzzy inference system (ANFIS). *The Journal of the Textile Institute* 103 (11): 1205–1212.

43 Ucar, N. and Ertugrul, S. (2002). Predicting circular knitting machine parameters for cotton plain fabrics using conventional and neuro-fuzzy methods. *Textile Research Journal* 72 (4): 361–366.

44 Bilisik, K. and Demiryurek, O. (2011). Analysis and tensile-tear properties of abraded denim fabrics depending on pattern relations using statistical and artificial neural network models. *Fibers and Polymers* 12 (3): 422–430.

45 Zeydan, M. (2008). Modelling the woven fabric strength using artificial neural network and Taguchi methodologies. *International Journal of Clothing Science and Technology* 20 (2): 104–118.

46 Hadizadeh, M., Tehran, M.A., and Jeddi, A.A.A. (2010). Application of an adaptive neuro fuzzy system for prediction of initial load/extension behavior of plain-woven fabrics. *Textile Research Journal* 80 (10): 981–990.

47 Bhattacharjee, D. and Kothari, V.K. (2007). A neural network system for prediction of thermal resistance of textile fabrics. *Textile Research Journal* 77 (1): 4–12.

48 Fayala, F., Alibi, H., Benltoufa, S., and Jemni, A. (2008). Neural network for predicting thermal conductivity of knit materials. *Journal of Engineered Fibers and Fabrics* 3 (4): 53–60.

49 Majumdar, A. (2011). Modelling of thermal conductivity of knitted fabrics made of cotton–bamboo yarns using artificial neural network. *The Journal of the Textile Institute* 102 (9): 752–762.

50 Tokarska, M. (2004). Neural model of the permeability features of woven fabrics. *Textile Research Journal* 74 (12): 1045–1048.

51 Brasquet, C. and Le Cloirec, P. (2000). Pressure drop through textile fabrics-experimental data modeling using classical models and neural networks. *Chemical Engineering Science* 55: 2767–2778.

52 Brasquet, C.F. and Le Cloirec, P. (2003). Modelling of the flow behavior of activated carbon cloths using a neural network approach. *Chemical Engineering and Processing* 42: 645–652.

53 Hatua, P., Majumdar, A., and Das, A. (2013). Predicting the ultraviolet radiation protection by polyester-cotton blended woven fabrics using nonlinear regression and artificial neural network models. *Photodermatology, Photoimmunology and Photomedicine* 29: 182–189.

54 Hatua, P., Majumdar, A., and Das, A. (2014). Modeling ultraviolet protection factor of polyester-cotton blended woven fabrics using soft computing approaches. *Journal of Engineered Fibres and Fabrics* 9: 99–106.

55 Rawal, A., Majumdar, A., Anand, S., and Shah, T. (2009). Predicting the properties of needlepunched nonwovens using artificial neural network. *Journal of Applied Polymer Science* 112: 3375–3338.

22

Fiber-Shaped Electronic Devices

Yang Zhou, Jian Fang, Yan Zhao, and Tong Lin

Deakin University, Institute for Frontier Materials, Geelong Waurn Ponds Campus, Pigdons Road, Geelong, VIC 3216, Australia

22.1 Introduction

Wearable electronic devices are designed to be soft, flexible, and even stretchable, to adapt to bodies' shape and movement [1–4]. They are very useful for health monitoring, activity tracking, military garments, and energy conversion. Early wearable electronic devices were prepared by mounting electronic components on clothes or accessories. Some wearable electronics such as smart watches and wristbands have been commercialized for health monitoring. Since based on silicon electronics, they are rigid and unwashable, hence unsuitable for use in textiles. To improve device flexibility, great attention has been paid toward integrating functional components (e.g. integrated circuits, electrodes) into flexible plastic films. These flexible devices, however, still have poor wear comfort feature due to impermeable nature of the substrates to air and moisture. Meanwhile, these flexible devices are vulnerable to withstand twisting or wrapping [5, 6]. An alternative way to fabricate wearable electronics is to incorporate electronic functions into fibers and then weaving or knitting the electronic fibers into fabrics or textiles. Wearable electronic textiles prepared in this way possess high air permeability. They allow to prepare multifunctional devices simply by assembling fibers of different functions into the same fabric [7–9].

Considerable efforts have been made over recent decade to prepare fiber-shaped electronics, including energy generators [10, 11], energy storage devices [12–14], electrochromic fibers [15], and transistors [16]. The fiber-based energy generators can generate electricity from solar energy or mechanical vibration. They are expected to supply power to personal electronic devices and develop self-powered smart textiles. Electrochromic fibers, which can change color upon being charged with an electric potential, were reported with the purposes to prepare textiles with displaying feature [17]. Figure 22.1 shows an overview of fiber-shaped electronic devices.

Although the significant progress have been made in wearable fiber-based electronics, most of the existing fiber-shaped devices show inferior performance

Handbook of Fibrous Materials, First Edition. Edited by Jinlian Hu, Bipin Kumar, and Jing Lu.

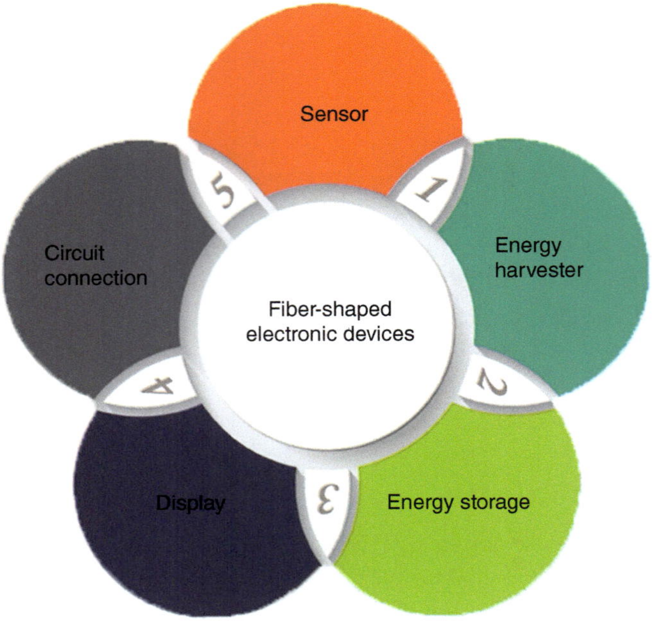

Figure 22.1 Overview of applications of fiber-based electronics.

when compared with their conventional counterparts. Challenges still exist in making flexible fiber-shaped devices. The small size and non-flat structure of fibers increase complicity during device fabrication. The working environment of wearable electronics requires the devices withstanding repeated squeezing or bending deformations caused by body movement. The durability against wearing and washing is critical.

In this chapter, we summarize typical device structures that have been employed in fiber-shaped electronics, along with an introduction of electrode materials and recent progresses in fiber electronics. The development trend of fibrous smart electronics is discussed as well.

22.2 Fiber-Shaped Electronic Devices

Four different structures of fiber-shaped electronic devices have been reported. They all consist of electrodes and functional layers (also called "active layers" in some papers). According to structure features, these devices are named as fiber electrode wrapped fiber devices (FEWFDs), twisted fiber devices (TFDs), sheath–core single fiber devices (SCSFDs), and parallel coil fiber devices (PCFDs). Figure 22.2 illustrates the typical structures of the devices, and the details about each devices are described below.

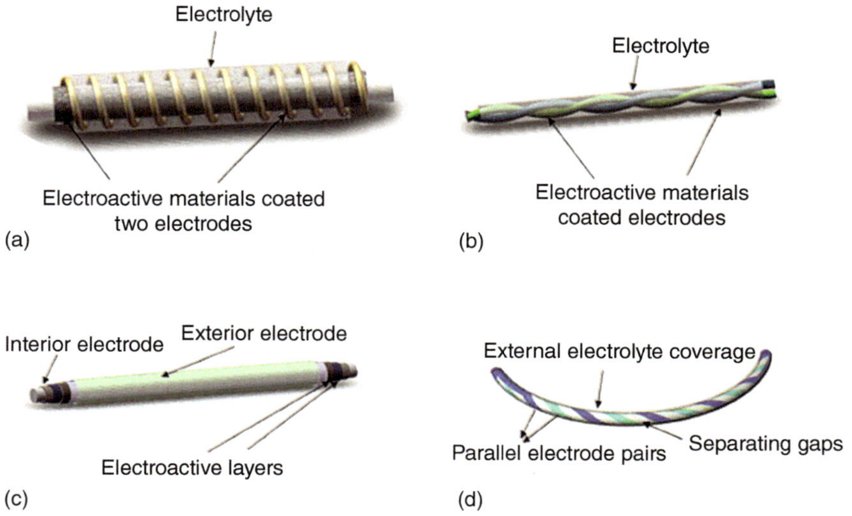

Figure 22.2 Typical structures of the fiber-shaped electronic devices: (a) FEWFDs, (b) TFDs, (c) SCSFDs, and (d) PCFDs.

22.2.1 Twisted Fiber Devices (TFDs)

TFDs are prepared by twisting two fibrous electrodes together. The fibrous electrodes can be made either by the electrode materials in bulk (e.g. carbon fiber or metal wire) or conducting coating on a polymer fiber. The active layer is loaded on either of the electrodes or the both. The twisting feature endows the TFD devices with sufficient specific surface area and contact points for occurrence of electrochemical reactions, which is beneficial for performance optimization. TFDs do not require an external substrate to hold electrodes. Hence, the fabrication process and dimension of TFDs can be simplified and minimized, respectively. The performance and stability of TFD devices are significantly affected by twisting parameters (e.g. twisting tightness, pitches), fiber diameters, and thickness of the active layers. Zhang et al. [18] reported the influence of twisting features on the performance of a TFD solar cell, where Ti wire and multiwalled carbon nanotube (MWCNT) fiber were used as two electrodes. It has been revealed that a TFD with an optimal pitch distance of 750 μm can provide the highest efficiency. Reducing pitch distance increases the possibility of blocking the incident light by MWCNT fibers, whereas increasing pitch distance will also reduce the contact area between two electrodes. It is mentioned as well that either over-twisted or under-twisted electrodes can affect device stability. A further investigation on tightness has been conducted by Chen et al. [19], where they reported that the twist tightness played a crucial role in deciding the performance of fiber-shaped solar cells. Increasing the tightness would increase the twisting stress and the possibility of short circuit within the device, whereas the efficiency will be reduced when the twists become loosed due to the decreased intimate contact. The major concern of TFDs is the susceptibility to physical deformation when being used. For example, frequent fold, rub, or bending could result in loosing or untwisting the structure, or motion between two electrodes. To obtain better stability, fibrous electrodes for TFDs usually have similar diameters and mechanical properties.

22.2.2 Fiber Electrode Wrapped Fiber Devices (FEWFDs)

FEWFDs are prepared by wrapping a fibrous electrode on another central fibrous electrode. The two electrodes in FEWFDs are set asymmetrically, and the mechanical stretchability and elasticity of the devices can be turned through the central fiber. Electroactive materials are deposited onto either or both fibrous electrodes. For example, Wang et al. [20] deposited a layer of TiO_2 on Ti wire electrode to improve the power conversion efficiency (PCE) of a fiber solar cell. The TiO_2 layer provided the pathway for charge transport, and it served as a separator to avoid short circuit, which could be caused by the close contact between two electrodes. In most cases, electron and charge transport occurs within the effective area defined by two electrodes and the surrounded electrolyte. The effective area in the FEWFDs can be tuned by varying the angle and density of the wrapping fibers, which are key parameters to affect performance. Chen et al. [19] discussed the influence of various screw pitches on the performance of a FEWFD solar cell. The device with 0.8 mm pitch distance has the largest current density, whereas the device shows the lowest current density when the pitch distance was 1.4 mm. They explained that lower screw distance increased the number of contact point, which facilitates charge transport and increases the short circuit current density. In addition, increasing the coil density could block the incident light captured by the central electrode, resulting in decrease in light harvesting performance. Besides, exfoliation of coating layer might occur when high friction was introduced during wrapping the electrode.

Bending could lead to disconnection of the electrode. As shown in Figure 22.3, disconnection occurs between two electrodes after repeated bending. The PCE reduced from 0.48% to 0.14% after 50 cycles of bending the device.

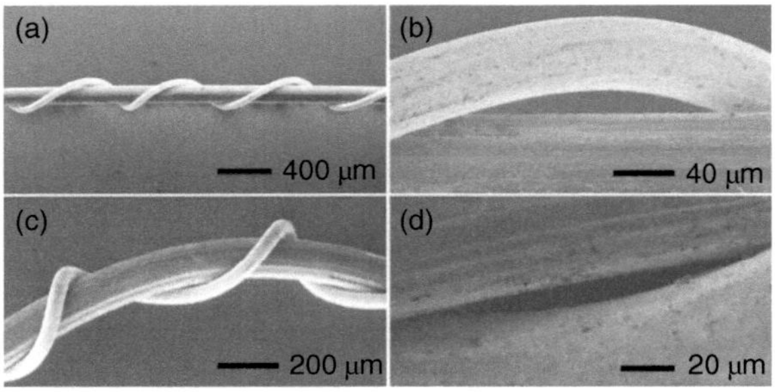

Figure 22.3 Fiber-shaped solar cell with FEWFD structure before bending at (a) low and (b) high magnifications, respectively. The twisted, wire-shaped polymer solar cell with the silver wire as the electrode after bending at (c) low and (d) high magnifications, respectively. Source: Zhang et al. 2014 [18]. Reprinted with permission of John Wiley & Sons.

The wrapping electrode offer a possibility to hold liquid electrolyte in the device. Liu et al. [21] reported a FEWFD solar cell using liquid electrolyte. The capillary effect originated from the gap between the two fiber-shaped electrode enables liquid to be stored in the space.

22.2.3 Sheath–Core Single Fiber Devices (SCSFDs)

SCSFDs have a multiple layer structure. They have a similar configuration to sandwich structured flat electronic device. In general, the two electrodes are separated by active layers in between. Unlike FEWFDs and TFDs where two fibers are used as electrodes, SCSFDs have a sheath–core structure that reduces device dimension because all components are integrated into a fibrous substrate. Because of the intimate contact among the materials, SCSFDs have higher utilization and better physical stability than FEWFDs and TFDs [22]. Zhang et al. [23] reported a SCSFD-structured DSSC with Ti wire and carbon nanotube (CNT) film electrodes. They indicated that the compact contact between two electrodes facilitated ion diffusion toward counter electrode. This allows electrons accumulated at the counter electrode to be captured timely by triiodide ions. Since the electrolyte is infiltrated between the two electrode layers, leakage of the electrolyte substance is eliminated. In addition, the compact electrode layers avoid potential physical disconnection between the electrode and the active layer. Peng [22] highlighted superior volume utilization in SCSFDs by suggesting that the cross-sectional area of a SCSFD ($2\pi r^2$) is two times larger than that of TFDs ($4\pi r^2$) when other parameters were controlled same (see the illustration in Figure 22.4).

Chen et al. [24] indicated that the SCSFD supercapacitor showed better performance than those with TFD structure. Based on Nyquist plot results, they concluded that the internal resistance of the SCSFD was lower than that of TFD. The galvanostatic charge–discharge plot also suggested that SCSFD required less time for a charging–discharging cycle.

A limitation for SCSFDs is the low transparency of the external electrode layer. This affects device performance when used for solar cells, touch screen, and display. Since incident light is blocked by multiple layers of active substances, SCSFD-structured solar cells allow less light to reach electroactive layer [25, 26].

22.2.4 Parallel Coil Fiber Devices (PCFDs)

PCFDs were later developed than other types of fiber devices. They combine the structure features of FEWFDs and SCSFDs. However, PCFDs differ to FEWFDs

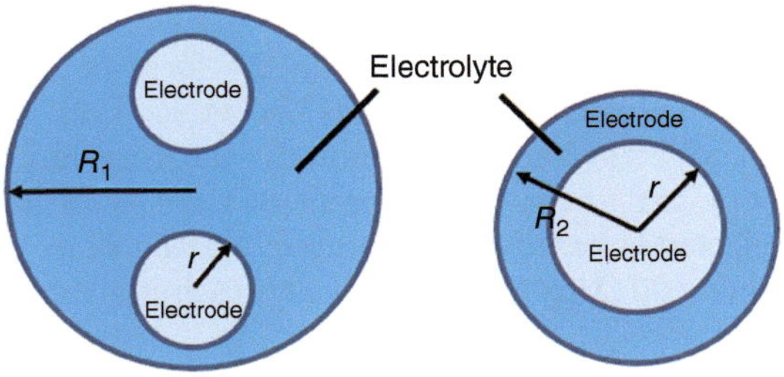

Figure 22.4 Cross-sectional comparison between fiber-shaped supercapacitor with twisting (left) and coaxial (right) structure with identical amount of electrode materials. Source: Peng 2015 [22]. Reproduced with permission of Springer.

and SCSFDs in that the fiber in PCFDs functions like a substrate. Two electrodes covered around the fiber substrate in the shape of parallel coils. The electrodes on the fiber substrate can be formed through a coating or deposition technology. This makes the parallel coils look like a structured sheath. Therefore, PCFDs have higher structure stability than FEWFDs and TFDs. Important structure parameters, features, and production techniques of abovementioned structures are summarized in Table 22.1.

22.3 Electrode Materials

Electrode materials play an essential role in deciding device properties. The electrode materials are often required to have the following features:

(1) Be electrically conductive to complete the circuit.
(2) Provide sufficient surfaces for connection with electroactive materials.
(3) Be able to form effective contact with active substances.
(4) Be stable in both physical and chemical properties.

The commonly used electrode materials for fibrous electronic devices include metal/metal oxides, carbon nanomaterials, and conducting polymers.

22.3.1 Metals and Metal Oxides

Metals are the mostly used electrode material because they are widely available and have high conductivity. For making fiber-shaped electronic devices, the metal electrode can be prepared either in the form of metallic wires or in a thin coating on a fiber substrate. The techniques to fabricate metal electrodes for fiber electronics include wrapping metal wires on the fibrous substrate or twisting with another fiber electrode. In some cases, the metallic wires are coated with a layer of electroactive materials on surface such as TiO_2 to enhance its light absorbing performance in DSSC [20].

For substrates that are intrinsically poor in conductivity, a wide range of coating techniques can be used, including sputter coating, thermal evaporation, electrodeposition, and dipcoating. Sputter coating and thermal evaporation are suitable mainly for flat surfaces, whereas wet chemical methods could suit both curve and flat surfaces.

Apart from metallic wires, indium tin oxide (ITO) is a widely employed owing to high transparency. Thermal evaporation, sputter coating, chemical vapor deposition (CVD), spray pyrolysis, and plating have been used to apply ITO on substrates. Most coating methods can be performed at room temperature, and they are cost effective and environment friendly as well [30, 31].

Metal and metal oxides in the form of nanowire possess both good transparency and conductivity. Lee et al. [32] reported the preparation of ITO nanowires, and the conductive coating has an optical transmittance up to 90% and sheet resistance of 89 Ω/sq. Kim et al. [33] prepared a transparent electrode based on Ag nanowires. The sheet resistance of the coating was less than 100 Ω/sq, and the optical transmittance was around 90%. By optimizing the

Table 22.1 Structure parameters and feature of the fiber-shaped electronic devices.

	Structure parameters	Preparation	Advantages	Disadvantages	References
TFDs	Twisting tightness Screw pitch	Twisting two fiber electrodes manually or via a twisting machine	High specific surface area Miniature in size Large incident area All-direction incident light received	Possibility of short circuit Ineffective contact Twisting stress	[18, 19, 27]
FEWFDs	Wrapping angle Screw pitch Wrapping tightness	Preparing central and sheath electrodes. Wrapping the sheath electrode onto the central electrode	More effective contact between electrodes Capillary effect	Wrapping may block incident light Gaps exist in both straight and bending states	[19, 28, 29]
SCSFDs	Coating thickness Substrate	Layer-by-layer covering the fiber substrate with the electrode electroactive materials and electrode	Intimate contact Compact size Structural stability	Exterior layer may block incident light Possibility of exfoliation	[22, 23, 25, 27]
PCFDs	Pattern design Coating thickness Substrate	Making pre-patterned on fiber substrate, then coating the patterned substrate with electrode materials, and finally removing the template	Parallel electrode Compact size Structural stability	Possibility of exfoliation	[17]

nanowire diameter and length, the nanowire coatings could be transparent in near-infrared region [34].

In addition, some transition metal oxides like zinc oxide (ZnO), manganese dioxide (MnO_2), and ruthenium dioxide (RuO_2) were used as electrode materials. They show great potential in making anode in lithium batteries and supercapacitors owing to the reversible redox reaction and high charge storage capacity [35]. When they are shaped into nanostructures, their electrical conductivity can be substantially improved. The synthesis of such transition metal oxides requires low working temperature and can be applied on a variety of surface especially on textiles and fabrics [36–38].

22.3.2 Carbon-Based Materials

Carbon-based materials such as CNTs, carbon nanofibers, and graphene have high carrier mobility, conductivity, and mechanical properties [39, 40]. Despite they have relatively lower electrical conductivity than metals, their chemical stability is much higher, and the conductivity can be improved by addition of conducting components [41].

The mechanical properties of CNTs were reported to have a tremendous effect on device performance. Liu et al. [28] reported that the PCE of a fibrous polymer solar cell can be improved from 0.8% to around 2% when silver wire electrode was replaced with CNT fiber. Firstly, high flexibility of CNTs led to superior performance by providing better-fitted contact area. The stability was improved since CNT films maintained their continuous morphology and adhered to substrate when the fiber substrate was under bending status. Secondly, the possibility of short circuit caused by the penetration of rigid metal wires on functional polymer layers can be reduced. The outstanding mechanical stability of CNTs allows them to be used to develop elastic electrodes. For example, a rubber fiber with a thin layer of CNTs sheet wrapped was used to prepare a fiber-shaped supercapacitor. Such a device can be stretched over 400% of its original length and still maintains the specific capacitance at 79.4 F/g [15].

CNT sheets are usually prepared by layer cross-stacking of aligned CNTs [42]. The number of layers and stack direction were used to modify the properties. When CNTs were doped with nitrogen, defects developed on the walls. The presence of the defects facilitates the diffusion of ions in batteries [43]. This improves battery performance where doped CNTs were used. It also enhances the capacitance when used as electrode material for supercapacitors [44].

Since electrode transparency is an important factor for wearable solar cell and organic light-emitting diodes (LEDs), the transparency of carbon-based materials has been improved by many methods. For example, optimization of rod coating or spin coating methods allowed to prepare an electrode with the transparency as high as 85% [45–47], and electrode from hybrid CNTs, fin-like Fe/CNTs, for instance, have an optical transmittance of 88% [48].

Graphene possesses high strength (0.1–0.5 GPa), excellent electrical conductivity (10^4 S/m), and wide potential window. Graphene has attracted intensive interest for making electrode [49–52]. Graphene-based materials can be used either in a conventional two-dimensional form or three-dimensional structures (e.g. aerogels and foams).

22.3.3 Conducting Polymers

Conducting polymers possess large conjugated structure. The electrical conductivity, solubility, optical properties, and stability of conducting polymers can be tailored by modifying the polymer structure. For example, the electrical conductivity and solubility of poly(3,4-ethylenedioxythiophene) (PEDOT) can be improved by doping the polymer with polystyrene sulfonate (PSS), which enables the easy applying of PEDOT on various substrates through dipcoating. Hou et al. [53] prepared dimethyl sulfoxide (DMSO)-doped PEDOT electrode, which had a high conductivity of 109 S/cm. Besides, the morphology of conducting polymers (e.g. nanofibers, nanotubes) also affects performance. Huang et al. [54] reported that polypyrrole (PPy) nanofibers with a diameter of 60–100 nm exhibited excellent conductivity (20–130 S/cm) at ambient temperature. Oh et al. [55] synthesized PPy nanowires via electrodeposition, and the PPy nanowires in aqueous dispersion had excellent electrochemical activities and strong emission. Table 22.2 summarizes typical electrode materials reported for wearable electronic devices.

Table 22.2 Summary of typical electrode materials used in wearable electronic devices.

Materials	Type	Features	References
Ti wire	Metal	Suitable for TiO_2 modification	[56–60]
Stainless wire	Metal	Anticorrosion	[61, 62]
Aligned CNTs sheet	Carbon-based	Good electrical and thermal conductivity High Young's modulus	[63–65]
Graphene/graphene oxide	Carbon-based	High specific surface area High theoretical specific capacitance Thermal conductivity	[66–68]
Carbon	Carbon-based	Good creep resistance Scalable production	[14, 69]
PANI/PEDOT	Conducting polymers	Redox reactions	[68, 70–72]
MnO_2	Metal oxide	High charge storage capacity for solar cell	[14]
ZnO	Metal oxide	Morphology can be precisely controlled Piezoelectricity	[73, 74]
$LiMn_2O_4/Li_4Ti_5O_{12}$	Metal oxide	High working voltage Little volume changes Excellent ion transport	[75, 76]

22.4 Applications

Fiber-shaped electronic devices have shown application potential in diverse fields, including energy generators, energy storage devices, electrochromic fibers, and transistors. The device structure and application performances of the fiber-shaped electronic devices are detailed below.

22.4.1 Energy Harvesters

22.4.1.1 Thermal Energy Harvesters

Thermoelectric devices convert temperature differences into electrical signal based on temperature gradient. Although the output voltage of thermoelectric devices remains small, the ability to convert heat energy, which is ubiquitous but usually cannot be effectively utilized, to electricity is favorable. Being as wearable electronics, thermoelectric devices can utilize the intrinsic temperature gradient from the human body, which is a promising feature over other electricity generator.

A thermoelectric device consists of thermoelectric material and two electrodes that used to conduct generated current. As early as 2008, Yadav et al. [77] reported a yarn-based thermoelectric generator as shown in Figure 22.5, where Bi_2Te_3 and Sb_2Te_3 served as n- and p-type semiconductors, respectively. Silicon fiber with a diameter of 710 μm was used as substrate. Silver and nickel were deposited via thermal deposition in an alternating pattern onto fiber substrates as metal hosts for heating. This fiber generator exhibited a maximum power of 2 nW for seven couples when ΔT was 6.6 K. Investigation on structure–performance relationship was conducted via finite element analysis (FEA), indicating that the power generated per couple increased firstly and then decreased along with the increase of segment length as depicted in Figure 22.5d. Before a maximum value was reached, the improvement of power output was

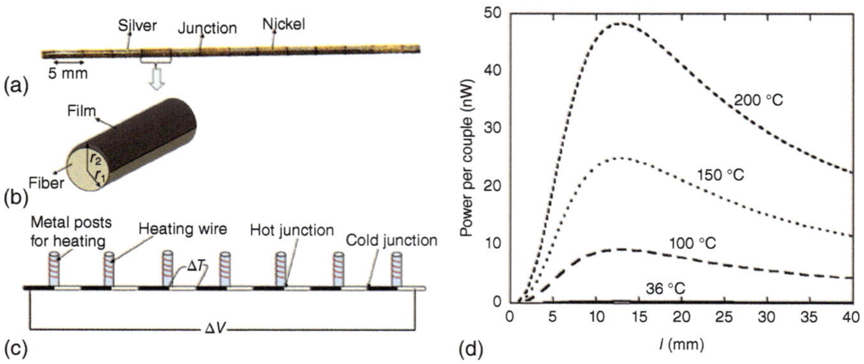

Figure 22.5 (a) Schematic of alternative coating of thermoelectric fibers. (b) Illustration of fiber with thin film deposited on one side. (c) Schematic of experimental setup for applying a temperature gradient and measuring the induced open-circuit voltage. (d) Power per couple vs. the segment length for different hot junction temperatures. Source: Yadav et al. 2008 [77]. Reprinted with permission of Elsevier.

attributed to the increase of temperature gradient caused by elongated segments. Afterward, the increasing resistance caused by excess length dominates, leading to a decline in final output.

It is known that the integration of electronics into textile will result in lower performance than theoretical value. To obtain an in-depth understanding of thermoelectric effects under practice scenario, Lee et al. [78] compared the output signals under various stitches, as shown in Figure 22.6. Flexible yarns were obtained by twisting basic n-type Bi_2Te_3 and p-type Sb_2Te_3 components. The yarns were then woven into zigzag stitch, garter stitch, and plain weave structures. The zigzag structure was easy to fabricate but had the lowest performance of 0.166 μW per couple since the requirement of insulating yarns reduced thermoelectric couples within a certain area. The garter stitch did not

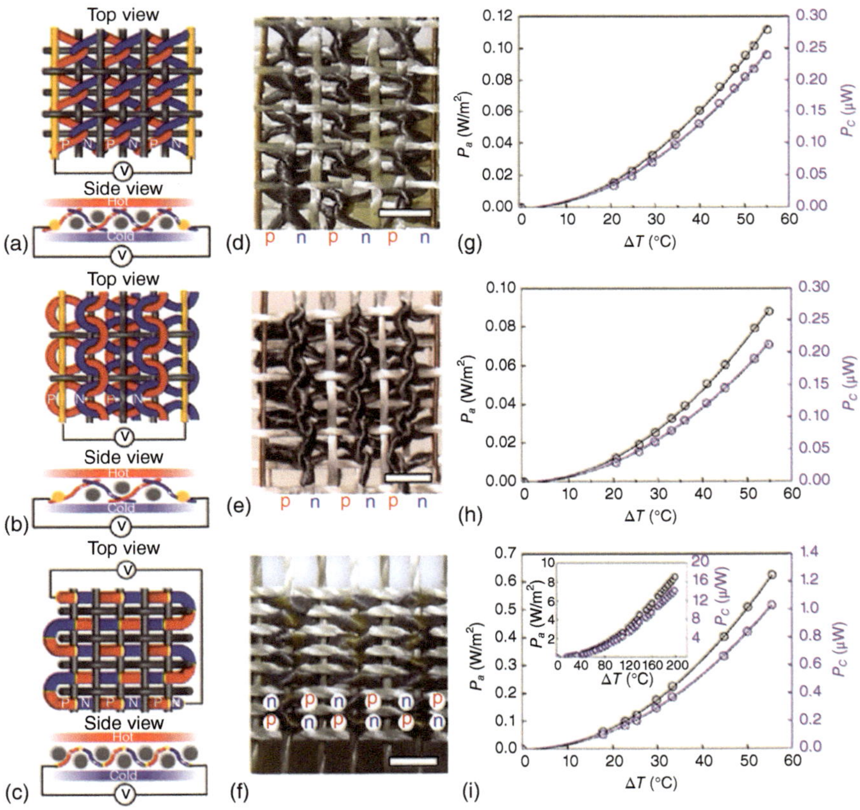

Figure 22.6 Structure and performance of TE textiles. (a–f) Schematic illustrations (a–c) and photographs (d–f) of realized zigzag stitch, garter stitch, and plain weave TE textiles, respectively. The scale bars are 2 mm long. (g–i) The output power per textile area (P_a) and per TE couple (P_c), measured as a function of the temperature difference (ΔT) for zigzag stitch, garter stitch, and plain weave TE textiles, respectively. The inset of (i) shows the output power of a plain weave TE textile for a ΔT of up to 200 °C. Source: Lee et al. 2016 [78]. Reprinted with permission of John Wiley & Sons.

need the presence of insulating yarn. The insulating yarns in garter stitch are used to control junction location during hand weaving process. The plain wave structure was used to provide thermoelectric materials connected in-series. The resulting thermoelectric textile has an outstanding output power up to 8.56 W/cm^2 when a temperature difference of 200 °C was applied toward textile thickness direction.

22.4.1.2 Solar Cells

The research on fiber-shaped solar cells focuses on three cell types: dye-sensitized solar cells, polymer solar cells, and perovskite solar cells. Although the exact mechanism varies from type to type, they all convert solar energy into electricity via photovoltaic effect. In general, dye-sensitized solar cells exhibit higher PCE than polymer cell while requiring the use of liquid electrolyte. Hence, most fiber-shaped dye-sensitized solar cells employ quasi-solid electrolyte or a tube infiltrated with electrolyte to protect the device. The key merit of polymer solar cell is the all-solid-state device feature where no liquid electrolyte is used. The all-solid-state feature is specifically suitable for wearable electronics. As an emerging type of solar cell, the possession of both solid state and high efficiency has made perovskite solar cells in the spotlight. As all of them are subgroups under solar cell, we do not intentionally distinct them in Section 22.4.1.2. In general, fiber-shaped solar cells consist of stainless steel or Ti wires coated with a layer of titanium dioxide (TiO_2) working electrode, and CNT sheet counter electrode [79]. Electroactive materials such as sensitizer and electrolyte surround the two electrodes in the form of gel or solid states. Table 22.3 lists recently reported fiber-shaped solar cells and their performance features.

In 2002, Baps et al. [81] reported a fiber-shaped solar cell with SCSFD structure where a TiO_2-coated stainless wire served as internal electrode and conducting polymer as external electrode. The device can generate a potential around 0.3 V. This prototype has inspired people to transform conventional planar device into a fiber. Later on, investigation on fiber-based solar cells has been burgeoning over the past decades. Liu et al. [82] reported a fiber-shaped polymer solar cell with a PCE of 0.6%. Fan et al. [83] reported a TFD-structured DSSC in 2008. The working electrode was made of TiO_2-coated stainless steel fiber, and the counter electrode was a polymer protective layer coated with Pt wire. The twisting structure enabled that incident light from all directions and angles can be captured by electrodes. Polymer protective layer was employed to release local stress between two electrodes via elastic deformation. Hence, the back-transfer of carrier can be reduced. The optimal thickness of protective layer was achieved at 3.5 μm. This device with 5 cm length exhibited 610 mV open circuit voltage and 0.06 mA short-circuit current. The emergence of fiber-shaped perovskite solar cell was in 2014. Qiu et al. reported a SCSFD-structured fibrous solar cell [25]. Stainless steel wire was adopted as internal electrode, followed by the deposition of perovskite electroactive material $CH_3NH_3PbI_3$. Aligned CNT sheet with high transparency was used as external electrode. The all-solid-states perovskite fiber showed promising future for wearable applications with a PCE of 3.3%, which was higher than fiber-shaped dye-sensitized solar cells and polymer cells reported on that time.

Table 22.3 Recent development on fiber-shaped solar cells.

Year	Structure	Electrode	PCE (%)	Feature	References
2013	SCSFD	Ti/TiO_2/CNT	4.10	Stability during deformation Ultrathin	[56]
2013	SCSFD	Ti/TiO_2/CNT	2.6	High thermal stability and flexibility Quasi-solid-state	[57]
2014	SCSFD	Ti/TiO_2/CNT	6.83	Core–sheath structured CNT/graphene nanoribbon fibrous electrode	[58]
2014	FEWFD	Ti/TiO_2/CNT	5.64	Investigation on a set of carbon-based materials	[59]
2014	FEWFD	Ti/TiO_2/CNT	7.13	Elastic and stretchable	[65]
2014	TFD	Ti/TiO_2/P3HT: PCBM/PEDOT: PSS/CNT	1.78	Sealing free Scalable production	[18]
2014	SCSFD	Stainless steel/ TiO_2/$MAPbI_3$/ OMeTAD/CNT	3.3	All solid flexible	[27]
2015	SCSFD	Stainless steel/ ZnO/$MAPbI_3$/ OMeTAD/CNT	3.8	Facile preparation Obelisk-like ZnO layer	[80]
2015	FEWFD	Ti/TiO_2/CNT	5.47	Stretchable Ultradurable	[11]
2015	FEWFD	Ti/TiO_2/CNT	5.22	Stable under both stretching and bending	[60]

Afterward, numerous efforts have been made to improve PCE and mechanical properties. The PCE of fiber-shaped solar cells can be improved by optimization of electrode in transparency and conductivity. For example, fiber-shaped solar cell with a PCE of 7.1% was produced recently [84]. The aligned CNTs electrode exhibited more than 80% transmittance and a conductivity of 500 S/cm. Low-dimensional metals such as Au, Ag, and Cu nanowires, which possess both high transparency and conductivity, have been incorporated into fiber substrate [85–87]. Based on scattering effect, Li et al. [88] reported a strategy to capture more light by controlling the thickness of TiO_2 layer.

Additionally, considerable efforts have been made to enhance electrode properties. For instance, the alignment of CNTs can enhanced both mechanical and electrical properties [89–91]. Alignment of CNTs leads to lower contact resistance, less defects, efficient load transfer, and larger surface area when compared with unaligned CNTs [92]. Additionally, aligned CNTs were condensed into thin films, serving as both effective hole extraction layers that eliminated the hole

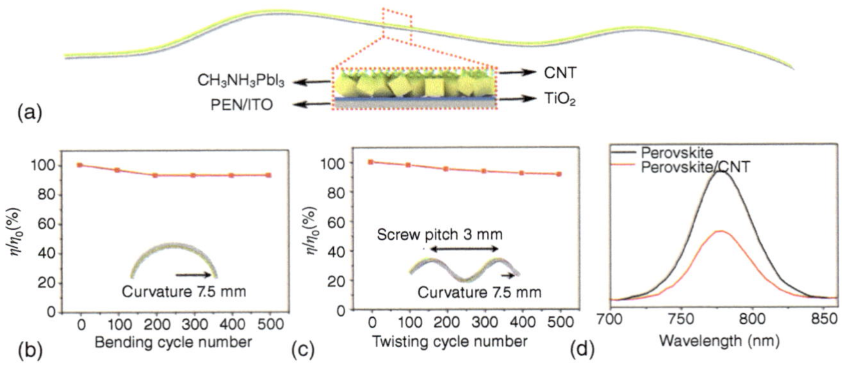

Figure 22.7 (a) Schematic of fiber perovskite solar cell, (b) performance decay under bending states, (c) performance decay under twisting states, and (d) fluorescence spectrum of devices with pristine perovskite and perovskite/CNT electrodes. Source: Qiu et al. 2016 [10]. Reprinted with permission of Royal Society of Chemistry.

transport layer. Benefited from aligned CNTs, Qiu et al. [10] reported a perovskite solar cell with high PCE up to 9.49% (Figure 22.7). It was stable under both bending and twisting operation modes. Significant decrease of fluorescence intensity in non-CNT contained perovskite indicated that CNTs facilitate the effective ion extraction process.

Apart from CNTs, silicon-based p-i-n photodiode junction fiber can be produced via high-pressure chemical vapor deposition (HPCVD) [93]. The fiber device possesses a coaxial p-i-n structure where amorphous silicon n^+, i, and p^+ layers were deposited, followed by crystallization process. The p-i-n photodiodes endowed the fibers with high quantum efficiency, which is desirable in solar cell. Zhang et al. [94] produced a SCSFD-structured solar cell based on $CuInSe_2$ (CIS) and Cds deposited Mo wire. The wire electrode was further coated by ZnO and ITO layers. The solar cell has a PCE of 2.31% under various working conditions. Due to the core–sheath structure where each layer was stacked firmly onto substrates, only 8% efficiency loss was observed when operating at 360° bending angle.

22.4.1.3 Mechanical Energy Harvesters

Mechanical energy harvesters can generate electricity from movement, which are particularly applicable for smart textiles because of the regular motion of wearers. Among many different types of mechanical energy harvesters, piezoelectric devices have attracted the majority of research interest. A piezoelectric device normally contains a layer of piezoelectric material and two electrodes on both sides. Piezoelectric outputs are generated under external force when the piezoelectric material experiences physical deformation. Qin et al. [74] reported a TFD-structured nanogenerator as elucidated in Figure 22.8a. Two fibers coated with ZnO and ZnO/Au were twisted together to form a piezoelectric device. Relative sliding and deflection of two fibers resulted in charge generation, where the uncoated ZnO produces potential across width direction and ZnO with gold coating is used for charge collection and transport. Figure 22.8b and c depicted

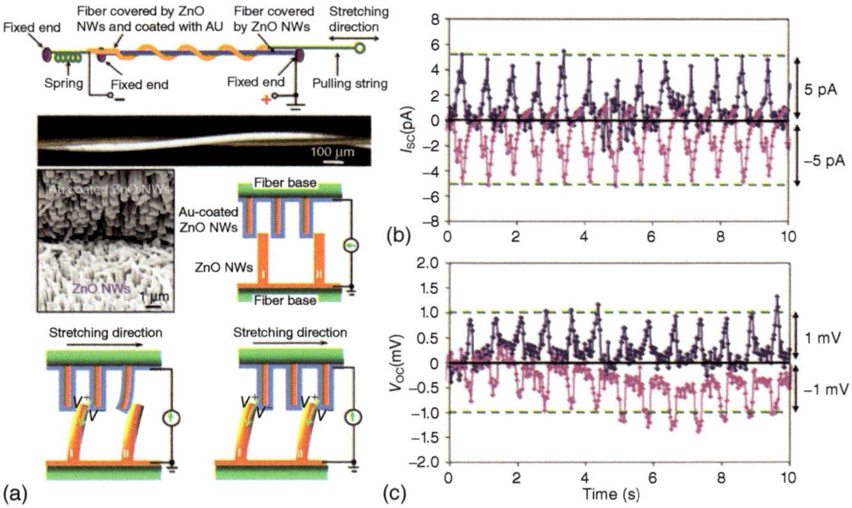

Figure 22.8 (a) Schematic of ZnO nanowires fiber-shaped piezoelectric device and electricity generating processes; (b) the short-circuit output current and (c) the open-circuit output voltage of a double-fiber nanogenerator measured by applying an external pulling force at a motor speed of 80 rpm. Source: Qin et al. 2008 [74]. Reprinted with permission of Springer Nature.

the short-circuit current and the open-circuit output voltage of the twisting nanogenerator. When as-prepared fiber generators were integrated into textile fabrics, an output density of 20–80 mW/m^2 was observed.

Liao et al. [95] reported a fiber-based piezoelectric device with the help of foldable ZnO paper (Figure 22.9). External force moved the Au-coated ZnO layer toward back and down direction, resulting in the formation of piezoelectric field. Flexible paper substrates can be folded at any angle and provide steady physical contact along with enlarged effective working area. The device showed an output voltage of 17 mV and current density of 0.09 $\mu A/cm^2$. Figure 22.9 can be further enhanced when operated under multifiber mode. A LED was lighten up by the power generated from a combination of 600 fibers.

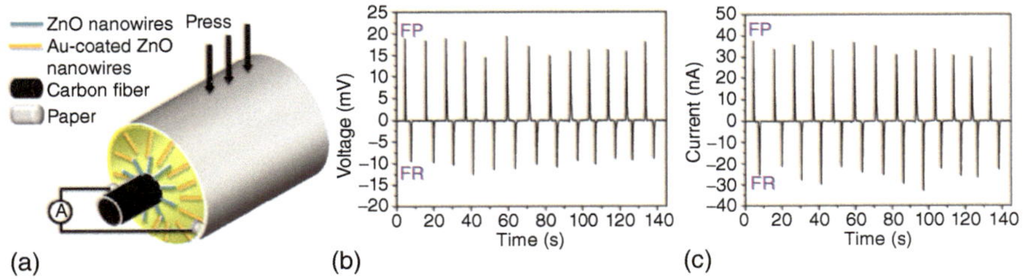

Figure 22.9 (a) Schematic of ZnO paper-based piezoelectric device, (b) output voltage, and (c) current of multifiber-based piezoelectric generator subjected to repeated cycles of FP and FR under forward-connected mode. Source: Liao et al. 2014 [95]. Reprinted with permission of Springer Nature.

Hybrid polyvinylidene fluoride (PVDF)/ZnO dual component device was prepared by incorporating both ZnO nanowires and PVDF polymer onto a fiber substrate. ZnO was considered as the main piezoelectric generator and served as semiconducting electrodes, which also improved piezoelectric coefficient of PVDF. PVDF was coated for multiple purposes. It served as a protective layer, piezoelectric generator, and surface contact enhancer. Hybrid feature endowed the device with better contact between two piezoelectric materials along with improvement of output signal and stability. It has been further demonstrated that electrical outputs could be generated by the releasing and folding processes of human elbow after attaching fiber generator to a human arm. A 2 cm length fiber device was able to generate a voltage output of 0.1 V, current density of $10\,nA/cm^2$, and power density of $16\,\mu W/cm^3$ [96].

22.4.2 Energy Storage

22.4.2.1 Supercapacitors

Fiber-shaped supercapacitors are more portable and ready to be woven into textiles when compared with their planar counterparts. A typical fiber-shaped supercapacitor consists of two electrodes and electrolyte, which acted as a separator and an ion transport domain. Recent studies on fiber-shaped supercapacitors are summarized in Table 22.4.

Single fiber-shaped supercapacitor was reported by Bae et al. [73] in 2011. They prepared a SCSFD that could convert energy and store it simultaneously. ZnO nanowires and graphene were deposited on the central plastic wire and copper mesh, respectively, as two electrodes, followed by the infiltration of PVA/H_3PO_4 gel electrolyte. The calculated specific capacitance reached to $0.4\,mF/cm^2$ and 0.025 mF/cm. Since then, giant leaps have been made toward two directions: improvement of capacitance and enhancement of mechanical properties. Electrode optimization is the mostly used method to improve specific capacitance. CNTs and their combination with metal, conducting polymer, or other carbon-based materials were reported to improve the capacitance of the devices. For example, graphene and CNT hybrid electrode have shown promising properties in making wire-shaped electronic devices, where graphene sheets served as conducting bridges and decreased contact resistance [99]. CNT sheets were drawn out from CVD-synthesized CNT arrays and stacked together with the addition of graphene oxide solution. The resulting CNTs exhibited multiwalled feature with a diameter of 10 μm. After reduction, strong interconnection between CNTs and graphene was established owing to π—π bonds. As-prepared energy device exhibited specific capacitance of 31.50 F/g, which is superior to bare CNT electrode.

Conducting polymers such as polyaniline (PANI) and PEDOT can provide extra specific surface area, effective pathway for charge transport, and pseudocapacitance when introduced into CNTs, leading to a significant increase in specific capacitance. Cai et al. [64] reported a PANI/CNT fiber supercapacitor with specific capacitance of 274 F/g. Figure 22.10 shows the influence of PANI on electrochemical properties. Two redox peaks were observed in the interval between 0.3 and 0.4 V, indicating pseudocapacitance feature. Symmetrical

Table 22.4 Recent progress on fiber-shaped supercapacitors.

Year	Structure	Electrode	Maximum performance	Features	References
2013	TFD	CNT/PANI	Specific capacitance of 274 F/g	Stable under bending Good weavability	[64]
2013	SCSFD	Ti/P3HT:PCBM/PEDOT:PSS/CNT	Energy density of 1.61×10^{-7} Wh/cm^2	Integration of solar cell and supercapacitor	[65]
2014	FEWFD	CNT/PANI	Energy density of 12.75 Wh/kg and power density of 1494 W/kg	Color indication	[15]
2014	SCSFD	CNT/PANI	Specific capacitance of 79.4 F/g after stretching at a strain of 300% and 100.8 F/g after bending	High stretchability over 400%	[70]
2015	FEWFD	CNT/PEDOT:PSS	Specific capacitance of 30.7 F/g at 0.5 A/g	Working at up to 350% strain	[71]
2016	SCSFD	MWCNT/PANI	Specific capacitance of 11 F/g	High output voltages up to 1000 V In-series feature	[72]
2016	SCSFD	Thermoplastic polyurethane/CNT	Specific capacitance of 24 F/g	Shape-memory features	[97]
2016	PCFD	GO/PEDOT:PSS/Vitamin C	Energy density of 27.1 $\mu Wh/cm^2$ and power density of 66.5 $\mu W/cm^2$	Hollow fiber electrodes	[68]
2016	Hybrid	Rubber/CNT	Specific capacitance of 2.38 mF/cm	Against deformation Twistable sandwich structure	[98]
2016	TFD	Microporous carbon/CNT CNT/PEDOT:PSS/MnO_2	Energy density of 11.3 mWh/cm^3 and power density of 2.1 W/cm^3	High performance	[14]

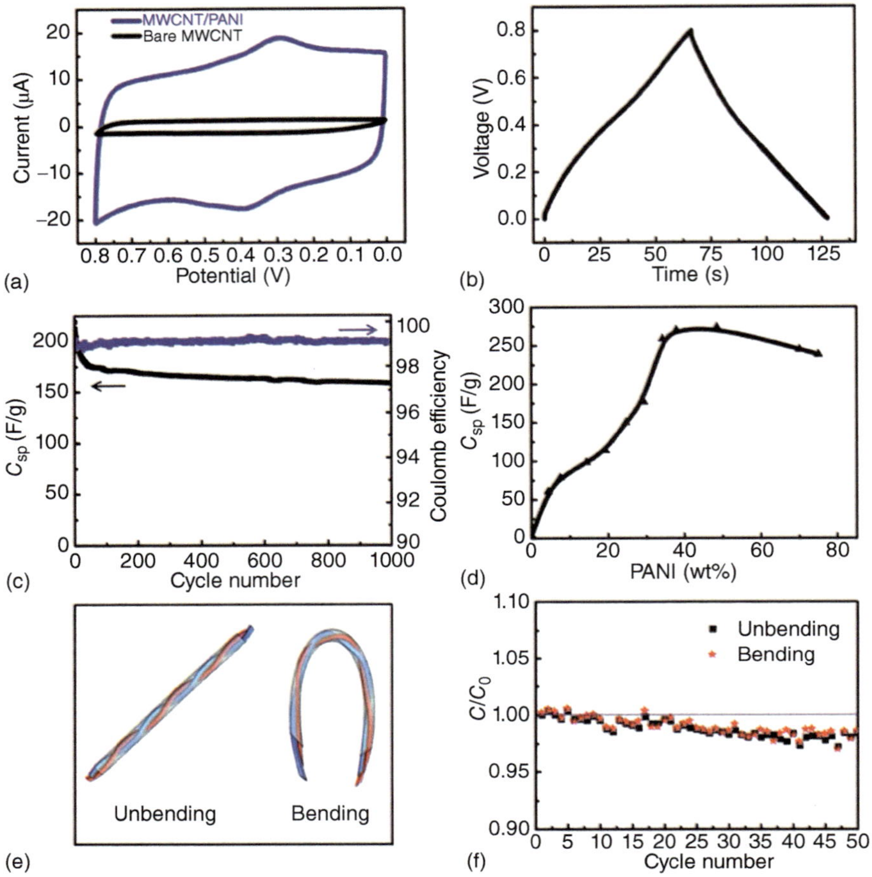

Figure 22.10 Electrochemical properties of supercapacitor wires. (a) Cyclic voltammogram of a supercapacitor with 24% PANI weight. (b) Galvanostatic charge–discharge curves of a supercapacitor with 24% PANI weight. (c) Dependence of specific capacitance and Coulomb efficiency on cycle number of a supercapacitor with 34% PANI weight. (d) Dependence of specific capacitance on the PANI weight percentage. (e) Schematic of the unbending and bending morphologies. (f) Dependence of the specific capacitance of a supercapacitor with 34% PANI weight on bending cycle number. Source: Cai et al. 2013 [64]. Reprinted with permission of Royal Society of Chemistry.

galvanostatic charge–discharge curves and stability test proved that the device exhibited improved stability. This significant improvement was attributed to the synergy of CNTs and PANI, where CNTs provided physical stability and effective contact and PANI afforded charge transfer pathway and high specific surface area. The weight percentage of PANI plays a crucial role in specific capacitance. Specific capacitance was increased along with the increase of PANI weight until a turning point was reached. The reverse trend after the turning point can be explained as excess PANI covered the surface of aligned CNTs and deteriorate electrical properties of CNTs. Meanwhile, stable mechanical properties of CNTs ensured the device can be operated under bending conditions.

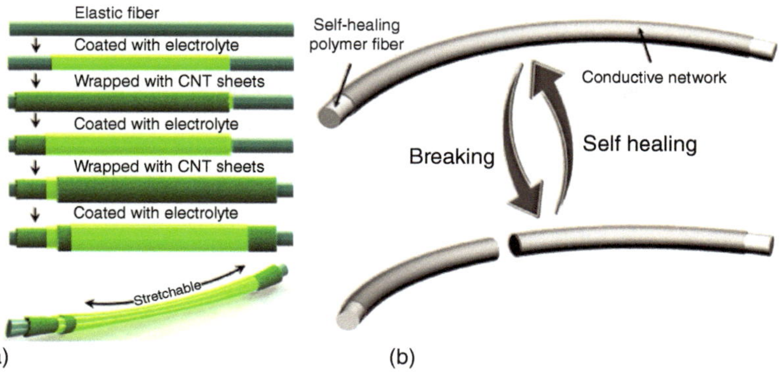

Figure 22.11 (a) Schematic of preparation process of a stretchable supercapacitor. Source: Reprinted with permission from Yang et al. 2013 [100]. (b) Schematic of self-healing supercapacitors. Source: Sun et al. 2014 [101]. Reprinted with permission of John Wiley & Sons.

Apart from specific capacitance, supercapacitors also require robust mechanical properties, which ensure devices can withstand various physical deformation. To date, most fiber-shaped supercapacitors have achieved the ability to work under bending states with little performance loss. However, other features such as stretchability are still highly desired. Yang et al. [100] have prepared SCSFD-structured supercapacitor with high stretchability based on an elastic fiber (Figure 22.11a). As-prepared device managed to achieve 18 F/g specific capacitance after being stretched by 75% for 100 cycles. Other features, such as self-healing and long lifetime fibrous devices, have been investigated as well [101]. As shown in Figure 22.11b, a self-healing polymer was used as a substrate, which was wrapped by CNTs and Ag nanowires. As-prepared device showed excellent performance (specific capacitance of 140.0 F/g) and could recover up to 92% of its original specific capacitance after repeated breaking–healing cycles.

22.4.2.2 Fiber-Shaped Li-Ion Battery

Fiber-shaped lithium-ion batteries are formed by two electrode and electrolyte. The development of fibrous Li-ion battery follows similar directions with fiber supercapacitors. Current fiber-shaped Li-ion batteries originated from a prototype reported by Kwon et al. [102] in 2012. A cable-shaped Li-ion battery was produced by employing a hollow Ni–Sn anode with multi-helix structure and $LiCoO_2$ cathode. Hollow structure reduced cell resistance and enlarged specific surface area, resulting better electrolyte permeability and electrical conductivity. A comparison between non-hollow and hollow structured devices was depicted in Figure 22.12. Device with hollow electrode can be charged and discharged repeatedly with less capacity loss.

Ren et al. [103] fabricated a wire-shaped battery with FEWFD structure that achieved a specific capacitance of 13.31 F/g, where lithium wire and MWCNTs were adopted as two electrodes. Later, they made a further improvement by coating silicon onto the exterior of MWCNTs as a composite anode [104]. When the weight percentage of Si increased from 0% to 38.1%, the specific capacities increased from 82 to 1670 mAh/g owing to remarkable specific capacity of Si

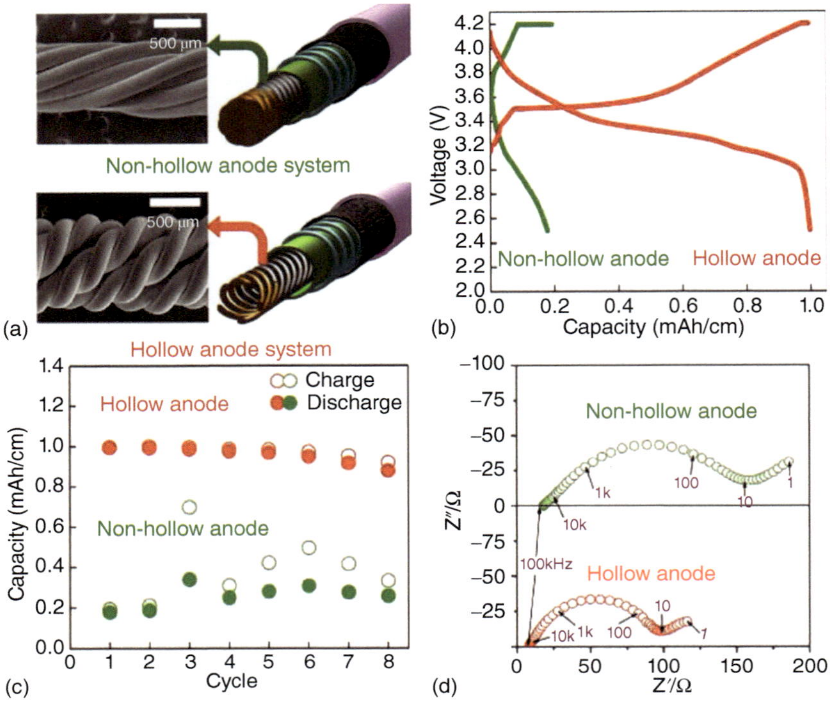

Figure 22.12 Electrochemical properties of the cable battery with hollow anode compared with a device with a dense anode. (a) Images of the cable batteries with dense and hollow anode system. (b) First charge and discharge profiles of cable batteries. (c) Capacity retention of cable batteries. (d) AC impedance spectra of cable battery after one cycle in the frequency range from 100 kHz to 1 Hz. Source: Kwon et al. 2012 [102]. Reprinted with permission of John Wiley & Sons.

[105]. The power density was further improved to 10 217.74 W/kg by employing $LiMn_2O_4$ (LMO)/CNT and polyimide (PI)/CNT hybrid fibers as cathode and anode, respectively, which was the highest reported value in this area and comparable with planar thin-film lithium-ion batteries (Figure 22.13) [75].

A typical method to improve the mechanical properties of fiber batteries is to use a core–sheath structure with a strong substrate. For example, self-healing devices have been produced by applying self-healing polymers to reconstruct their morphology after deformation [106–108]. Zhao et al. [106] produced a self-healing Li-ion battery with an energy density of 32.04 Wh/kg. The discharge voltage decreased from 1.58 to 1.45 V after three times cutting and self-healing processes. Specific capacity decreased from 28.2 to 17.2 mAh/g after the fifth self-healing process. Aside from self-healing features, lithium-ion batteries with high stretchability have also been fabricated. A conventional solution to enhance stretchability is to introduce elastomeric polymers into the device as a substrate [100, 109, 110]. However, the addition of elastic substrates will increase both size and weight of devices. A new solution to this issue is to use spring-like CNT fibers with coiled loops as electrodes instead of elastic polymers, as shown in

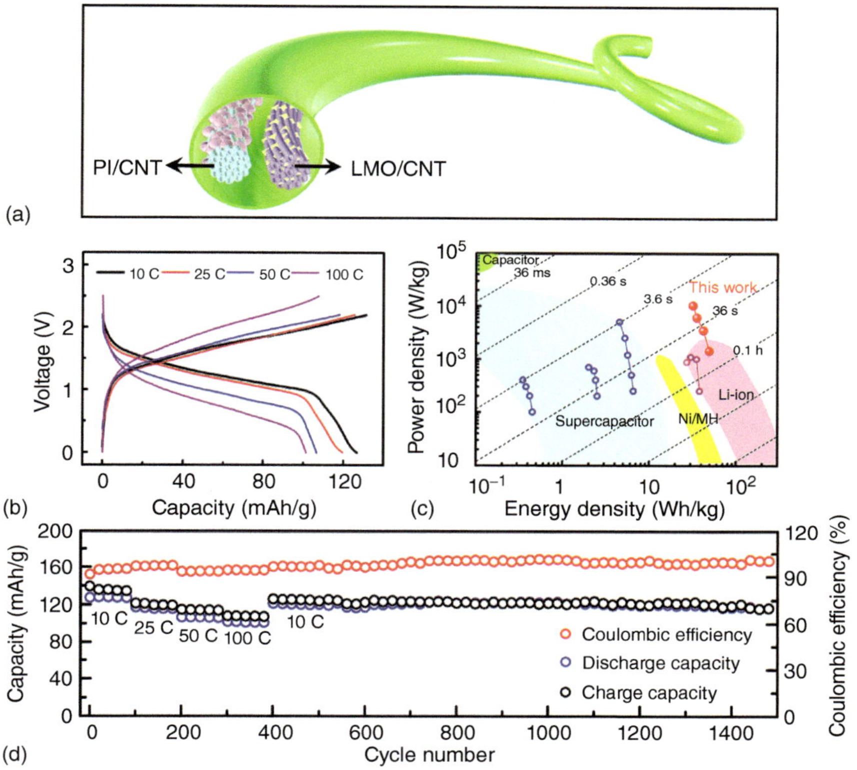

Figure 22.13 Schematic and electrochemical performance. (a) Schematic of a simplified structure showing only the two hybrid electrodes. (b) Charge and discharge curves under increasing current rates (1 C = 183 mA/g). (c) Energy and power densities compared with previous energy storage systems. (d) Rate performance at increasing current rates from 10 to 100 C and long-term stability test at a current rate of 10 C. Source: Zhang et al. 2016 [75]. Reprinted with permission of Royal Society of Chemistry.

Figure 22.14 [111]. By twisting CNT fibers into springs, an elongation of 305% and tensile stress of 82.7 MPa were attained. The volume and weight of this device can be reduced by 400% and 300%, respectively, when compared with devices that contain traditional polydimethylsiloxane (PDMS) or rubber. At a 100% strain level, the capacity of devices remained 85% of its original capacity.

22.4.3 Electrochromic Fibers

Electrochromic materials can change their optical properties by switching among various redox states that present different colors under an applied voltage. Electrochromic fibers are expected to be used for fashionable clothing, safety, and communication purposes. Early electrochromic fiber was made of polydiacetylene (PDA) and CNTs. It could switch color between blue and red by electrical stimulation [112]. The fiber turned from blue to red when an electric potential was applied, and the blue color restored once the applied voltage was

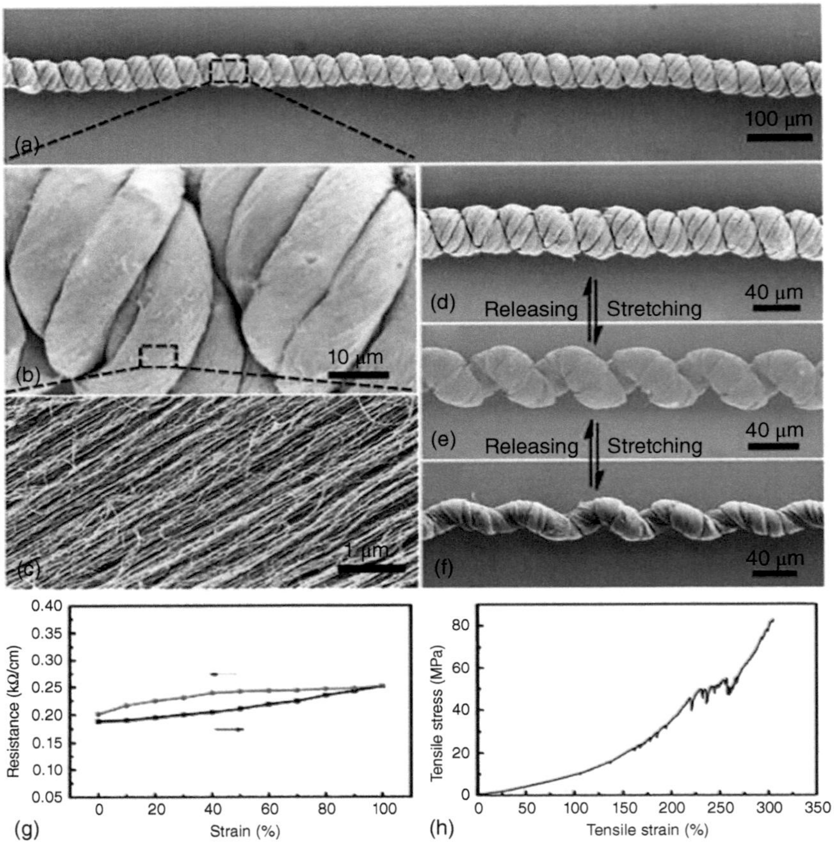

Figure 22.14 (a–c) Scanning electron microscopy (SEM) images of a spring-like fiber at different magnifications; (d–f) SEM images of a fiber at different strains 0% (d), 50% (e), and 100% (f). (g) Evolution of resistances of the fiber during a stretching and releasing process with a strain of 100%. (h) Tensile stress–strain curve of a fiber. Source: Zhang et al. 2014 [111]. Reprinted with permission of John Wiley & Sons.

removed. The CNT/PDA fibers possess high electrical conductivity and fast switching rate of 1×10^3 S/cm and one second, respectively. Color changing was attributed to the interaction between CNTs and PDA. Current induced electrons hopped among nanotubes within a fiber and led to the polarization of –COOH groups and PDA backbones. As a result, delocalization of π-electrons was decreased by above polarization, which corresponded to color changes.

Electrochromic fibers where color variation originated from switch among multiple redox states were later reported. Such electrochromic fibers were prepared by electrodepositing electrochromic materials onto fiber substrates. Li et al. [61] reported a set of electrochromic fibers with multiple colors when woven into clothes (Figure 22.15). A thinner stainless steel wire electrode was wrapped onto a coarse stainless steel wire, followed by electrodeposition of PEDOT, poly(3-methylthiophene) (P3MT), and poly(2,5-dimethoxyaniline)

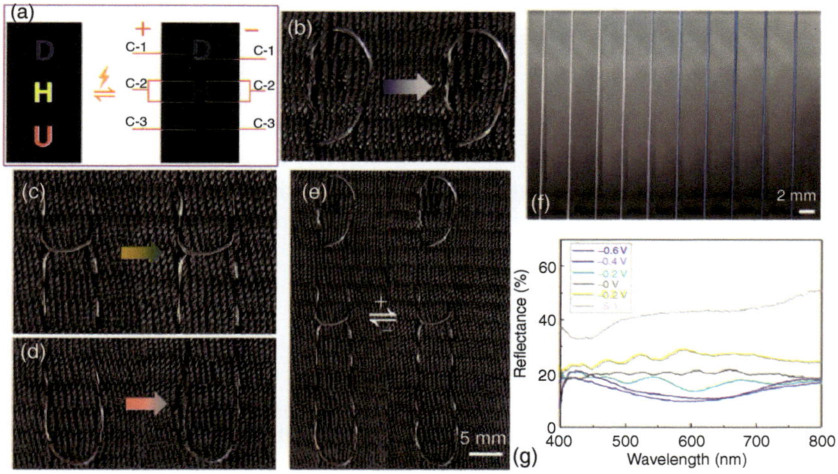

Figure 22.15 Illustration of electrochromic application of composite fibers: (a–e) digital photographs of electrochromic performance of fibers in various forms, (f) color gradient of electrochromic fiber, and (g) reflectance of composite fibers at different working potential. Source: Li et al. 2014 [61]. Reprinted with permission of American Chemical Society.

(PDMA). Multiple colors can be observed due to intrinsic optical properties of deposited polymers. Color response in millisecond at a relatively low voltage was observed. By manipulating the operating potential, color gradient ranging from gray to blue occurred, which was caused by the different intercalation level of dopant.

Aside from aesthetic enhancement, electrochromic fibers can be utilized as a potential monitor since electrochromic behavior is dependent on the potential applied. In addition, electrochromic polymer has pseudocapacitance features intrinsically and can be made as dual functional fiber. PEDOT-based dual functional fiber was reported as shown in Figure 22.16A [17]. Au was selectively coated onto fibrous substrate with the help of a helix-shaped mask, followed by electrodeposition of PEDOT. Uncoated blank space separated the continuous Au layer, resulting two electrodes integrated onto a single fiber substrate. Electrochromic performance can be observed on helix electrodes directly

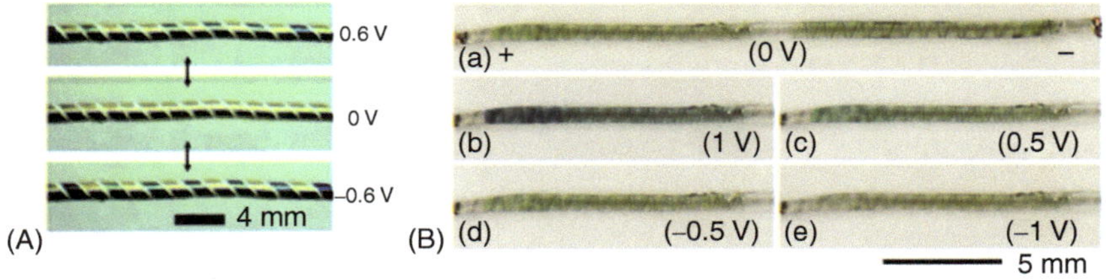

Figure 22.16 (A) Chromatic transitions during the charge–discharge process. Source: Zhou et al. 2016 [17]. Reprinted with permission of American Chemical Society. (B) An electrochromic fiber-shaped supercapacitor at various potential. Source: Chen et al. 2014 [15]. Reprinted with permission of John Wiley & Sons.

without color overlap, which occurred in a traditional core–sheath structure, and supercapacitor worked well without interference. Chen et al. [15] utilized this feature to monitor the working statues of a fiber-shaped supercapacitor. PANI-coated CNT electrodes were deposited onto elastic substrate, and the resulting substrate was further woven into a supercapacitor network, where PANI acted as both monitor and pseudocapacitor. The fiber exhibited various colors depending upon the potential of supercapacitor (Figure 22.16B). The direction and working potential can be known through different colors.

22.4.4 Transistors

Transistors are semiconductor components that amplify or switch electronic signals. Fiber-shaped organic field-effect transistors (OFETs) and organic electrochemical transistor (OECT) have become the mainstream of research interest among various types of transistors. Figure 22.17 presented a comparison between OFET and OECT [113]. For an OFET, the conductivity of polymer is subjected to the potential applied on gate electrode. More charges are accumulated within the polymer channel with increased conductivity when the device is under "ON"

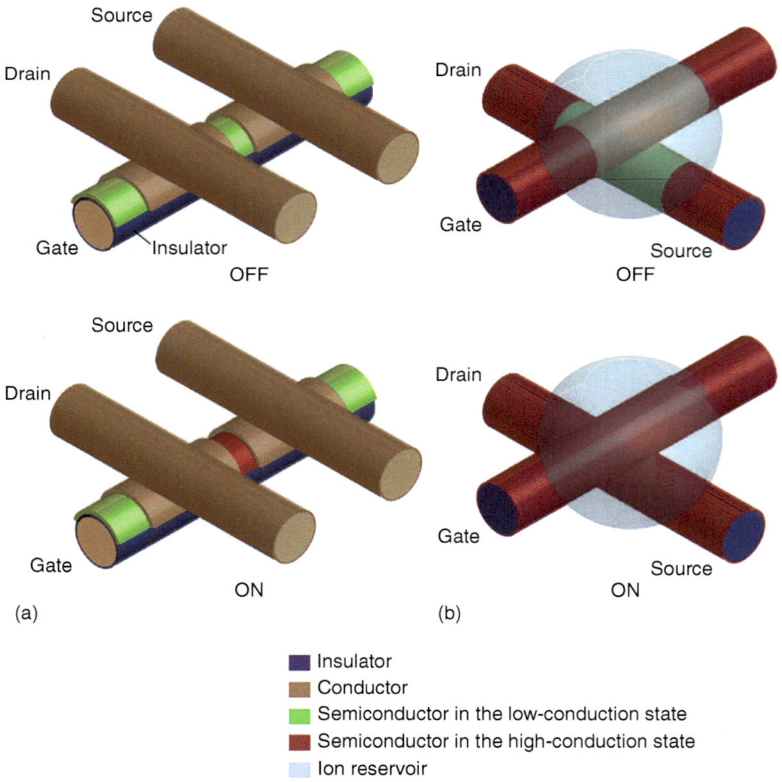

Figure 22.17 Working schematic of (a) OFET and (b) OECT with respect to their "ON" and "OFF" states. Source: De Rossi 2007 [113]. Reprinted with permission of Springer Nature.

mode. When it is switched to "OFF" mode, the number of accumulated charges reduces, leading to a decline in conductivity. In an OECT, the control of gate electrode results in conductivity change via the injection/extraction of ions from electrolyte.

Advantages of OECT include high current in channel and low cost. Additionally, an OECT does not require smooth substrate surface, making it more practical to be applied in wearable applications. However, restricted by the working mechanism, which is originated from the doping/de-doping features of conducting polymers, this type of transistor can only be operated in depletion mode. Meanwhile, the response time of OECT is longer than OFET since ion transport processes in ionic liquid are slower. Hamedi et al. [16] combined OECT and OFET by adopting poly(3-hexylthiophene) (P3HT) and imidazolium ionic liquid as electroactive materials. Conducting channel was established by the field-effect current at the interface of P3HT/electrolyte, followed by a significant response made by doping/de-doping process of P3HT. It possessed advantages of both OFET and OECT and can be operated at a low potential with large current density.

22.4.5 Fiber-Shaped Electroluminescent Device

Electroluminescent materials can emit light in response to the potential applied or current passed through. Fiber-shaped electroluminescent device can be woven into clothes for better information conveying functions. For example, Zhang et al. [114] prepared a fiber-shaped polymer light-emitting device by constructing a CNT/electroluminescent polymer (EP)/CNT core–sheath fiber. When the potential applied reached to a threshold, the EP switched to n-doped and p-doped states at cathode and anode, respectively, exhibiting a uniform electroluminescent effect (Figure 22.18).

The minimum working potential of resulting light-emitting device was 8.8 V, and the maximum light intensity reached at 505 cd/m^2, respectively. As-prepared device also exhibited remarkable performance under bending states where 91.2% of brightness was retained after 100 bending cycles.

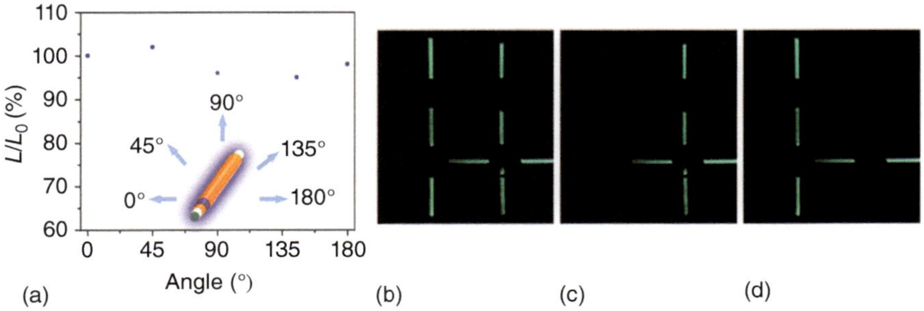

Figure 22.18 (a) The dependence of luminance on the observation angle for a "luminescent fiber." Here L_0 and L correspond to the luminance measured at 0 and at other observation angles, respectively. (b–d) Images of woven "luminescent fibers" being selectively lightened. Source: Zhang et al. 2015 [114]. Reprinted with permission of Royal Society of Chemistry.

22.4.6 Fiber-Shaped Sensors

Sensors are another typical application area of fiber-shaped electronic device. Ge et al. [115] produced a multifunctional sensor fabric, which could response to normal pressure, lateral strain, and flexion. A polyurethane fiber was used as substrate, followed by the wrapping of a nylon fiber. This basic fiber component was further treated with 3-triethoxysilylpropylamine in order to form hydrogen bonds between the interface. Ag nanowires and piezoresistive rubber were used afterward to construct a core–sheath structure. Figure 22.19 illustrated the working principle and performance of the mechanical sensor. The resistance of sensor

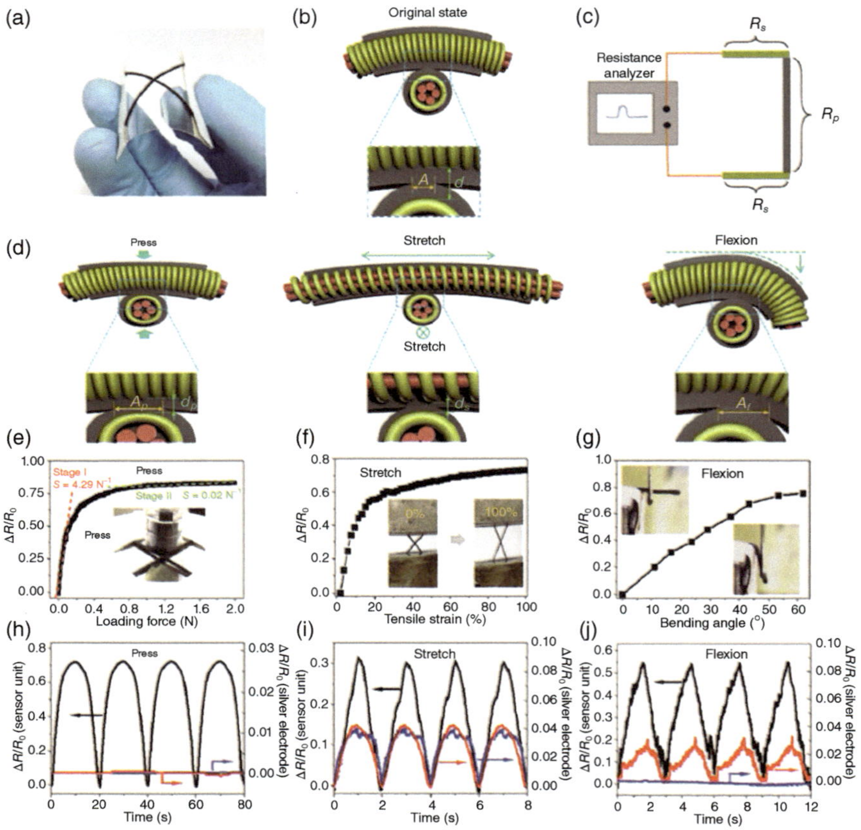

Figure 22.19 Mechanical–electric properties of the sensor unit. (a) Optical image of a typical sensor unit. (b) Schematic illustration of the cross contact point of the sensor unit. *A* is the contact area, and *d* is the thickness of the piezoresistive rubber layers between the silver electrodes. (c) The equivalent circuit of the sensor unit. (d) Schematic illustration of the shape deformation at the contact point of the sensor unit under press, stretch, and flexion. A_p and d_p are the contact area and thickness under press, d_s is the thickness under stretch, and A_f is the contact area under flexion. (e–g) The plots of relative resistance change ($\Delta R/R_0$) of the sensor unit as a function of loading force, tensile strain, and bending angle, respectively. Source: Reprinted with permission from Ge et al. 2016 [115].

will be changed under various operation modes and can be viewed as a function of input signal (e.g. loading force, tensile strain, and bending angle). They have further achieved force mapping by weaving a number of fiber sensors into a fabric, which can detect the position of the object placed on the fabric and quantify mechanical stress through pressure, strain, and flexion changes.

22.4.7 Characterization Techniques and Key Parameter

It is crucial to be aware of characterization techniques and corresponding parameters when describing fiber-shaped electronic devices that applied in various area. These parameters will be used to evaluate and compare the performance of a fiber-shaped electronic device with peers. Hence, a summary of characterization methods and evaluation parameters based on application types is presented as shown in Table 22.5.

Table 22.5 Summary of characterization techniques and key parameters.

Application	Characterization techniques	Evaluation parameters
Thermal energy harvesters	Voltage–temperature and current–temperature curves	Power density, figure-of-merit, net thermal voltage
Solar cells	Photoluminescence spectra, voltage–current curve	Fill factor, power conversion efficiency, open circuit voltage, short circuit current
Mechanical energy harvesters	Voltage–strain and current–strain curves, open-circuit voltage responses, short-circuit current plots	Power density, open-circuit voltage, short-circuit current
Supercapacitors	Cyclic voltammetry curves, galvanostatic charge–discharge curves, impedance spectroscopy	Energy density, power density, specific capacitance
Li-ion batteries	Cyclic voltammetry curves, charge–discharge curves, impedance spectroscopy	Energy density, power density, specific capacity, impedance
Electrochromic fibers	UV–vis spectrum, chromaticity	Contrast ratio, coloration efficiency, color chromaticity, response time
Transistors	Current–voltage plot, impedance spectroscopy	Capacitance, response time, driving voltage, current density
Electroluminescent fibers	Current–voltage plot, chromaticity	Light intensity, color chromaticity, illumination efficiency
Sensors	Strain–current plot, impedance spectroscopy	Sensitivity

22.5 Conclusions

During the past decade, the performance of fiber-based electronic devices has been significantly improved through material innovation and novel structure design. Meanwhile, the functionality of devices has evolved from single function to multiple functions. However, with great challenges in this area, continuous efforts are needed to make breakthrough innovation in the near future, especially in the following directions:

Improve performance through new materials and structures: The development of new materials and structures is a major driving force to facilitate the realization of wearable electronics, including optimization of microstructure, electrolyte, dopants, and modification of molecular chains, etc. For example, previous studies have demonstrated that aligned CNT sheets have remarkable electrical properties compared with CNTs without treatment.

Enhance interface for better stability and durability: Good stability and durability are mainly originated from structure optimization. Devices with high integration level will intrinsically have better stability. Future trends are high-level integration of electronic components within a fiber device and miniature of device dimensions. Single fiber-based electronics that can replace existing multifiber structures are desired. Making fiber electronics in micro- or nanoscales will improve the specific area and let people assemble as much as possible fiber components given certain surface areas.

Produce fibrous devices on large scale: To achieve commercialization of wearable devices, scalable fabrication and device assembly must be developed. High temperature and voltage should be avoided during manufacturing processes. The development of large-scale production of carbon-based materials might become a promising topic in terms of electrode fabrication.

Integration with normal textiles, washing, and ironing: The ultimate aim of wearable electronics is to integrate electronic components into textiles unperceivably, which requires extensive studies on woven ability, post treatment, and safety issues such as electrolyte leakage. Hence, scientists must seek balance between comfortability, performance, and safety. Some fiber-shaped devices are based on stainless steel wires or hard materials, which are not suitable to be worn. Taking a step further, we should treat wearable electronics as same as normal textiles, which means launderability, and the ability to undergo ironing and folding cannot be omitted.

In summary, we are still far from the truly realization of wearable electronics. More systematic awareness of wearable electronics, including materials, structures, fabrications, etc., should be established. However, given the joint efforts that have been made so far, it is believed that the ideal wearable electronics are at the corner.

References

1 He, X. et al. (2016). A highly stretchable fiber-based triboelectric nanogenerator for self-powered wearable electronics. *Advanced Functional Materials* 27 (4): 1604378.

2 Yu, N. et al. (2016). High-performance fiber-shaped all-solid-state asymmetric supercapacitors based on ultrathin MnO_2 nanosheet/carbon fiber cathodes for wearable electronics. *Advanced Energy Materials* 6 (2): 1501458.

3 Jung, Y.H. et al. (2016). Stretchable twisted-pair transmission lines for microwave frequency wearable electronics. *Advanced Functional Materials* 26 (26): 4635–4642.

4 Wang, J. et al. (2016). Sustainably powering wearable electronics solely by biomechanical energy. *Nature Communications* 7: 12744.

5 Yun, M.J. et al. (2015). Insertion of dye-sensitized solar cells in textiles using a conventional weaving process. *Scientific Reports* 5: 11022.

6 Zou, D.C. et al. (2010). Fiber-shaped flexible solar cells. *Coordination Chemistry Reviews* 254 (9–10): 1169–1178.

7 Xu, Y. et al. (2016). An all-solid-state fiber-shaped aluminum–air battery with flexibility, stretchability, and high electrochemical performance. *Angewandte Chemie* 128 (28): 8111–8114.

8 Fan, F.R., Tang, W., and Wang, Z.L. (2016). Flexible nanogenerators for energy harvesting and self-powered electronics. *Advanced Materials* 28 (22): 4283–4305.

9 Stoppa, M. and Chiolerio, A. (2014). Wearable electronics and smart textiles: a critical review. *Sensors* 14 (7): 11957–11992.

10 Qiu, L.B. et al. (2016). An all-solid-state fiber-type solar cell achieving 9.49% efficiency. *Journal of Materials Chemistry A* 4 (26): 10105–10109.

11 Li, H.P. et al. (2015). Stable hydrophobic ionic liquid gel electrolyte for stretchable fiber-shaped dye-sensitized solar cell. *ChemNanoMat* 1 (6): 399–402.

12 Zhang, Z.T. et al. (2015). A colour-tunable, weavable fibre-shaped polymer light-emitting electrochemical cell. *Nature Photonics* 9 (4): 233–238.

13 Xu, Y.F. et al. (2015). Flexible, stretchable, and rechargeable fiber-shaped zinc-air battery based on cross-stacked carbon nanotube sheets. *Angewandte Chemie International Edition* 54 (51): 15390–15394.

14 Cheng, X.L. et al. (2016). Design of a hierarchical ternary hybrid for a fiber-shaped asymmetric supercapacitor with high volumetric energy density. *Journal of Physical Chemistry C* 120 (18): 9685–9691.

15 Chen, X. et al. (2014). Electrochromic fiber-shaped supercapacitors. *Advanced Materials* 26 (48): 8126–8132.

16 Hamedi, M. et al. (2009). Fiber-embedded electrolyte-gated field-effect transistors for e-textiles. *Advanced Materials* 21 (5): 573–577.

17 Zhou, Y. et al. (2016). Electrochromic/supercapacitive dual functional fibres. *RSC Advances* 6 (111): 110164–110170.

18 Zhang, Z.T. et al. (2014). Weaving efficient polymer solar cell wires into flexible power textiles. *Advanced Energy Materials* 4 (11): 1301750.

19 Chen, T. et al. (2012). Designing aligned inorganic nanotubes at the electrode interface: towards highly efficient photovoltaic wires. *Advanced Materials* 24 (34): 4623–4628.

20 Wang, H. et al. (2009). Hydrothermal growth of large-scale macroporous TiO_2 nanowires and its application in 3D dye-sensitized solar cells. *Applied Physics A: Materials Science and Processing* 97 (1): 25–29.

21 Liu, Y. et al. (2010). Synthesis of TiO_2 nanotube arrays and its application in mini-3D dye-sensitized solar cells. *Journal of Physics D: Applied Physics* 43 (20): 205103.

22 Peng, H. (2015). *Fiber-Shaped Energy Harvesting and Storage Devices.* Springer.

23 Zhang, S. et al. (2011). Single-wire dye-sensitized solar cells wrapped by carbon nanotube film electrodes. *Nano Letters* 11 (8): 3383–3387.

24 Chen, X. et al. (2013). Novel electric double-layer capacitor with a coaxial fiber structure. *Advanced Materials* 25 (44): 6436–6441.

25 Qiu, L. et al. (2014). Integrating perovskite solar cells into a flexible fiber. *Angewandte Chemie International Edition* 53 (39): 10425–10428.

26 Chen, T. et al. (2012). An integrated "energy wire" for both photoelectric conversion and energy storage. *Angewandte Chemie International Edition* 51 (48): 11977–11980.

27 Chen, T. et al. (2012). Polymer photovoltaic wires based on aligned carbon nanotube fibers. *Journal of Materials Chemistry* 22 (44): 23655–23658.

28 Liu, D. et al. (2012). Solid-state, polymer-based fiber solar cells with carbon nanotube electrodes. *ACS Nano* 6 (12): 11027–11034.

29 Yang, Z.B. et al. (2014). Stretchable, wearable dye-sensitized solar cells. *Advanced Materials* 26 (17): 2643–2647.

30 Marikkannan, M. et al. (2015). Effect of ambient combinations of argon, oxygen, and hydrogen on the properties of DC magnetron sputtered indium tin oxide films. *AIP Advances* 5 (1): 017128.

31 Kudryashov, D., Gudovskikh, A., and Zelentsov, K. (2013). Low temperature growth of ITO transparent conductive oxide layers in oxygen-free environment by RF magnetron sputtering. *Journal of Physics: Conference Series.* 012021.

32 Lee, J.Y. et al. (2008). Solution-processed metal nanowire mesh transparent electrodes. *Nano Letters* 8 (2): 689–692.

33 Kim, T. et al. (2014). Highly transparent Au-coated Ag nanowire transparent electrode with reduction in haze. *ACS Applied Materials & Interfaces* 6 (16): 13527–13534.

34 Hu, L.B. et al. (2010). Scalable coating and properties of transparent, flexible, silver nanowire electrodes. *ACS Nano* 4 (5): 2955–2963.

35 Wang, R.R. et al. (2014). A novel method to enhance the conductance of transitional metal oxide electrodes. *Nanoscale* 6 (7): 3791–3795.

36 Bae, J. et al. (2011). Fiber supercapacitors made of nanowire-fiber hybrid structures for wearable/flexible energy storage. *Angewandte Chemie International Edition* 50 (7): 1683–1687.

37 Wei, Z. et al. (2013). Facile synthesis of Sn/TiO_2 nanowire array composites as superior lithium-ion battery anodes. *Journal of Power Sources* 223: 50–55.

38 Ren, J. et al. (2013). Twisting carbon nanotube fibers for both wire-shaped micro-supercapacitor and micro-battery. *Advanced Materials* 25 (8): 1155–1159.

39 Sun, D.M. et al. (2013). A review of carbon nanotube- and graphene-based flexible thin-film transistors. *Small* 9 (8): 1188–1205.

40 Zeng, W. et al. (2014). Fiber-based wearable electronics: a review of materials, fabrication, devices, and applications. *Advanced Materials* 26 (31): 5310–5336.
41 Hou, S. et al. (2011). Transparent conductive oxide-less, flexible, and highly efficient dye-sensitized solar cells with commercialized carbon fiber as the counter electrode. *Journal of Materials Chemistry* 21 (36): 13776–13779.
42 Zhang, M., Atkinson, K.R., and Baughman, R.H. (2004). Multifunctional carbon nanotube yarns by downsizing an ancient technology. *Science* 306 (5700): 1358–1361.
43 Reddy, A.L.M. et al. (2010). Synthesis of nitrogen-doped graphene films for lithium battery application. *ACS Nano* 4 (11): 6337–6342.
44 Zhang, Z. et al. (2016). Nitrogen-doped core-sheath carbon nanotube array for highly stretchable supercapacitor. *Advanced Energy Materials* 27 (4): 1604378.
45 Zhang, D.H. et al. (2006). Transparent, conductive, and flexible carbon nanotube films and their application in organic light-emitting diodes. *Nano Letters* 6 (9): 1880–1886.
46 Wang, X. et al. (2008). Transparent carbon films as electrodes in organic solar cells. *Angewandte Chemie International Edition* 47 (16): 2990–2992.
47 Dan, B., Irvin, G.C., and Pasquali, M. (2009). Continuous and scalable fabrication of transparent conducting carbon nanotube films. *ACS Nano* 3 (4): 835–843.
48 Jiang, D. et al. (2016). Embedded fin-like metal/CNT hybrid structures for flexible and transparent conductors. *Small* 12 (11): 1521–1526.
49 Hu, Y. et al. (2014). All-in-one graphene fiber supercapacitor. *Nanoscale* 6 (12): 6448–6451.
50 Cong, H.P. et al. (2012). Wet-spinning assembly of continuous, neat, and macroscopic graphene fibers. *Scientific Reports* 2: 613.
51 Kim, K.M. et al. (2016). Shape-engineerable composite fibers and their supercapacitor application. *Nanoscale* 8 (4): 1910–1914.
52 Chen, D., Tang, L.H., and Li, J.H. (2010). Graphene-based materials in electrochemistry. *Chemical Society Reviews* 39 (8): 3157–3180.
53 Hou, S.C. et al. (2012). Flexible conductive threads for wearable dye-sensitized solar cells. *Journal of Materials Chemistry* 22 (14): 6549–6552.
54 Huang, K. et al. (2005). Multi-functional polypyrrole nanofibers via a functional dopant-introduced process. *Synthetic Metals* 155 (3): 495–500.
55 Oh, J.H. et al. (2009). Solution-processed, high-performance n-channel organic microwire transistors. *Proceedings of the National Academy of Sciences of the United States of America* 106 (15): 6065–6070.
56 Sun, H. et al. (2013). Winding ultrathin, transparent, and electrically conductive carbon nanotube sheets into high-performance fiber-shaped dye-sensitized solar cells. *Journal of Materials Chemistry A* 1 (40): 12422–12425.
57 Sun, H. et al. (2014). Quasi-solid-state, coaxial, fiber-shaped dye-sensitized solar cells. *Journal of Materials Chemistry A* 2 (2): 345–349.

58 Fang, X. et al. (2014). Core-sheath carbon nanostructured fibers for efficient wire-shaped dye-sensitized solar cells. *Advanced Materials* 26 (11): 1694–1698.
59 Pan, S.W. et al. (2014). Carbon nanostructured fibers as counter electrodes in wire-shaped dye-sensitized solar cells. *Journal of Physical Chemistry C* 118 (30): 16419–16425.
60 Deng, J. et al. (2015). Elastic perovskite solar cells. *Journal of Materials Chemistry A* 3 (42): 21070–21076.
61 Li, K.R. et al. (2014). Red, green, blue (RGB) electrochromic fibers for the new smart color change fabrics. *ACS Applied Materials & Interfaces* 6 (15): 13043–13050.
62 Lee, M.R. et al. (2009). Solar power wires based on organic photovoltaic materials. *Science* 324 (5924): 232–235.
63 Liu, L., Ma, W., and Zhang, Z. (2011). Macroscopic carbon nanotube assemblies: preparation, properties, and potential applications. *Small* 7 (11): 1504–1520.
64 Cai, Z.B. et al. (2013). Flexible, weavable and efficient microsupercapacitor wires based on polyaniline composite fibers incorporated with aligned carbon nanotubes. *Journal of Materials Chemistry A* 1 (2): 258–261.
65 Zhang, Z.T. et al. (2014). Integrated polymer solar cell and electrochemical supercapacitor in a flexible and stable fiber format. *Advanced Materials* 26 (3): 466–470.
66 Li, X. et al. (2013). Large-area flexible core–shell graphene/porous carbon woven fabric films for fiber supercapacitor electrodes. *Advanced Functional Materials* 23 (38): 4862–4869.
67 Meng, Y. et al. (2013). All-graphene core-sheath microfibers for all-solid-state, stretchable fibriform supercapacitors and wearable electronic textiles. *Advanced Materials* 25 (16): 2326–2331.
68 Qu, G.X. et al. (2016). A fiber supercapacitor with high energy density based on hollow graphene/conducting polymer fiber electrode. *Advanced Materials* 28 (19): 3646–3652.
69 Hou, S.C. et al. (2012). Flexible, metal-free composite counter electrodes for efficient fiber-shaped dye-sensitized solar cells. *Journal of Power Sources* 215: 164–169.
70 Zhang, Z.T. et al. (2015). Superelastic supercapacitors with high performances during stretching. *Advanced Materials* 27 (2): 356–362.
71 Chen, T. et al. (2015). High-performance, stretchable, wire-shaped supercapacitors. *Angewandte Chemie International Edition* 54 (2): 618–622.
72 Sun, H. et al. (2016). Electrochemical capacitors with high output voltages that mimic electric eels. *Advanced Materials* 28 (10): 2070–2076.
73 Bae, J. et al. (2011). Single-fiber-based hybridization of energy converters and storage units using graphene as electrodes. *Advanced Materials* 23 (30): 3446–3449.
74 Qin, Y., Wang, X.D., and Wang, Z.L. (2008). Microfibre-nanowire hybrid structure for energy scavenging. *Nature* 451 (7180): 809–813.
75 Zhang, Y. et al. (2016). A fiber-shaped aqueous lithium ion battery with high power density. *Journal of Materials Chemistry A* 4 (23): 9002–9008.

76 Ren, J. et al. (2014). Elastic and wearable wire-shaped lithium-ion battery with high electrochemical performance. *Angewandte Chemie* 126 (30): 7998–8003.

77 Yadav, A., Pipe, K.P., and Shtein, M. (2008). Fiber-based flexible thermoelectric power generator. *Journal of Power Sources* 175 (2): 909–913.

78 Lee, J.A. et al. (2016). Woven-yarn thermoelectric textiles. *Advanced Materials* 28 (25): 5038–5044.

79 Zou, D.C. et al. (2012). Macro/microfiber-shaped electronic devices. *Nano Energy* 1 (2): 273–281.

80 He, S.S. et al. (2015). Radically grown obelisk-like ZnO arrays for perovskite solar cell fibers and fabrics through a mild solution process. *Journal of Materials Chemistry A* 3 (18): 9406–9410.

81 Baps, B., Eber-Koyuncu, M., and Koyuncu, M. (2002). Ceramic based solar cells in fiber form. *Key Engineering Materials* 937–940.

82 Liu, J., Namboothiry, M.A.G., and Carroll, D.L. (2007). Fiber-based architectures for organic photovoltaics. *Applied Physics Letters* 90 (6): 063501.

83 Fan, X. et al. (2008). Wire-shaped flexible dye-sensitized solar cells. *Advanced Materials* 20 (3): 592–595.

84 Qiu, L.B. et al. (2016). Fiber-shaped perovskite solar cells with high power conversion efficiency. *Small* 12 (18): 2419–2424.

85 Rathmell, A.R. et al. (2010). The growth mechanism of copper nanowires and their properties in flexible, transparent conducting films. *Advanced Materials* 22 (32): 3558–3563.

86 Yang, Y. et al. (2016). Facile fabrication of stretchable Ag nanowire/polyurethane electrodes using high intensity pulsed light. *Nano Research* 9 (2): 401–414.

87 Maurer, J.H.M. et al. (2016). Templated self-assembly of ultrathin gold nanowires by nanoimprinting for transparent flexible electronics. *Nano Letters* 16 (5): 2921–2925.

88 Li, Z.D. et al. (2016). Electrophoretic deposition of graphene-TiO_2 hierarchical spheres onto Ti thread for flexible fiber-shaped dye-sensitized solar cells. *Materials and Design* 105: 352–358.

89 Craddock, J.D. and Weisenberger, M.C. (2015). Harvesting of large, substrate-free sheets of vertically aligned multiwall carbon nanotube arrays. *Carbon* 81: 839–841.

90 Zhai, H.Y. et al. (2015). A sensitive electrochemical sensor with sulfonated graphene sheets/oxygen-functionalized multi-walled carbon nanotubes modified electrode for the detection of clenbuterol. *Sensors and Actuators B: Chemical* 210: 483–490.

91 Ternon, C. et al. (2015). Carbon nanotube sheet as top contact electrode for nanowires: highly versatile and simple process. *Journal of Nanoscience and Nanotechnology* 15 (2): 1669–1673.

92 Chen, H. et al. (2010). Controlled growth and modification of vertically-aligned carbon nanotubes for multifunctional applications. *Materials Science and Engineering R: Reports* 70 (3–6): 63–91.

93 He, R.R. et al. (2013). Silicon p-i-n junction fibers. *Advanced Materials* 25 (10): 1461–1467.

94 Zhang, L. et al. (2012). Flexible fiber-shaped $CuInSe_2$ solar cells with single-wire-structure: design, construction and performance. *Nano Energy* 1 (6): 769–776.
95 Liao, Q. et al. (2014). Flexible piezoelectric nanogenerators based on a fiber/ZnO nanowires/paper hybrid structure for energy harvesting. *Nano Research* 7 (6): 917–928.
96 Lee, M. et al. (2012). A hybrid piezoelectric structure for wearable nanogenerators. *Advanced Materials* 24 (13): 1759–1764.
97 Deng, J.E. et al. (2015). A shape-memory supercapacitor fiber. *Angewandte Chemie International Edition* 54 (51): 15419–15423.
98 Choi, C. et al. (2016). Twistable and stretchable sandwich structured fiber for wearable sensors and supercapacitors. *Nano Letters* 16 (12): 7677–7684.
99 Sun, H. et al. (2014). Novel graphene/carbon nanotube composite fibers for efficient wire-shaped miniature energy devices. *Advanced Materials* 26 (18): 2868–2873.
100 Yang, Z.B. et al. (2013). A highly stretchable, fiber-shaped supercapacitor. *Angewandte Chemie International Edition* 52 (50): 13453–13457.
101 Sun, H. et al. (2014). Self-healable electrically conducting wires for wearable microelectronics. *Angewandte Chemie International Edition* 53 (36): 9526–9531.
102 Kwon, Y.H. et al. (2012). Cable-type flexible lithium ion battery based on hollow multi-helix electrodes. *Advanced Materials* 24 (38): 5192–5197.
103 Ren, J. et al. (2013). Batteries: twisting carbon nanotube fibers for both wire-shaped micro-supercapacitor and micro-battery (Adv. Mater. 8/2013). *Advanced Materials* 25 (8): 1224–1224.
104 Lin, H.J. et al. (2014). Twisted aligned carbon nanotube/silicon composite fiber anode for flexible wire-shaped lithium-ion battery. *Advanced Materials* 26 (8): 1217–1222.
105 Weng, W. et al. (2014). Winding aligned carbon nanotube composite yarns into coaxial fiber full batteries with high performances. *Nano Letters* 14 (6): 3432–3438.
106 Zhao, Y. et al. (2016). A self-healing aqueous lithium-ion battery. *Angewandte Chemie International Edition* 55 (46): 14382–14386.
107 Borre, E. et al. (2016). Light-powered self-healable metallosupramolecular soft actuators. *Angewandte Chemie International Edition* 55 (4): 1313–1317.
108 Huang, Y. et al. (2015). Magnetic-assisted, self-healable, yarn-based supercapacitor. *ACS Nano* 9 (6): 6242–6251.
109 Li, X., Gu, T.L., and Wei, B.Q. (2012). Dynamic and galvanic stability of stretchable supercapacitors. *Nano Letters* 12 (12): 6366–6371.
110 Zhao, C. et al. (2013). Intrinsically stretchable supercapacitors composed of polypyrrole electrodes and highly stretchable gel electrolyte. *ACS Applied Materials & Interfaces* 5 (18): 9008–9014.
111 Zhang, Y. et al. (2014). Flexible and stretchable lithium-ion batteries and supercapacitors based on electrically conducting carbon nanotube fiber springs. *Angewandte Chemie International Edition* 53 (52): 14564–14568.
112 Peng, H.S. et al. (2009). Electrochromatic carbon nanotube/polydiacetylene nanocomposite fibres. *Nature Nanotechnology* 4 (11): 738–741.

113 De Rossi, D. (2007). Electronic textiles: a logical step. *Nature Materials* 6 (5): 328–329.

114 Zhang, Z.T. et al. (2015). Flexible electroluminescent fiber fabricated from coaxially wound carbon nanotube sheets. *Journal of Materials Chemistry C* 3 (22): 5621–5624.

115 Ge, J. et al. (2016). A stretchable electronic fabric artificial skin with pressure-, lateral strain-, and flexion-sensitive properties. *Advanced Materials* 28 (4): 722–728.

23

Fibers for Optical Textiles

Dana Křemenáková, Jiri Militky, and Rajesh Mishra

Technical University of Liberec, Faculty of Textile Engineering, Department of Material Engineering (KMI), Building B, 2nd Floor, 46117 Liberec, Czech Republic

23.1 Introduction

Optical fibers are one category of specialty fibers originally developed for light transmission (fiber optic) and information (optical cables). Guiding of light through water stream was shown by John Tyndall (1870). History of classic optical fiber development and their basic functions are described, e.g. in the book [27].

Conventional end-emitting inorganic or polymeric optical fiber (POF) is a dielectric wave guide that transmits light or infrared radiation along its axis by total internal reflection at the interface between two media with different refractive index. The fiber consists of a core surrounded by a cladding. In order to transmit the optical signal, the refractive index of the core should be higher than the cladding (see Figure 23.1).

Inorganic optical fibers are mainly used to transmit data over long distances at very high bit rates. The advantage of using optical fibers in telecommunications is their flexibility, no electromagnetic interference, and low attenuation rate. In data networks, the multimode optical fibers with a diameter of 50/125 μm and in the United States 62.5/125 μm are most commonly used. In telecommunications single-mode fibers with a diameter of 9/125 μm are used. The advantages of optical fibers over metal conductors are as follows: low attenuation (less amps on the optical path), resistance to electromagnetic interference and cross-talk, electrical insulation, transmission security, and high bandwidth.

The well-known POF with polymethylmethacrylate (PMMA) core were introduced in the 1960s of last century by Du Pont. The commercial product was first marketed by Mitsubishi Rayon.

In the past several decades, concurrent with the successive improvements in glass fibers, POF have become increasingly popular, owing to their growing utility [2–7]. The classical POF are made of polycarbonate (PC), PMMA, or polystyrene (PS). Modern POF based on "perfluorinated" polymers are used to replace conventional inorganic optical fibers in particular for the transmission of information. For the production of simple "light pipes," extrusion of polymers

Handbook of Fibrous Materials, First Edition. Edited by Jinlian Hu, Bipin Kumar, and Jing Lu.

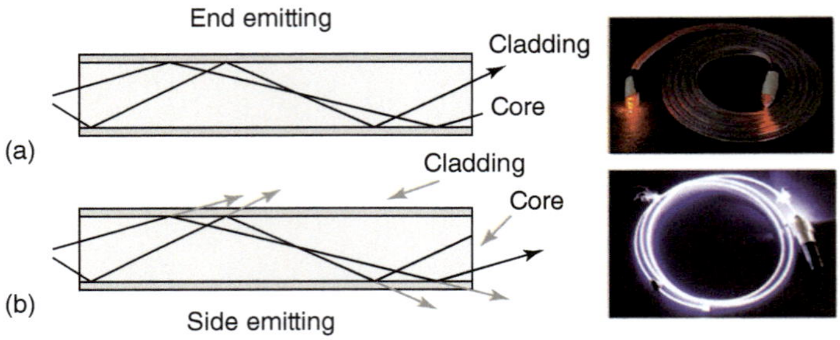

Figure 23.1 Light passing through (a) end-emitting POF and (b) side emitting optical fibers (SEPOF).

and for transmitting data subsequent drawing is used. Classic "light pipes" have a core–jacket structure. The core consists of a polymer with higher refractive index, and the cladding is formed by a polymeric layer having a lower refractive index, thereby ensuring conduction of light inside the fiber and minimal loss of radiation from the side.

In addition, POF serves as a complement for glass fibers in short-haul communication links because they are easy to handle, flexible, and economical, although they are not used for very long distances because of their relatively high attenuation. These characteristics make them especially suitable as a means of connection between a large net of glass optical fiber and a residential area, where distances to cover are generally less than 1 km. Nowadays, with the PMMA core optical fibers, transmissions at 156 Mbit/s over distances up to 100 m can be carried out [8], and transmission speeds of 500 Mbit/s over 50 m can be reached [9]. To achieve higher transmission speeds, graded-index POF can be used [10].

The large diameter of POF (typically 0.25–1 mm) allows low precision plastic connectors to be used, which reduces the total cost of the system. In addition, POFs stand out for their greater flexibility and resistance to impacts and vibrations, as well as for the greater coupling of light from the light source to the fiber. Because of these merits, varied applications with POF have been developed and commercialized, from their use as a simple light transmission guide in displays to their utilization as sensors and telecommunication cables. Because of their large diameter, POF are easier to install and align than their glass counterparts [2–7]. POF found its use as a high-capacity transmission medium only recently, thanks to their improved transparency and bandwidth [6, 7].

However, still it is difficult to draw conclusions whether the POF can replace the silica fiber to be the main transmission medium in communication networks; it depends on the further development of production technology in the future. From the current situation, there is still a big gap in attenuation, bandwidth, and working wavelength for POF compared with silica fiber, and silica fiber is also not less than POF in flexibility, connection performance, and cost. For example, silica fiber can be wound even knotted on the fingers; the assembly of site connector is quick and easy. As for cost, the price of POF is higher than silica fiber.

In the field of sensors, numerous types of products based on POF have been commercialized [11], for example, scanning heads, shape-defect detectors used in bottling plants, and liquid-level detectors. In addition, by using conventional POF, it is possible to make sensors measure distance, position, shape, color, brightness, opacity, density, turbidity, etc. [11, 12]. These sensors can serve to control the various manufacturing parameters in automated production processes with robots.

Gradually, the use of optical fibers was extended to the special textiles (displays, safety fabrics, special effects for apparel fabrics, etc.). For these applications, either by a suitable arrangement (micro-bends) or structures (side-emitting plastic optical fiber [SEPOF]), the side emission from surface of the fibers was obtained (see Figure 23.1).

POF integrated in weaving patterns can be described by a sequence of locally bent and straight sections. For this case the transmission loss is based on the distance between threads, thickness of threads, and weaves. Local bends are unwanted in the cases of using SEPOF for achieving active visibility at higher distances from light source. It is therefore necessary to prepare line illumination hybrid structure (LIHS) combining SEPOF in the core wrapped by the special textile-based outer layer, e.g. in form of tube from textile yarns.

Comprehensive information about properties and applications of POF in the textile field are presented in the book [1] and special chapter [28]. In this chapter the basic types of POF and their potential use in textile structures are discussed. One of main aims here is description of basic properties of SEPOF and their not so known applications in LIHS enabling them to be used in textile structures and enhancing side illumination intensity. Applications of LIHS for safety textiles, fashion purposes, and for special line illumination purposes are shown.

23.2 Principles of Fibers Optics

Standard, i.e. *end-emitting optical fiber*, is a special form of a relatively long distance (hundreds meters to some kilometers) optical wave guide and consists typically of a cylindrical core and a cladding (see Figure 23.2), both made from highly transparent materials. The core refractive index n_1 should be slightly higher than that of the cladding n_2 to avoid loss of illumination power due to side emission. Depending on ultimate applications, on the surface, there is additional protective layer called as jacket.

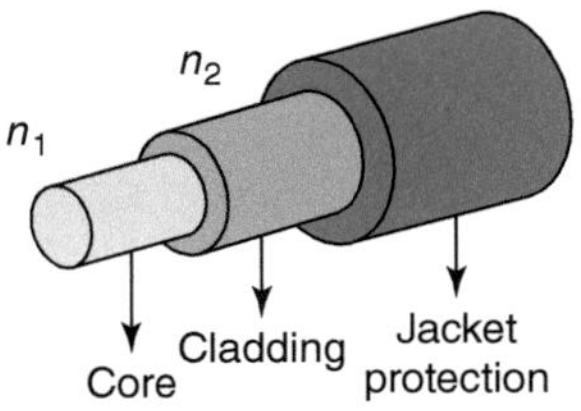

Figure 23.2 Typical structure of optical fiber.

For purposes of line lighting in short distances (few meters), the so-called SEPOF are used. The simple way how to obtain side emission is to change magnitude of refractive indexes, i.e. using slightly smaller core refractive index n_1 than that of the cladding n_2. Still there are a limited number of companies only producing SEPOF based on this simple idea. Second possibility is to use the end-emitting optical fiber with modification enabling side emission. There are in fact possibilities to promote scattering of core by adding particles or changing super molecular structure; creating imperfections (grooves) in cladding; and using structures with sharp bends (as in woven fabrics). In fact all these modifications are a little bit against aims of producers to suppress attenuation, i.e. side emission. The SEPOF will be probably in future produced according to their main goal, i.e. side emission, with more even and approximately constant decay of illumination power along their working length.

The properties and functionality of optical fibers are dependent on the refractive indexes of core and cladding. The refractive index $n = n(\lambda)$ of medium can be expressed as ratio of the velocity of a light wave having wavelength λ in vacuum C_v and velocity of a light wave in a medium $C_m(\lambda)$

$$n(\lambda) = \frac{C_v}{C_m(\lambda)} \tag{23.1}$$

The speed of light in vacuum is about 300 000 000 m/s. Typical refractive index of optical fiber cladding is $n_2 = 1.46$, and typical value for the core is $n_1 = 1.48$ [1]. The higher the refractive index, the slower rate of light travels in the given environment.

For transparent and clear substances, refractive index can be regarded as constant in the whole range of visible light. In this case, the refractive index is always greater than 1. The higher the refractive index of a material, the slower the speed of light, and the lower the refractive index, the higher the speed of light (see Figure 23.3).

Factors that affect the amount of light that is transmitted through a medium are absorption and scattering of the light. The absorption of radiation results in heat generation in a material. Sum of absorption and scattering is known as attenuation.

When light passes from a medium with a lower refractive index to a medium with a higher index, it is bent toward the normal plane between media. Otherwise light bends in the opposite direction (away from the normal plane, see Figure 23.4).

If the angle of incidence in this case is greater than the critical angle N_c, all the light energy is reflected back into the medium with a higher refractive index. This phenomenon known as total internal reflection is used to guide the light in optical fibers (Figure 23.4).

The reflection and refraction of incident ray in optical fibers vary accordingly with the material refractive index. For instance, higher refractive index would result in higher reflection than refraction of the incident light ray. The fraction of the incident light ray power R (reflectance) at direction normal to interface z

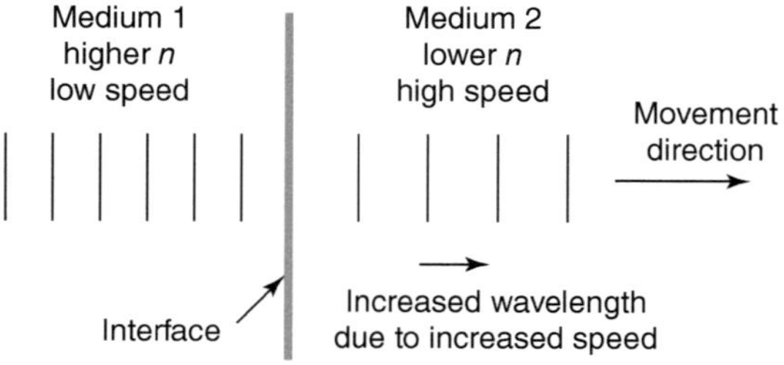

Figure 23.3 Effect of refractive index on speed of light rays. Source: Adapted from Crisp 2001 [42].

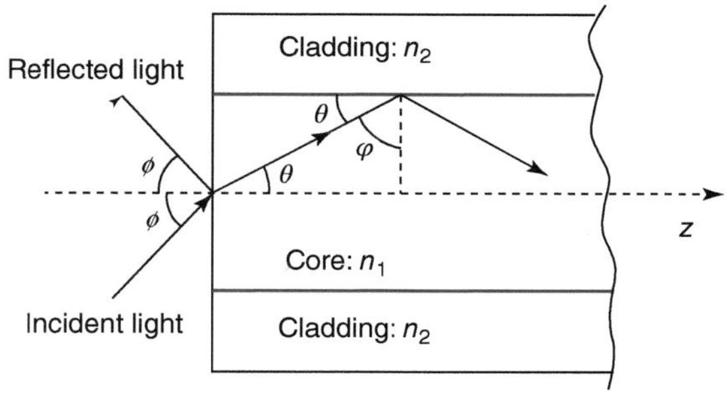

Figure 23.4 Internal reflection in optical fibers. Source: Azadeh 2009 [84]. Reproduced with permission of Springer.

($\varphi = 0°$ in Figure 23.4) reflected from the interface can be described by Fresnel's law [60]:

$$R = \left(\frac{n_1 - n_2}{n_1 + n_2}\right)^2 \tag{23.2}$$

In the standard end-emitting optical fibers, refractive index of core n_1 is always higher than refractive index of cladding n_2, i.e. $n_1 > n_2$. Below a certain angle N_c, the total reflection then takes place at the boundary between the core and the cladding. If light enters under higher angle, the partial refraction (side emission) occurs, and only part of the total light energy is reflected back to core. Figure 23.4 shows that total internal reflection (generally electromagnetic wave) occurs in the optical fibers at the boundary between the core and the sheath. The minimum angle φ_m (see Figure 23.4) when there is total internal reflection of light is equal to

$$\varphi_m = \arcsin\left(\frac{n_2}{n_1}\right) \quad \text{or} \theta_m = \arccos\left(\frac{n_2}{n_1}\right) \quad \text{or}$$
$$\left(\frac{n_2}{n_1}\right) = \sqrt{(1 - \sin^2(\varphi_m))} \tag{23.3}$$

From Figure 23.4 it is obvious that $\theta = 90 - \varphi$ and $\sin(\theta) = \cos(\varphi)$. According to Snell's law, the upper critical angle $\phi = N_c$ at input into the optical fiber is equal to [60]

$$N_c = \arcsin\left(\frac{\sqrt{n_1^2 - n_2^2}}{n_0}\right) \tag{23.4}$$

Here n_0 is refractive index of surroundings (for air $n_0 = 1$). The acceptance angle ($2N_c$) corresponds to the vertex angle of the largest cone of rays that can enter the core of the fiber.

Assuming that there is a total internal reflection (i.e. input angle ϕ is below the critical angle N_c), the whole electromagnetic radiation is captured in core and propagates along the fiber axis. If the angle of incidence on the optical fiber is higher, the part of the light is refracted, and there appears so-called leakage or partial side emission. Side emission occurs if the input angle of light incidence ϕ

is higher than critical angle N_c. This effect can be obtained by the increase of n_2 or decrease of n_1 or by the change of input angle of incident light ϕ.

The angle φ of the reflected light at the boundary between the core and the cladding at given incidence angle ϕ is given by Snell's law:

$$\phi = \arccos\left(\frac{n_0 \sin(\varphi)}{n_1}\right) \tag{23.5}$$

The number of reflections of light beam in the optical fiber is related to its length L and a core diameter d according to the equation [3]:

$$K = \frac{L}{d}\sqrt{\frac{n_2^2}{n_1^2} - 1} \tag{23.6}$$

The sine of the maximum incident angle N_c is defined as the *numerical aperture* NA $= n_0 \sin(N_c)$. The angle N_c is often denoted to as the acceptance angle, and twice the acceptance angle is denoted as the aperture angle. The NA is defined by relation

$$\mathrm{NA} = \sqrt{n_1^2 - n_2^2} \tag{23.7}$$

The NA is a very important parameter of optical fibers, since it indicates their capacity for accepting and guiding light. The size of the NA is solely dependent on the difference in the refractive indexes of the core and cladding material. For standard PMMA fiber $n_1 = 1.49$ and $n_2 = 1.40$; thus NA $= 0.50$ and $N_c = 30°$.

Compared with other fiber types [6], POF has the largest numerical aperture and the largest core diameter. This is one of the most important advantages of POF, since the connection technology that can be used for POF fibers is more economical to apply than that used for glass fibers.

The behavior of light rays in optical fiber is influenced by the profile of the refractive index of the core and cladding. In a step-index profile fiber (see Figure 23.5), the refractive index is constant across the entire cross section of the core and cladding, while the light rays propagate along straight lines in the core and are completely reflected at the core/cladding boundary [6].

It is visible that the individual light rays are passing by different trajectories and have different transit times. Graded-index profile optical fibers can minimize these differences. These fibers have cladding with a constant refractive index, and core follows the parabolic refractive index profile (see Figure 23.5). The mathematical expression of this profile has the following form [6]:

$$n^2(r) = \begin{cases} n_{1\max}^2\left[1 - 2\Delta\left(\frac{r}{r_c}\right)^2\right]; & 0 < r < r_c \\ n_{1\max}^2[1 - 2\Delta] = n_2^2; & r > r_c \end{cases} \tag{23.8}$$

where $n_{1\max}$ is maximum refractive index of core, r_c is core radius, and $\Delta = \frac{n_1^2 - n_2^2}{2n_1^2}$.

The numerical aperture of the step-index profile fiber remains constant over the entire core; the graded-index profile fiber exhibits a decreasing acceptance angle from the center of the core to the cladding.

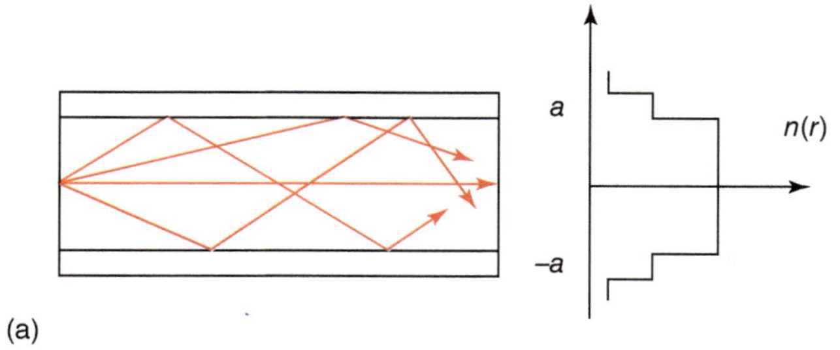

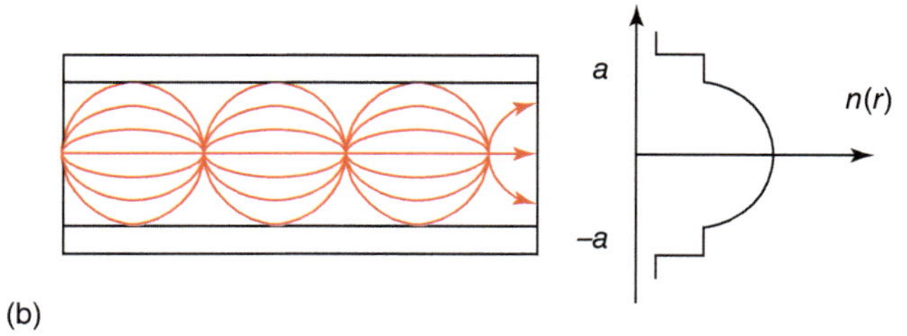

Figure 23.5 (a) Step-index and (b) graded-index profile optical fiber.

Figure 23.6 Light passing through single-mode optical fibers.

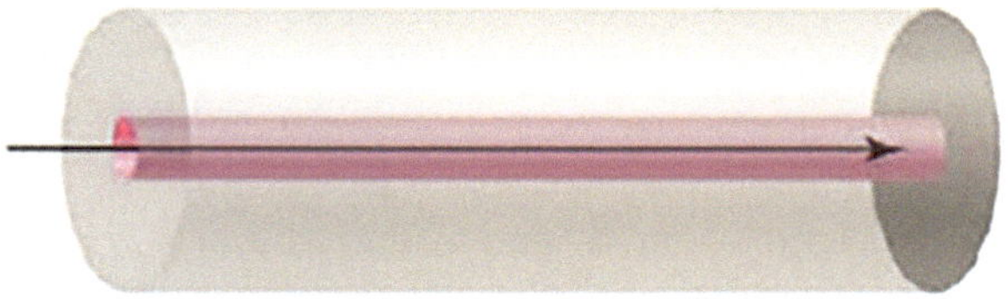

According to the mode of light propagation (the number of transmitted light rays), the optical fibers are divided into single mode (see Figure 23.6) and multimode (see Figure 23.7) [1–6].

Single-mode optical fibers have a comparatively small core diameter, typically around 8–10 μm, ensuring to guide light by direct way only. They can be analyzed by using the wave model of light (see Figure 23.8).

This type of fibers allows only the lowest order or the fundamental mode and is suitable for the applications preferably at 1300 nm wavelength. Single-mode optical fibers have lower signal loss and higher information capacity (i.e. bandwidth) than multimode optical fibers. These fibers are capable of transferring higher amounts of optical data due to lower fiber dispersion [13].

Single-mode fiber has low transmitted light loss about 0.25 dB/km. They are relatively cheaper. Electromagnetic radiation is ideally emitted from only one end, which is also used for the data transmission. The disadvantage of single-mode optical fiber is just keeping only one light ray.

Multimode optical fibers have a large core (see Figure 23.7). They can be analyzed by geometric ray-tracing model (see Figure 23.8).

Step-index multimode optical fiber guides light rays by more modes simultaneously. The individual light rays enter the fiber at different angles, thereby each

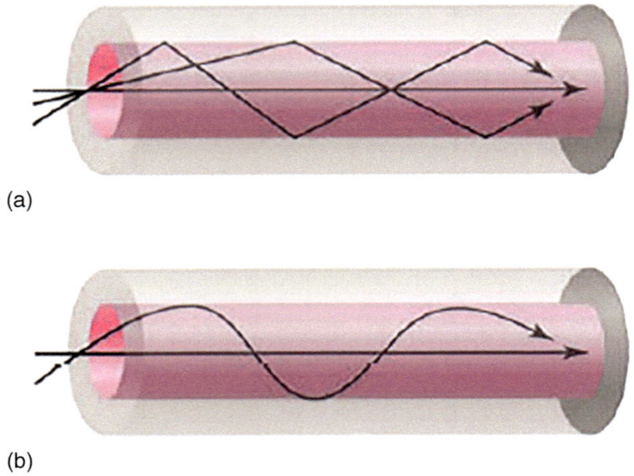

Figure 23.7 Light passing through multimode optical fibers. (a) Step index and (b) graded index.

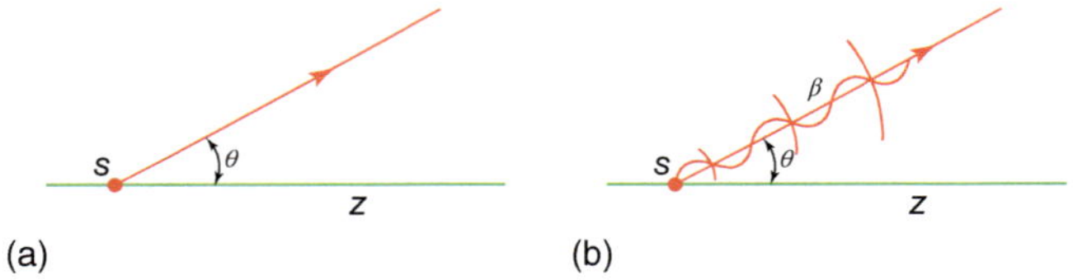

Figure 23.8 Different models of light propagation. (a) Geometric model of light propagation and (b) wave model of light propagation.

reflected along a different path. They are used primarily for short connections due to the scattering of light rays.

Graded multimode optical fibers guide light rays by more modes simultaneously, but the light is refracted in them gradually, thereby forming a coil circumscribing the interior of the fiber.

Optical fibers for special purposes have noncircular core or sheath. Typically the elliptical or rectangular cross section is used. Multimode optical fibers are generally suitable for illumination purposes, because their larger diameter allows obtaining higher light flux intensity.

In dependence on core diameter, the aperture of glass and POFs differ (see Figure 23.9).

The boundary between single-mode and multimode optical fiber types are determined by the structural parameter V, which is given by

$$V = \frac{2\pi\rho}{\lambda}\sqrt{(n_1^2 - n_2^2)} \tag{23.9}$$

where ρ is fiber density and λ is wavelength of rays passing through optical fiber. Typical wavelength is 650 nm for PMMA and 850 or 1310 nm for perfluorinated polymer core.

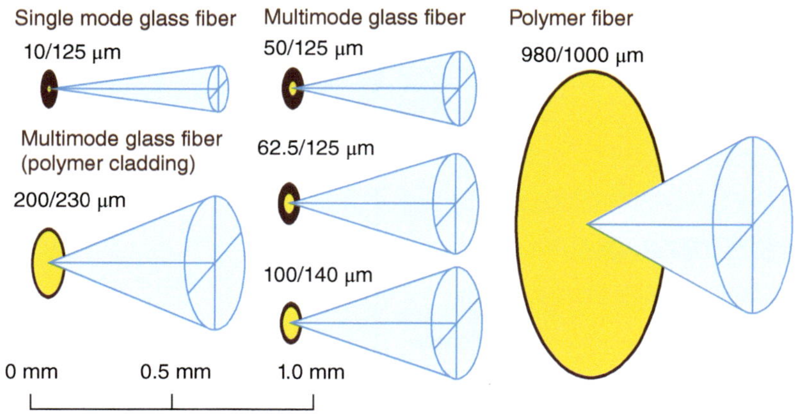

Figure 23.9 Aperture angle and core diameter of single and multimode glass fibers and polymer optical fibers. Source: Daum et al. 2002 [6]. Reproduced with permission of Springer.

When the size of the parameter V is higher than 2.405, it can be proved that for silica core the optical fiber is multimode [14, 15]. For larger values of V is a number of modes equal to

$$N = V^2/2 \tag{23.10}$$

Each mode represents one solution of Maxwell's equations. For typical plastic optical fibers, the number of modes is of the order of 10^4–10^6.

Single-mode plastic optical fibers exhibit much higher attenuation than glass single-mode optical fibers [14, 15].

Due to the transmission loss, the power of radiation P (W) emitted in any direction decays exponentially along the fiber axis with increasing distance from the light source of the fiber as observed by Zajkowski [16, 17], while the percentage of light emitted per unit length is uniform over the entire fiber length.

The simple exponential model for prediction of this attenuation is proposed. Illumination power $P(L)$ for straight optical fiber is decreasing with increasing distance from source L according to relation [18]:

$$P(L) = P(0)10^{-\alpha L/10} \tag{23.11}$$

where $P(0)$ is illumination intensity of source and α is attenuation coefficient of optical fiber. Coefficient α describes attenuation dependent on fiber length:

$$\alpha = \frac{-10}{L} \log\left(\frac{P(L)}{P(0)}\right) \tag{23.12}$$

Unit of attenuation coefficient α is dB/m. Decibel (dB) is logarithm of ratio between two illuminating powers P_1 and P_2:

$$\text{dB} = 10 \log\left(\frac{P_2}{P_1}\right)$$

i.e. if $P_2 = 10P_1$, then P_2 is higher by about 10 dB. Illuminating power is dependent on the fiber radius and on the bending radius. High level of illumination loss should be due to light transmission through bent fiber [18].

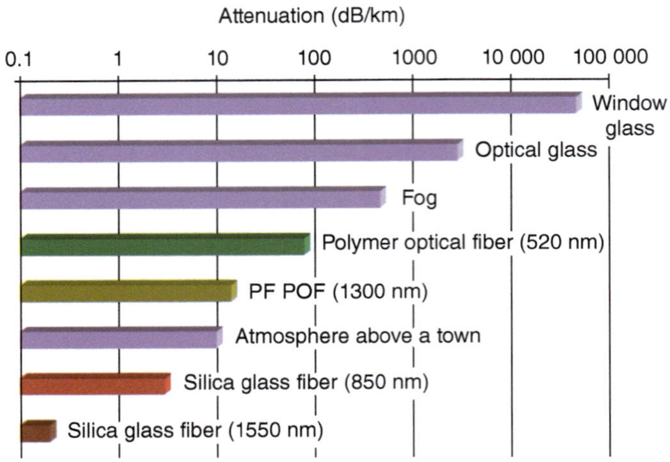

Figure 23.10 Attenuation of various media. Source: Daum et al. 2002 [6]. Reproduced with permission of Springer.

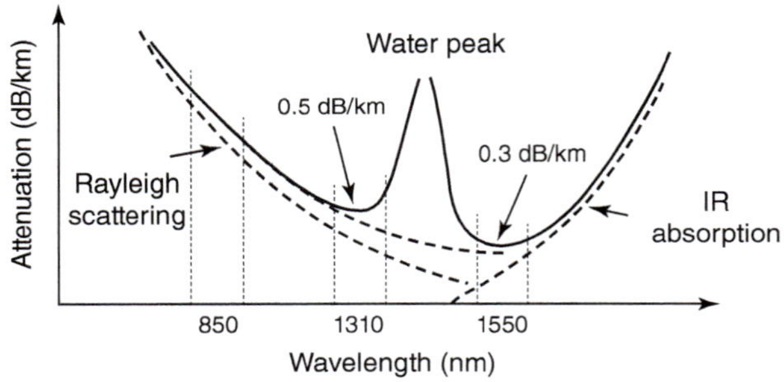

Figure 23.11 Attenuation profile of silica optical fiber.

The attenuation coefficient in the different media is shown in Figure 23.10.

Due to their high level of attenuation, POFs are suitable for short distances (approx. 100 m).

Attenuation profile of silica optical fiber is shown in Figure 23.11.

For the purpose of optical communication, the silica optical fiber uses three bands around the wavelengths of 850, 1310, and 1550 nm, which lies in the infrared region.

Attenuation profile of PMMA optical fiber is shown in Figure 23.12.

The attenuation profile of PMMA step-index fiber has three attenuation minima at 520, 570, and 650 nm [6]. Due to the molecular vibrations, attenuation increases considerably with the wavelength. If a light-emitting diode (LED) is used at 650 nm, a level of attenuation substantially higher than 130 dB/km is obtained [6].

The attenuation of POF is caused by intrinsic and extrinsic losses (see Table 23.1).

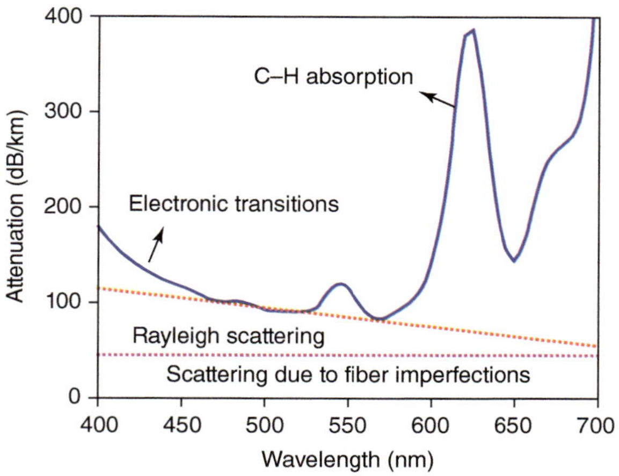

Figure 23.12 Attenuation profile of PMMA optical fiber.

Table 23.1 The attenuation of POF.

Type	Mechanism	Origin
Intrinsic	Absorption	• Molecular vibrations of C–H • Electronic transitions (UV, IR range)
	Rayleigh scattering	• Density or refractive index fluctuations • Orientation fluctuations • Composition fluctuations
Extrinsic	Absorption	• Transition metals • Organic contaminants • Absorbed water
	Scattering	• Dust, micro-voids, and fractures • Fluctuations in core diameter • Orientation birefringence • Core–cladding boundary • Micro- and macro-voids

The intrinsic sources of attenuation are caused by the material mainly. Rayleigh scattering caused particularly by fluctuations in density, composition, and molecular orientation of the polymer material changes is proportional to λ^{-4}.

The primary sources of absorption are vibrations and electronic transitions of molecular groups.

Vibrations of molecular groups such as carbon–hydrogen (C—H), carbon–deuterium (C—D), carbon–fluorine (C—F), carbon–chlorine (C—Cl), carbon–bromine (C—Br), oxygen–hydrogen (O—H), carbon double-bond oxygen (C=O), carbon–oxygen (C—O), and carbon–carbon (C—C) are responsible for the absorption loss. Molecular groups such as carbon double-bond carbon (C=C) and C=O are responsible for the optical loss by light absorption owing to electronic transitions [1–6].

The extrinsic sources of attenuation are caused by external contamination in the fiber core and physical imperfections in the fiber. The presence of impurities such as polymerization initiators, additives, and polymerization by-products are the most likely organic contaminants in POFs, which can cause a significant increase of the attenuation by light absorption or scattering. The water absorption by POF core can induce an increase in its attenuation due to OH group vibration absorption and a minor scattering loss due to agglomeration of water molecules. During the fiber drawing process, many imperfections such as dusts, micro-voids, micro-cracks, core diameter fluctuation, orientation birefringence, and core–cladding boundary imperfections appeared in the fiber. These imperfections can increase the optical loss by light scattering, independent of wavelength. The extrinsic sources of attenuation can be suppressed by proper manufacturing of POF. In the case of SEPOF, scattering of light rays that are wanted and targeted adding of particles into core is often used to enhance this.

Special class of optical fibers is microstructured optical fibers (photonic crystals). In contrast to the standard optical fibers, the microstructured optical fibers are made from one material, and light guidance is achieved by introducing microscopic air holes running along the entire fiber length [48]. These fibers can have a cladding formed from microstructural elements, usually holes. Basic preference of microstructured optical fibers is the dependence of optical characteristics on the geometry of the microstructure. The size of the microstructural elements is of the same order as the wavelength of the radiation passing through them. Other types of special POF are described in the book [1].

23.3 Materials of POF

The fundamental requirements for selecting materials for POF fabrication are that the polymers should be (i) nearly completely transparent, (ii) resistant to high temperatures, (iii) able to be drawn into a fiber, and (iv) sufficiently flexible. Although the temperature requirement depends on utilization of POF, in most cases, it is preferred that POF glass transition T_g should be above 80 °C.

Polymeric optical fibers can be produced from the preform, like silica optical fibers, only at a much lower temperature (e.g. 200 °C). Preforms can be produced by using a hollow tube of sheath material, which is filled with the liquid mixture of monomer and suitable reactive substance enabling polymerization of the core. Alternatively, a dopant dispersed in a tube or within tubes can be used. These processes can be tuned with respect to the acquisition of different refractive index profiles. An alternative to preforms is extrusion processes. These techniques were originally used for PMMA optical fiber, but there are also variants for perfluorinated materials. In the extruder the two distinct polymer materials can be inserted only. By the control of the diffusion process, the nonconstant refractive index profiles can be produced. Typical operation is hot drawing changing super molecular structure and optical properties [1–6].

23.3.1 Core

The most widely used polymers for optical fibers core include *PMMA* (Figure 23.13), which has a high transmission rate and low attenuation. Amorphous homopolymer of PMMA has a refractive index of $n = 1.492$ measured at wavelength of 589.2 nm. Refractive index is systematically decreasing with increasing of wavelength especially in infrared region.

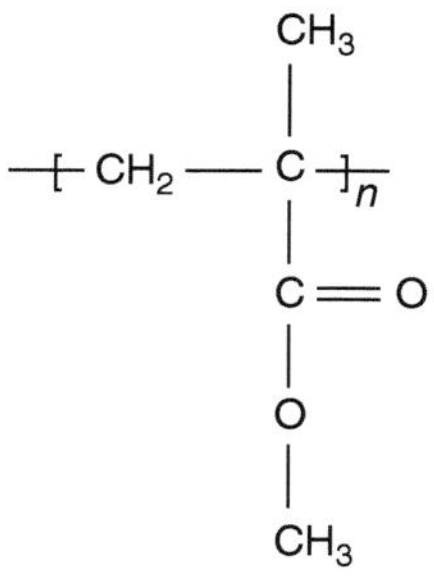

Figure 23.13 Structure of PMMA.

This index can be changed using special dopants. PMMA has a relatively high T_g of around 105 °C. It can be drawn in the temperature range from 175 to 220 °C. The polymer is stable in air at temperatures of 85 °C. At higher temperatures, the presence of moisture leads to rapid degradation and durability constraints.

Contact angle of water for PMMA is 80° and bandwidth 0.003 GHz km. Typical core diameter is 250–1000 μm, and numerical aperture is NA = 0.47.

The PMMA core has two attenuation, absolute low value of 70 dB/km at 522 and 570 nm (green region) and the relative minimum around 650 nm (red region) [41]. Each methyl methacrylate unit has a total of eight C—H bonds. The vibrations of these bonds or precisely their harmonic waves are a main cause for the losses in PMMA fibers.

PMMA is thermoplastic polymer and can be spun from melt. Solvents for PMMA are aromatic or chlorinated hydrocarbons, esters and ketones as benzene, toluene, dichloromethane, chloroform, tetrahydrofuran, dimethylformamide, ethyl acetate, and acetone.

Selected properties of PMMA are summarized in Table 23.2.

Over a greater temperature range up to 120 °C *PC* (Figure 23.14), core may be used, which is also mechanically more robust and less sensitive to moisture and has a larger ductility (10.8% compared with 6.3% for PMMA).

Table 23.2 Selected properties of PMMA [19, 20, 28, 32].

Property	Typical value	Unit	Test
Refractive index	1.491	—	DIN 53291
Density	1180	kg/m^3	DIN 53479
Tensile strength	80 (72)	MPa	DIN 53455
Impact strength	15	kJ/m^2	ISO 179/1D
Bending strength	115 (105)	MPa	DIN 53452
Initial modulus	3300	MPa	DIN 53457
Glass temperature, T_g	105	°C	[55]
Thermal expansion		K^{-1}	DIN 53752-A
• Linear	7.0×10^{-5} (0–50 °C)		
• Volume	2.72×10^{-4} ($<T_g$)		[20]
	5.80×10^{-4} ($>T_g$)		
Starting temperature of shrinkage	>80	°C	[20]
Water absorption	30	mg	DIN 53495

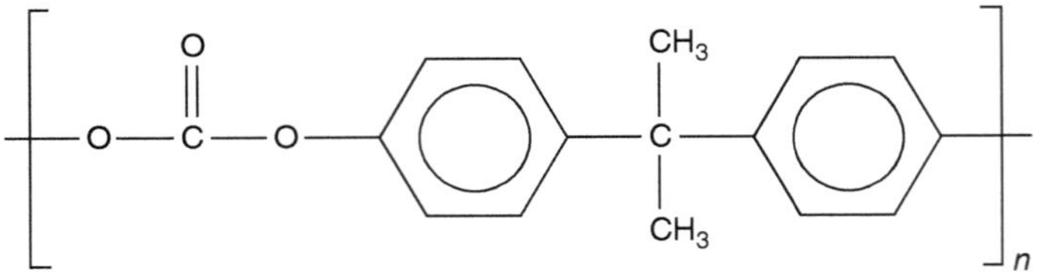

Figure 23.14 Structure of polycarbonate (PC).

PC has a refractive index of $n = 1.582$, and bandwidth is 0.0015 GHz km. Typical core diameter is 500–1000 μm, and numerical aperture is NA = 0.78. The PC core has minimal attenuation of 600 dB/km at 670 nm, density 1200 kg/m^3, and glass transition temperature $T_g = 145\,°C$ [1]. The bisphenol A type PC POF core can be used in industrial applications where the temperature is an important parameter. PC core POF can be used for creation of textile integrated illumination elements [21]. A problem with PC is that it has large birefringence.

More hydrophobic *PS* (Figure 23.15) core has a minimum attenuation 330 dB/km at 580 nm, i.e. in the red region.

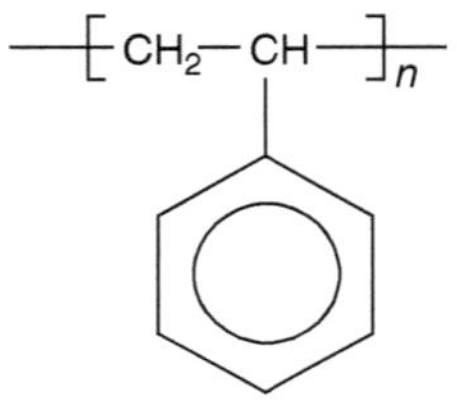

Figure 23.15 Structure of polystyrene (PS).

PS is soluble in aromatic and halogenated hydrocarbons and in aliphatic ethers, esters, and ketones. From the mechanical standpoint, the PS fibers are superior to fibers of PMMA, although they have higher attenuation [41]. PS has a refractive index of $n = 1.592$, and bandwidth is 0.0015 GHz km. Typical core diameter is 500–1000 μm, and numerical aperture is NA = 0.73 [41]. The glass transition temperature of PS is around 100 °C. This material is now not of any practical significance [6].

One of the lowest attenuation 16 dB/km at 1310 nm has amorphous *fluorinated polymer* cyclic transparent optical polymer (*CYTOP*) (Figure 23.16) [39–41]. This material poly(perfluorobutenyl vinyl ether) no longer contains hydrogen.

CYTOP has similar excellent chemical, thermal, electrical, and surface properties as conventional fluoropolymers. This material obeys high optical transparency and good solubility in specific fluorinated solvent due to amorphous morphology. CYTOP has a refractive index of $n = 1.353$, bandwidth 0.0005 GHz km, density 2030 kg/m^3, $T_g = 108\,°C$, and contact angle with water 110°. Typical core diameter is 110–1000 μm, and numerical aperture is NA = 0.40 [41].

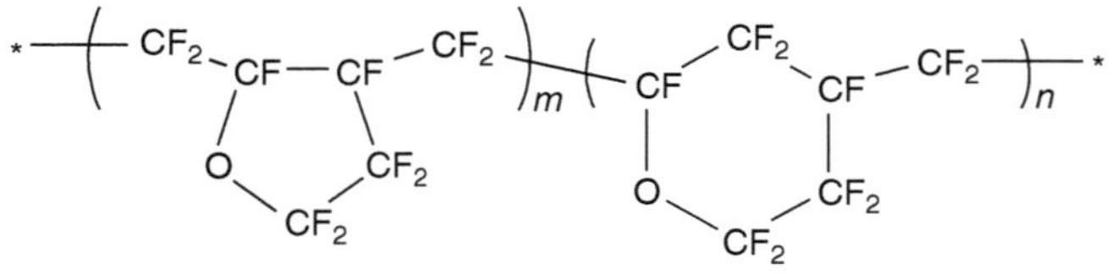

Figure 23.16 Structure of CYTOP.

Core attenuation generally increases with the decrease of its diameter. Attenuation of PMMA core at 570 nm is 70 dB/km for a diameter of 0.5 mm and 130 dB/km for a diameter of 0.25 mm. Production process is difficult to be controlled for smaller diameters of POF. The temperature gradient is greater, resulting in a high number of defects.

Attenuation can be induced by bending and tensile or torsion deformation [35, 36]. The initial tensile modulus for POF is almost 2 orders of magnitude lower than that of silica optical fibers (2.1 GPa for PMMA [37], 1.55 GPa for PS, and 2.55 GPa for PC [35]). Therefore, the POF of diameter 1 mm are still sufficiently flexible and the minimum bend radius for POF is smaller than for conventional glass optical fibers.

POF light transfer rate also depends on the degree of elongation (drawing). When stretched by 10% the attenuation increases by less than 0.1 dB [37]. Also cyclic bending causes changes in attenuation (for 1 mm thick PMMA, it is 0.15 dB after 1000 bends around bending radius of 50 mm [38]).

The maximum operating temperature of the POF is 80–100 °C. Above this threshold, there is a loss of rigidity and transparency. The operating temperature can be increased up to 125 °C or even to 135 °C using a cladding of cross-linked polyethylene or polyolefin elastomers.

Resistance to high temperatures strongly depends on the humidity. For example, if the POF is maintained at 85 °C with 85% RH (air relative humidity) for 1000 hours, it is increasing attenuation of 0.02 dB/km. If the relative humidity is around 90%, it increases the attenuation of more than about 0.03 dB/m. Fluorinated POF cores do not absorb water, so that their attenuation is virtually unchanged with the increase of the moisture.

Standard core materials (PMMA, PS, and PC) generally have a high attenuation, which limits their applicability for information transfer.

It is possible to reduce the absorption losses of POF by using different materials in which less or no C—H bonds are present. A heavier atom X replacing hydrogen atoms H leads to lowering of stretch and bending fundamental vibrations of C–X and moving the attenuation bands to a larger wavelength. The selected atoms X are deuterium (heavy hydrogen with the atomic mass 2), fluorine (atomic mass 19), and chlorine (atomic mass 35 or 37). Generally, the materials for POF can be compounds containing hydrogen, compounds with partial substitution of hydrogen, and compounds with complete substitution of hydrogen [6].

One way to suppress attenuation was to use deuterated polymers [6]. In deuterated polymer, hydrogen is substituted by heavier deuterium. This isotope has two times higher atomic mass than hydrogen. Behavior of deuterium is the same as hydrogen so that it is simple to use so-called heavy water (D_2O) instead of water H_2O for synthesis. It was possible to prepare deuterated material with a minimum attenuation of 20 dB/km at 680 nm. Disadvantages of deuterated polymers are a gradual increase of attenuation due to replacement of deuterium by hydrogen as a result of atmospheric water vapor absorption and a high price.

Problems with the attenuation and wetting ability of the core can be solved by using halogenated polymers. Fully halogenated polymers, e.g. polytetrafluoroethylene (PTFE), are highly crystalline, which increases optical attenuation. Usually the partially halogenated polymers blended with PMMA or amorphous fluorinated polymers of CYTOP type can be used for avoiding unwanted wetting and lowering of attenuation.

For producing core of step-index POF, *bulk polymerization* technique is preferred because of the purity of an obtainable polymer. Continuous or batch extrusion and melt spinning followed by hot drawing are mostly widely used for POF preparation [1–6]. Production of graded-index POF mainly uses the preform hot drawing technology [1–6].

23.3.2 Cladding

Two of the main requirements for cladding materials are low refractive index (but close to the index of the core) and good film forming ability. In addition, the application of cladding can provide a good mechanical and thermal resistance to the core. A variety of cladding polymers have been developed since the invention of glass and POFs. Majority of perfluoropolymers and partially fluorinated polymers alone are unsuitable for the POF core but are good cladding materials. Actually, there are two classes of fluoropolymers, which are widely used as cladding materials: copolymers of fluoroolefins and poly(fluoroalkyl acrylates) (PFA). The PFA are preferred because of their amorphous nature, high transparency, good adhesion properties, etc. As a result durable optical properties can be obtained. Another main advantage of PFA is their monomers easily coating (by solution) on the core and can be easily photopolymerized.

In general, there are already a number of monomers of PFA mainly for fiber-optic applications especially as cladding materials available in the market. Some monomers of poly(fluoroalkyl methacrylates) are listed in Table 23.3.

However, the formulation of cladding materials depends on individual optical fiber producers. Copolymers of fluorinated methyl and ethyl esters of acrylic and methacrylic acids are usually used as cladding materials for the PMMA core.

23.3.3 Jacket

Selection of jacket depends on the intended field of POF use. The basic mechanical and thermal characteristics of jacket affect the overall life of the POF. Selected suitable types of jackets and their properties are listed in Table 23.4 [6].

Table 23.3 Monomers of poly(fluoroalkyl methacrylates).

Monomer	Alkyl group in methacrylate	T_g (°C)	n
Methyl methacrylate	$—CH_3$	105	1.489
2,2,2-Trifluoroethyl methacrylate	$—CH_2—CF_3$	69	1.418
2,2,3,3-Tetrafluoropropyl methacrylate	$—CH_2—CF_2—CHF_2$	68	1.417–1.422
2,2,3,3,3-Pentafluoropropyl methacrylate	$—CH_2—CF_2—CF_3$	70–77	1.1395
2,2,3,3,4,4,4-Heptafluorobutyl methacrylate	$—CH_2—CF_2—CF_2—CF_3$	65	1.383

Table 23.4 Polymers suitable for jackets.

Polymer	Maximum working temperature (°C)	Density (kg/m³)
Polyvinylchloride (PVC)	70	1200–1500
Polyethylene (PE)		
• Low density	70	1300–1600
• High density	80	950–980
Polypropylene	90	910
Polyamide 6 (PA 6)	80–90	1100–1150
Polyurethane (PU)	90–100	1150–1200
Copolymer ethylene-vinyl acetate (EVA)	120	1300–1500
Perfluoroethylenepropylene (PFEP)	180	2000–2300
Polytetrafluoroethylene (PTFE)	260	2000–2300

Polyethylene (PE) jacket also serves to protect the POF, to liquids such as water, NaOH, sulfuric acid (34.6%), or motor oil. Attenuation remains constant if the POF having POE jacket is immersed in these fluids at 50 °C for 1000 hours [6].

23.4 Side-Emitting POF

For a variety of textile applications (e.g. the so-called active safety textiles) the side emission should be ensured. In side-emitting POF, i.e. *SEPOF*, the light leaks out from their surface. For illumination purposes, light emission along the fiber is necessary. In this case, multimodal fibers are most likely to be used, since their larger diameter allows higher emission intensities to be obtained. Side emission occurs if the light incidence angle is higher than critical angle N_c. This effect can be obtained by the increasing of cladding refractive index, decreasing of core refractive index, or changing of incident light angle. For optical characterization of SEPOF, it is favorable to use so-called *attenuation profile* (see Section 5.1). The *attenuation profile* is the dependence of light intensity of omnidirectional rays passing from optical fibers through their sides measured as function of distance from light source. Higher level and smaller decay (slope) of attenuation profile is better for more even side emission. Detailed discussion about attenuation profile characterization and shapes for different types of side-emitting fibers is presented by Křemenáková et al. [29].

There are some possibilities on how to obtain side emission by using standard or modified end-emitting POF. Various types of patented side-emitting fibers and wave guides including methods of their preparation are discussed in Ref. [26]. In all cases the aim is to avoid total internal reflection by using multiple micro-bending of the fiber, dispersion of fluorescent additives or particles into the core or cladding material, creating asymmetries in the core/cladding geometry, partial destruction or grooving of cladding, etc. [22–26]. Emission from side-emitting optical fibers referring to a fiber design with scattering cladding is described in work [22, 23]. Various commercial side emission optical fibers and light guides and methods for their manufacture were developed and patented [21–26].

23.4.1 SEPOF Based on Difference Between Refractive Indexes

There exist a lot of producers of SEPOF, e.g. from China, England, the United States, and Australia offering wide range of diameters, different types of cladding, different composition of core, and different flexibility. Majority of these fibers are based on changing relation between refractive indexes of core n_1 and cladding n_2, i.e. following inequality $n_1 < n_2$. For specific applications it is necessary to select proper diameter mainly because the illumination intensity and height of attenuation profile (see Figure 23.19) depends critically on this parameter.

As example, the two typical SEPOF fibers, *standard* and *flexi* (Grace POF Co., Ltd. China), are compared. The fiber Grace standard is composed from PMMA core and PC cladding of low irregular thickness below 10 μm (see Figure 23.17). These fibers are relatively stiff and brittle. Offered diameters of these fibers are from 0.25 to 1 mm.

Fibers of type Grace flexi are formed of a PMMA core and polytetrafluorethylene cladding of thickness about 0.2 mm, which can be easily detached (see Figure 23.18). Grace flexi fibers are produced in diameters of 2–14 mm or more. These fibers have lower stiffness and higher flexibility in spite of higher diameter, which is advantageous for their incorporation into textiles.

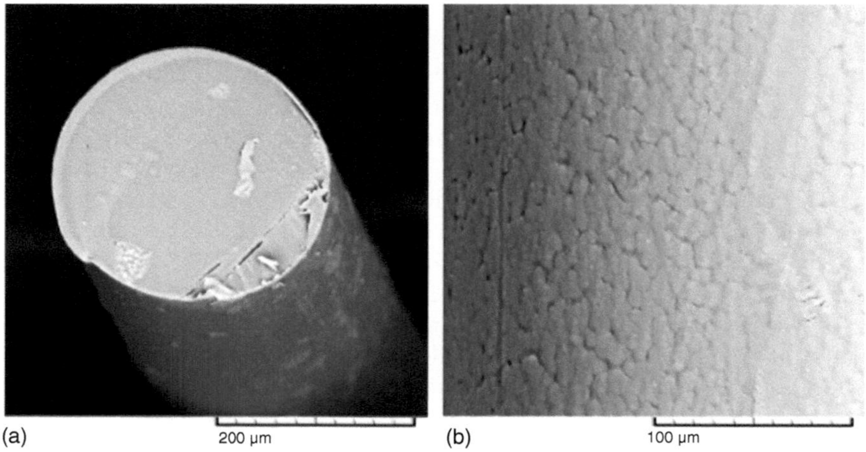

Figure 23.17 SEPOF Grace standard. (a) Cross section – magnification 350× and (b) surface – magnification 700×.

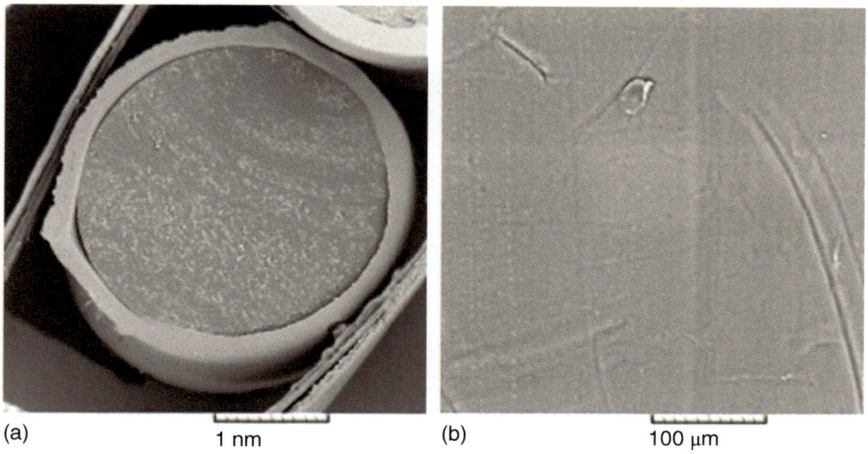

Figure 23.18 SEPOF Grace flexi. (a) Cross section – magnification 40× and (b) surface – magnification 400×.

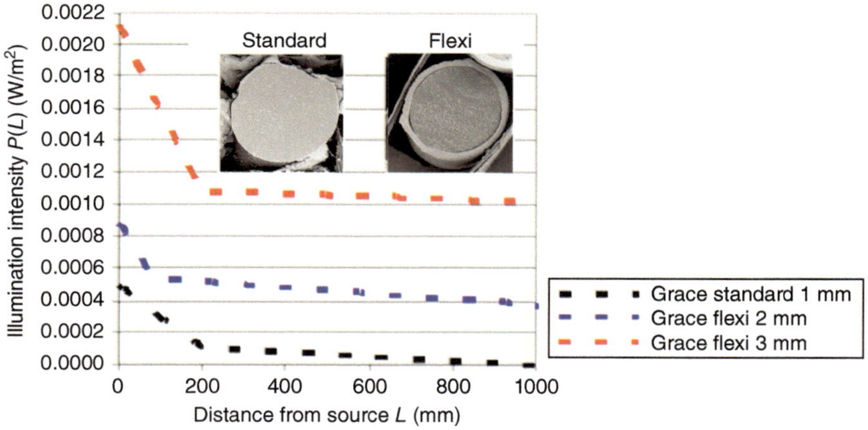

Figure 23.19 Influence of type and diameter of SEPOF on their illumination intensity.

The attenuation profiles of these fibers (see Section 5.1) for different diameters are compared in Figure 23.19.

It is clearly visible that the illumination intensity, i.e. height of attenuation profile, can be simply enhanced by increasing of SEPOF diameter.

Second important parameter is applied voltage, i.e. electric power necessary for light generation. In Figure 23.20, attenuation profiles of three portable power sources are compared for the same LED. Here the light intensity is expressed in lux (lm) units.

It is clear that the proper voltage is very important to obtain required illumination intensity and therefore lighting efficiency.

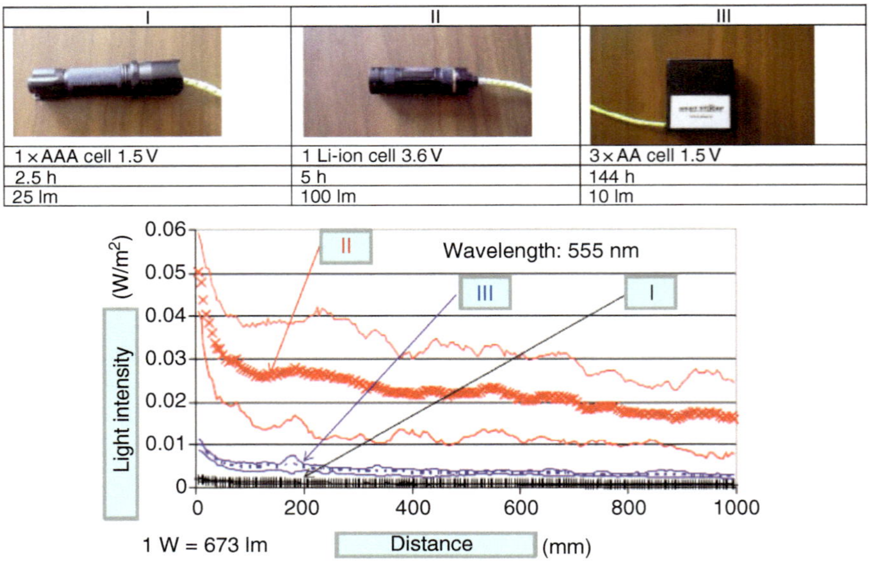

Figure 23.20 Influence power source voltage on SEPOF illumination intensity.

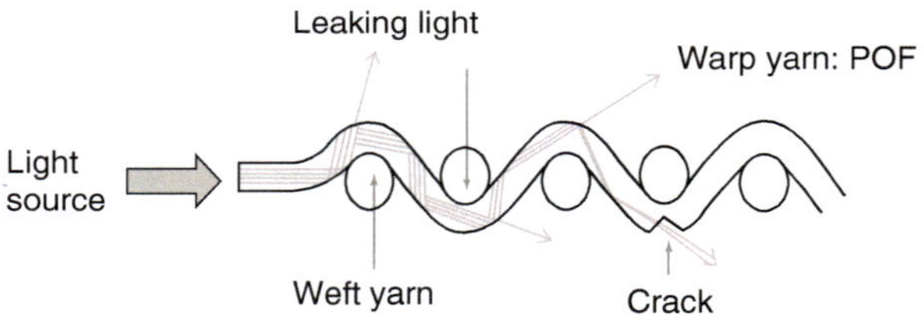

Figure 23.21 Local side emitting due to creating of woven fabric. Source: Adapted from Masuda et al. 2006 [43].

23.4.2 SEPOF Based on Multiple Micro-bending

For end-emitting optical fibers, micro-bending is usually unwanted phenomenon causing increase of attenuation. Micro-bending is here commonly described as a random variable with a distribution of spacing and amplitude; radius of bent is typically less than 1 mm [74, 75]. Micro-bending due to the incorporation of POF fibers into woven or knitted fabrics, respectively, will induce their side emission. Incorporation of end-emitting POF into textile structures thus can cause local side illumination (see Figure 23.21). When creating micro-bending due to weaving of end-emitting POF to fabrics, the bend angle exceeds over critical angle N_c, and the light is reflected out.

By the creation of woven/knitted structures, optical fibers are subjected to bending, and the angle of incidence of light rays at the interface between the core and the cladding is locally changed so that there may be appearing local side emission (instead of internal reflections in end-emitting optical fibers). It is known that the intensity of the transmitted light generally decreases with increasing bending angle of optical fibers. Transmission loss increases exponentially with increase in the ratio between the bending radius and the radius of the

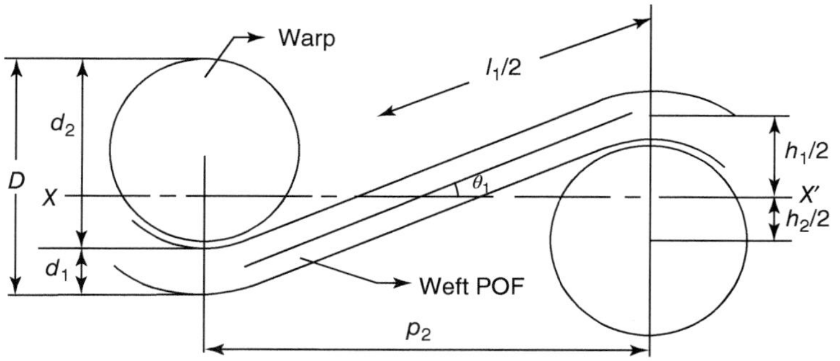

Figure 23.22 Peirce geometrical model of cloth. Source: Adapted from Peirce 1937 [64] and Behera et al. 2012 [70].

fiber [62]. For optical fibers inserted into textile structures, the influence of the bending on the macroscale is relatively small. On the other hand, the bending caused by the integration of fibers into the fabric structure, i.e. a local bending in binding points between warps and wefts (mesoscale), can lead to significant losses in light transmission or provide side emission.

Usually at least the local side radiation is required, so the incorporation of end-emitting optical fiber as part of the fabric structure is advantageous (local side radiation due mesoscale bends at crimps ensuring the high curvature). For SEPOF this local increase of side radiation in areas of highest curvature is usually not so beneficial. That is why the side-emitting fibers are inserted into the textile structures with the smallest deviations from the straight state (textile structure is serving as wrapping).

A separate problem is that flexibility and elasticity of optical fibers is limited, and at higher crimps due to weave, some damage may occur. The incorporation of optical fibers into the fabric so that at least part of them are on surface (as part of the structure) is protected by a patent [69]. Patent pending [68] protects the use of LIHS, wherein no part of SEPOF is on the fabric surface.

For characterization of POF shape and meso-bending in the fabric, the various geometric models can be used. Most common Peirce geometric model [64, 70] is shown in Figure 23.22. This model defines the arrangement of threads (yarns) in the woven fabric elementary unit (cell) for the case of direct inextensible but flexible materials. Threads have a circular cross section and consist of straight and bent segments.

The basic parameters of this model are as follows: the diameter of the warp threads d_2; the diameter of the weft thread d_1 warp sett $S_o = 1/p_2$, where p_2 is the distance between warp yarns; weft sett $S_u = 1/p_1$, where p_1 is the distance between the weft threads; and the fabrics thickness D and h_1, h_2, i.e. maximum displacement of thread axis normal to the plane of cloth (crimp height) of warp (2) and weft (1) in Figure 23.22.

Based on this two-dimensional model of the unit cell of fabric, a number of geometric parameters of fabrics, such as the angle of inclination of the thread, the thickness of the fabric, etc., can be calculated [70]. For a radius of curvature of the weft threads r_c, it can be then derived

$$r_c = \frac{1/S_o}{2\,\mathrm{arc}\,t_g\left(\frac{D-d_2}{1/S_u}\right)} + \frac{D}{2} \tag{23.13}$$

The fabric thickness is equal to $D = h_2 + d_1 \cong d_1 + d_2$. Trajectory of optical fiber integrated in weaving patterns can be simply described by Novikov waviness (see Ref. [73]). Warp weaves amplitude h_o and weft weave amplitude h_u are distance between fabric axis and yarn warp/weft axis (see Figure 23.10). Novikov relative warp waviness e_o and relative weft waviness e_u are ratio between warp/weft weave amplitude h_o/h_u and mean warp/weft diameter D_{str} [73]:

$$e_1 = h_1/d_{\mathrm{str}} \quad e_2 = h_2/d_{\mathrm{str}} \quad e_1 + e_2 = 1 \quad d_{\mathrm{str}} = (d_1 + d_2)/2 \tag{23.14}$$

Value of this ratio for the case of POF in the warp can be varied from 0 (Phase 1 – straight) to 1 (Phase 9 – maximum bending of warp thread, axis of weft thread is lying on the straight line).

The radius of curvature of the optical fiber built in warp can be determined by using Eq. (23.13). It is only necessary to replace the parameter of warp by parameters of weft and vice versa. The physical simulation system of fabrics created from metal wires replacing the wefts and putting optical fibers as warps was proposed in Ref. [65]. The bending radius r_b of locally bent optical fiber in this simulated system is equal to

$$r_b = \frac{D}{4} + \frac{p_2^2}{16D} \tag{23.15}$$

Numerical aperture NA_b of bent POF can be calculated from relation [66]:

$$\mathrm{NA}_b = \sqrt{\mathrm{NA}^2 - \frac{n_2^2 r_c}{r_b}\left(2 + \frac{r_c}{r_b}\right)} \tag{23.16}$$

where NA is numerical aperture of straight POF, n_2 is refractive index of cladding, and r_c is core radius. The ratio of transmitted light energies is equal to the square of the ratio of NA, and therefore the light transmission efficiency E_t of bent POF is equal to

$$E_t = \left(\frac{\mathrm{NA}_b}{\mathrm{NA}}\right)^2 \tag{23.17}$$

The light emission efficiency E_e of bent POF without including intrinsic losses can be calculated from relation $E_t + E_e = 1$ [67]. For the same POF, light emission efficiency is then determined by the core radius and bending radius of POF only [65]. Based on the results of simulation experiments, it was found that the side-emitting intensity is directly proportional to the diameter of POF and indirectly proportional to the bending radius of POF. In accordance with Eqs. (23.15) and (23.16), the following model for prediction of side-emitting intensity I_e was proposed [65]:

$$I_e = \frac{A(r_c)}{r_b^2} + \frac{B(r_c)}{r_b} + C \tag{23.18}$$

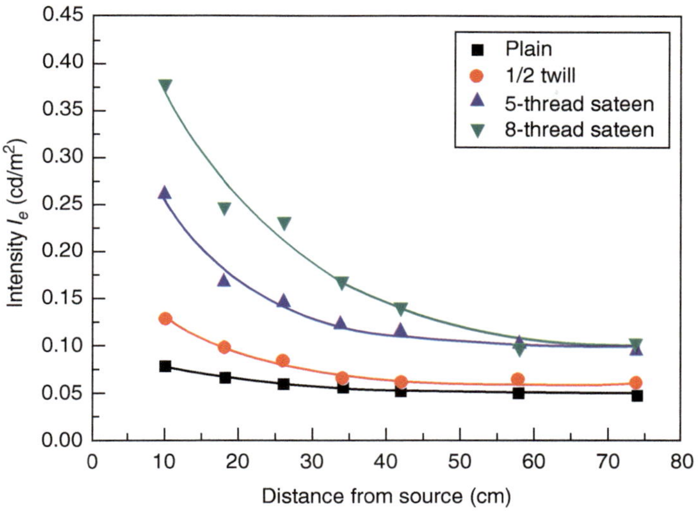

Figure 23.23 Side-emitting intensity of woven structures with POF fibers in weft. Source: Adapted from Wang et al. 2013 [65].

where constants $A(r_c)$ and $B(r_c)$ depend on the radius of core and constant C is connected with light source quality and POF diameter. The minimum bending radius r_{bm} of the POF before the break can be calculated from knowledge of deformation at yield point ε_y according to the relation [66]:

$$r_{\text{bm}} = \frac{r_c}{\varepsilon_y} \tag{23.19}$$

It was found that for PMMA core POF having diameters 250 and 500 μm, minimum bending radius r_{bm} were equal to 2.6 and 4.46 mm. Side-emitting intensity I_e of different woven structures having POF from PMMA with diameter 250 mm in weft is shown in Figure 23.23.

It is obvious that looser weaves provide higher side emission intensity. In the case of sateen weaves, there was a significant improvement in the side emission intensity of radiation at small radii of curvature (see Figure 23.24).

Generally, it is therefore necessary to appropriately select the bending radius (dependent on weave type) and the radius of the POF with respect to initial modulus, strength, and parameters at the yield point, respectively. Light transmission losses of bent optical fiber may be unacceptably high for a SEPOF as well [44].

The most significant loss of illumination intensity of about 55% are in the area where POF passes from straight shape to bent form. Since the weft yarns in the fabric are relatively more bent than the warp yarn, the fabrics with POF in the warp obey more side emission for the same weave.

Increased lateral emissions can be achieved also by using fluorescent dye stuffs in the cladding of the POF [31]. During textile processing of POF, their plastic elongation occurs frequently, which negatively affects transmission of light. For weaving of POF, the rapier weaving machines are particularly suitable.

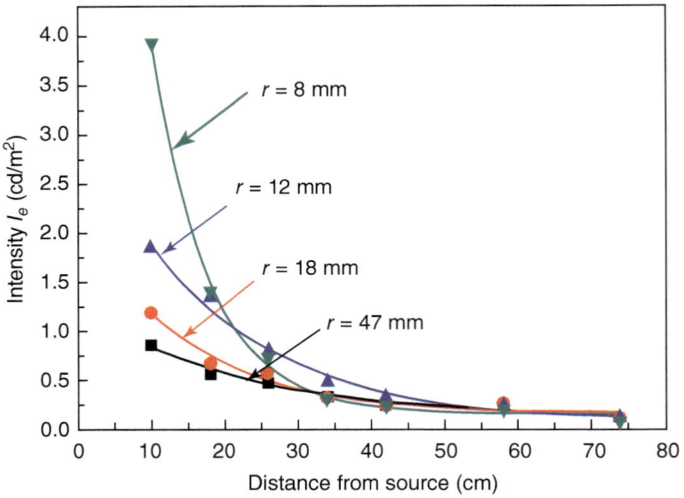

Figure 23.24 Influence of radii of curvature *r* on the side-emitting intensity of sateen weaves. Source: Adapted from Wang et al. 2013 [65].

23.4.3 Modifications of POF Structure

To provide or enhance side emission, it is possible to use the following techniques modifying standard end-emitting POF:

1) During the manufacture of fibers, appropriate "micro" particles (ZnO, Al_2O_3, etc.) are placed into the polymer core or cladding (see Figures 23.26 and 23.27).
2) The cladding of the POF is chemically or mechanically damaged (see Figures 23.31, 23.33 and 23.34).
3) The micro-cuts or notches are added in the cladding layer (see Figures 23.28 and 23.29).

For industrial purposes these techniques can be combined with the selection of materials of core and cladding according to principles of side emission, i.e. $n_2 > n_1$.

For some textile applications, the special polymers comprising alternating layers of material having a high refractive index difference (2D photonic crystals) are proposed [49]. In their cross section these fibers contain a periodic sequence of micron-sized layers of different materials. The so-called Bragg fibers combining PMMA core, multilayered PPMA/PC Bragg reflector, and PMMA cladding are shown in Figure 23.25 [49].

Photonic bandgap (PBG) fibers guide light using PBG effect rather than total internal reflection. Intensity of side-emitted light can be controlled by choosing the number of layers in the microstructured region surrounding the optical fiber core. Such fibers always emit a certain color sideways without necessity of surface corrugation or micro-bending.

Hollow core Bragg fiber consists of an air-filled core surrounded by a periodic sequence of high and low refractive index layers forming a so-called Bragg reflector [50]. Typical feature of reflector is the presence of regions of high reflector

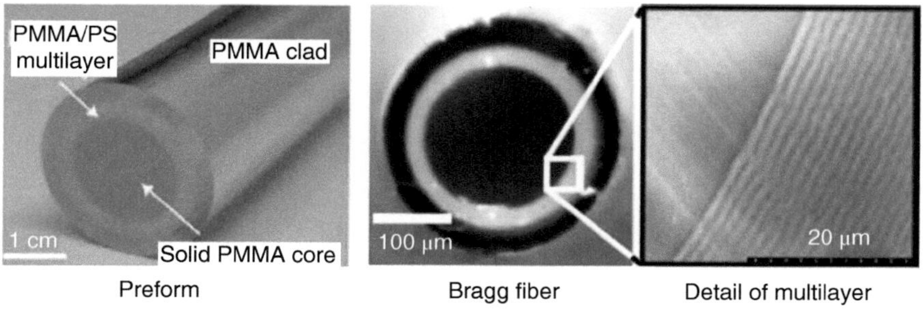

Figure 23.25 Solid core Bragg photonic fiber.

efficiency caused by the interference effects inside a periodic multilayer. Light with frequency inside of a reflector bandgap can be confined in the fiber hollow core through reflections from a surrounding periodic reflector. By changing the number of reflector layers, it is possible to control the rate of potential side emission. For light transmission the number of reflector layers should be sufficiently high to suppress radiation loss. For illumination purposes a relatively small number of reflector layers should be used.

The SEPOF containing particles of Al_2O_3 in core are schematically illustrated in Figure 23.26, and SEPOF containing particles of ZnO in cladding are in Figure 23.27.

Diffusing jacket on SEPOF (Figure 23.26) is usually used as a protection against humidity and mechanical damage. Metallic particles such as iron, copper, cobalt, nickel, manganese, and chromium have the effect of absorbing the radiation,

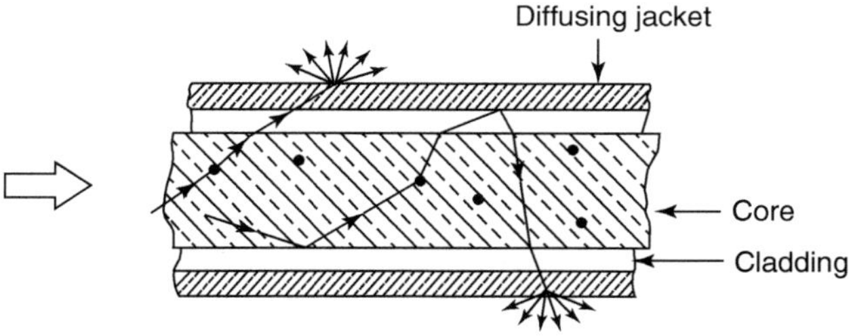

Figure 23.26 SEPOF with particles in core [25].

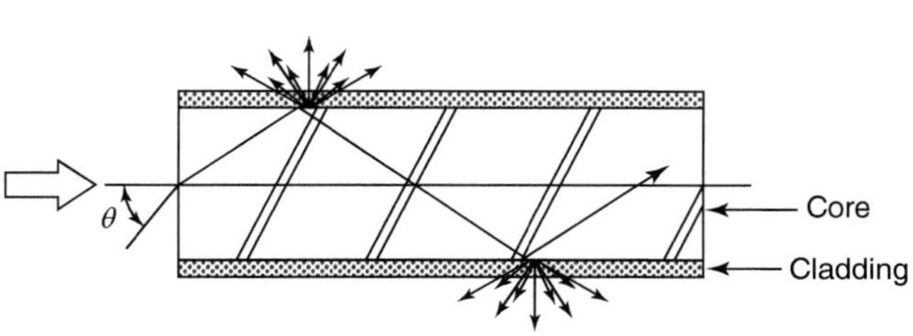

Figure 23.27 SEPOF with particles in cladding [47].

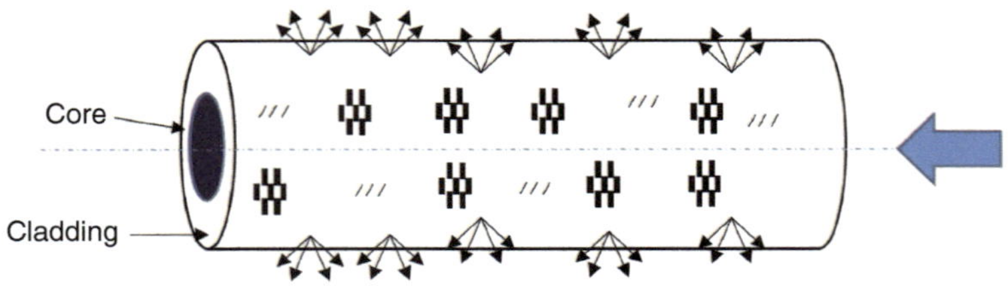

Figure 23.28 SEPOF with micro-cuts or notches in the cladding. Source: Spigulis et al. 1996 [22]. Reproduced with permission of Society of Photo-Optical Instrumentation Engineers (SPIE).

increasing attenuation, and, consequently, decreasing the overall radiation intensity in the fiber [33, 51, 52].

SEPOF with micro-cuts or notches in the cladding is shown in Figure 23.28.

Notches can be created by local surface melting by using CO_2 infrared laser micromachining [71]. The size of one notch created on POF having the PMMA core (diameter 240 mm) and fluorinated PMMA cladding (diameter 10 mm) was as follows: height $h = 98.3$ mm, angle $2\delta = 148.2°$.

Structure of typical notch created by local laser surface melting is shown in Figure 23.29.

Combination of surface notches created by laser and micro-bending in woven structure for creation of light side-emitting fabrics was investigated in work [71].

Enhanced side illumination of SEPOF having PMMA core and PTFE cladding by combination of TiO_2 particles (diameter of 0.5 mm) attached on cladding surface and subsequent CO_2 laser treatment was investigated in work [72].

It is visible (see Figure 23.30) that surface of cladding is damaged especially in places where agglomerates of TiO_2 particles are present.

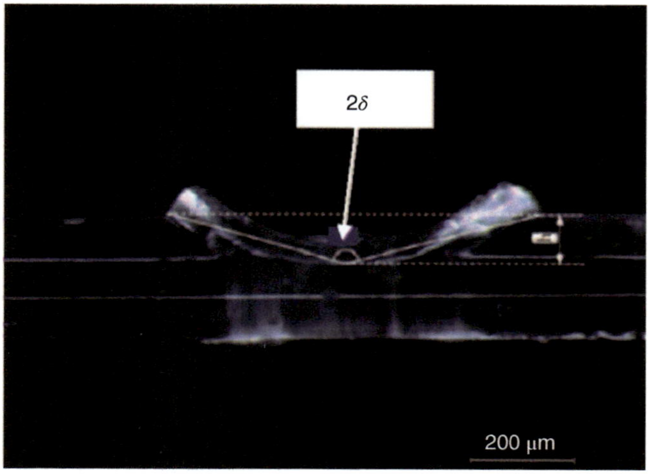

Figure 23.29 Micrograph of polymer optical fiber with one notch. Source: Shen et al. 2013 [71]. Reproduced with permission of SAGE.

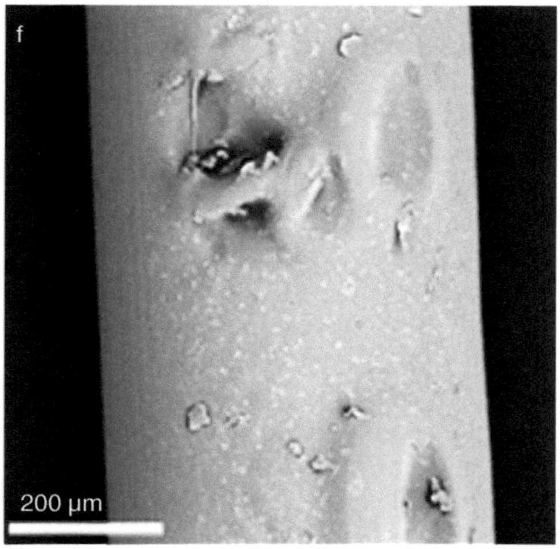

Figure 23.30 Surface scanning electron microscopy (SEM) images of SEPOF with TiO_2 particles after CO_2 laser treatment (intensity 150 μs pixel time). Source: Huang et al. 2015 [72]. Reproduced with permission of De Gruyter.

The formation of micro-mirrors inside a POF with PMMA core through cutting, masking, metallic coating, and polymer redeposition is described in work [51]. It was found that power loss at a single micro-mirror is about 0.7 dB, and approximately 16 mirrors are uniformly embedded along a 20 mm.

Zajkowski [15, 16] experimentally studied the intensity of side-emitting optical fibers for materials with different values of refractive index of core and cladding. The intensity of the total side emission was strongly decreased along the direction of fiber axis.

Spigulis et al. [22] described the side emission of optical fiber containing additives causing scattering in the cladding. Zarian et al. [54] described an optical fiber that was manufactured using the notches in claddings extended to the core. Suitably defined nonconstant changes in the distance between the notches permitted to obtain a constant intensity of the side emission along the fiber.

Sillyman et al. [53] used paraboloid reflector system to control the numerical aperture for delivery of light to ensure a constant intensity of the lateral radiation along the fiber.

Harlin et al. [21] reported that the main area of side-emitting optical fiber utilization is illumination in the visible spectrum. There are also more advanced applications, e.g. photochemical reactions induced by a side-emitting optical fiber, a photomedical therapy [56], the disinfection of waste water [57] and the photocuring of the polymers. In these applications the UV radiation is generally used. UV radiation obeys stronger attenuation during transmission through the fibers than visible light.

Various types of commercial SEPOF were compared according to their emission spectrum and their suitability for applications in the spectral range from UV-A to the visible blue [55]. The SEPOF based on particles in core and notches in cladding were compared. The higher levels of intensity obey SEPOF fibers with micro-notches, but intensity distribution is more uneven. PMMA core SEPOF with micro-notches has significantly greater loss in emission at longer distances from the light source compared with fibers containing silicon core. The higher uniformity of side emission can be achieved also by appropriate surface treatment [55].

23.4.4 Local Side Emission

The local side emission of end-emitting POF and enhancement of side emission intensity can be obtained by partial damage of cladding layer. Simple method is mechanical damage by abrasion. The extent of damage to SEPOF by abrasion on Zweigle yarn abrasion device covered by emery paper 80 after 20 cycles is shown in Figure 23.31.

The changes of attenuation profiles due to abrasion are shown in Figure 23.32. For comparison, the end-emitting POF of the same diameters subjected to the same abrasion degree is presented as well.

It is visible that improvement of side illumination intensity in the case of end-emitting POF is relatively small and far from intensity of SEPOF. In the case of SEPOF, abrasion is effective, and local illumination is much higher (see Figure 23.32).

The visual changes of side emission of end-emitting POF with fluoropolymer cladding due to abrasion, etching in ethyl acetate (20 minutes) and combined etching abrasion were published in work [26]. The best results, i.e. maximum of side emission, were obtained by abrasion followed by solvent etching (see Figure 23.33).

The repeated bending is in fact responsible for local mechanical damage of cladding as well. The changes of SEPOF Grace standard attenuation profile due to repeated bending on Flexometer device to prescribed percentage of cycles till fracture (25%, 50%, and 75%) are shown in Figure 23.34.

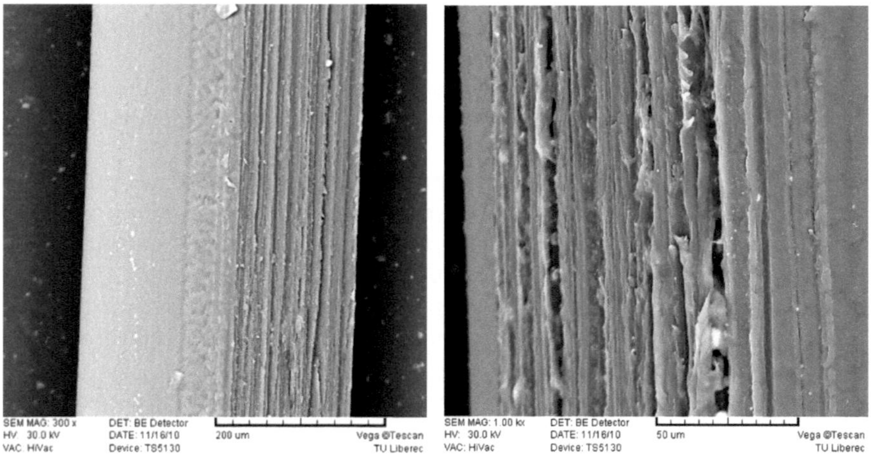

(a) (b)

Figure 23.31 Abrasion of SEPOF type Hypo by using of emery paper graininess 80 (after 20 cycles) at different magnifications.

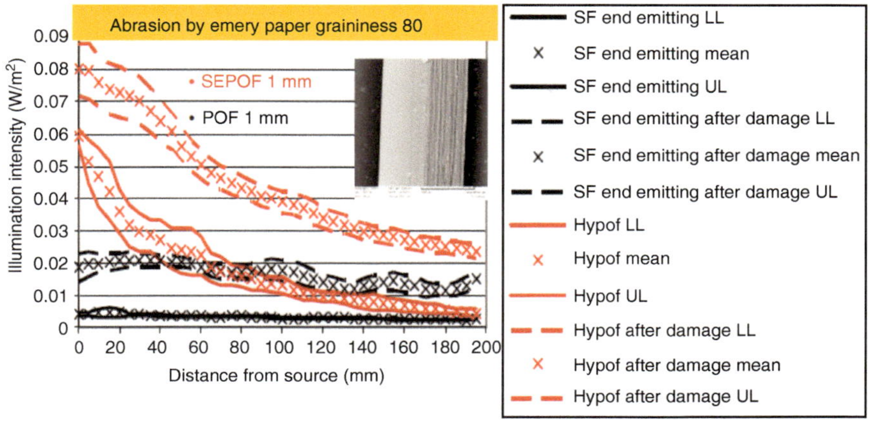

Figure 23.32 Changes of attenuation profiles of end-emitting POF and SEPOF before and after abrasion.

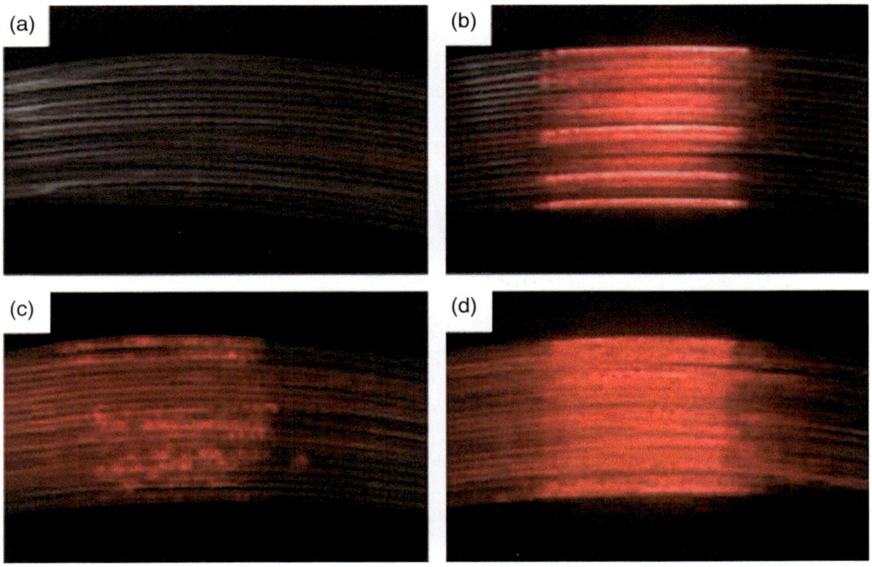

Figure 23.33 Side light emission effect on end-emitting POF: (a) without treatment; (b) abraded by sandpaper; (c) etched with solvent; (d) abraded and then solvent-etched. Source: Im 2007 [26]. Reproduced with permission of Springer.

For comparative purposes the SEPOF Grace standard with a diameter of 0.75 mm was subjected to mechanical and chemical damage and influence of UV radiation under the following conditions:

1) Twenty cycles of abrasion on Zweigle yarn abrasion device covered by emery paper 80
2) Treatment in tetrachloroethylene for one hour

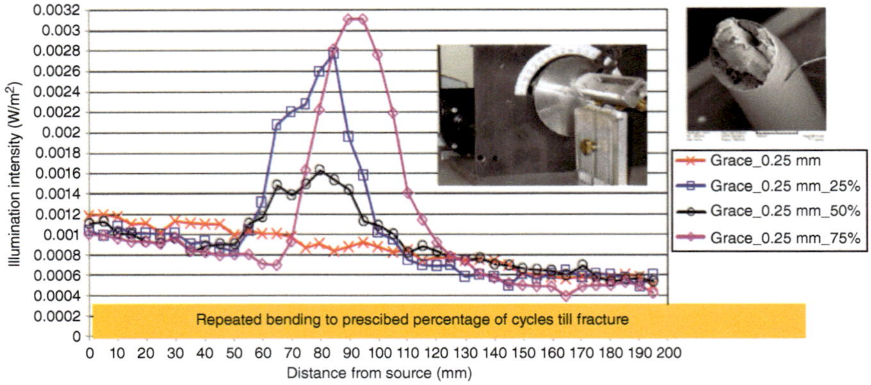

Figure 23.34 Attenuation profiles of fibers damaged by repeated bending.

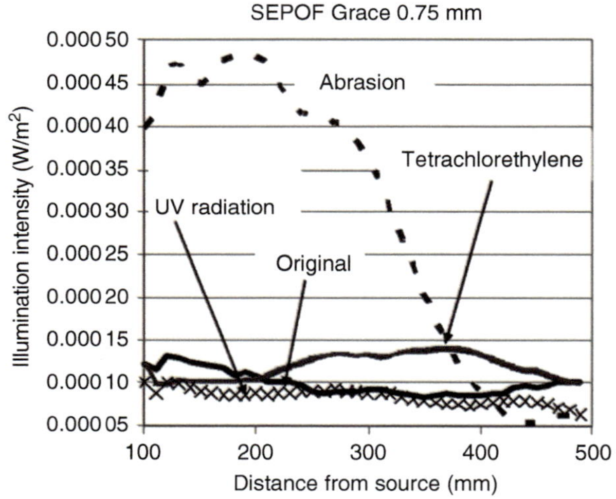

Figure 23.35 Attenuation profiles of fibers damaged by various manner.

3) UV exposure for seven days on the instrument ATLAS UV 340th

Comparison of attenuation profiles for original and damaged samples is shown in Figure 23.35.

The abrasion looks as very simple and in practice can be combined with increasing of temperature. Few cycles only have visible effect. Very simple method for local enhancement of POF emission is the combination of pressure and temperature, i.e. pressing at elevated temperatures (see Figure 23.36).

23.5 Properties of POF

Like glass fibers, POF is characterized by a wide range of different parameters. The properties characterizing the quality of light propagation as attenuation and

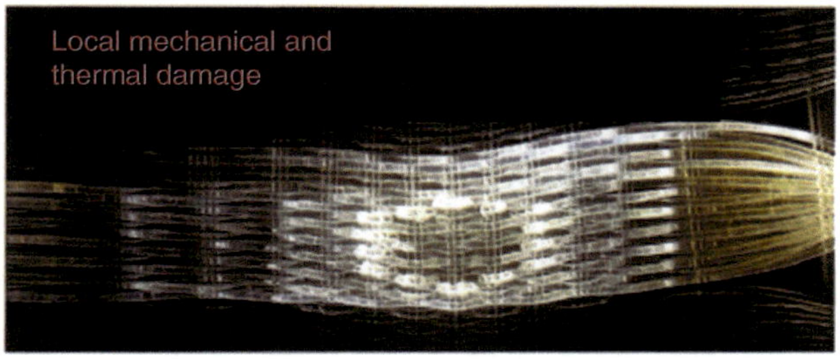

Figure 23.36 Local enhancement of POF side emission by combination of pressure and temperature.

bandwidth are essential for applications of POF. For the practical use, it is necessary to have POF with other characteristics such as the following:

- Durability of the fiber in conditions of use
- Thermal resistance
- Resistance to bending and repeated flexing
- Mechanical stability (flexibility, stress limits for tensile and compressive stress, etc.)
- UV radiation resistance
- Resistance against chemicals, water vapor, etc.

Some of these characteristics are dependent critically on the jacket and can be tuned in connection with areas of application. Some properties, e.g. repeated flexing, are important for SEPOF mainly, and some are necessary for all kind of POF. General properties of POF including measurements and characterization are comprehensively described in many books [1–6]. Here we are focused on optical, mechanical, and thermal properties of POF and mainly SEPOF.

23.5.1 Optical Attenuation

Since the development of POFs, the primary importance has been given to understand and to reduce their optical transmission loss.

Loss of light intensity generally depends on the following: light wavelength, fiber type, fiber structure (i.e. crystallinity and orientation), impurities and accompanying substances (dopants), outer geometric shape (micro-bends, macro-bends, surface damage), and mainly on distance from the source. For classical end-emitting optical fibers, the goal is to suppress attenuation because it is the source of information loss or loss of transferred illumination power. For SEPOF side emission is wanted and can be supported. The standard approach to characterize attenuation, dispersion, and bandwidth of end-emitting optical fibers is comprehensively described in books [1–6], and therefore it is not discussed here.

Special measurement and evaluation system was constructed for evaluation of SEPOF attenuation in straight state and creation of the *attenuation profile* $P(L)$, i.e. dependence of illumination intensity on distance from light source L till maximum length in the range of 1–10 m [44]. Main principle is the measurement of illumination intensity by using integrating cylinder. The light emitted by the POF falls on the inner surface of the cylinder, which has a high reflectivity and is opaque; thus incident rays are scattered randomly in all directions (into the interior of the cylinder). For sufficiently big dimensions of this cylinder, the random light scattering extends substantially to the statistically uniform illumination of its internal surface.

Irradiation of inner surface $E(A)$ (W/m^2) depends on the light power (illumination intensity) $P(L)$ (W/m^2) escaping from measured surface area A of the POF to the inner surface of the cylinder at distance L from light source:

$$E(A) = \eta P(L)\left(\frac{\pi d^2}{4} l\right)^{-1} \tag{23.20}$$

where η (–) is efficiency coefficient, l is length, and d is inner diameter of integrating cylinder. The sensor of light power has active area S_s. Then the illumination intensity $P(L)$ is calculated from measured radiant flux Φ_V (W) by using the following equation:

$$P(L) = \frac{\pi d^2}{4} \frac{\Phi_V}{\eta S_s} l \tag{23.21}$$

This principle was used for construction of online computer-controlled measurement and evaluation system for creation of the *attenuation profile* (Figure 23.37). System is composed of radiant flux light sensor (THORLABS PM 1000 SB), step driver, control unit, measuring channel, input/output rolls, and illumination unit based on LED.

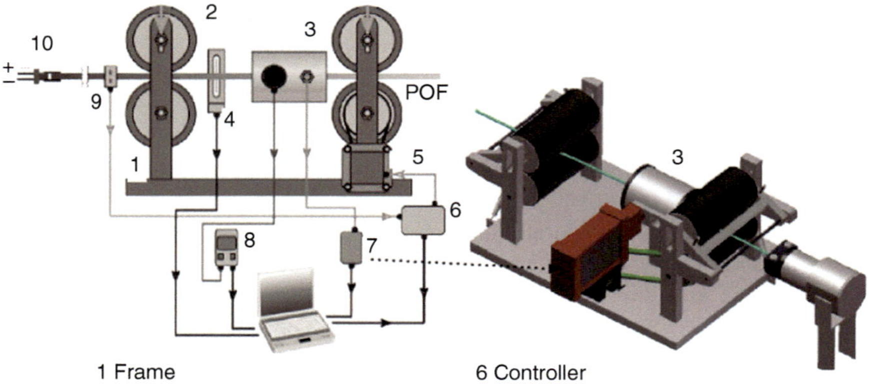

1 Frame
2 Rubber roller
3 Optical integration cylinder
4 Raster camera
5 Stepping motor
6 Controller
7 Spectrophotometer
8 Power measuring unit
9 End of fiber detector
10 Illumination unit

Figure 23.37 Device for measurement of POF light intensity in straight state. Source: Adapted from Lédl and Václavík 2009 [45].

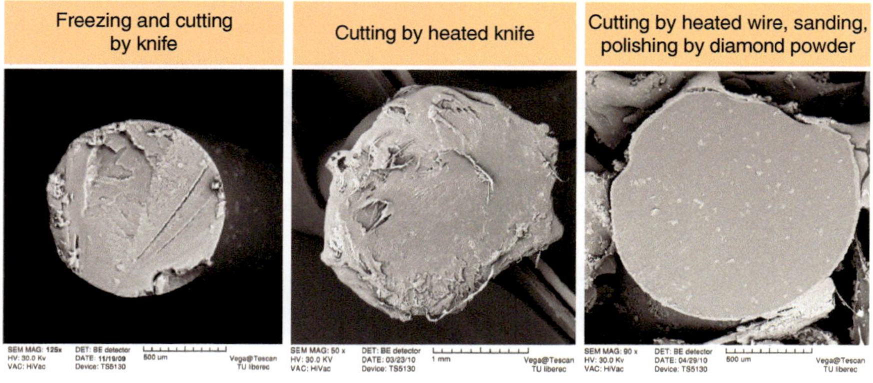

Figure 23.38 Preparation of POF ends. Source: Adapted from Křemenáková et al. 2013 [29].

The illumination of POF is realized by illumination unit composed of LED connected with power source. Electric power supply is 3 W (3 V voltage and current of 1 A). At 10% conversion efficiency, the light output power is of 300 mW. Moreover, about 30% radiates outward from the fiber, so that the radiant flux at the input into the fiber is 100 mW.

The POF or tubular textile structure is located between the feed rollers, which guide them to the integrating cylinder. The tow rollers are driven by step motor. The actual measurement is performed in predefined step lengths. Step lengths are processed by a step motor that drives the two rollers. The measurement process is controlled by a computer program created in MATLAB.

The computer unit is used also for the calculation of *attenuation profile* including statistical analysis.

The POF end connected with illumination unit was prepared by cutting with heated wire or freeze knife and then by polishing with diamond powder (Figure 23.38).

Typical mean attenuation profile of SEPOF Grace standard having diameter 0.25 mm is shown in Figure 23.39 [29]. Due to relative high variability of results,

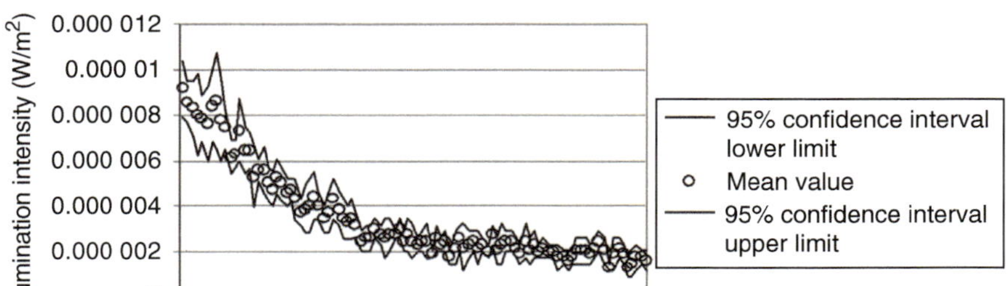

Figure 23.39 Mean attenuation profile and lines of 95% confidence interval for SEPOF Grace standard.

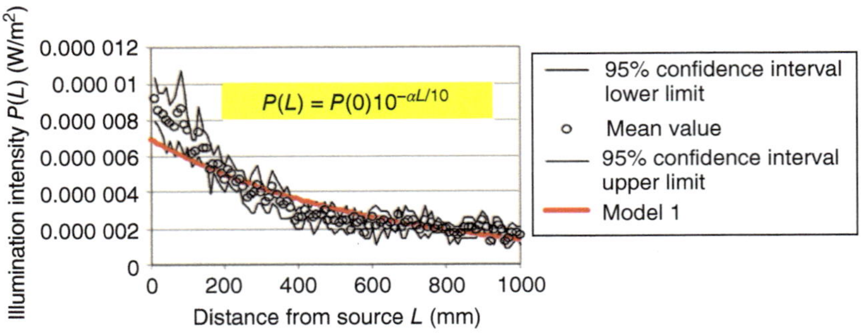

Figure 23.40 Mean attenuation profile for SEPOF Grace standard with added line of $P(L)$ from Eq. (23.11).

20 attenuation profiles are usually measured. Robust estimator of mean illumination intensity (points in Figure 23.4) and corresponding 95% confidence intervals (lines in Figure 23.39) is then calculated by robust Horn procedure suitable for small sample sizes [46].

Experimental mean attenuation profiles can be used for obtaining parameters of exponential model describing the $P(L)$ function (see Eq. (23.11)) by the standard nonlinear or linearized regression using least-squares criterion [46]. By logarithmic transformation of Eq. (23.11), the straight line type dependence $\log P(L) = -\alpha L/10 + \log P(0)$ results. Slope of this straight line k can be used for calculation of mean attenuation rate $\alpha = -10k$, and intercept q can be used for calculation of illumination intensity on the fiber input $P(0) = 10^q$. By using regression straight line parameters k and q, coefficient of determination R^2 and parameters $P(0)$ and α can be calculated [46]. The experimental variances of $P(L)$ can be used as weights in regression analysis. For refining the estimators of α and $P(0)$, the nonlinear regression should be used. The $P(L)$ curve based on Eq. (23.11) is added into experimental attenuation profile in Figure 23.40.

It is clearly visible that the model from Eq. (23.11) is not sufficient for mean intensity profile experimental course approximation. It was found that at short distances from light source, illumination intensity *is* strongly decreasing especially for POF with higher diameter (higher than 1 mm). For better expression of $P(L)$ behavior, the so-called LLF2 model was proposed. LLF2 is in fact linear spline, i.e. linear piece-wise function composed from two different straight sections [46]. This model is based on the assumption that in short distances from light source, there are some nonuniformities in side emission due to accommodation to aperture and critical angle. In second phase the illumination intensity is slowly decreasing with distance L from light source (system is accommodated). Local slopes of LLF2 are in fact sensitivity coefficients a_1, a_2. Corrected illumination intensity on the fiber input is $P_{cor}(0)$. LLF2 model is described by the following equation (linear regression spline with one knot):

$$\mathrm{LLF2} = P_{cor}(0) + a_1 L + a_2 (L - L_c)_+ \tag{23.22}$$

where function $(x)_+ = 0$ if x is negative and if x is positive, function $(x)_+ = x$. L_c is the distance of transition between first and second phase. By using modified

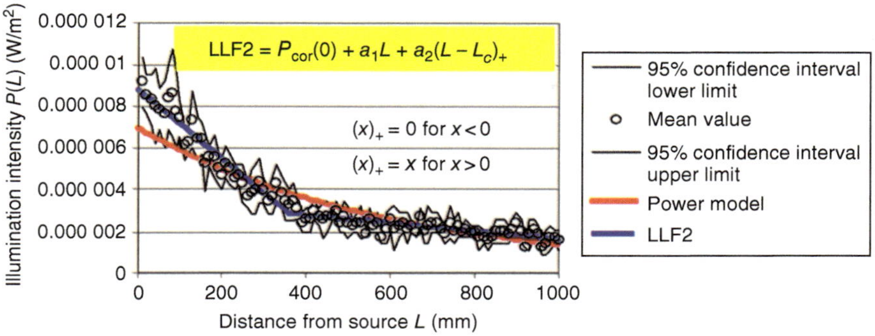

Figure 23.41 Mean attenuation profile for SEPOF Grace standard with added line of $P(L)$ from Eq. (23.11) and line corresponding to LLF2.

linear regression [46], parameters of LLF2 for experimental mean attenuation profiles can be calculated. The *LLF2* curve from Eq. (23.22) is added into experimental attenuation profile in Figure 23.41.

As attenuation profile characteristics of POF, the sensitivity coefficients a_1, a_2, corrected illumination intensity on the fiber input $P_{\text{cor}}(0)$, and distance of transition between first and second phase L_c can be used. Based on comprehensive tests, it was found that the main factor influencing the intensity of the side emission of SEPOF is their diameter. For straight SEPOF, 20–40% of illumination intensity is lost at short distances from light source. The intensity of radiation is here falling sharply. After a certain length L_c the illumination intensity decreases slowly.

The intensity of radiation at the entrance to the fiber $P_{\text{cor}}(0)$ and slope a_2 of model (23.22) obey power law dependence on SEPOF diameter [29].

For expression of acceptable working length L_p of POF, illuminating power P_{L_p} must be on the end of this length corresponding to sufficient value of attenuation coefficient α_{L_p}:

$$P_{L_p} = P(0)10^{-\alpha_{L_p}L_p/10} \tag{23.23}$$

Illuminating power P_{L_p} can replace $P(L)$ in Eq. (23.23), and then working length of optical fiber L_p can be calculated:

$$L_p = \frac{10}{\alpha_{L_p}} \log(P(0)/P_{L_p}) \tag{23.24}$$

Working length of optical fiber are typically calculated for attenuation coefficient α_{L_p} from 10 to 20 dB.

23.5.2 Mechanical Properties

Several authors have studied the mechanical properties of POF. Most of these studies have been focused on the attenuation induced by bends and tensile or torsion stresses and fiber fatigue [35, 36, 58, 59]. In contrast to silica optical fibers,

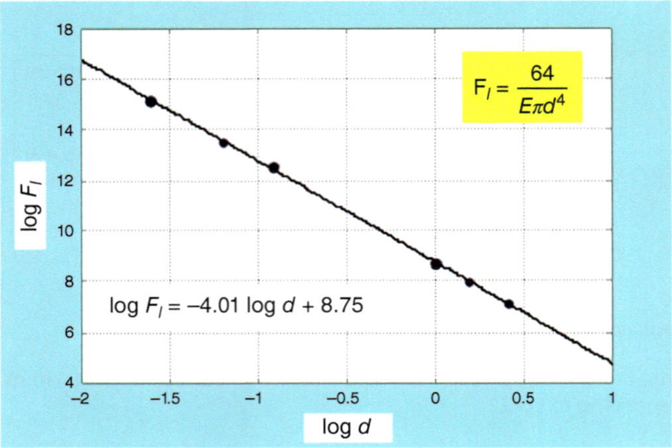

Figure 23.42 Dependence of SEPOF flexibility on their diameter *d*.

POFs are made of polymeric materials. One of main difference is that initial modulus E for a POF (about 3 GPa) is nearly 2 orders of magnitude lower than that of a silica optical fiber (about 73 GPa).

The flexibility F_l of optical fibers with diameter d can be calculated as characterization of their usefulness:

$$F_l = \frac{64}{E\pi d^4} \tag{23.25}$$

For this reason, e.g. even a diameter of POF will be approximately two times higher than diameter of silica fibers, it will have the same flexibility sufficient to be installed according to typical configurations. For the same reason, the minimum bending radius for POF is smaller, since polymers are more ductile and much flexible than silica or glass. For SEPOF brand Hypof, dependence of flexibility on fiber diameter is approximately linear (see Figure 23.42).

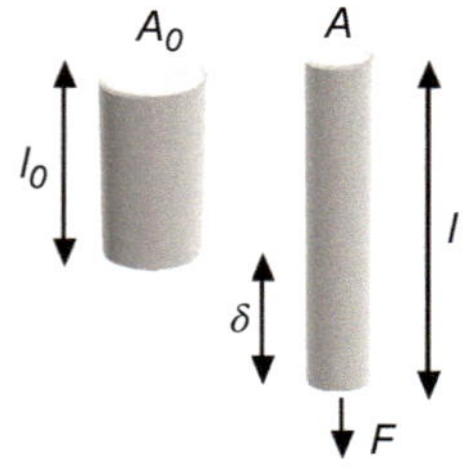

Figure 23.43 Ideal geometry o fiber extension.

Stress–strain curves of POF are created by using tensile testing machines (dynamometers), which are at disposal in every material testing laboratory. Initial undeformed straight fiber has cross-sectional area A_o (diameter d_o) and length l_o. After applying the load F, the fiber is extended to the length l and shortened to the area A or diameter d (see Figure 23.43).

The engineering stress σ, engineering strain ε, and draw ratio λ are defined by well-known equations:

$$\sigma = \frac{F}{A_o} \quad \varepsilon = \frac{l - l_o}{l_o} = \frac{\delta}{l_o} \quad \lambda = \frac{l}{l_o} = 1 + \varepsilon \tag{23.26}$$

The differential of draw ratio is then defined as $d\lambda = dl/l$. After integration the so-called true strain ε_t can be obtained:

$$\varepsilon_t = \int_{l_o}^{l} \frac{dl^*}{l^*} = \ln(\lambda) = \ln(1+\varepsilon) \tag{23.27}$$

The important characteristics of material deformation is so-called Poisson ratio ν, defined as ratio of the relative transversal deformation ε_T and relative longitudinal extension ε:

$$\nu = -\frac{\varepsilon_T}{\varepsilon} \quad \text{where } \varepsilon_T = \frac{d - d_o}{d_o} \tag{23.28}$$

For liquids and rubber the volume is not changed during deformation, i.e. $V/V_o = 1$ (incompressible material) and $\nu = 0.5$. For majority of fiber forming polymers, volume is increased due to extension (i.e. $V/V_o > 1$) and then $\nu < 0.5$ (obviously $0.2 \leq \nu \leq 0.45$). Poisson ratio of PS is 0.325–0.33, Poisson ratio of PC is 0.37, and Poisson ratio of PMMA is 0.34–0.4. For sufficiently small deformations, the values computed for constant ν are fully acceptable, and the true stress σ_t can be then expressed in the form

$$\sigma_t = \frac{F}{A} = \frac{\sigma}{(1-\nu\varepsilon)^2} \tag{23.29}$$

It is interesting that for linear true stress/true strain dependence $\sigma_t = E\varepsilon_t$, the engineering stress–strain diagram is a convex increasing curve defined as

$$\sigma = E(1-\nu\varepsilon)^2 \ln(1+\varepsilon) \approx \frac{E\varepsilon(1-\nu\varepsilon)^2}{1+\varepsilon} \tag{23.30}$$

Equation (23.30) can be used for small deformation range till yield point. The initial modulus for true stress/strain dependence is equal to initial modulus of the engineering stress–strain curve. It was experimentally proved that Poisson ratio decreased with increasing tensile elongation. It is interesting that the stress–strain relation of silica optical fibers was described by relation [76]:

$$\sigma = E\varepsilon\left(1 + \frac{0.75\alpha\varepsilon}{2}\right) \tag{23.31}$$

where value for the nonlinearity constant α is about 6. For small ε, it is Eq. (23.30) quadratic polynomial according to ε as Eq. (23.31).

There are big differences between mechanical characteristic of POF obtained by different authors [61]. One reason is that the mechanical properties of POF depend strongly upon the polymer molecular mass and structure of fiber influenced by drawing process parameters as the drawing ratio and temperature. For PMMA polymers the following typical mechanical parameters are tabulated: impact strength (Charpy) 20 kJ/m^2, shear modulus 1.7 GPa, initial modulus 3.3 GPa, and mechanical loss factor ($\tan\delta$) 0.08 at 25 °C and 10 Hz [20].

The mean initial modulus $\overline{E}$, mean deformation to break $\overline{D}_B$, and tensile strength $\overline{\sigma}_B$ with corresponding standard deviation CV_σ for SEPOF brand Hypof for different fiber diameters are given in Table 23.5 [30, 34].

It is interesting that the tensile strength is decreasing for the increasing of fiber diameter. It can be partially caused by different draw ratio. Preparation of

Table 23.5 Tensile characteristics of Hypof.

d (mm)	$\overline{E}$ (GPa)	$\overline{D}_B$ (%)	$\overline{\sigma}_B$ (MPa)	CV_σ (%)
0.2	3.3	21.80	107.6	7.4
0.3	3.4	42.1	91.5	2.4
0.4	2.8	22.4	78.6	9.3
1	3.4	4.8	88.2	8.6
1.2	3.3	5.8	84.7	4.5
1.5	3.1	8.7	82.9	1.2

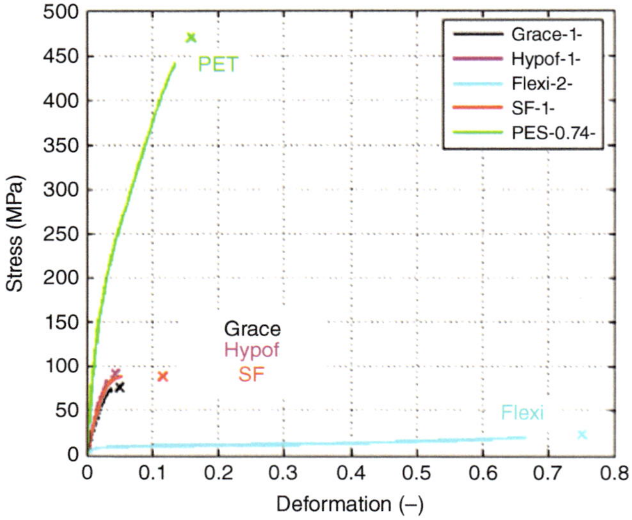

Figure 23.44 Stress–strain curves of SEPOF and polyester fibers.

thicker fiber requires generally lower draw ratio. The deformation to break is ideally increasing for increasing of fiber diameter.

The stress–strain tensile curves of different SEPOF (Grace, Hypof, SF, and Grace Flexi) and standard polyester fiber of the similar diameter are shown in Figure 23.44.

It is visible that SEPOF have very low tensile strength in comparison with standard textile fibers. SEPOF Flexi is characterized by lowest tensile strength but very high deformation to break. The breaking stress distribution of polymeric fibers follows often three-parameter Weibull distribution (see Eq. (23.34)).

For service lifetime predictions of optical fibers, the static fatigue measurements are commonly used. In the tensile tests the time to failure of a fiber due to axial constant applied stress σ_a is measured. The mandrel bending tests involve winding a portion of optical fiber around a precision ground mandrel, and the two-point bend technique involves bending short portions of fiber due to insertion into a precision bore glass tube. Comparison of these techniques and statistical analysis of life time data is presented in the work [57]. The optical fiber lifetime t_f as a function of constant applied stress σ_a and fiber strength σ_b is expressed by empiric relation:

$$t_f = B\frac{\sigma_b^p}{\sigma_a^n} \tag{23.32}$$

where B is constant dependent on environment, n is considered as material property (stress corrosion parameter), and p depends on the nature of defects causing failure (for crack like flaws, $p = n - 2$). The n value for bulk silica is around 40 and around 20 and 25 for fibers (see Ref. [57]).

For evaluation of flex fatigue of POF, SEPOF, and LIHS, the special device was constructed [30, 34]. Basic principle is the evaluation of number of repeated flexing cycles till break FC of fibrous materials. Fiber is fixed in upper movable clamp and guided through slot of diameter 0.2 till 3 mm. Upper clamp should be adjusted to the selected maximum flexing angle from 0° to 140° (see Figure 23.45). On the free end of fiber (adjustable in range 8–27 mm), prescribed load is mounted. Upper clamp is driven by electrical drive with adjustable frequency of flexing per minute.

For analysis of FC data, it is necessary to know the suitable theoretical distribution. The computationally assisted exploratory data analysis methods including techniques used here are described in the book [63].

The estimation of proper distribution of FC can be realized by using so-called quantile–quantile (Q–Q) plot. Classical Q–Q plot is based on comparison of empirical quantile function $Q(P_i) \approx \text{FC}_{(i)}$ with chosen theoretical quantile function $\text{QT}(P_i)$. The probability estimator $P_i = i/(n+1)$ and so-called order statistics [63].

$$\text{FC}_{(1)} < \text{FC}_{(2)} < \cdots < \text{FC}_{(N)}$$

The sample values arranged in the increasing order are used. It was found that for flex fatigue of polymeric fibers, three-parameter Weibull distribution is useful

(a)

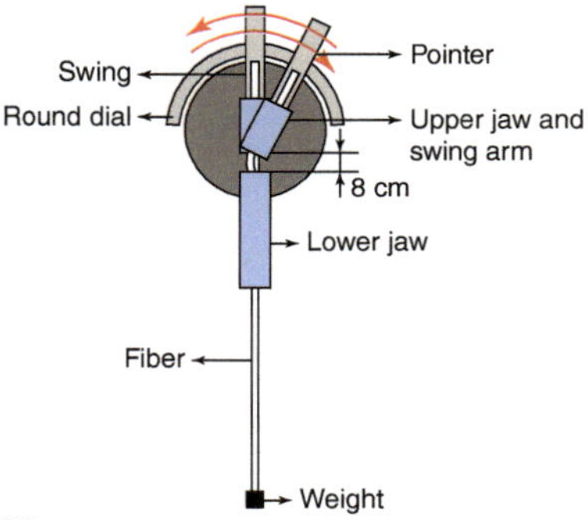

(b)

Figure 23.45 Device for flex fatigue measurements.

[63]. Distribution function has simple form:

$$F(\mathrm{FC}) = 1 - \exp\left[-\left(\frac{\mathrm{FC} - A}{B}\right)^{C}\right] \tag{23.33}$$

Here A is threshold, i.e. lowest number of repeated flexing cycles till break, B is scale parameter, and C is shape parameter. For quick and rough parameter estimates of three-parameter Weibull models, the moment-based method can be used. The main idea of this method is very simple. Based on the selected three sample moments and corresponding theoretical moments for number of repeated flexing cycles till break, the three nonlinear equations can be created. Their complexity is based on the suitable selection of moments [63].

Moment method based on so-called Weibull moments is very simple (see Ref. [30]). The Weibull sample moments m_r are defined as

$$m_r = \sum_{i=0}^{N-1} (1 - i/N)^r [\mathrm{FC}_{(i+1)} - \mathrm{FC}_{(i)}] \tag{23.34}$$

For $i = 0$ is formally $\mathrm{FC}_{(0)} = 0$. Shape parameter C can be estimated from relation

$$C = \frac{\ln(2)}{\ln(m_1 - m_2) - \ln(m_2 - m_4)} \tag{23.35}$$

For estimation of the lower limiting number of repeated flexing A is valid

$$A = \frac{m_1 m_4 - m_2^2}{m_1 + m_4 - 2m_2} \tag{23.36}$$

and estimate of scale parameter B is in the form

$$B = \frac{m_1 - A}{\Gamma(1 + 1/C)} \tag{23.37}$$

where $\Gamma(x)$ is Gamma function.

This very simple technique can be used for the rough estimation of threshold A in three-parameter Weibull models.

Then Q–Q plot can be simply constructed for three-parameter Weibull distribution [63]. After rearrangements the linear dependence, $y = ax + b$ in Q–Q plot occurs. Here,

$$y = \ln[-\ln(1 - p_i)], \quad x = \ln(FC_{(i)}) - A$$

$$a = C \text{ and } b = -\ln(B)C$$

In this case it is necessary to know in advance the estimator of lowest number of repeated flexing cycles till break A calculated from Eq. (23.3).

The bending fatigue tests were realized on device shown in Figure 23.45 for SEPOF brand Hypof with different diameters. For bending fatigue measurements, the following conditions were selected:

- Frequency of flexing – 116 (min^{-1})
- Free-end load – 5 (g).
- Fiber length – 8 (mm)

Table 23.6 Standard characteristics of flex fatigue.

Fiber diameter d (mm)	FC_M (cycles)	S_{FC} (cycles)	CV_{FC} (%)
0.2	2454.3	1284.2	52.32
0.3	7904	7091.1	89.7
0.4	1034.8	554.1	53.5
1	89.0	37.7	42.4
1.2	41.7	14.2	34.1
1.5	6.2	2.5	40.6

Table 23.7 Calculated parameters of Weibull distribution.

Fiber diameter (mm)	A (cycles)	B (cycles)	C (–)
0.2	1028	3887.9	3.21
0.3	1377.3	6314.3	0.93
0.4	455.5	599.3	1.093
1	16.12	80.65	2.1
1.2	30.69	13.32	1.1
1.5	2.12	4.52	1.81

- Maximum bending angle – 140°

For comparative purposes the initial modulus, deformation to break and tensile strength was evaluated by Instron device at rate of deformation 0.5 min^{-1} (see Table 23.5).

In Table 23.6 the arithmetic means of numbers of repeated flexing cycles till break FC_M of Hypof fibers corresponding standard deviations S_{FC} and variation coefficients CV_{FC} calculated by standard manner (as moment estimators) are given.

The rough estimators of Weibull distribution parameters calculated from Eqs. (23.35), (23.36), and (23.37) are given in Table 23.7.

It is visible that the higher fiber diameter leads to the decrease of lowest number of repeated flexing cycles till break A and corresponding scale parameter B.

The mean number of repeated flexing cycles $E(\text{FC})$ and corresponding standard deviation SD(FC) can be now computed from the following equations:

$$E(\text{FC}) = A + B\Gamma\left(1 + \frac{1}{C}\right) \tag{23.38}$$

$$\text{SD(FC)} = E(\text{FC})\left(\frac{\Gamma\left(1 + \frac{2}{C}\right)}{\Gamma^2\left(1 + \frac{1}{C}\right)} - 1\right)^{\frac{1}{2}} \tag{23.39}$$

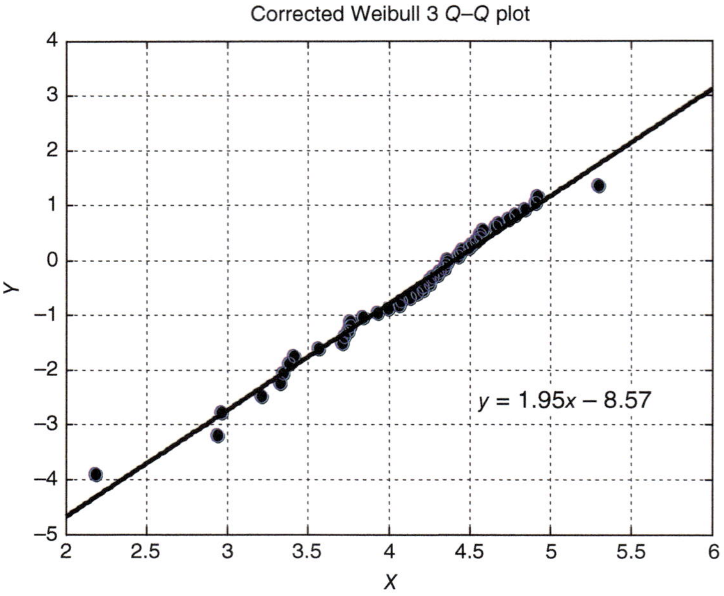

Figure 23.46 Weibull Q–Q plot for Hypof 1 mm.

The typical Q–Q plot for Hypof (diameter of 1 mm) is shown in Figure 23.46. It is visible that the linearity is relatively good. This plot for all investigated fibers was approximately linear, which is proving the usefulness of Weibull distribution.

It is interesting to compare mean number of repeated flexing cycles till break FC_M with fiber flexibility F_l (see Eq. (23.25)) calculated from measured initial modulus. The dependence of $\log(F_l)$ on $\log(FC_M)$ is shown in Figure 23.47.

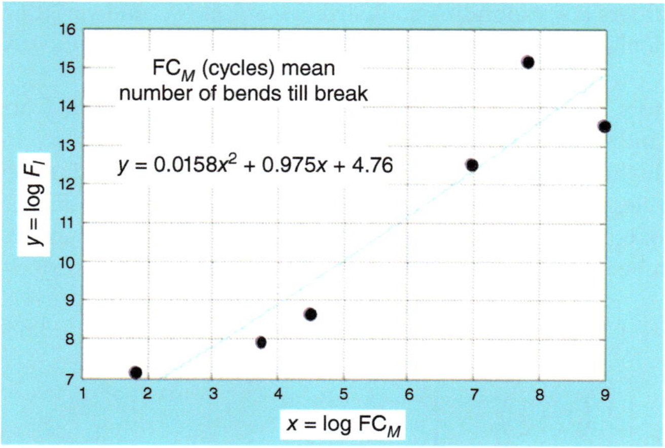

Figure 23.47 The dependence of $\log(F_l)$ on $\log(FC_M)$.

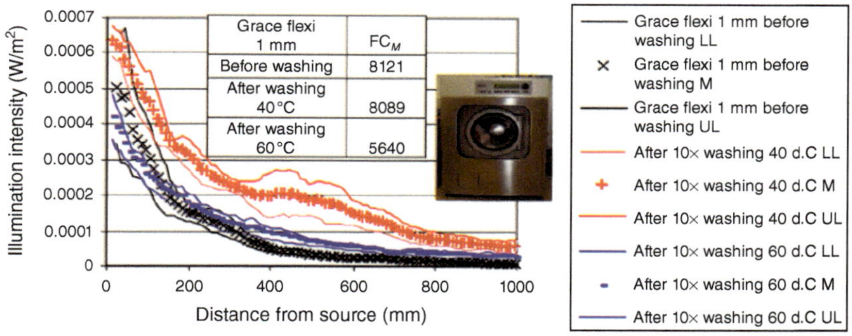

Figure 23.48 Change of attenuation and FC_M of SEPOF due to repeated washing.

It is visible that the flex fatigue characteristic FC_M is the increasing function of flexibility parameter. From corresponding quadratic fitting, the empiric equation

$$\ln(F_l) = 0.0158[\ln(FC_M)]^2 + 0.975\ \ln(FC_M) + 4.76$$

was found.

The fatigue characteristic FC_M can be used for characterization of changes of mechanical response to external influences. For SEPOF brand Grace flexi, this characteristic *is* used as indication of *changes* in mechanical characteristics due to repeated washing (see Figure 23.48).

It is clear that the washing is changing structure (loss of FC_M) and amount of illumination intensity as well (in details see Ref. [29]).

The transmission rate through a POF also depends on some mechanical properties [35]. For example, if the POF is elongated till 10% of its length, the attenuation increases about less than 0.1 dB. Regarding other factors that can change the POF optical properties, cyclic bending can also cause variations in the attenuation, up to a certain limit 0.15 dB for 1000 bending times with a bending radius of 50 mm, in the case of 1 mm PMMA POF [80].

23.5.3 Thermal Properties

Standard POF can be used in working conditions at temperatures up to 80–110 °C. At higher temperatures the POFs lose their transparency. The working temperature can be increased to 125 °C or to 135 °C by using jackets made from more temperature-resistant materials [2–6].

It is known that amorphous polymers including PMMA exhibit anisotropic thermal expansion dependent on the polymer processing history. The thermal expansion coefficient is lower in the direction of the fiber axis than in the perpendicular direction. Such anisotropic behavior of polymers can also be expected in the fiber core material PMMA. Changes of POF surface structure composed of PMMA core and fluoropolymer cladding are due to thermal exposition.

In connection with the absorption of humidity, the thermal aging that occurs at high temperatures is the essential mechanism influencing the durability of POFs.

Thermal aging leads to the decomposition effects as a consequence of the polymer chain splitting by the energy of heat.

With PMMA there is particular risk of depolymerization, i.e. splitting off of end groups and the loosening of monomer components from the end of the chain.

During the depolymerization, free radicals are formed. They can react with oxygen and promote degradation by thermooxidation. The consequence is brittleness and disintegration, with a direct effect on the mechanical and optical properties of the POF.

Takezewa et al. [78] have investigated the effects of high-temperature (150 °C) exposure of cross-linked PMMA-based POF and found that oxidative degradation products of the core polymer cause an increase of the optical loss by light absorption. Schartel et al. [79] have studied the thermooxidative degradation behavior of PMMA-based silica POF. They have found that the optical transmission stability of POF is most likely to be governed by the thermooxidative stability of the core.

The resistance of some POF to higher temperatures strongly depends on the air relative humidity RH as well. For example, if the PMMA core POF is maintained at 85 °C with 85% of relative humidity for 1000 hours, the attenuation will increase in 0.02 dB/km. If the relative humidity level is around 90%, the attenuation increases more than 0.03 dB/m [33].

Model for the deterioration of the optical transmission under high temperature and humidity was proposed in Ref. [6]. According to this model, the loss of optical transmission of silica POF (cables) occurs in four phases: first phase characterized by a definite loss of transmission within a short period (25–50 hours), second phase characterized by a very slow deterioration of transmission, third phase characterized by rapid decline in the transmission, and finally fourth phase characterized by increase of the transmission to a higher level when the humidity is varied while the temperature is kept constant. This behavior is due to the strong OH^- absorption band in the visible range. The anisotropic nature of PMMA core fibers due to the orientation of polymer chains was found [77]. Therefore the more expansion in the transversal direction than in the longitudinal direction of POF can be predicted. The anisotropic behavior of expansion of polymers can be visible when they are heated to their T_g and then cooled down.

Therefore the observed significant increase of the POF diameter and a contraction of their length should be due to plasticizing effects of water in the core of POF. It is known that water in POF reduced their T_g by increasing the chain mobility. As a result the expansion occurred in a faster rate than it would take place when exposing to dry conditions. The plasticizer effect is irreversible because the observed expansion (diameter and contraction of length) and brittle nature of the cooled POF are an indication of the occurrence of molecular disorder [32].

23.6 Illumination Systems Using POF

Direct application of POF for embedding into textiles or creation of line lighting systems is not so easy. POFs are created from polymeric materials that are sensitive to surface damage, their mechanical properties are not as good as properties of synthetic fibers, durability in standard conditions of use is limited, they are degraded under UV radiation and sensitive to moisture, and their comfort properties including hand (touch) and drape are generally bad. Some negative characteristics of POF can be suppressed by proper selection of jacket, but it is not sufficient for their effective application especially for direct embedding into the textile structures.

In bent state, POF structure is subjected to plastic deformations and repeated multiaxial deformation during use, which can deteriorated their properties. Each optical fiber should have delivery of light, which is serious limitation for using multiple fibers in wider areas.

For SEPOF it is possible to have sufficient side illumination in straight state, and all kinds of deformation appeared due to necessities to form textile products mainly. There is a continuous effort to increase the side emission by using dopants as fluorescent or luminescent dyestuffs. Direct dyeing of POF is complicated due to changes of their structure and negative influence on their properties. One solution on how to avoid these problems with SEPOF is to use so-called LIHS composed from SEPOF in core and cover (textile tube) from textile material [18, 29, 44]. This layer can be prepared by weaving, knitting, braiding, etc. [18, 29, 44]. Textile structures containing embedded side-emitting POF, i.e. LIHS, have the following benefits:

- Enhancing illumination intensity by selection of textile layer composition and surface dopants
- Protection of LIHS against weather including UV and combination of temperature/moisture
- Enabling LIHS standard maintenance during use as washing
- Suppression of mechanical damage as abrasion and sensitivity to repeated multiaxial deformation
- Simplification of attachment of LIHS into textiles by standard techniques as sewing

Typical structure of LIHS is shown in Figure 23.49.

It is visible that beside of the textile tube covering the SEPOF, the tape is created. This tape is enabling the simple attachments to supporting material in the case of side illumination or sewing into seems in the case of textile products. Comparison

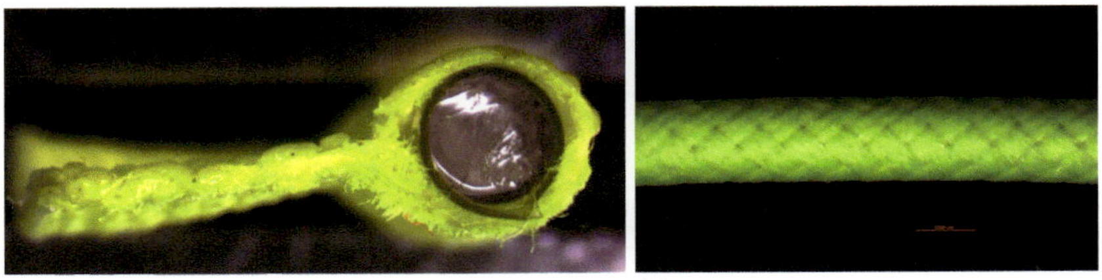

(a) (b)

Figure 23.49 Device for flex fatigue measurements. (a) Picture and (b) schematic arrangement.

of SEPOF and LIHS containing the same SEPOF active illumination is shown in Figure 23.50.

For quantitative comparison of SEPOF and LIHT side illumination, the attenuation profiles were created. Illumination intensity as function of distance from source for POF and the same POF embedded into textile tube are shown in Figure 23.51.

It is visible that LIHS (SEPOF in textile cover) is producing higher side emission intensity in comparison with SEPOF only.

LIHS can be incorporated into a variety of textile structures as labels, strips and bands, cords, special patches, woven fabrics, and knitted structures.

Each structure containing LIHS needs to have lighting system and power supply. The resulting complex system can be constructed as easily attachable to various fabrics and easily removable. Components of side illumination system containing LIHS (SEPOF and textile cover) are shown in Figure 23.52.

The light source is critical for obtaining sufficient side emission for creation of safety textiles, fashion purposes, and line lighting.

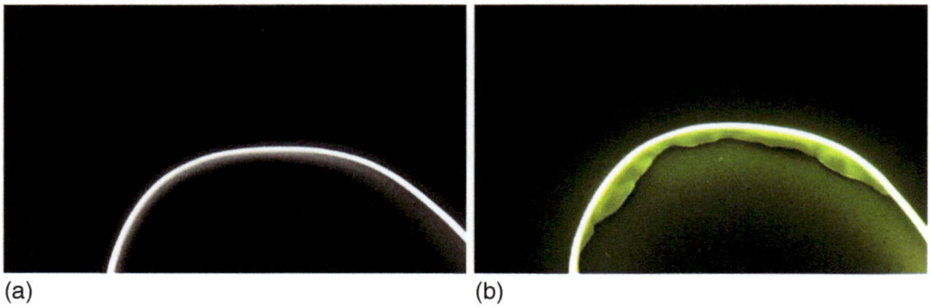

Figure 23.50 Active illumination by (a) SEPOF and (b) LIHS.

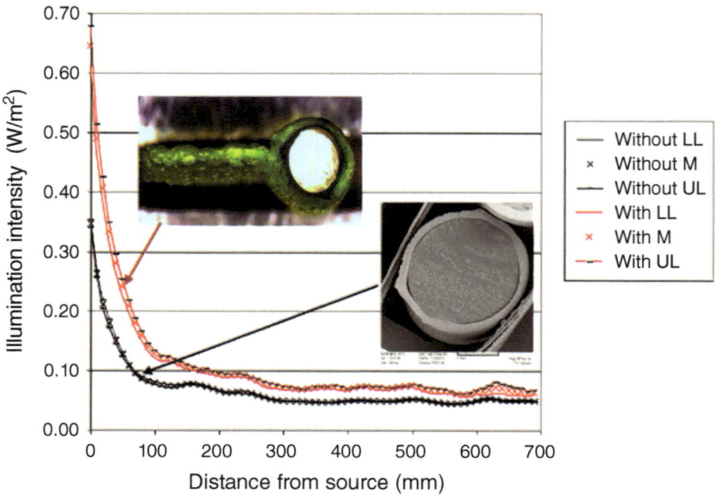

Figure 23.51 Side illumination intensity of SEPOF and LIHS.

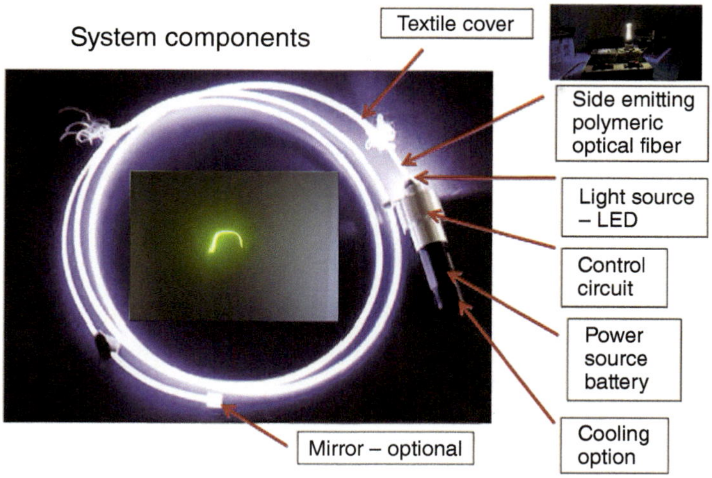

Figure 23.52 System components for side illumination.

As effective source of light with directional output, the LEDs are used for all kind of applications. The comparison of light outputs from incandescent bulb, compact fluorescent light bulb, and LED is shown in Figure 23.53.

The comparison of some characteristics of these sources of illumination is shown in Table 23.8.

The use of LED leads to considerable savings in the use of electrical power and, in addition, lowers maintenance costs. The LED lifetime is the time in which the brightness of 50% of a sampling of LEDs decreases by 30% of its incipient value,

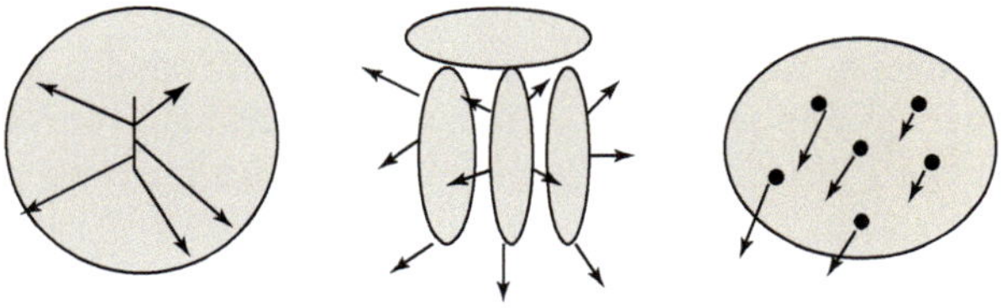

Figure 23.53 Light output from typical sources of illumination [81]. (a) Incandescent, (b) CFL, and (c) LED. Source: Held 2009 [81]. Reproduced with permission of Taylor & Francis.

Table 23.8 Selected characteristics of typical sources of illumination.

Characteristics	Incandescent bulb	CFL	LED
Life (h)	750	7 500	60 000
Equivalent number of bulbs	80	8	1
Electric power (W) for 1100 lm	100	20	9
Effectivity (lm/W)	11	55	122

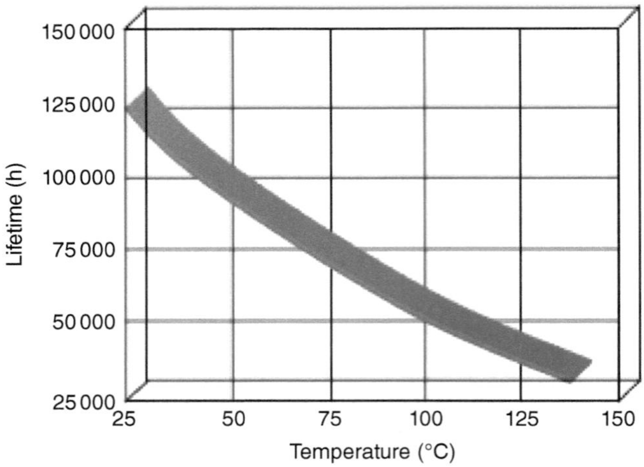

Figure 23.54 Influence of temperature on the lifetime of LED.

that is, drops down to 70% of the brightness in the beginning. The LED lifetime is approximately exponentially decreasing with increase of temperature as is shown in Figure 23.54.

During lighting by using LED, temperature is rising till 60–80 °C in two to three minutes, and after this, it is nearly constant. This temperature can be reduced by selection of cover of illumination unit and its construction. The best portable powering structures have maximum temperature due to lighting significantly lower than 50 °C.

Based on the comprehensive testing and refinements, one-chip LED with power in the range of 1–3 W was selected as light source. These LED delivers optical power in the range of 40–120 lx/m, which is sufficient to illumination by SEPOF till length of several meters. At the same time such LED have small size, and the heat generation is manageable despite the smaller size of the light source unit. The typical attachment of LED for two branches illumination purpose is shown in Figure 23.55.

Selection of energy source depends on the LED used and on the application area. Recent very quick development of portable energy sources based on hard or flexible miniature batteries for wearable electronics including sensing, heating, and lighting applications is promising for future. It will be interesting to try to check the energy harvesting from mechanical vibrations, thermoelectric effect, solar power, etc. Some of currently available portable powering structures for LIHS are shown in Figure 23.56.

There are plenty of different types of rechargeable batteries in current market suitable for use in powering structure for LIHS. Only few of them are here mentioned.

Lithium-ion (Li-ion) batteries were the good choice of battery for powering consumer electronics due to their high energy and power density and stable electrochemical performance. Batteries have high capacity (mAh) and energy density, but they are bulky and nonflexible. For the fabrication of flexible batteries,

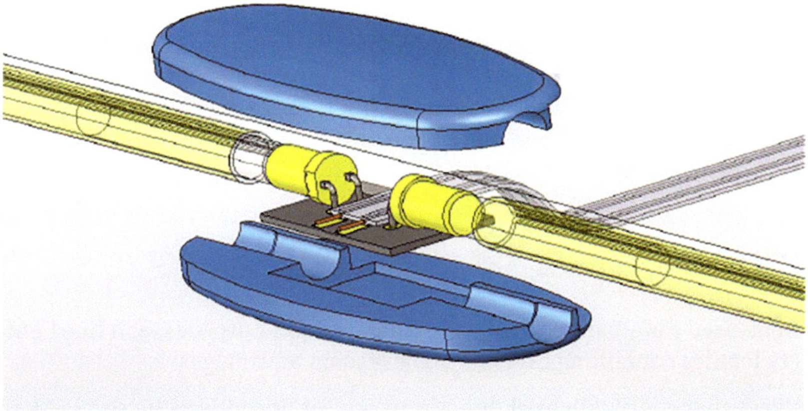

Figure 23.55 Attachment of LED for two branches of SEPOF.

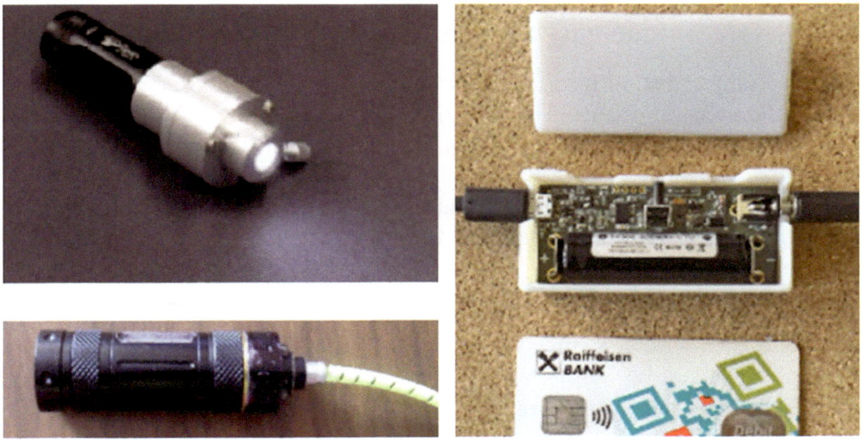

Figure 23.56 Some portable powering structure for LIHS.

a thin-film configuration is adopted where single layers of anode, separator, and cathode are stacked together and sealed within a flexible encapsulation.

Main limitations of these batteries are as follows:

- Requires protection circuit to maintain voltage and current within safe limits
- Subject to aging, even if not in use – storage in a cool place at 40% charge reduces aging effect
- Expensive to manufacture – about 40% higher in cost than nickel–cadmium
- Fire hazard because electrolyte is organic solvent

Alternates are LiPol batteries with high-energy density (20% more than classical Li-ion batteries). Their flexible cover causes high reduction of total mass.

Low self-discharge nickel–metal hydride (Ni–MH) rechargeable batteries Eneloop Plus have a thermistor built-in that cuts the power in case the batteries are overheating. This makes them especially suitable for toys and devices, which generate an increased amount of heat. AA size has 1900 mAh minimum capacity. The discharge rate holds 90% up to one year and approx. 70% after five years. Number of recharge: 1800 times.

Lithium phosphorus oxynitride electrolyte (LiPON), deposited as amorphous thin (2 μm) film is used in all-solid-state micro-batteries. The capacity loss is negligible during 100 charge and discharge cycles [82].

23.7 LIHS Applications

LIHS can be used for creation of safety textiles, fashion purposes, and line lighting. Safety textiles containing LIHS have these main advantages:

- The object shape is highlighted directly, which significantly reduce the risk of incorrect interpretation of their size.
- Side emission is not dependent on the conditions of external exposure (typically vehicle headlights); these active lighting systems operate well in darkness.
- There is no glare disturbing, e.g. road users.

Due to full integration into the fabric, SEPOF are protected against environmental influences (moisture, UV radiation). It is simple to use reflective or fluorescence colors for dyeing or coating outer textile layer enhancing diffusion scattering.

The simple removal of power/light source enables treatment of textiles according to standard procedures. Based on the practical testing, the following care rules were specified:

- Gentle machine wash, moderate mechanical treatment, rinses, and spin drying
- Do not bleach with chlorine
- Do not iron
- Can be cleaned with perchlorethylene,
- Do not tumble dry

The examples of application of systems containing LIHS for clothing and accessories are shown in Figure 23.57.

Line lighting systems based on LIHS have the following advantages:

- Can be used for line illumination of complicated ways and significantly reduce the local variation of light intensity
- Low cost of lighting on the level of illumination standards
- Not local over lighting
- Intensity of light is controlled by voltage
- Simple connections to permanent or portable electricity sources
- Simple branching of line lighting
- Maintenance according to standard procedures

Line lighting systems based on LIHS can be used for

- setting of limits (parking barriers, end of carpets, stairs, etc.),

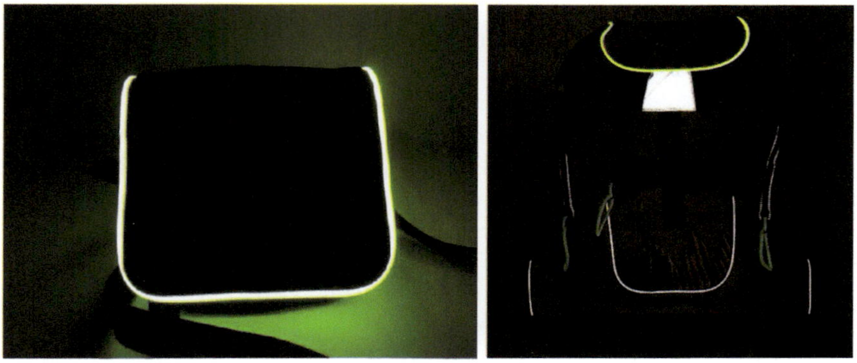

Figure 23.57 Systems using LIHS for accessories.

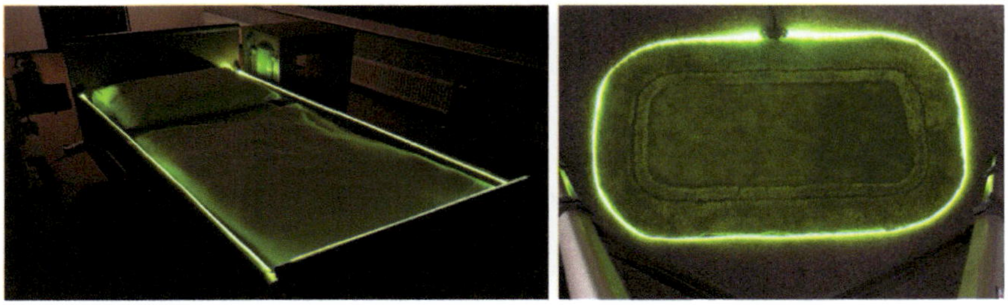

Figure 23.58 Systems using LIHS for line illumination.

- emergency lighting (hospitals, hotel, dormitories, supermarkets),
- lighting in corridors, lifts, edge visualization, etc.

The example of line lighting is in Figure 23.58.

23.8 Conclusion

The emphasis of POF is focused on reduction of attenuation coefficient and increase in thermal stability mainly. For example, PC, silicone resin, cross-linked acrylic acid, and copolymer can be used to increase the thermal stability up to 125–150 °C. Plastic optical fibers applied in illumination and decorations are as follows [83]:

- End-emitting optical fiber POF are mainly using the lighting effect of the end face of fibers to achieve decorative lighting. The fibers have many applications, such as, using endpoints to form various patterns and simulate star blinking effect – as if going into fairyland, making a variety of technological models to create an atmosphere for entertainment, and making Christmas tree more attractive and advertising signs graceful, in order to increase the noble mystery of artifacts and jewelry.
- Side-emitting optical fiber SEPOFs are easy to be used to set off the outline of patterns of objects and produce a variety of artistic design.

- Flash point optical fiber is used to make special effect. These fibers can be made into fiber-optic curtains, fiber-optic waterfall, and other fiber-optic craft products.

Recently, the POF and SEPOF technology has been developing quickly; the applications have been recognized in many branches. The LIHS are still in the first stage of industrial realization, and it will be necessary to optimize their composition and functionality for various targeted applications including surface effects, doping, and use of special dye stuffs. The system of creation of illumination intensity profiles (see Figure 23.45) will be simple tool for evaluation of new effects and optimization of LIHS-based systems. Probably the most important element for efficient low cost active illumination system creation is the battery. Limits are flexibility, power, size, weight, and price.

References

1 Bunge, C.A. et al. (eds.) (2017). *Polymer Optical Fibres Fibre Types, Materials, Fabrication, Characterisation and Applications*. Elsevier Amsterdam.
2 Koike, Y. (2015). *Fundamentals of Plastic Optical Fibers*. Weinheim: Wiley-VCH.
3 Bäumer, S. (ed.) (2010). *Handbook of Plastic Optics*. Weinheim: Wiley-VCH.
4 Harmon, J.P. and Noren, G.K. (eds.) (2001). *Optical Polymers Fibers and Waveguides*. Washington, DC: American Chemical Society.
5 Kuzyk, M.G. (2007). *Polymer Fiber Optics Materials, Physics, and Applications*. Boca Raton, FL: CRC Press.
6 Daum, W. et al. (2002). *POF - Polymer Optical Fibers for Data Communication*. Berlin: Springer-Verlag.
7 Thévenaz, L. (ed.) (2011). *Advanced Fiber Optics Concepts and Technology*. Lausanne: EPFL Press.
8 Nyu, T. et al. (1995). Experiments on 156 Mbps-100 m transmission using 650 nm LED and SI POF. In: *Proceedings of the 4th International Conference on Plastic Optical Fibres and Applications POF'95*, 119–121. Boston, MA.
9 Numata, K., Furusawa, S., and Mirikura, S. (1999). Transmission characteristics of 500 Mbs optical link using 650 nm RC-LED and POF. In: *Proceedings of the 8th International Conference on Plastic Optical Fibres and Applications POF'99*, 74–77. Chiba, Japan.
10 Koike, Y., Ishigure, T., and Nihei, E. (1995). High-bandwidth graded index polymer optical fiber. *Journal of Lightwave Technology* 13: 1475.
11 Matias, I.R. et al. (eds.) (2017). *Fiber Optic Sensors Current Status and Future Possibilities*. Springer International Publishing Switzerland.
12 Bartlett, R.J. et al. (1998). Plastic optical fibers sensors and devices. In: *Proceedings of the 7th International Conference on Plastic Optical Fibres and Applications POF'98*, 245–246. Berlin, Germany.
13 Schilk, A.J. and Bowyer, T.W. (1995). The use of scintillating plastic optical fibers for the detection of radioactive contamination. In: *Proceedings of*

the 4th International Conference on Plastic Optical Fibres and Applications POF'95, 171–176. Boston, MA.
14 Marcuse, D. (1974). *Theory of Dielectric Waveguides*, 75–80. San Diego, CA: Academic Press.
15 Snyder, A.W. and Love, J.D. (1983). *Optical Waveguide Theory*, 2e. London: Chapman and Hall.
16 Zajkowski, M. (2002). Emission of flux light in 'side light' fiber optic. *Proceedings of SPIE* 5125: 322–327.
17 Zajkowski, M. (2005). Luminous flux emission calculation analysis in side light illumination optical fibres. *Proceedings of SPIE* 5775: 440–445.
18 Křemenáková, D., Militký, J., Meryová, B., and Lédl, V. (2012). Testing and characterization of side emitting polymer optical fibers. In: *International Symposium TBIS*. Ueda, Japan: Shinshu University.
19 Stickler, M. and Rhein, T. (2005). Polymethacrylates. In: *Ullamann's Encyclopedia of Industrial Chemistry*, Weinheim: Wiley-VCH Verlag GmbH & Co. KGaA. 10.1002/14356007.a21 473.
20 Wunderlich, W. (1989). Physical constants of PMMA. In: *Polymer Handbook* (eds. J. Brandrup and E.H. Immergut). Wiley.
21 Harlin, A., Makinen, M., and Vuorivirta, A. (2003). Development of polymeric optical fiber fabrics as illumination elements and textile displays. *AUTEX Research Journal* 3: 1–8.
22 Spigulis, J. et al. (1996). The 'glowing' optical fiber designs and parameters. *Proceedings of SPIE* 2967: 231–236.
23 Koncar, V. (2005). Optical fiber fabric displays. *Optics & Photonics News* 16 (4): 40–44.
24 Franklin, J.B., Smith, G.B., and Joseph, K.J. (2004). Side-scattering polymer light guide and method of manufacture. US Patent No. 0196648.
25 Joseph, E.K., Franklin, J.B., and Smith, G.B. (2006). Side-scattering light guides. US Patent No. 2006140562.
26 Im, M.H. (2007). Modification of plastic optical fiber for side-illumination. In: *Human-Computer Interaction, Part II* (ed. J. Jacko), 1123–1129. Berlin, Heidelberg: Springer-Verlag.
27 Šesták, J. and Militký, J. (2013). Selected properties of functional materials. In: *Background of Fiber Optics*, Chapter 5 (eds. D. Křemenáková et al.). Pilzen: OPS Kanina Publ. House.
28 Huang, J. et al. (2013). Selected properties of functional materials. In: *Review on Polymeric Optical Fibers*, Chapter 6 (eds. D. Křemenáková et al.). Pilzen: OPS Kanina Publ. House.
29 Křemenáková, D. et al. (2013). Selected properties of functional materials. In: *Characterization of Side Emitting Polymeric Optical Fibres Illumination Intensity*, Chapter 7 (eds. D. Křemenáková et al.). Pilzen: OPS Kanina Publ. House.
30 Militký, J. and Křemenáková, D. (2013). Selected properties of functional materials. In: *Flex Fatigue of Side Emitting Polymeric Optical Fiber*, Chapter 8 (eds. D. Křemenáková et al.). Pilzen: OPS Kanina Publ. House.
31 Mishra, R. et al. (2013). Selected properties of functional materials. In: *Surface Modification of Polymer Optical Fibers for Enhanced Side Emission*

Behavior, Chapter 9 (eds. D. Křemenáková et al.). Pilzen: OPS Kanina Publ. House.
32 Appajaiah, A. (2005). Climatic stability of polymer optical fibers (POF). BAM-Dissertationsreihe Band 9 Berlin.
33 Zubia, J. and Arrue, J. (2001). Plastic optical fibers: an introduction to their technological processes and applications. *Optical Fiber Technology* 7: 101–140.
34 Militký, J. and Křemenáková, D. (2012). Flex fatigue of polymeric optical fibers. In: *41st TRS – Textile Research Symposium*. Guimarães, Portugal: University of Minho (12–14 September 2012).
35 Guerrero, H. et al. (1998). Mechanical properties of polycarbonate optical fibers. *Fiber and Integrated Optics* 17: 231.
36 Zubia, J. et al. (1997). Theoretical analysis of the torsion induced optical effect in plastic optical fibers. *Optical Fiber Technology* 3: 162.
37 Blyler, L. (1999). Material science and technology for POF. In: *Proceedings of the 8th Conference on Plastic Optical Fibers and Applications POF'99*, 196–200. Chiva, Japan.
38 Information Gatekeepers (1993). *Plastic Optical Fiber POF Data Book*, 47–49. Information Gatekeepers.
39 Anonym. *CYTOP™ Technical Information*. AGC Chemicals Europe http://www.agcce.com/cytop-technical-information/.
40 Liu, H.Y. et al. (2002). Thermal stability of gratings in PMMA and CYTOP polymer fibers. *Optics Communications* 204: 151–156.
41 De Torres, M.C.P. (2013). Dispositivos fotonicos de medida basados en fibra optica de polımero. Thesis Universidad de Alcala, Chapter 2.
42 Crisp, J. (2001). *Introduction to Fiber Optics*. Oxford: Butterworth-Heinemann.
43 Masuda, A. et al. (2006). Optical properties of woven fabrics by plastic optical fibers. *Journal of Textile Engineering* 52: 93–97.
44 Křemenáková, D. et al. (2012). Characterization of side emitting polymeric optical fibres. *Journal of Fiber Bioengineering & Informatics* 5 (4): 423–431.
45 Lédl, V. and Václavík, J. (2009). Device for Evaluating the Side Emitting Optical Fiber Illumination Intensity. Report TU Liberec December 2009.
46 Meloun, M. and Militky, J. (2012). *Statistical Data Analysis*. New Delhi: Woodhead.
47 Yokogawa, H. et al. (2000). Side face illuminating optical fiber. US Patent 6, 154, 595.
48 King, J.C. et al. (1996). All-silica single-mode optical fiber with photonic crystal cladding. *Optics Letters* 21: 1547–15489.
49 Gauvreau, B. et al. (2008). Color-changing and color-tunable photonic bandgap fiber textiles. *Optics Express* 16 (20): 15677–15693.
50 Johnson, S.G. et al. (2001). Low-loss asymptotically single-mode propagation in large core OmniGuide fibers. *Optics Express* 9: 748.
51 Appajaiah, A., Kretzschmar, H.J., and Daum, W. (2007). Aging behavior of polymer optical fibers: degradation characterization by FTIR. *Journal of Applied Polymer Science* 103: 860–870.

52 Takezawa, Y. et al. (1992). Empirial estimation method of intrinsic loss spectra in transparent amorphous polymers for plastic optical fibres. *Journal of Applied Polymer Science* 46: 1835–1841.
53 Sillyman, S. (2004). a kol.: Source numerical aperture control for efficient light emission from notched, side-lighting fiber optics. *Proceedings of SPIE* 5529: 70–78.
54 Zarian, J.R. et al. (1999). Side lighting optical conduit. US Patent 5, 987, 199, 16 December 1999.
55 Endruweit, A. et al. (2008). Spectroscopic experiments regarding the efficiency of side emission optical fibres in the UV-A and visible blue spectrum. *Optics and Lasers in Engineering* 46: 97–105.
56 Spigulis, J. and Pfafrods, D. (1997). Clinical potential of the side-glowing optical fibers. *Proceedings of SPIE* 2977: 84–88.
57 Matthewson, M.J. and Kurkjian, C.R. (1987). Static fatigue of optical fibers in bending. *Journal of the American Ceramic Society* 70: 662–668.
58 Arrue, J. et al. (1998). Light power behavior when bending plastic optical fibers. *IEEE Proceedings - Optoelectronics* 145 (6): 313.
59 Measures, R.M. (2001). *Structural Monitoring with Fiber Optic Technology.* San Diego, CA: Academic Press.
60 Hecht, E. (2017). *Optic.* Boston, MA: Pearson.
61 Peters, K. (2011). Polymer optical fiber sensors – a review. *Smart Materials and Structures* 20: 013002 (17 pp).
62 Endruweit, A., Long, A.C., and Johnson, M.S. (2006). Textile composites with embedded optical fibers. In: *8th International Conference on Textile Composites (TEXCOMP-8).* Nottingham, UK.
63 Meloun, M. and Militký, J. (2011). *Experimental Data Treatment.* New Delhi: Woodhead Publishing.
64 Peirce, F.T. (1937). The geometry of cloth structure. *Journal of the Textile Institute* 28: T45–T196.
65 Wang, J. et al. (2013). Effect of weave structure on the side-emitting properties of polymer optical fiber jacquard fabrics. *Textile Research Journal* 83: 1170–1180.
66 Fu, H.J. and Chen, M.X. (1993). Effect of bending on optical fiber numerical aperture. *Opt. Commun.* 65: 59–61.
67 Yao, S. et al. (2011). Transmission efficiency of bending fiber with small radius. *Acta Optica Sinica* 31 (01): 06001.
68 Křemenáková, D. et al. Textile sheath containing side-emitting optical fibre. EP 2917391, pending.
69 Givoletti, M. (2001). Textile product containing optical fibres. EP 1307687B1.
70 Behera, B.K. et al. (2012). *Modeling of Woven Fabrics Geometry and Properties: Woven Fabrics,* Chapter 2 (ed. H.Y. Jeon). Rjeka: Intech.
71 Shen, J. et al. (2013). Light-emitting fabrics integrated with structured polymer optical fibers treated with an infrared CO_2 laser. *Textile Research Journal* 83 (7): 730–739.
72 Huang, J. et al. (2015). Evaluation of illumination intensity of plastic optical fibers with TiO_2 particles by laser treatment. *AUTEX Research Journal* 15 (1): 13–18.

73 Chepelyuk, E. et al. (2010). Geometric disposition of threads in single-layer woven structures. *International Journal of Clothing Science and Technology* 22: 35–48.
74 Barnoski, M.K. et al. (2001). *Fundamentals of Optical Fiber Communication.* Academic Press.
75 Jay, J.A. (2010). *An Overview of Macrobending and Microbending of Optical Fibers.* New York: Corning Inc.
76 Mallinder, F.P. and Proctor, B.A. (1964). Elastic constants of fused silica as a function of large tensile strain. *Physics and Chemistry of Glasses* 5 (4): 91–103.
77 Dugas, J. and Maurel, G. (1992). Mode-coupling processes in polymethyl methacrylate-core optical fibers. *Applied Optics* 31: 5069–5079.
78 Takezewa, Y. et al. (1991). Analysis of thermal degradation for plastic optical fibers. *Journal of Applied Polymer Science* 42: 2811–2817.
79 Schartel, B. et al. (1999). Chemiluminescence: a promising new testing method for plastic optical fibers. *Journal of Lightwave Technology* 17 (11): 2291–2296.
80 Boston Optical Fiber (2000). *Raytela Polymer Optical Fiber Cord, Toray Industries*. "OptiMega and OptiGiga".
81 Held, G. (2009). *Introduction to Light Emitting Diode Technology and Applications.* Boca Raton, FL: CRC Press.
82 Wang, X. et al. (2014). Flexible energy-storage devices. *Advanced Materials* 26 (28): 4763–4782.
83 Mao, Q. (2010). Plastic optical fiber and its application prospect, Communication technology and standards, Expert forum, 9 China.
84 Azadeh, M. (2009). *Fiber Optic Engineering*. Dordrecht: Springer.

24

Fibers as Energy Materials

Jiadeng Zhu, Esra Serife Pampal, Yeqian Ge, Jennifer D. Leary, and Xiangwu Zhang

North Carolina State University, College of Textiles, Department of Textile Engineering, Chemistry and Science, 1020 Main Campus Drive, Raleigh, NC 27695-8301, USA

24.1 Introduction to Fibers as Energy Materials

Renewable energy resources and their devices have received a lot of attention recently, caused by the limited availability and negative environmental impacts of fossil and nuclear fuels. It is also mandatory to develop an efficient storage system to retain all of the energy generated by renewable sources such as solar and wind power [1–3]. In contrast to those intermittent sources, rechargeable batteries and supercapacitors are increasing in importance, and they have become ubiquitous in our daily lives by powering numerous electronic devices, including cell phones, laptops, rechargeable vehicles, etc. [4–6].

Fibrous materials are considered promising candidates for fabricating energy-related systems due to their unique structure and excellent properties [7–9]. In particular, they can be employed as high-performance electrodes and separators in supercapacitors and secondary rechargeable batteries, transparent conductive electrodes or active materials in solar cells, and electrocatalysts for oxygen reduction in fuel cells, to name a few examples [10–15].

This chapter introduces the fundamental principles of energy storage and conversion systems and the applications of fibrous materials in rechargeable batteries and supercapacitors, followed by detailed discussion. It concludes with a comprehensive and up-to-date review of current and future trends for fibers as energy materials.

24.2 Fundamental Principles

Among the various dimensional energy materials, fiber-based 1D materials have attracted much attention due to the unique structure of those fibrous materials that can provide a continuous pathway for either electrons or ions and enhance the kinetic properties of the system [16, 17].

Handbook of Fibrous Materials, First Edition. Edited by Jinlian Hu, Bipin Kumar, and Jing Lu.

24.2.1 Basic Terms Related to Energy

Several basic terms related to energy are explained here to give a basic understanding on the required properties of fibers for use as energy materials. Generally, energy density represents the amount of energy stored in a system (e.g. batteries, fuel cells, supercapacitors) per unit volume (volumetric energy density) or mass (gravimetric energy density). Power density is the amount of power per unit volume or mass. A measure of the density of an electric current is called current density, which is the amount of current per unit area or mass. Capacity and capacitance are the system's ability to store electric charge for batteries and supercapacitors, respectively. Gravimetric specific capacity or capacitance specifies the amount of capacity or capacitance per unit mass. Coulombic efficiency is generally considered as the efficiency of charge or electron transportation in a system promoting an electrochemical reaction.

24.2.2 Principles of Lithium-Ion Batteries

A lithium-ion battery (LIB) is an energy storage system that stores electrical energy based on lithium-ion chemistry. It involves a reversible intercalation/deintercalation of lithium ions during the discharge/charge process. In the first-generation LIBs (Figure 24.1), graphite (C) and layered lithium cobalt dioxide ($LiCoO_2$) have been used as anode and cathode, respectively [18]. The electrolyte, acting as a conductive medium to shuttle lithium ions between the active electrodes, is 1 M lithium hexafluorophosphate ($LiPF_6$) dissolved in a mixture of ethylene carbonate, dimethyl carbonate, and diethyl carbonate with a volume ratio of 1 : 1 : 1. Microporous polyolefin membranes are typically utilized as separators for lithium batteries of this type.

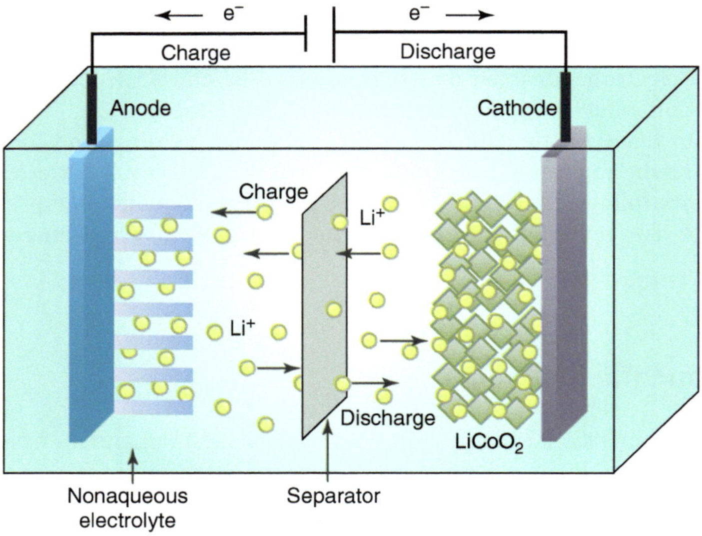

Figure 24.1 Schematic representation of a first-generation Li-ion cell. Source: Reproduced with permission from Roy and Srivastava [18]. Copyright 2014, Royal Society of Chemistry.

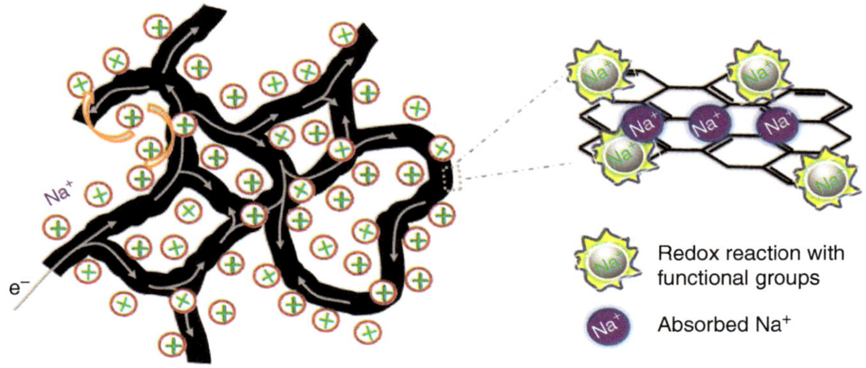

Figure 24.2 Schematic illustration of sodium storage in carbon nanofibers. Source: Reproduced with permission from Wang et al. [19]. Copyright 2013, Elsevier.

The lithium ions move from the anode (C) to the cathode ($LiCoO_2$) during discharge and return back while charging. The reversible reactions during the discharge process can be written as

$$\text{Cathode:}\quad Li_{1-x}CoO_2 + xLi^+ + xe^- \longrightarrow LiCoO_2 \tag{24.1}$$

$$\text{Anode:}\quad Li_xC_6 \longrightarrow xLi^+ + 6C + xe^- \tag{24.2}$$

It is important to note that fibrous electrode material can be beneficial to the electrochemical performance of the devices by shortening the diffusion distance of Li/Na ions and further improving the effective transfer pathway of electrons (Figure 24.2) [19].

24.2.3 Principles of Supercapacitors

Supercapacitors, also called ultracapacitors, are energy devices that store and release energy through charge separation at the electrochemical interface between electrodes and electrolytes. They exhibit high power density, longer cycle life, and rapid charge/discharge ability [20, 21].

There are two main types of supercapacitors. One type, called the double-layer capacitor, achieves energy storage by separation of charge in a Helmholtz double layer, as shown in Figure 24.3 [20]. Generally, the electrodes used for double-layer capacitor are porous carbon materials with high specific surface area. Compared with the infinite planar one, the behavior of electrical double-layer is more complicated since the ion transportation in a confined system can be influenced by many factors, such as the space constrain inside the pores, the tortuous mass transfer path, and the wettability of the pore surface by the electrolyte [20].

The second type is known as a pseudocapacitor, and its electrical energy is obtained by faradaic redox reactions with charge transfer. It can be seen from Figure 24.4 that fast and reversible faradaic reactions occur on the electrode materials when a potential is applied, producing charges and resulting in charge transfer [21]. Pseudocapacitors usually suffer from a severe capacitance decay/fading due to those redox reactions. It is, therefore, important to design a stable structure that can be maintained during cycling.

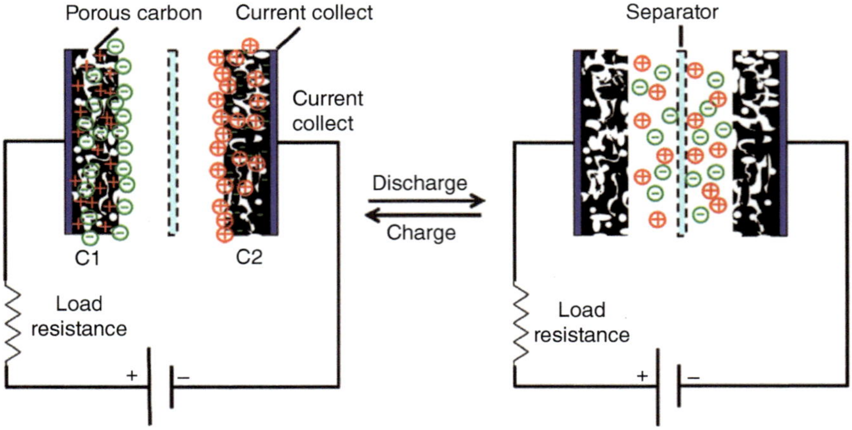

Figure 24.3 Schematic representation of a double-layer capacitor by using porous carbon electrodes. Source: Reproduced with permission from Wu and Xu [20]. Copyright 2014, Royal Society of Chemistry.

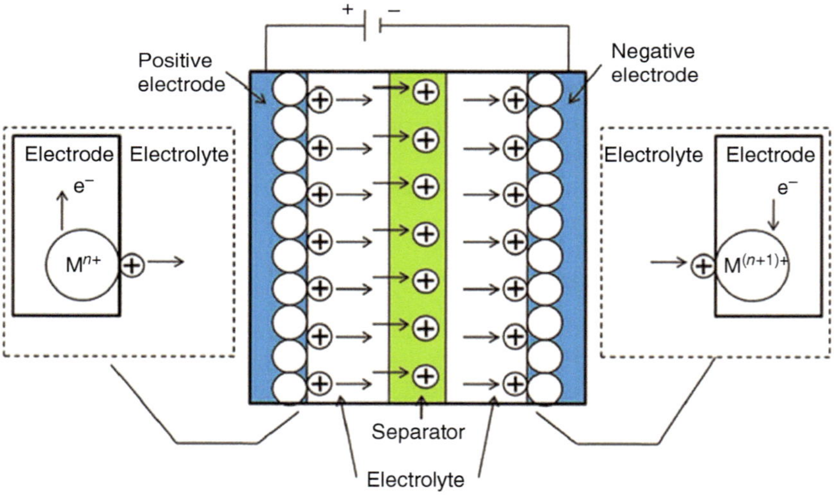

Figure 24.4 Schematic illustration of a pseudocapacitor (M: a metal atom). Source: Reproduced with permission from Shi et al. [21]. Copyright 2014, Royal Society of Chemistry.

24.3 Characterization, Structure, and Fabrication of Fibrous Energy Materials

Some common characterization techniques have been introduced to investigate the physical and chemical properties of energy materials with fibrous structures to pursue better properties. When incorporated into electrodes for electrochemical systems, materials are expected to possess good cycling life and high energy/power density. In some cases, flexibility is also one of the important factors. One-dimensional fiber structure gives rise to high surface area, short ion diffusion distance, and electron transfer path in electronic devices, thus capturing many researchers' attention. Besides, as separators, membranes are desired to have excellent ion permeability, good wettability, and chemical stability. Meanwhile, they are also characterized as having large porosity, interconnected open pore structure, high electrolyte permeability, and excellent electrolyte uptake. Therefore, fibrous mats are quite promising in contrast to commercial polyolefin membranes [22, 23]. In short, the structure and fabrication of energy materials are extremely critical in this field.

24.3.1 Commonly Used Characterization Techniques for Fibrous Energy Materials

It is well known that both physical and chemical properties of fibrous materials play important roles in determining their final applications. The morphology and diameters of fibrous energy materials can be observed by using scanning electron microscopy (SEM) and transmission electron microscopy (TEM). Their crystal structure and surface areas can be investigated by X-ray diffraction (XRD) and Brunauer–Emmett–Teller (BET), respectively. Thermogravimetric analysis (TGA) and differential scanning calorimetry (DSSC) can be performed to acquire their thermal properties. The composition of fibers can be characterized by using X-ray photoelectron spectroscopy (XPS), elemental analysis (EA), and energy-dispersive X-ray spectroscopy (EDS). Raman spectroscopy is a useful tool for characterizing the carbonaceous structure in carbon-based fibers.

24.3.2 Fibrous Structures of Electrodes

A variety of fibrous structures have been explored in the pursuit of obtaining the perfect construction, including carbon fibers, fiber composites, core–shell and tubular structure fiber, and porous fibers [24–26]. Here, we introduce some typical fibrous structures.

24.3.2.1 Continuous Fibers

The continuous fiber network of carbon fiber possesses large surface area and good mechanical properties, extending the electron transfer pathway. Therefore, carbon fibers have great potential to be applied as electrodes for advanced energy storage systems [27]. Zhu et al. have produced continuous nitrogen-doped carbon nanofibers (N-CNFs) derived from polyacrylonitrile (PAN), which were subsequently used as anode materials for sodium-ion batteries by electrospinning and thermal treatment. The prepared N-CNFs have a diameter of 150–250 nm [24]. It provided an efficient and uninterrupted electron transport and enhanced ion transport and storage capability, shown in Figure 24.5.

24.3.2.2 Carbon Fibers Containing Nanoparticles

It is well known that the conventional method of electrode preparation by mixing active material with conductive agent and binder can reduce the mass loading

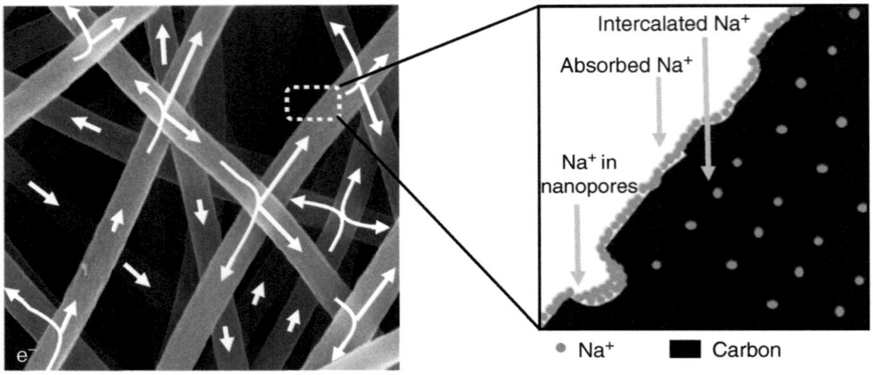

Figure 24.5 Schematic illustration of sodium storage mechanism in N-CNFs. Source: Reproduced with permission from Zhu et al. [24]. Copyright 2015, Elsevier.

and specific capacity of the electrode. On the other hand, by impregnating active material into carbon fiber matrix by electrospinning, the electrode can be used as flexible and binder free. Especially, while silicon is used as the anode, it has a large volume change during charge–discharge process, which is suppressed by using carbon fiber matrix as a buffer to confine the volume change [28]. Ji and Zhang have introduced a framework of Si/carbon nanofiber (Si/CNF), where Si nanoparticles were embedded in carbon nanofibers (CNFs). This composite was obtained by electrospinning from the mixture of Si nanoparticles and PAN solution with subsequent heating process. This anode material demonstrated the advantage of using carbon matrix in terms of its cycle life as well as the active Si (high lithium-storage capacity) [29].

What's more, some active materials such as metal oxides have low electrical conductivity, but when combined with CNF, the electrical conductivity can be significantly enhanced. Ge et al. have prepared copper-doped $Li_4Ti_5O_{12}$/CNF composite by electrospinning with subsequent process, where poly(vinylpyrrolidone) (PVP) was used as the carbon source. The prepared composite demonstrated consistent fibrous structure in the diameter of 200–600 nm (Figure 24.6). Within the carbon fiber matrix, copper-doped $Li_4Ti_5O_{12}$ particles with a grain size of 50 nm were uniformly distributed [30].

Core–shell structure is another approach to combine inorganic components with CNF. Li et al. designed Si/C–C core–shell nanofiber structure by dual nozzle coaxial electrospinning with carbonization process and used these as anode in LIBs. As shown in Figure 24.7, the core was filled with Si/C composite and covered with carbon as the shell. The carbon shell served as buffer to confine the large volume expansion of Si/C core during the lithiation process [31].

24.3.2.3 Porous and Tubular Fibers

Generally, porous fiber structure and tubular fiber structure are made by eliminating the sacrificial components in the fiber to create hollow space, which could provide high surface area and thus enhance the electrochemical properties of

Figure 24.6 SEM image of $Li_{3.95}Cu_{0.05}Ti_5O_{12}$/CNFs. Source: Reproduced with permission from Ge et al. [30]. Copyright 2014, Elsevier.

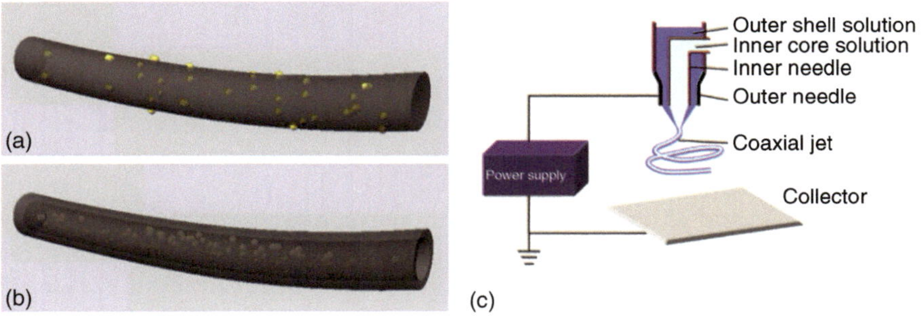

Figure 24.7 Schematic image of (a) Si/C composite nanofibers and (b) Si/C–C core–shell nanofiber, (c) dual nozzle coaxial electrospinning setup. Source: Reproduced with permission from Li et al. [31]. Copyright 2014, Elsevier.

energy materials. Basically, inorganic substances such as SiO_2 [32] and Ni particles as well as certain polymers such as poly(methyl methacrylate) (PMMA) [33] and poly-L-lactic acid [34] are favorable choices for designers. Sacrificial components are removed by dissolution in suitable solvents or decomposed during heat treatment.

Ji et al. generated porous CNF using SiO_2 nanoparticles as sacrificial template, which were etched by hydrofluoric (HF) acid in the as-produced carbon/SiO_2 composite. After etching, the surface area was increased from 26.1 to 91.8 m^2/g, as seen in Figure 24.8 [32]. Lu et al. prepared porous CNF via centrifugal spinning of PAN/PMMA solution and used these in electric double-layer capacitor (EDLC) electrodes. The result indicated that when the PAN/PMMA weight ratio was 7/3, the specific surface area increased from 405 to 444 m^2/g as the carbonization temperature increased from 700 to 900 °C. The total pore volume also increased from 0.171 to 0.310 cm^3/g [35].

Figure 24.8 Schematic image of porous carbon nanofiber. Source: Reproduced with permission from Ji et al. [32]. Copyright 2009, Elsevier.

24.3.3 Fibrous Structures of Separators

It is well known that separators play an important role in the electrochemical devices [23]. They allow ions to be transported from one side to the other while preventing electrons from passing. Therefore, they should have good insulativity and penetrability.

The thickness and porosity of fiber membranes are two key factors for separators. Yanilmaz et al. prepared silica/PAN membranes as the separator for LIBs. The thickness of the prepared membranes was around 65 μm. In addition, it could reduce the pore size from 478 to 275 nm by adding 27 wt% SiO_2 in the fiber composite [36]. In Figure 24.9, Raghavan et al. designed sandwich structure of poly(vinylidene fluoride-*co*-hexafluoropropylene) and PAN and converted it into polymer electrolytes [37]. The obtained membranes had uniform morphology

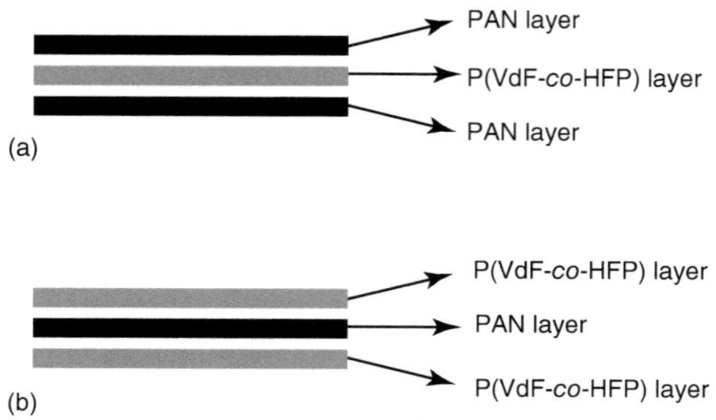

Figure 24.9 Schematic of electrospun trilayer nonwoven mats. (a) PAN/P(VdF-co-HFP)/PAN and (b) P(VdF-co-HFP)/PAN/ P(VdF-co-HFP). Source: Reproduced with permission from Raghavan et al. [37]. Copyright 2010, Elsevier.

with a diameter of 320–490 nm and a high porosity of 82–85%, demonstrating a high electrolyte uptake.

24.3.4 Fabrication Approaches

Several methodological approaches have been explored to produce fibrous materials in the nanoscale or microscale in order to achieve high surface area and high kinetic for energy storage applications. In fact, for the usage in energy storage systems, fibers have been achieved overwhelmingly by two main approaches: electrospinning and centrifugal spinning [25, 38, 39].

Electrospinning is a relatively mature and common technique to produce ultrafine fiber with diameter in the range of 40–2000 nm [40, 41]. The electrospinning setup revealed in Figure 24.10 consists of a high voltage supply, spinning nozzle, syringe, and collector. Polymer solution is stretched from the needle tip by electrostatic force to form a "Taylor cone" and then stacked on the collector [23]. This technique supplies many advantages. First of all, the process is simple and agile and can be used to tailor various one-dimensional structures, including structures of multicomponent nanofiber, core–shell nanofiber, and heterogeneous nanofiber [27]. Secondly, by using rotary collector, one can obtain an oriented and aligned nanofiber mat [42]. The method is widely adopted in the field of energy conversion and storage devices. Ji et al. prepared one-dimensional Ni/CNFs for lithium-ion storage by initially electrospinning the solution of $[Ni(OAc)_2]\cdot 4H_2O$ (30 wt%) and PAN (10 wt%) dissolved in dimethylformamide (DMF) and then carbonized the as-spun nanofiber mat. The final obtained CNFs showed that the Ni spherical nanoparticles could be uniformly distributed in CNFs and the prepared Ni/CNFs delivered favorable electrochemical performances [43].

Centrifugal spinning is an effective and high-speed route to create fibrous structures in the micro- or nanoscale at high throughput, deriving from cotton

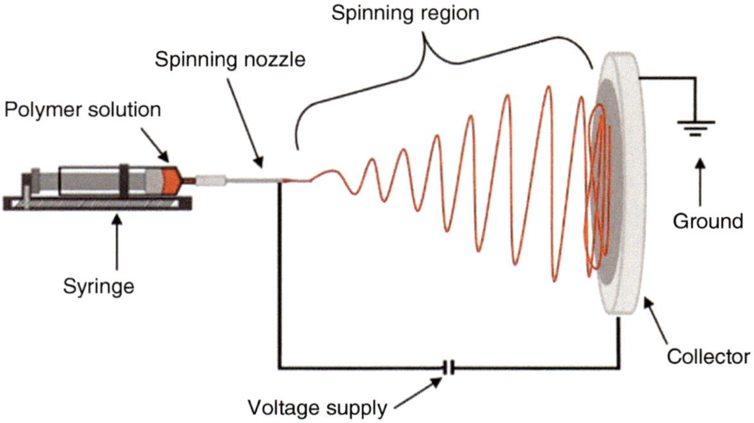

Figure 24.10 Schematic of a typical electrospinning setup. Source: Reproduced with permission from Lee et al. [23]. Copyright 2014, Royal Society of Chemistry.

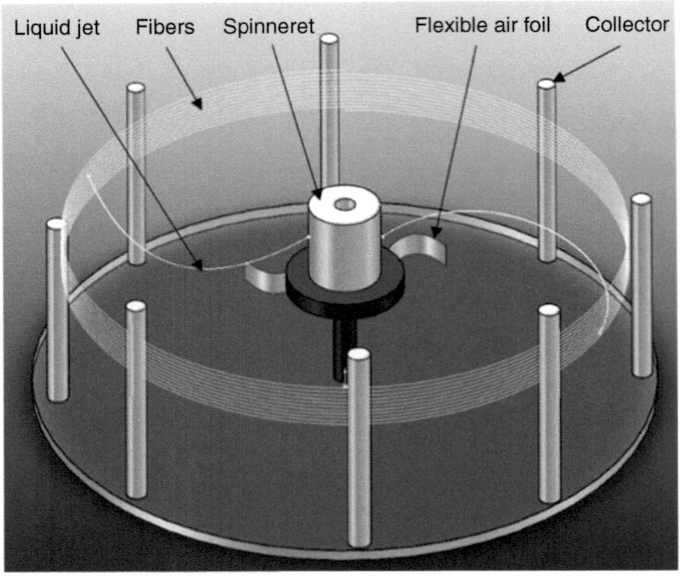

Figure 24.11 Schematic image of the centrifugal spinning system. Source: Reproduced with permission from Lu et al. [44]. Copyright 2013, Elsevier.

candy maker. In Figure 24.11, the centrifugal spinning setup shown is composed of a perforated Teflon spinneret fixed in the center of the spinning platform, with a motor underneath the spinneret to provide high-speed rotation, a speed controller connected to the motor to adjust the rotational speed, and a set of collectors in the outer perimeter of the spinning platform. During the processing, the polymer precursor is subjected to the action of centrifugal force and drawn to the collectors. The advantages of centrifugal spinning can be summarized as follows: (i) It is productive with high-speed rotation of 2000–13 000 rpm [45]; (ii) the fiber is generated by centrifugal force, which eliminates some safety issue from high voltage like electrospinning; (iii) the process has less impact on the environment, and it can produce environmentally sensitive or electrically sensitive polymers. There are an increasing number of publications about centrifugal spinning fibers for energy material preparation. Lu et al. produced PAN micro-/nanofiber by using centrifugal spinning approach. The resultant fiber was in the diameter of 407–1066 nm [44]. Dirican and Zhang fabricated porous carbon microfibers as anode materials for sodium-ion batteries by centrifugal spinning the solution of PAN/PMMA precursor and combining with thermal process. The prepared fiber was observed with many internal pores and surface roughness, which gave rise to greater sodium-ion storage ability [33].

Apart from the techniques mentioned above, self-assembly, chemical vapor deposition (CVD), and hydrothermal method or hybrid process are also adopted to produce fiber materials for energy storage use [46–48]. Fu et al. tailored aligned CNT–Si structure, which could be used as binder-free and flexible anodes for LIBs. Carbon nanotube (CNT) sheet was winded from CNT forests, which were grown on a quartz substrate using iron chloride ($FeCl_2$) as the catalyst in a horizontal tube furnace. Then silicon-active material was deposited onto 10-layer CNT sheet (2 mg/cm^3) by CVD processing [49].

24.4 Applications to Batteries, Supercapacitors, and Energy Harvesting

The growing interest in renewable energy sources also increases the need for devices that will store the energy produced from them. Therefore, the investigation on energy harvesting and energy storage components occupies huge space of the academic and industrial R&D programs. In this regard, the development of electrodes for either energy storage or conversion takes the position in these programs as a core part that can overcome the shortcomings and significantly improve the performances of such devices. Various nanostructures made of different materials have been proposed for this purpose [50]. Among them, fibrous 1D structures attract attention due to their unique properties that allow improved and efficient energy harvesting and storage [51].

24.4.1 Batteries

Since the commercialization of LIBs, graphite has been used as the negative electrode. Due to its layered structure, which enables reversible Li-ion insertion and extraction, a high and stable energy efficiency is achieved despite its limited capacity (372 mAh/g) [52]. Soft carbons obtained from carbonization of an organic compound, usually resins, have been used as an alternative; however these showed a lower reversible capacity retention. CNFs obtained from different precursors and by different methods attract the attention due to their 1D structure, controllable diameter, enhanced surface area-to-volume ratio, low density, and the possibility to be produced in a feasible process in a variety of structures and in combination with other compounds. Neat, porous, or doped CNFs obtained from PAN by electrospinning and subsequent carbonization have been most widely investigated as anodes for LIBs [53, 54]. Electrospun PAN-based CNFs obtained after carbonization at 1000 °C provided a capacity of 450 mAh/g at a current density of 30 mA/g, or 350 mAh/g at 100 mA/g [53]. The high price and the limited processability of PAN forced researchers to investigate more environmentally friendly and cost-effective alternatives. Among them, lignin, as an abundant natural polymer with high carbon content and thermoplastic character, has great potential to replace PAN to produce carbon fiber-based electrodes for energy applications [55]. Polyvinyl alcohol (PVA), polyvinyl chloride (PVC), PVP, poly(amic acid) (PAA), polyimide (PI), and polybenzimidazole (PBI) are other alternatives toward lower-cost CNF production [56–58]. Although neat CNFs can provide higher capacity than graphite, it still cannot meet the required increased demand for high capacity and long-lasting batteries for larger applications, such as electrical vehicles, grid energy storage, etc. [59]. Therefore, the research in the last few decades has focused on the development of 1D structures containing active materials that can store large amounts of Li ions.

With a theoretical capacity of 4212 mAh/g, silicon (Si) is the most investigated active material for anode application in LIBs in a wide range of structures and combinations [18]. Despite its high Li-ion storage capacity, it shows large capacity loss during cycling as a result of its volume expansion when alloying with Li ions. This volume change causes electrode disintegration and decrease of electrical contact between particles [59]. Thus, its combination with carbonaceous materials is most commonly used approach to suppress the volume expansion and to maintain the electrical conductivity during cycling. Among the structures proposed, fibrous 1D structures show promising electrochemical performances. Si/CNF composites obtained from electrospinning of Si/PAN solution and subsequent carbonization demonstrate an effective production method. They delivered an initial capacity of 1750 mAh/g that decreased to around 1000 mAh/g after 20 cycles [60]. However, optimization of the nanoparticle loading and other process parameters can improve the cycling stability of these electrodes. Si nanoparticles embedded within the holes of porous CNFs is one alternative approach to this goal [61]. Such structures covered with an additional carbon layer and used as freestanding anodes exhibited capacity of 600 mAh/g after 200 cycles at a current density of 100 mA/g. The stability of the composites can be further improved with the use of carbon-coated Si/CNF composite nanofibers. Such coating can be achieved by immersing the Si/CNF into sucrose solution followed by carbonization, as can be seen in Figure 24.12 [62]. Optimized carbon-layer thickness buffers the volume expansion of the Si nanoparticles maintaining the

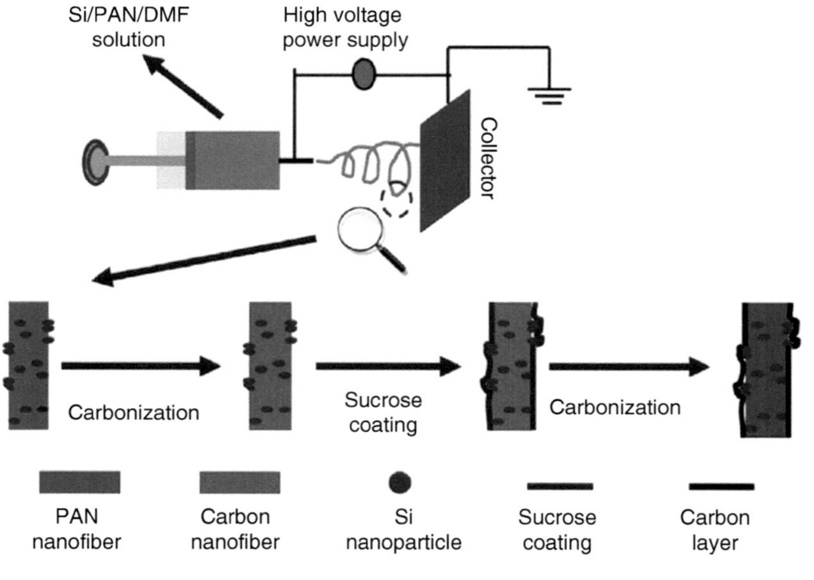

Figure 24.12 Schematic representation of Si/C/CNF nanofiber production and the way its structure changes during different production steps. Source: Reproduced with permission from Chen et al. [62]. Copyright 2015, Elsevier.

structural integrity of the electrode during cycling. These Si/C/CNF could give a capacity of 1215 mAh/g after 50 cycles at 6 A/g. SiO_2-coated Si/CNF composite could also improve the cycling stability and rate capability of the anodes [63]. Additionally, Lin et al. reported a study of twisted aligned multiwalled carbon nanotube (MWCNT) and MWCNT/Si fibers as the anode materials for LIBs, and the procedure is shown in Figure 24.13a [64]. The diameter of the twisted aligned MWCNT fiber was 30 μm, shown in Figure 24.13b,c. In contrast, the diameter of twisted aligned MWCNT/Si composite fiber increased to 60 μm due to the presence of a uniform Si coating (Figure 24.13d,e). The prepared aligned composite fiber could deliver a capacity of 1970 mAh/g with a current density of 1 A/g at the 50th cycle.

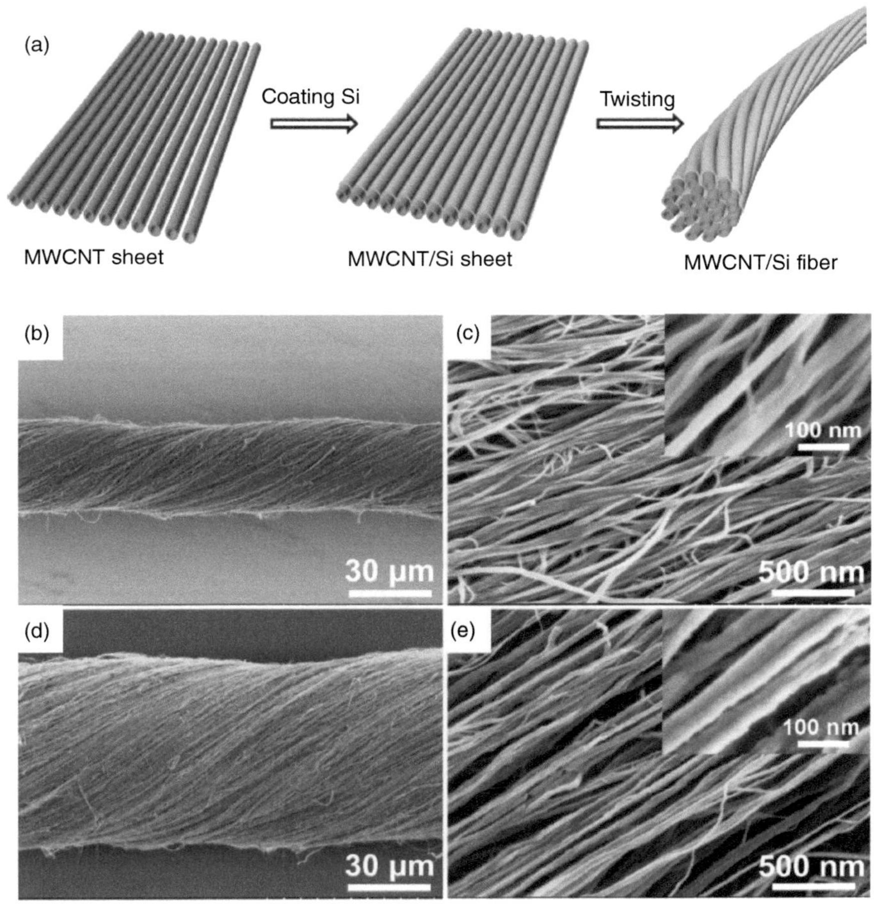

Figure 24.13 (a) Schematic image of the synthesis of aligned MWCNT/Si fiber; SEM images of the twisted aligned (b, c) pure MWCNT and (d, e) MWCNT/Si composite fibers with different magnifications. Source: Reproduced with permission from Lin et al. [64]. Copyright 2014, John Wiley & Sons.

Besides silicon, tin (Sn) has also attracted significant attention for LIB and NIB (sodium-ion battery) anodes application due to its low cost and high theoretical capacity for both chemistries, namely, 990 and 847 mAh/g for LIB and NIB, respectively. Similar to silicon, Sn also suffers from volume expansion due to the alloy formation with Li/Na ions during cycling, which further provokes electrode disintegration and shortens cycle life. Sn nanoparticles embedded into submicron carbon fibers could be produced via the cost-effective centrifugal spinning method [65]. Carbonized $SnCl_2$/PAN nanofibers obtained by this method exhibited a capacity of approximately 400 mAh/g after 100 cycles at 100 mA/g when used as anodes for LIBs. Notably, centrifugally spun SnO_2 microfibers, applied as anodes in NIBs, had a capacity of 150 mAh/g at 640 mA/g, which is three times higher the capacity of pure SnO_2 nanoparticles at the same current density [66]. Furthermore, SnO_2 microfibers show better rate performance with lower capacity decay at higher current densities. Sn/Cu/CNF core–shell structures with improved conductivity have been suggested as an alternate way to improve the cycling stability over prolonged cycling for both Li- and Na-ion chemistries [67]. Tin oxide with its rutile structure and theoretical capacity of 782 mAh/g can also be used as active material for energy storage, due to its lighter volume expansion compared with pure Sn [68]. Electrodeposited SnO_2 over porous CNFs as a freestanding anode in LIBs gave a capacity of 554 mAh/g after 100 cycles, and their cycling stability could be further improved by controlling carbon deposition [69]. Freestanding Sn/SnO_2/CNFs coated with carbon layer exhibited an excellent capacity of 712 mAh/g after 200 cycles at 800 mA/g in the LIB half-cell [70]. A similar structure used for a NIB anode also improved the capacity and the cycling stability of the cell, with 82.7% capacity retention at 50 mA/g (or 374 mAh/g) after 100 cycles [71]. Moreover, ZnO/SnO_2 heterostructures with optimized fiber diameter, particle size, and porosity showed excellent cycling stability of LIBs [72].

On the other hand, hematite iron oxide (α-Fe_2O_3) is a low-cost and non-toxic potential active material for LIB anodes with a theoretical capacity of 1007 mAh/g. Its composites with carbonaceous materials showed an improved electrochemical performance due to the increased electrical conductivity [73]. However, the electrochemical capacity of α-Fe_2O_3 nanoparticles cannot be fully utilized when embedded into CNF, due to the lack of space for Li-ion diffusion. Therefore, to facilitate the Li-ion transport to the pores of the active materials, 1D α-Fe_2O_3 nanofibers were coated with carbon as schematized in Figure 24.14 [74]. The space between the particles were covered with a carbon layer and embedded into 1D structure. This allowed facile and reversible insertion and

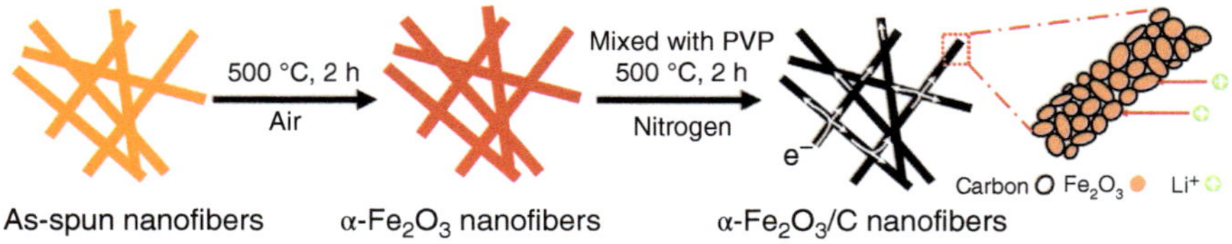

Figure 24.14 Schematic representation of porous 1D α-Fe_2O_3/CNF nanofiber production and its storage mechanism. Source: Reproduced with permission from Zhu et al. [74]. Copyright 2016, Elsevier.

de-insertion of Li ions during cycling, which enables full usage of the active material and enough space to buffer its volume expansion. This unique 1D structure showed capacity of 715 mAh/g (capacity retention of 91%) after 200 cycles at 100 mA/g, or almost double the capacity of α-Fe_2O_3 nanoparticles/CNF composite (400 mAh/g after 100 cycles at 100 mA/g).

The versatility of CNFs allows them to be used as conductive support for 3D structures. $Zn_xCo_{3-x}O_4$ (ZCO) nanocubes coated onto CNFs have been successfully applied as anodes for LIBs [75]. The combination of hydrothermal–thermal treatment of CNF mats with suitable chemicals and optimized process conditions allows formation of nanosized ternary metal oxide/CNF composites with enhanced electrochemical properties that can find application not only for energy storage but also for energy conversion devices.

Despite the extended research on fiber structures as anodes, which are usually carbon based, the versatile approaches for fiber/nanofiber production allow their application as cathodes as well [76]. Even though the structure of cathode materials is more complex, they do require high conductivity and reversible and stable reactivity with the ions [77]. In this regard, layered, spinel, and olivine single or multi-metal structures have been designed as fibrous structures and applied in either LIBs or NIBs [76–79].

$LiFePO_4$/CNF freestanding cathodes were obtained by electrospinning of PAN solution containing Li, Fe acetate, and phosphoric acid followed by carbonization [78]. This 1D structure enabled the formation of well-distributed $LiFePO_4$ particles throughout the CNF, which resulted in enhanced capacity and cycling stability of the cathode. The addition of CNT or graphene within the structure further improved the electrochemical performances due to the increased electrical conductivity between the particles [78, 79]. Zhang et al. prepared LiF/Fe/CNF cathodes from ferrocene/PAN solution via electrospinning [80]. Such cathodes, with a controllable fiber diameter and distribution of active particles, enable better contact between LiF and Fe, which provided high and stable capacity during cycling. On the other hand, the presence of carbon over the particles of the active material enables easy and fast Li-ion diffusion and electron transport. Embedding the multicomponent olivine of $Li_2Mn_{(1-x)}Cr_xSiO_4$ with controlled Mn-to-Cr ratio into a CNF matrix could significantly prevent its structural collapse during cycling [81]. Furthermore, the enhanced conductivity between the particles during cycling enables reaching capacity near the theoretical capacity of Li_2MnSiO_4 (330 mAh/g). The improved performances of Li_2MnSiO_4 cathode in LIB cells can be also obtained by Fe doping [82]. Recently, Lu et al. obtained Li-substituted Na-layered transition metal oxide fibers by centrifugal spinning of PAN/nitrile solutions with different content of Li and Na, as shown in Figure 24.15 [83]. The results indicated that the obtained materials had good cycling stability when performed as electrodes for NIBs.

Fiber cathodes for Li–S and Li–air batteries have also been widely investigated [84–87]. CNFs act as supports for sulfur or catalysts where carbon compensates for the low electrical conductivity of the active materials. This enables freestanding structures with high surface area and porosity. In the case of Li–S batteries, PAN-based CNFs can be doped with sulfur via a vapor deposition [88]. CNFs can act as interlayers between the cathode and the separator in Li–S batteries

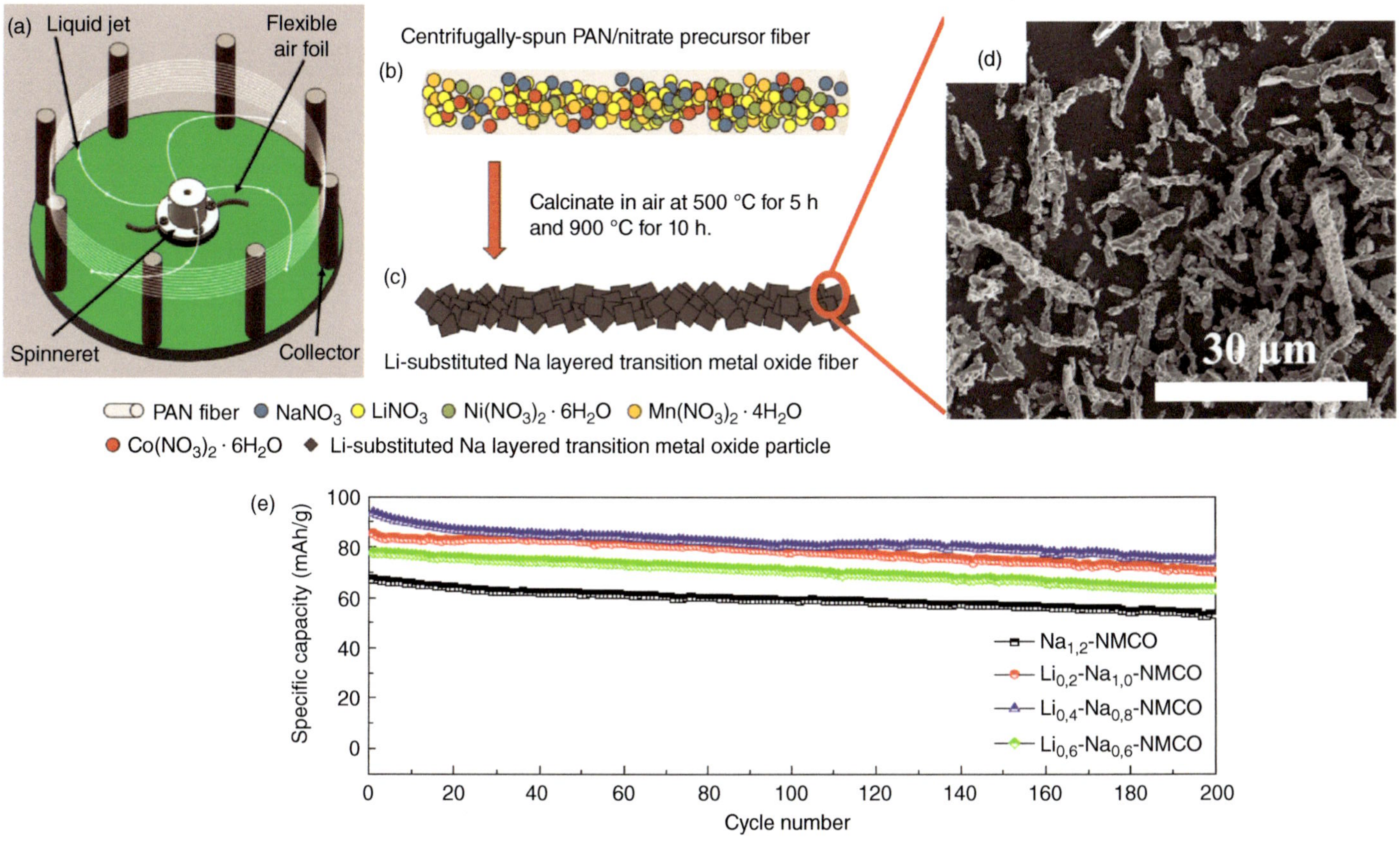

Figure 24.15 Schematic illustration of (a) centrifugal spinning device, (b) the fiber precursor, and (c) the fiber cathode, including its (d) SEM image and (e) cycling performances at 75 mA/g. Source: Reproduced with permission from Lu et al. [83]. Copyright 2015, Elsevier.

in order to prevent the shuttling effect by localizing the soluble polysulfides during cycling. That improves the electrical conductivity of the cathode and reduces the charge-transfer resistance [86]. The electrical conductivity of the cell can be enhanced by using CNFs deposited on polyvinylidene fluoride (PVDF) membrane [87], or CNF/rGO (reduced graphene oxide) membrane obtained by the filtration method. Different catalysts with high activity for oxygen reduction reaction (ORR) and the oxygen evolution reaction (OER) such as cobalt oxide (Co_2O_3), nickel–cobalt oxide ($NiCo_2O_4$), or MnO_2 can be in situ introduced into CNF [89, 90] or mixed with CNF in a suspension that gives a freestanding porous film after filtration method [91]. Well-distributed active particles in CNF showed lower overpotential and improved capacity and cyclic performance [91].

Polymer fibrous membranes with good thermal stability and chemical resistivity lead to facile migration of ions and decreased cell resistance [77]. In this regard, nanofiber membranes have been widely used as separators for energy storage systems [92, 93].

Silica nanopowder embedded in PVDF [93], PAN [36], or nylon 6,6 [94] nanofibers can be used as a separator with improved thermal stability, higher ionic conductivity, and lower interfacial resistance that results in an enhanced cycling performance of a LIB cell. As can be seen in Figure 24.16, when spun onto self-standing CNT sheets, the composite structure acted as both separator and interlayer for Li–S batteries, providing fast diffusion of both lithium ions and electrons, and reduced diffusion of polysulfides [95].

Centrifugally spun fibrous membranes either doped or from a blend of polymers can also be used as separators in energy storage devices [39, 96–98]. SiO_2/PAN membranes with optimized amount of SiO_2, obtained by centrifugal spinning, could significantly improve the ionic conductivity of the separator and consequently enhance the discharge capacity of LIBs (Figure 24.17) [39]. Furthermore, it offered low-cost fiber production due to much lower power consumption and high production rates.

24.4.2 Supercapacitors

Supercapacitors are energy storage devices that can provide high power density due to the highly porous electrode structure with controlled pore size distribution

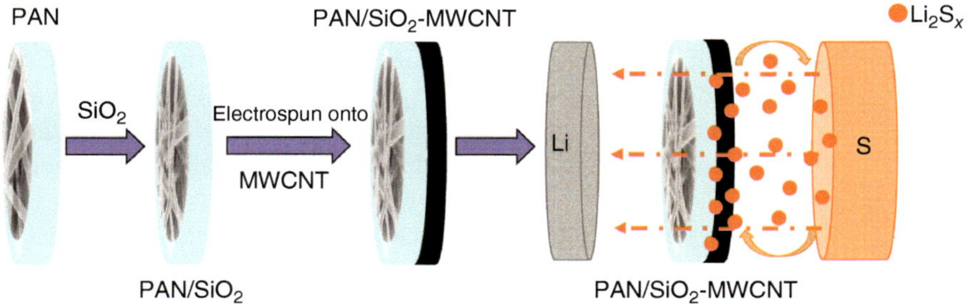

Figure 24.16 Schematic presentation of PAN/SiO_2-adapted separator for Li–S batteries. Source: Reproduced with permission from Zhu et al. [95]. Copyright 2016, Royal Society of Chemistry.

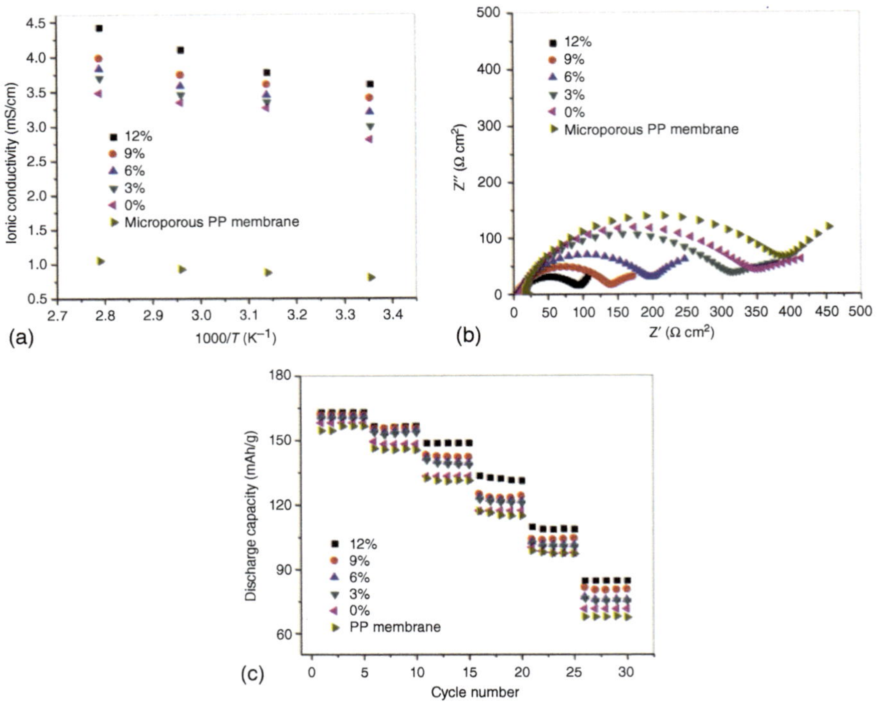

Figure 24.17 (a) Ionic conductivity, (b) electrochemical impedance spectra of SiO_2/PAN membranes, and (c) C-rate performance of Li/$LiFePO_4$ cells containing SiO_2/PAN membranes as separator. Source: Reproduced with permission from Yanilmaz et al. [39]. Copyright 2015, Elsevier.

via an electrostatic mechanism [99]. Thus, CNFs with increased surface area can be ideal for this application [100].

Lu et al. obtained CNFs with controlled porosity by centrifugal spinning of polymer blends and thermal treatment. The prepared CNFs that were carbonized at 900 °C could deliver a capacitance of approximately 100 F/g at 2 A/g. A high capacity retention of 74% could be achieved for the sample carbonized at 900 °C when current density increased from 0.1 to 2 A/g [35]. The production steps and the morphology of such CNFs are given in Figure 24.18. On the other hand, CNF doped with metal oxides such as RuO_2, Fe_2O_3, MnO_2, and NiO could serve as electrodes for hybrid capacitors with higher energy and power densities because of the synergetic effect between carbon's conductivity and metal oxide's pseudocapacitance [101, 102].

Conductive polymers can also give pseudocapacitance and can be deposited onto porous CNF or be used alone as electrodes for either symmetric or asymmetric supercapacitors [103, 104]. Studies showed that hollow fiber structures made of conductive polymers could be used as supercapacitor electrodes, which significantly improved the pseudocapacitance performance of the device. Electrodes were obtained from electrospun PAA fiber membrane as a template for in

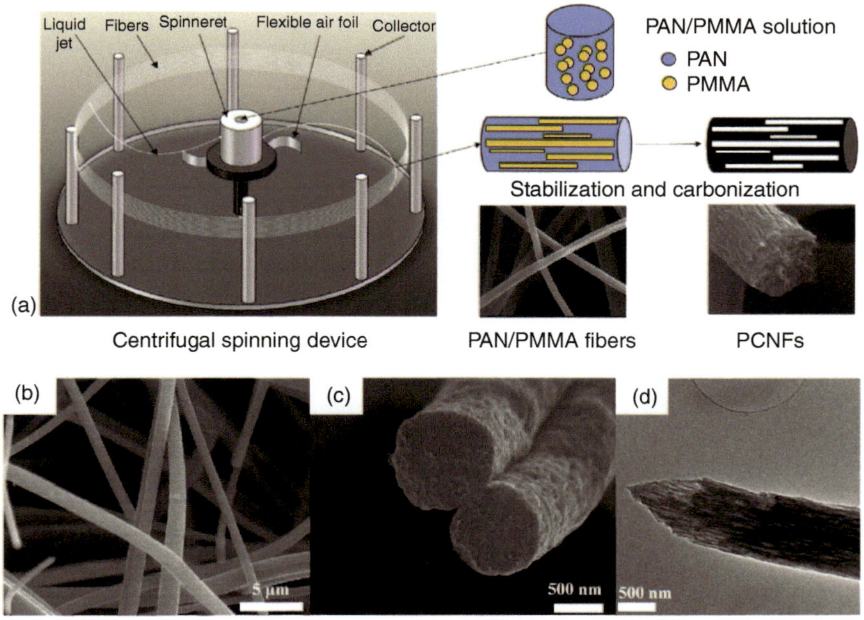

Figure 24.18 (a) Production method of porous CNFs and (b–d) their morphology: (b) plane-viewed SEM, (c) cross-sectional SEM, and (d) TEM images. Source: Reproduced with permission from Lu et al. [35]. Copyright 2015, Elsevier.

situ polymerization of polyaniline (PANI) (Figure 24.19) [103]. These electrodes delivered the capacitance of 500 F/g at 2 A/g using three-electrode system.

Pan et al. prepared a novel wearable energy device by using aligned CNT fiber textiles that were coated by PANI [104]. During the preparation, CNT fibers were first formed from a spinnable CNT array that were then woven into textiles (Figure 24.20a). The CV curves of pristine CNT and CNT/PANI composite textiles are shown in Figure 24.20b. It is obvious that the current density of CNT/PANI textiles was much higher than that of pristine CNT textiles. In Figure 24.20c, the composite textiles with 50 wt% PANI could deliver a specific capacitance of 272.7 F/g at a current density of 1 A/g.

24.4.3 Energy Harvesting Devices

Energy harvesting or the process of capturing and transforming the energy derived from an external force has attracted a great amount of attention in the last decade. Among the sources from which electrical energy can be obtained, photovoltaic, fuel cells, and nanogenerators give a chance for incorporation of fibrous structures as their constituents [105].

In dye-sensitized solar cells (DSSCs), the high specific surface area of nanofibrous structure enables improved absorption of photosensitizing dye and electrolyte uptake. Additionally, their improved interconnectivity lowers the grain boundaries that leads to enhanced charge conduction and

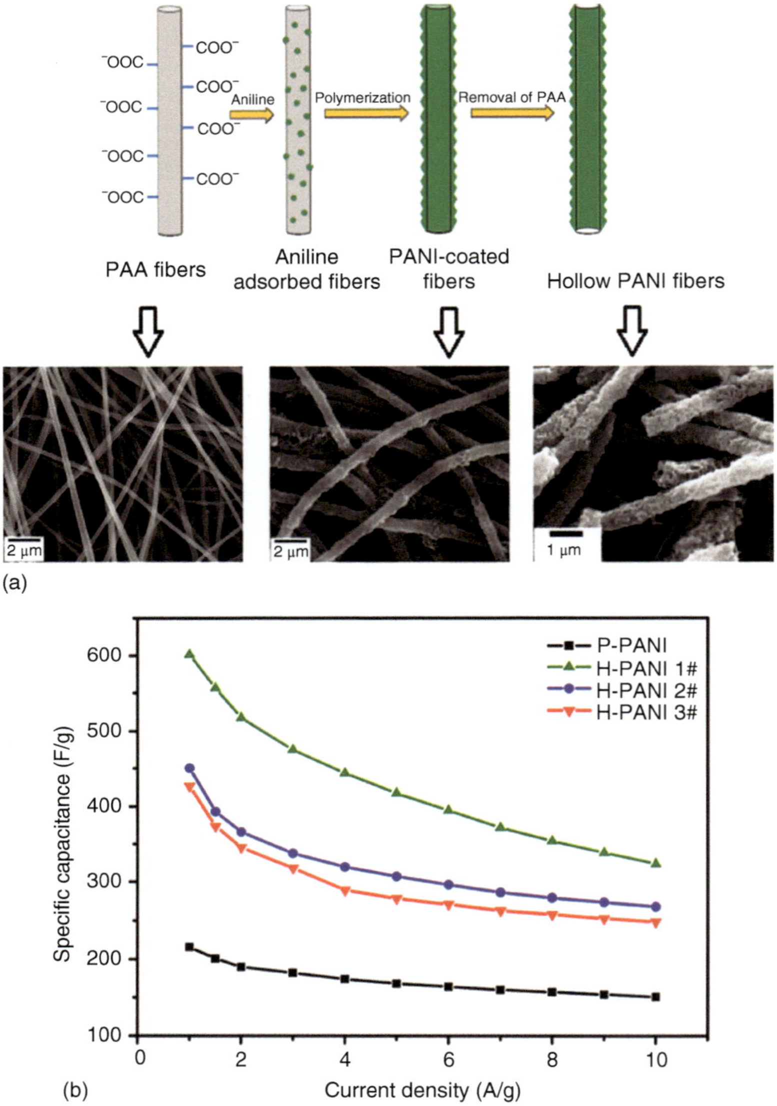

Figure 24.19 (a) Production method and (b) rate capability of a supercapacitor using hollow PANI electrode. Source: Reproduced with permission from Miao et al. [103]. Copyright 2013, American Chemical Society.

reduced charge–carrier recombination. For this purpose, inorganic titania and silica-based nanofibers have been used as photoanodes [106, 107]. Mohamed et al. synthesized one-dimensional $SnO_2@TiO_2$ nanofibers (NFs), which were developed into electrode for DSSCs [107]. The diameters of the prepared nanofibers cover a range from 160 to 360 nm (Figure 24.21a). The photocurrent–voltage curves (J–V) of the cells with $SnO_2@TiO_2$ NFs and pure

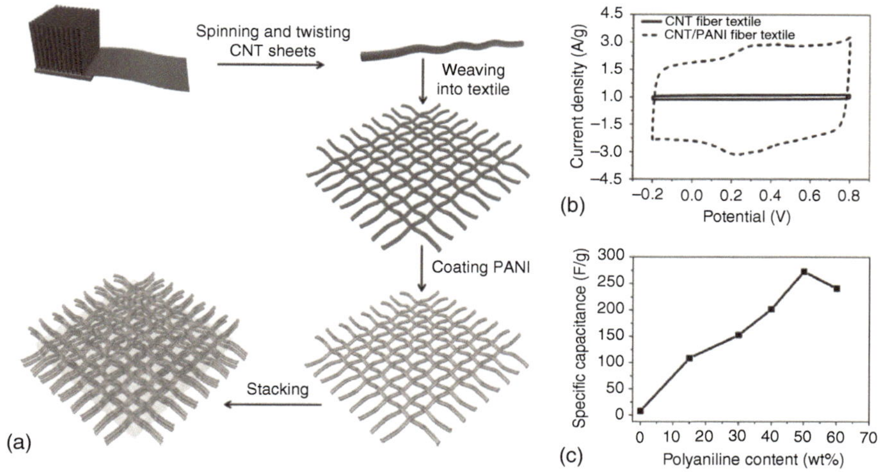

Figure 24.20 (a) Schematic image of preparation of CNT/PANI composite fiber-based textiles used for a supercapacitor; (b) CV curves of CNT fiber textile and CNT/PANI fiber textile; (c) Specific capacitances of CNT/PANI fiber textiles with various PANI weight percentages at a current density of 1 A/g. Source: Reproduced with permission from Pan et al. [104]. Copyright 2015, John Wiley & Sons.

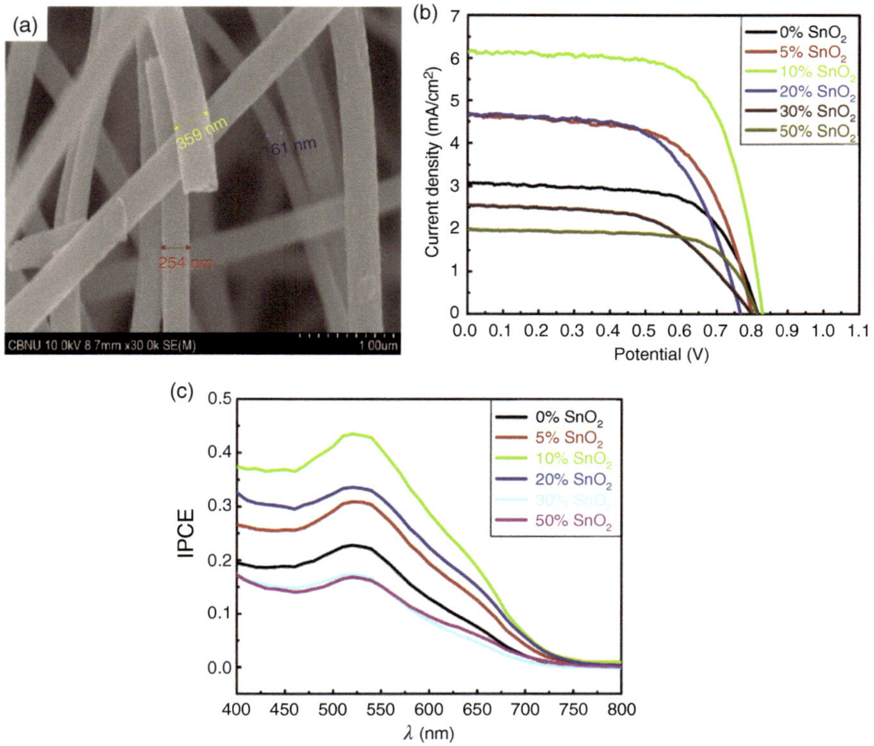

Figure 24.21 (a) SEM image of the SnO_2-TiO_2-NF photoanode containing 10% SnO_2 and comparison of performances of dye-sensitized SnO_2-TiO_2-NF composite photoanodes with different concentrations of SnO_2 (b) J–V curves and (c) IPCE spectra. Source: Reproduced with permission from Mohamed et al. [107]. Copyright 2016, Elsevier.

TiO_2 NFs in comparison are shown in Figure 24.21b. It can be seen that TiO_2 NFs with 10% SnO_2 have the highest current density (6.10 mA/cm^2) with a significant enhancement in the incident photon-to-current conversion efficiency (IPCE) (Figure 24.21c), which is probably because the existence of SnO_2 at a proper concentration (10%) can improve the electron transfer from the dye to the fluorine-tin-oxide (FTO) substrate by reducing the recombination between electrons and the excited dye molecules. Stable, ultrastrong, and flexible aligned CNT fibers were obtained by Chen et al. for DSSC [15]. During the preparation process, a highly aligned CNT fiber (Figure 24.22a,c) was first formed by using a CVD method. The obtained aligned CNT fiber was then dipped into a TiO_2 dispersion, followed by heating at 500 °C for 60 minutes. TiO_2 particles with size of ~25 nm were uniformly attached on the surface of the oriented CNT fiber (Figure 24.22b,d). A high-power conversion efficiency of 2.94% could be achieved by using this unique fiber-shaped DSSC.

In fuel cells, CNFs can serve as support for catalysts that enable ORR and OER. Nitrogen, Pt, Au, or Mn oxides are among the most widely used catalysts incorporated into CNF structures [108, 109]. As shown in Figure 24.23a, Pt-doped porous carbon nanofibers (Pt/N-PCNF) with an average diameter of 150 nm (Figure 24.23b) were successfully fabricated by the synthetic procedure [108]. It is obvious that the electrochemical surface area (ECSA) retention of Pt/N-PCNF is better compared with that of both the commercial one (JM20) and Pt-supported porous CNF (Pt/PCNF) (Figure 24.23c), which is due to the existence of functional anchoring groups as well as the enhanced Pt-orbital

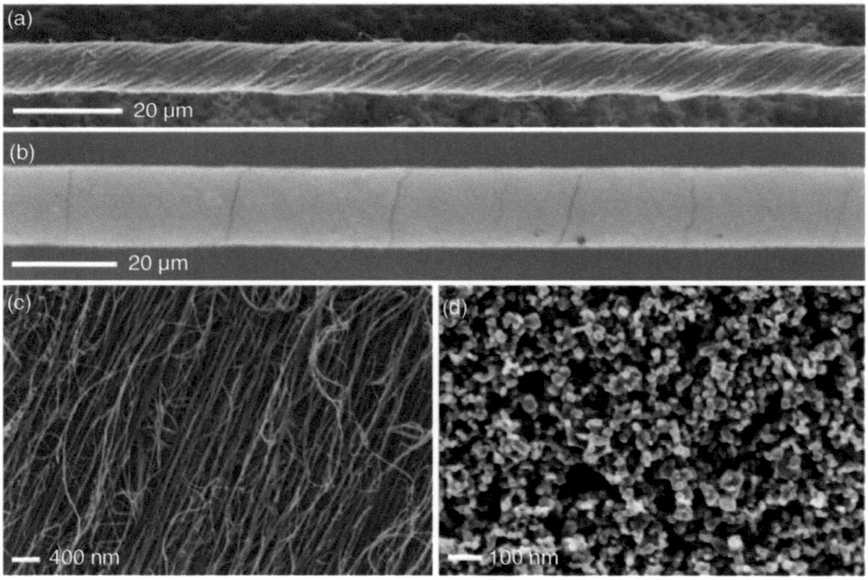

Figure 24.22 SEM images of twisted aligned (a, c) CNT and (b, d) CNT/TiO_2 fibers with different magnifications. Source: Reproduced with permission from Chen et al. [15]. Copyright 2012, American Chemical Society.

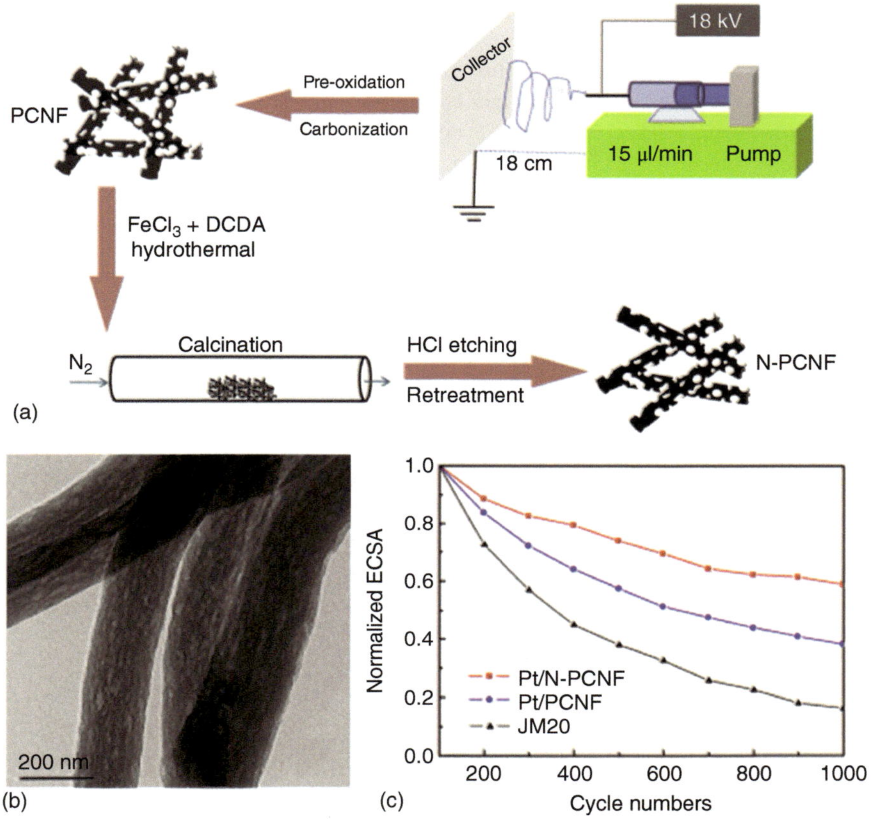

Figure 24.23 (a) Production method of Pt-supported N-doped porous CNF, (b) their morphology, and (c) their electrochemical surface area retention. Source: Reproduced with permission from Wang et al. [108]. Copyright 2016, Elsevier.

bonding strength. Bessel et al. used 5 wt% Pt supported on graphite nanofibers (Pt/GNF) as an electrode for fuel cell applications [109]. The results showed that the activities of the obtained Pt/GNF were comparable with that of about 25 wt% Pt on Vulcan carbon. This performance improvement is probably due to the metal particles, Pt, adopting specific crystallographic orientations when dispersed on the highly tailored GNF structures [109].

Nanogenerators are mechanical energy harvesting devices that use the piezoelectric effect to capture and convert energy [105]. In other words, the mechanical movement of a people or objects can be converted into an electric charge. These devices can use textile fibers, yarns, fabrics, or nanomembranes as electrodes due to their flexibility and mechanical stability [110–113]. A piezoelectric nanogenerator based on lead zirconate titanate (PZT) nanofibers with a diameter and length of about 60 nm and 500 μm was prepared by Chen et al. [114]. This nanogenerator provided a voltage of 1.63 V and a power of 0.03 μW, respectively. Meanwhile, the PZT nanofibers could be protected by embedding in soft polydimethylsiloxane (PDMS), prolonging the cycle life of this nanogenerator.

24.5 Future Trends

Fibers applied as energy materials have been intensively studied. Research interests mainly focus on (i) new materials' exploitation with fibrous structure, (ii) the combination of modification, and (iii) application developments.

Generally, electrode materials having fibrous structures are able to provide more active area and enhance the reaction kinetics. Additionally, if using CNFs as the matrix, it is possible to limit the volume change of the active materials during the cycling process, especially for Si, Sn, metal oxides, etc. To further improve the electrochemical properties in terms of cycling ability and rate capability, many additional modifications on fibrous energy materials have also been highlighted. For example, certain doping agents have been added into the substance that can increase the electrical conductivity and surface properties.

Moreover, some continuous fiber networks possess good mechanical property withstanding bending, stretching, compressing forces in a certain extent, and achieving freestanding electrodes, even flexible energy devices (flexible batteries and supercapacitors).

From another perspective, apart from randomly arranged fibers, they can also be fabricated into 1D aligned or even 2D woven textile architecture for energy use. For further developing, common textiles (such as cotton, polyester, nylon, carbon fiber, etc.) that incorporate activated materials (CNT, graphene, etc.) will be explored.

In summary, there has been great interest in flexible and wearable energy devices for applications in stretchable electronics by using fibrous materials. Enhancement of the performances and cost reduction of those energy storage systems should be achieved for future development.

References

1 Goldemberg, J. (2007). Ethanol for a sustainable energy future. *Science* 315 (5813): 808–810.
2 Zhu, J., Ge, Y., Kim, D. et al. (2016). A novel separator coated by carbon for achieving exceptional high performance lithium-sulfur batteries. *Nano Energy* 20: 176–184.
3 Nema, P., Nema, R.K., and Rangnekar, S. (2009). A current and future state of art development of hybrid energy system using wind and PV-solar: a review. *Renewable and Sustainable Energy Reviews* 13 (8): 2096–2103.
4 Lu, L., Han, X., Li, J. et al. (2013). A review on the key issues for lithium-ion battery management in electric vehicles. *Journal of Power Sources* 226: 272–288.
5 He, C., Wu, S., Zhao, N. et al. (2013). Carbon-encapsulated Fe_3O_4 nanoparticles as a high-rate lithium ion battery anode material. *ACS Nano* 7 (5): 4459–4469.
6 Jiang, H., Lee, P.S., and Li, C. (2013). 3D carbon based nanostructures for advanced supercapacitors. *Energy & Environmental Science* 6 (1): 41–53.

7 Fu, Y., Cai, X., Wu, H. et al. (2012). Fiber supercapacitors utilizing pen ink for flexible/wearable energy storage. *Advanced Materials* 24 (42): 5713–5718.

8 Zeng, W., Shu, L., Li, Q. et al. (2014). Fiber-based wearable electronics: a review of materials, fabrication, devices, and applications. *Advanced Materials* 26 (31): 5310–5336.

9 Wang, X., Liu, B., Liu, R. et al. (2014). Fiber-based flexible all-solid-state asymmetric supercapacitors for integrated photodetecting system. *Angewandte Chemie International Edition* 53 (7): 1849–1853.

10 Saetia, K., Schnorr, J.M., Mannarino, M.M. et al. (2014). Spray-layer-by-layer carbon nanotube/electrospun fiber electrodes for flexible chemiresistive sensor applications. *Advanced Functional Materials* 24 (4): 492–502.

11 Zhu, J., Yanilmaz, M., Fu, K. et al. (2016). Understanding glass fiber membrane used as a novel separator for lithium-sulfur batteries. *Journal of Membrane Science* 504: 89–96.

12 Zhu, J. (2016). Advanced Separator Selection and Design for High-performance Lithium-Sulfur Batteries. PhD Thesis, North Carolina State University, Raleigh, United States.

13 Lv, Z., Yu, J., Wu, H. et al. (2012). Highly efficient and completely flexible fiber-shaped dye-sensitized solar cell based on TiO_2 nanotube array. *Nanoscale* 4 (4): 1248–1253.

14 Liu, G., Peng, M., Song, W. et al. (2015). An 8.07% efficient fiber dye-sensitized solar cell based on a TiO_2 micron-core array and multilayer structure photoanode. *Nano Energy* 11: 341–347.

15 Chen, T., Qiu, L., Cai, Z. et al. (2012). Intertwined aligned carbon nanotube fiber based dye-sensitized solar cells. *Nano Letters* 12 (5): 2568–2572.

16 Wang, S.X., Yang, L., Stubbs, L.P. et al. (2013). Lignin-derived fused electrospun carbon fibrous mats as high performance anode materials for lithium ion batteries. *ACS Applied Materials & Interfaces* 5 (23): 12275–12282.

17 Wu, J., Qin, X., Miao, C. et al. (2016). A honeycomb-cobweb inspired hierarchical core–shell structure design for electrospun silicon/carbon fibers as lithium-ion battery anodes. *Carbon* 98: 582–591.

18 Roy, P. and Srivastava, S.K. (2015). Nanostructured anode materials for lithium ion batteries. *Journal of Materials Chemistry A* 3 (6): 2454–2484.

19 Wang, Z., Qie, L., Yuan, L. et al. (2013). Functionalized N-doped interconnected carbon nanofibers as an anode material for sodium-ion storage with excellent performance. *Carbon* 55: 328–334.

20 Wu, X.L. and Xu, A.W. (2014). Carbonaceous hydrogels and aerogels for supercapacitors. *Journal of Materials Chemistry A* 2 (14): 4852–4864.

21 Shi, F., Li, L., Wang, X.L. et al. (2014). Metal oxide/hydroxide-based materials for supercapacitors. *RSC Advances* 4 (79): 41910–41921.

22 Zhang, X., Ji, L., Toprakci, O. et al. (2011). Electrospun nanofiber-based anodes, cathodes, and separators for advanced lithium-ion batteries. *Polymer Reviews* 51 (3): 239–264.

23 Lee, H., Yanilmaz, M., Toprakci, O. et al. (2014). A review of recent developments in membrane separators for rechargeable lithium-ion batteries. *Energy & Environmental Science* 7: 3857–3886.

24 Zhu, J., Chen, C., Lu, Y. et al. (2015). Nitrogen-doped carbon nanofibers derived from polyacrylonitrile for use as anode material in sodium-ion batteries. *Carbon* 94: 189–195.

25 Ge, Y., Zhu, J., Lu, Y. et al. (2015). The study on structure and electrochemical sodiation of one-dimensional nanocrystalline TiO_2@C nanofiber composites. *Electrochimica Acta* 176: 989–996.

26 Dirican, M., Yildiz, O., Lu, Y. et al. (2015). Flexible binder-free silicon/silica/carbon nanofiber composites as anode for lithium-ion batteries. *Electrochimica Acta* 169: 52–60.

27 Peng, S., Li, L., Lee, J.K.Y. et al. (2016). Electrospun carbon nanofibers and their hybrid composites as advanced materials for energy conversion and storage. *Nano Energy* 22: 361–395.

28 Zhang, C.L. and Yu, S.H. (2014). Nanoparticles meet electrospinning: recent advances and future prospects. *Chemical Society Reviews* 43 (13): 4423–4448.

29 Ji, L. and Zhang, X. (2009). Electrospun carbon nanofibers containing silicon particles as an energy-storage medium. *Carbon* 47 (14): 3219–3226.

30 Ge, Y., Jiang, H., Fu, K. et al. (2014). Copper-doped $Li_4Ti_5O_{12}$/carbon nanofiber composites as anode for high-performance sodium-ion batteries. *Journal of Power Sources* 272: 860–865.

31 Li, Y., Xu, G., Yao, Y. et al. (2014). Coaxial electrospun Si/C-C core–shell composite nanofibers as binder-free anodes for lithium-ion batteries. *Solid State Ionics* 258: 67–73.

32 Ji, L., Lin, Z., Medford, A.J., and Zhang, X. (2009). Porous carbon nanofibers from electrospun polyacrylonitrile/SiO_2 composites as an energy storage material. *Carbon* 47 (14): 3346–3354.

33 Dirican, M. and Zhang, X. (2016). Centrifugally-spun carbon microfibers and porous carbon microfibers as anode materials for sodium-ion batteries. *Journal of Power Sources* 327: 333–339.

34 Ji, L. and Zhang, X. (2009). Fabrication of porous carbon/Si composite nanofibers as high-capacity battery electrodes. *Electrochemistry Communications* 11 (6): 1146–1149.

35 Lu, Y., Fu, K., Zhang, S. et al. (2015). Centrifugal spinning: a novel approach to fabricate porous carbon fibers as binder-free electrodes for electric double-layer capacitors. *Journal of Power Sources* 273: 502–510.

36 Yanilmaz, M., Lu, Y., Zhu, J., and Zhang, X. (2016). Silica/polyacrylonitrile hybrid nanofiber membrane separators via sol–gel and electrospinning techniques for lithium-ion batteries. *Journal of Power Sources* 313: 205–212.

37 Raghavan, P., Zhao, X., Shin, C. et al. (2010). Preparation and electrochemical characterization of polymer electrolytes based on electrospun poly(vinylidene fluoride-co-hexafluoropropylene)/polyacrylonitrile blend/composite membranes for lithium batteries. *Journal of Power Sources* 195 (18): 6088–6094.

38 Zhu, J., Chen, C., Lu, Y. et al. (2016). Highly porous polyacrylonitrile/graphene oxide membrane separator exhibiting excellent anti-self-discharge feature for high-performance lithium–sulfur batteries. *Carbon* 101: 272–280.

39 Yanilmaz, M., Lu, Y., Li, Y., and Zhang, X. (2015). SiO_2/polyacrylonitrile membranes via centrifugal spinning as a separator for Li-ion batteries. *Journal of Power Sources* 273: 1114–1119.
40 Du, J., Shintay, S., and Zhang, X. (2008). Diameter control of electrospun polyacrylonitrile/iron acetylacetonate ultrafine nanofibers. *Journal of Polymer Science Part B: Polymer Physics* 46 (15): 1611–1618.
41 Reneker, D.H. and Chun, I. (1996). Nanometre-diameter-fibres-of-polymer,-produced-by-electrospinning. *Nanotechnology* 7: 216–223.
42 Valizadeh, A. and Farkhani, S.M. (2014). Electrospinning and electrospun nanofibres. *IET Nanobiotechnology* 8 (2): 83–92.
43 Ji, L., Lin, Z., Medford, A.J., and Zhang, X. (2009). In-situ encapsulation of nickel particles in electrospun carbon nanofibers and the resultant electrochemical performance. *Chemistry A European Journal* 15 (41): 10718–10722.
44 Lu, Y., Li, Y., Zhang, S. et al. (2013). Parameter study and characterization for polyacrylonitrile nanofibers fabricated via centrifugal spinning process. *European Polymer Journal* 49 (12): 3834–3845.
45 Marano, S., Barker, S.A., Raimi-Abraham, B.T. et al. (2016). Development of micro-fibrous solid dispersions of poorly water-soluble drugs in sucrose using temperature-controlled centrifugal spinning. *European Journal of Pharmaceutics and Biopharmaceutics* 103: 84–94.
46 Li, Z., Liu, Z., Li, B. et al. (2014). MnO_2 nanosilks self-assembled micropowders: facile one-step hydrothermal synthesis and their application as supercapacitor electrodes. *Journal of the Taiwan Institute of Chemical Engineers* 45 (6): 2995–2999.
47 Zhang, R., Yang, X., Gao, Y. et al. (2016). In-situ preparation of $Li_3V_2(PO_4)_3$/C and carbon nanofibers hierarchical cathode by the chemical vapor deposition reaction. *Electrochimica Acta* 188: 254–261.
48 Waller, G.H., Lai, S.Y., Rainwater, B.H., and Liu, M. (2014). Hydrothermal synthesis of $LiMn_2O_4$ onto carbon fiber paper current collector for binder free lithium-ion battery positive electrodes. *Journal of Power Sources* 251: 411–416.
49 Fu, K., Yildiz, O., Bhanushali, H. et al. (2013). Aligned carbon nanotube-silicon sheets: a novel nano-architecture for flexible lithium ion battery electrodes. *Advanced Materials* 25 (36): 5109–5114.
50 Arico, A.S., Bruce, P., Scrosati, B. et al. (2005). Nanostructured materials for advanced energy conversion and storage devices. *Nature Materials* 4 (5): 366–377.
51 Lu, Y., Chen, C., and Zhang, X. (2015). Functional nanofibers for energy storage. In: *Handbook of Smart Textiles* (ed. X. Tao), 1–28. Springer Singapore.
52 Flandrois, S. and Simon, B. (1999). Carbon materials for lithium-ion rechargeable batteries. *Carbon* 37 (2): 165–180.
53 Kim, C., Yang, K.S., Kojima, M. et al. (2006). Fabrication of electrospinning-derived carbon nanofiber webs for the anode material of lithium-ion secondary batteries. *Advanced Functional Materials* 16 (18): 2393–2397.

54 Raza, A., Wang, J., Yang, S. et al. (2014). Hierarchical porous carbon nanofibers via electrospinning. *Carbon Letters* 15 (1): 1–14.

55 Chatterjee, S., Saito, T., Rios, O., and Johs, A. (2014). Lignin based carbon materials for energy storage applications. In: *Green Technologies for the Environment*, vol. 1186, 203–218. Washington, DC: American Chemical Society.

56 Bai, Y., Wang, Z., Wu, C. et al. (2015). Hard carbon originated from polyvinyl chloride nanofibers as high-performance anode material for na-ion battery. *ACS Applied Materials & Interfaces* 7 (9): 5598–5604.

57 Zou, L., Gan, L., Kang, F. et al. (2010). Sn/C non-woven film prepared by electrospinning as anode materials for lithium ion batteries. *Journal of Power Sources* 195 (4): 1216–1220.

58 Qie, L., Chen, W.-M., Wang, Z.-H. et al. (2012). Nitrogen-doped porous carbon nanofiber webs as anodes for lithium ion batteries with a superhigh capacity and rate capability. *Advanced Materials* 24 (15): 2047–2050.

59 Ji, L., Lin, Z., Alcoutlabi, M., and Zhang, X. (2011). Recent developments in nanostructured anode materials for rechargeable lithium-ion batteries. *Energy & Environmental Science* 4 (8): 2682–2699.

60 Li, S., Chen, C., Fu, K. et al. (2014). Comparison of Si/C, Ge/C and Sn/C composite nanofiber anodes used in advanced lithium-ion batteries. *Solid State Ionics* 254: 17–26.

61 Fu, K., Lu, Y., Dirican, M. et al. (2014). Chamber-confined silicon–carbon nanofiber composites for prolonged cycling life of Li-ion batteries. *Nanoscale* 6 (13): 7489–7495.

62 Chen, Y., Hu, Y., Shao, J. et al. (2015). Pyrolytic carbon-coated silicon/carbon nanofiber composite anodes for high-performance lithium-ion batteries. *Journal of Power Sources* 298: 130–137.

63 Dirican, M., Lu, Y., Fu, K. et al. (2015). SiO_2-confined silicon/carbon nanofiber composites as an anode for lithium-ion batteries. *RSC Advances* 5 (44): 34744–34751.

64 Lin, H., Weng, W., Ren, J. et al. (2014). Twisted aligned carbon nanotube/silicon composite fiber anode for flexible wire-shaped lithium-ion battery. *Advanced Materials* 26 (8): 1217–1222.

65 Jiang, H., Ge, Y., Fu, K. et al. (2014). Centrifugally-spun tin-containing carbon nanofibers as anode material for lithium-ion batteries. *Journal of Materials Science* 50 (3): 1094–1102.

66 Lu, Y., Yanilmaz, M., Chen, C. et al. (2015). Centrifugally spun SnO_2 microfibers composed of interconnected nanoparticles as the anode in sodium-ion batteries. *ChemElectroChem* 2 (12): 1947–1956.

67 Kim, J.-C. and Kim, D.-W. (2014). Electrospun Cu/Sn/C nanocomposite fiber anodes with superior usable lifetime for lithium- and sodium-ion batteries. *Chemistry - An Asian Journal* 9 (11): 3313–3318.

68 Nitta, N. and Yushin, G. (2014). High-capacity anode materials for lithium-ion batteries: choice of elements and structures for active particles. *Particle and Particle Systems Characterization* 31 (3): 317–336.

69 Dirican, M., Yanilmaz, M., Fu, K. et al. (2014). Carbon-enhanced electrodeposited SnO_2/carbon nanofiber composites as anode for lithium-ion batteries. *Journal of Power Sources* 264: 240–247.

70 Shen, Z., Hu, Y., Chen, Y. et al. (2016). Controllable synthesis of carbon-coated Sn–SnO_2–carbon-nanofiber membrane as advanced binder-free anode for lithium-ion batteries. *Electrochimica Acta* 188: 661–670.

71 Dirican, M., Lu, Y., Ge, Y. et al. (2015). Carbon-confined SnO_2-electrodeposited porous carbon nanofiber composite as high-capacity sodium-ion battery anode material. *ACS Applied Materials & Interfaces* 7 (33): 18387–18396.

72 Luo, L., Xu, W., Xia, Z. et al. (2016). Electrospun ZnO–SnO_2 composite nanofibers with enhanced electrochemical performance as lithium-ion anodes. *Ceramics International* 42 (9): 10826–10832.

73 Ji, L., Toprakci, O., Alcoutlabi, M. et al. (2012). α-Fe_2O_3 nanoparticle-loaded carbon nanofibers as stable and high-capacity anodes for rechargeable lithium-ion batteries. *ACS Applied Materials & Interfaces* 4 (5): 2672–2679.

74 Zhu, J., Lu, Y., Chen, C. et al. (2016). Porous one-dimensional carbon/iron oxide composite for rechargeable lithium-ion batteries with high and stable capacity. *Journal of Alloys and Compounds* 672: 79–85.

75 Chen, R., Hu, Y., Shen, Z. et al. (2016). Controlled synthesis of carbon nanofibers anchored with $Zn_xCo_{3-x}O_4$ nanocubes as binder-free anode materials for lithium-ion batteries. *ACS Applied Materials & Interfaces* 8 (4): 2591–2599.

76 Pampal, E.S., Stojanovska, E., Simon, B., and Kilic, A. (2015). A review of nanofibrous structures in lithium ion batteries. *Journal of Power Sources* 300: 199–215.

77 Whittingham, M.S. (2004). Lithium batteries and cathode materials. *Chemical Reviews* 104 (10): 4271–4302.

78 Toprakci, O., Toprakci, H.A.K., Ji, L. et al. (2012). Carbon nanotube-loaded electrospun $LiFePO_4$/carbon composite nanofibers as stable and binder-free cathodes for rechargeable lithium-ion batteries. *ACS Applied Materials & Interfaces* 4 (3): 1273–1280.

79 Toprakci, O., Toprakci, H.A.K., Ji, L. et al. (2012). $LiFePO_4$ nanoparticles encapsulated in graphene-containing carbon nanofibers for use as energy storage materials. *Journal of Renewable and Sustainable Energy* 4 (1): 013121.

80 Zhang, S., Lu, Y., Xu, G. et al. (2012). LiF/Fe/C nanofibres as a high-capacity cathode material for Li-ion batteries. *Journal of Physics D: Applied Physics* 45 (39): 395301.

81 Zhang, S., Lin, Z., Ji, L. et al. (2012). Cr-doped Li_2MnSiO_4/carbon composite nanofibers as high-energy cathodes for Li-ion batteries. *Journal of Materials Chemistry* 22 (29): 14661–14666.

82 Zhang, S., Li, Y., Xu, G. et al. (2012). High-capacity $Li_2Mn_{0.8}Fe_{0.2}SiO_4$/carbon composite nanofiber cathodes for lithium-ion batteries. *Journal of Power Sources* 213: 10–15.

83 Lu, Y., Yanilmaz, M., Chen, C. et al. (2015). Lithium-substituted sodium layered transition metal oxide fibers as cathodes for sodium-ion batteries. *Energy Storage Materials* 1: 74–81.
84 Zeng, L., Yao, Y., Shi, J. et al. (2016). A flexible $S_{1-x}Se_x$@porous carbon nanofibers ($x \leq 0.1$) thin film with high performance for Li-S batteries and room-temperature Na-S batteries. *Energy Storage Materials* 5: 50–57.
85 Liang, G., Wu, J., Qin, X. et al. (2016). Ultrafine TiO_2 decorated carbon nanofibers as multifunctional interlayer for high-performance lithium-sulfur battery. *ACS Applied Materials & Interfaces* 8 (35): 23105–23113.
86 Zhang, Z., Wang, G., Lai, Y., and Li, J. (2016). A freestanding hollow carbon nanofiber/reduced graphene oxide interlayer for high-performance lithium-sulfur batteries. *Journal of Alloys and Compounds* 663: 501–506.
87 Wang, Z., Zhang, J., Yang, Y. et al. (2016). Flexible carbon nanofiber/polyvinylidene fluoride composite membranes as interlayers in high-performance lithium sulfur batteries. *Journal of Power Sources* 329: 305–313.
88 Fu, K., Li, Y., Dirican, M. et al. (2014). Sulfur gradient-distributed CNF composite: a self-inhibiting cathode for binder-free lithium-sulfur batteries. *Chemical Communications* 50 (71): 10277–10280.
89 Kim, D.S. and Park, Y.J. (2014). Buckypaper electrode containing carbon nanofiber/Co_3O_4 composite for enhanced lithium air batteries. *Solid State Ionics* 268: 216–221.
90 Xue, H., Mu, X., Tang, J. et al. (2016). A nickel cobaltate nanoparticle-decorated hierarchical porous N-doped carbon nanofiber film as a binder-free self-supported cathode for nonaqueous Li–O_2 batteries. *Journal of Materials Chemistry A* 4 (23): 9106–9112.
91 Zhang, G.Q., Zheng, J.P., Liang, R. et al. (2011). α-MnO_2/carbon nanotube/carbon nanofiber composite catalytic air electrodes for rechargeable lithium-air batteries. *Journal of the Electrochemical Society* 158 (7): A822–A827.
92 Alcoutlabi, M., Lee, H., Watson, J.V., and Zhang, X. (2012). Preparation and properties of nanofiber-coated composite membranes as battery separators via electrospinning. *Journal of Materials Science* 48 (6): 2690–2700.
93 Yanilmaz, M., Lu, Y., Dirican, M. et al. (2014). Nanoparticle-on-nanofiber hybrid membrane separators for lithium-ion batteries via combining electrospraying and electrospinning techniques. *Journal of Membrane Science* 456: 57–65.
94 Yanilmaz, M., Dirican, M., and Zhang, X. (2014). Evaluation of electrospun SiO_2/nylon 6,6 nanofiber membranes as a thermally-stable separator for lithium-ion batteries. *Electrochimica Acta* 133: 501–508.
95 Zhu, J., Yildirim, E., Aly, K. et al. (2016). Hierarchical multi-component nanofiber separators for lithium polysulfide capture in lithium-sulfur batteries: an experimental and molecular modeling study. *Journal of Materials Chemistry A* 4 (35): 13572–13581.
96 Yanilmaz, M. and Zhang, X. (2015). Polymethylmethacrylate/polyacrylonitrile membranes via centrifugal spinning as separator in Li-ion batteries. *Polymers* 7 (4): 629–643.

97 Zhang, X. and Lu, Y. (2014). Centrifugal spinning: an alternative approach to fabricate nanofibers at high speed and low cost. *Polymer Reviews* 54 (4): 677–701.

98 Lv, R., Zhu, Y., Liu, H. et al. (2016). Poly(vinylidene fluoride)/poly(acrylonitrile) blend fibrous membranes by centrifugal spinning for high-performance lithium ion battery separators. *Journal of Applied Polymer Science* 134 (8): 4451511–7.

99 Luo, X., Wang, J., Dooner, M., and Clarke, J. (2015). Overview of current development in electrical energy storage technologies and the application potential in power system operation. *Applied Energy* 137: 511–536.

100 Wei, K. and Kim, I.S. (2014). Application of nanofibers in supercapacitors. In: *Electrospun Nanofibers for Energy and Environmental Applications* (eds. Ding, B. and Yu, J.), 163–181. Springer.

101 Lai, F., Miao, Y.-E., Huang, Y. et al. (2015). Flexible hybrid membranes of $NiCo_2O_4$-doped carbon nanofiber@MnO_2 core–sheath nanostructures for high-performance supercapacitors. *The Journal of Physical Chemistry C* 119 (24): 13442–13450.

102 Wang, J.-G., Yang, Y., Huang, Z.-H., and Kang, F. (2011). Coaxial carbon nanofibers/MnO_2 nanocomposites as freestanding electrodes for high-performance electrochemical capacitors. *Electrochimica Acta* 56 (25): 9240–9247.

103 Miao, Y.-E., Fan, W., Chen, D., and Liu, T. (2013). High-performance supercapacitors based on hollow polyaniline nanofibers by electrospinning. *ACS Applied Materials & Interfaces* 5 (10): 4423–4428.

104 Pan, S., Lin, H., Deng, J. et al. (2015). Novel wearable energy devices based on aligned carbon nanotube fiber textiles. *Advanced Energy Materials* 5 (4): 1401438.

105 Shi, X., Zhou, W., Ma, D. et al. (2015). Electrospinning of nanofibers and their applications for energy devices. *Journal of Nanomaterials* 16 (1): 122.

106 Macdonald, T.J., Tune, D.D., Dewi, M.R. et al. (2015). A TiO_2 nanofiber–carbon nanotube-composite photoanode for improved efficiency in dye-sensitized solar cells. *ChemSusChem* 8 (20): 3396–3400.

107 Mohamed, I.M.A., Dao, V.-D., Yasin, A.S. et al. (2016). Synthesis of novel SnO_2@TiO_2 nanofibers as an efficient photoanode of dye-sensitized solar cells. *International Journal of Hydrogen Energy* 41 (25): 10578–10589.

108 Wang, Y., Jin, J., Yang, S. et al. (2016). Nitrogen-doped porous carbon nanofiber based oxygen reduction reaction electrocatalysts with high activity and durability. *International Journal of Hydrogen Energy* 41 (26): 11174–11184.

109 Bessel, C.A., Laubernds, K., Rodriguez, N.M., and Baker, R.T.K. (2001). Graphite nanofibers as an electrode for fuel cell applications. *The Journal of Physical Chemistry B* 105 (6): 1115–1118.

110 Sung, M.-T., Chang, M.-H., and Ho, M.-H. (2014). Investigation of cathode electrocatalysts composed of electrospun Pt nanowires and Pt/C for proton exchange membrane fuel cells. *Journal of Power Sources* 249: 320–326.

111 Delmondo, L., Salvador, G.P., Muñoz-Tabares, J.A. et al. (2016). Nanostructured Mn_xO_y for oxygen reduction reaction (ORR) catalysts. *Applied Surface Science* 388: 631–639.

112 Pu, X., Li, L., Song, H. et al. (2015). A self-charging power unit by integration of a textile triboelectric nanogenerator and a flexible lithium-ion battery for wearable electronics. *Advanced Materials* 27 (15): 2472–2478.

113 Jost, K., Dion, G., and Gogotsi, Y. (2014). Textile energy storage in perspective. *Journal of Materials Chemistry A* 2 (28): 10776–10787.

114 Chen, X., Xu, S., Yao, N., and Shi, Y. (2010). 1.6 V nanogenerator for mechanical energy harvesting using PZT nanofibers. *Nano Letters* 10 (6): 2133–2137.

25

Fiber-Based Sensors and Actuators

Xiaomeng Fang, Kony Chatterjee, Ashish Kapoor, and Tushar Ghosh

Wilson College of Textiles, North Carolina State University, 1020 Main Campus Drive, Raleigh, NC 27606, USA

25.1 Introduction

Wearable electronic devices are increasingly becoming a part of our daily lives. Commercial wearable devices currently available range from smart glasses with audiovisual integration to athletic apparel for real-time biometric tracking. As the field of flexible electronics progresses, there is significant research underway to integrate wearable and other electronic capabilities into textiles. The motivation is obvious since textiles are deployed over a vast number of surfaces including the human body. The increasing use of textiles for clothing and beyond is due to its unique and desirable properties derived mostly from their hierarchical structure with fibers as building blocks. Fiber-based "wearables" are inherently advantageous because they combine breathability, conformability, strength, and stability of textiles with electrical functionalities. Arguably, the future seems to point toward a system with electronic capabilities integrated into fiber-based textile substrates, or electronic textiles. Sensors and actuators are likely to become integral part of this emerging system. A sensor detects a change in any stimulus into a measurable signal, similar to human sensory organs [1, 2]. Actuators are used to convert a form of input energy into mechanical motion. As an illustration, the concept of an autonomous closed loop system for human thermophysiological comfort with sensors and actuators is shown in Figure 25.1.

This chapter focuses particularly on the working principles, fabrication processes, and applications of fiber-based sensors and actuators reported in the literature. The operating principles, performance, and the remaining challenges associated with these technologies are described. While sensory and actuation properties could be built into a monolithic fiber, it can also be developed in an assembly of fibers. For this discussion, both types are included with a particular focus on sensors and actuators that have been designed in the form of fibers. Additionally, a less restrictive definition of a fiber being a flexible cylindrical body with a high aspect ratio and diameter in the range of hundreds of microns is used.

Handbook of Fibrous Materials, First Edition. Edited by Jinlian Hu, Bipin Kumar, and Jing Lu.

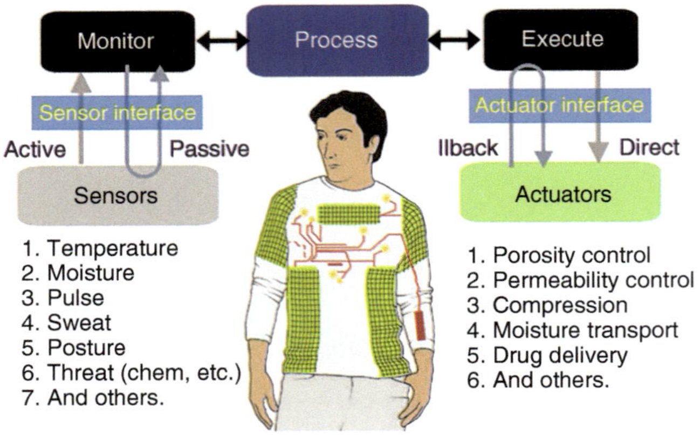

Figure 25.1 Illustration of a closed loop of system of integrated sensors and actuators.

25.2 Fibers as Actuators and Sensors

The traditional electromagnetic, pneumatic, and hydraulic actuators, currently used extensively in many applications, are not particularly suitable for wearable or flexible electronic applications. Material-based actuation technologies with high power to mass ratio such as piezoelectric ceramics, shape memory alloys (SMAs), and electroactive polymers (EAPs) seem to offer properties that are more attractive when flexibility and wearability are required. Biological muscles work for billions of cycles to generate about 150–300 kPa stress and 25% strain driven by complex mechanisms with a very short response time (milliseconds) [3, 4]. Hence, they have always been used as a benchmark to evaluate artificial material-based actuators. Interestingly, the natural muscle consists of hierarchical, aligned in parallel, robust, and resilient fibers [3, 5]. Hence, fiber-shaped actuators have the potential to mimic biological muscles. It is also important to note that some materials that are used as actuators are also capable of sensing modalities and hence can act as self-sensing actuators.

Sensors in the form of fibers in a textile substrate can be used in intimate contact with the human body via clothing or as sensors in large areas of deployment, for example, on tents, parachutes, etc. Fiber sensors are inherently advantageous because they can help retain the desirable properties of textiles – breathability, conformability, and strength – while simultaneously offering a sensory response. In the broadest sense, a sensor is a device that can detect changes in its environment and provide a usable output suitable for processing. In addition to the selective sensitivity toward a specific stimulus, important sensor characteristics include fast response, repeatability, reliability, and low hysteresis [6]. Apart from fiber-optic sensors, the concept of fiber sensors is relatively new. In this chapter, sensors are categorized based on their sensing modalities, i.e. strain sensing, humidity and temperature sensing, and chemical sensing. The discussion that follows covers the fiber-sensor concepts proposed in the literature mainly in the context of flexible electronics and electronic textiles. It is important to remember that very few of these have resulted in the commercial production of devices except for fiber-optic gyroscopes, temperature, pressure, and vibration sensors.

25.3 Fundamental Principle and Types of Fiber Actuators

Fiber-based actuators convert input energy, in the form of thermal, electrical, magnetic, or others into mechanical motion. These actuators differ from traditional rigid actuators by exhibiting a high degree of conformability and offering flexibility in their material selection and synthesis [7].

To comprehensively summarize relevant technologies and developments, this chapter includes most of the fiber-shaped devices with a large aspect ratio as well as those in tubular structures with relatively large dimensions (beyond macro), different from traditional fibers. The information about actuators in this chapter is classified and organized based on the materials used in the device.

25.3.1 Shape Memory Materials

These constitute a class of smart materials that can remember their original shape/size after significant and seemingly inelastic deformation when subjected to a stimulus such as temperature or magnetic field. Shape memory materials include shape memory alloys (SMA) and shape memory polymers (SMP). The most investigated SMA is the nickel–titanium (NiTi) alloy known as Nitinol. Typical SMA fiber actuators are relatively larger than traditional textile fibers with diameters in the range of 100–400 μm. Elastic and flexible SMA filaments have been plied with commonly used textile fibers and subsequently woven and knitted into fabrics with controllable properties [8]. SMPs are generally considered more suitable for textiles because of their flexibility, lightweight, and better processability. SMPs have a much higher actuation strain of 400–800% with substantially low densities (0.9–1.25 g/cm^3 compared with 6–8 g/cm^3 for SMAs) [9].

SMPs can be actuated using thermal, chemical, and optical stimuli. Thermally activated SMPs are the most common and are generally designed to maintain a temporary shape below a transition temperature (T_{trans}) and actuate to their original shape as the temperature increases above T_{trans}. Polymer chains tend to stay in their random coiled status as long as they are constrained by cross-links or temporary bonding. In case of chemical cross-links, SMPs are phase separated into hard and soft segments that act as the frozen and reversible phases, respectively [10]. Many polymeric materials could be used as segments in SMPs, for example, the most commonly used is polyurethane (PU) based.

The fabrication processes for SMP fibers can be similar to those of traditional polymeric fibers, such as wet spinning [11, 12], melting spinning [13, 14], and electrospinning [15]. In addition to monofilaments, SMP fibers can also be processed into yarns and woven/knitted/nonwoven fabrics [16]. However, due to

Table 25.1 Summary of shape memory polymer-based fiber actuators.

Materials and dimension	Actuation stimuli	Maximum linear strain	Applications
PU[a)]-based SMP Diameter: 200–500 μm [16–18]	Heat, 50 °C	N/A	• To improve textile aesthetics with "lively" effects • Breathable fabrics
PU-MWCNT[b)]composite fiber Diameter: ~250 μm [13]	Heat, 70 °C	Fixed strain to 100% and shape recovery ratio up to 90%	• Smart textiles • Biomedical materials • High-performance actuators, e.g. microgrippers
Triblock liquid-crystal polymers Diameter: ~200 μm [19]	Heat	Strain up to 500%	• Smart textiles • Surgical technology • Artificial muscles
PU[a)]-based SMP Diameter: N/A [15]	Heat	Fixed strain to 100% and shape recovery ratio up to 90%	• Smart textiles

a) Polyurethane.
b) Multiwalled carbon nanotube.

their unusual tactile properties, SMP fibers are generally incorporated with other types of fibers. In SMP-based textiles, the fabric structure also has a considerable influence on the shape memory effect [15].

Table 25.1 describes some SMP fiber actuators proposed in the literature. Typical SMP fiber diameters range from 200 to 500 μm. Nanoscale SMP fibers have also been fabricated with diameters ranging from 50 to 700 nm via electrospinning [20]. SMP fiber actuators offer a vast array of applications for textile- and clothing-related products, such as interactive shape changeable fabrics for windows or wall hangings, breathable fabrics, shoes, and crease- and shrink-resistant apparels. SMP fibers/fabrics are also used in biomedical application such as implants, scaffolds for tissue constructs, and sutures [10]. Shape memory material-based fiber actuators (refer Figure 25.2) offer new vistas for smart textiles, fashion garments, robotics, and biomedical product design [9].

25.3.2 Piezoelectric Materials

Piezoelectric effect is an electromechanical phenomenon wherein materials produce an electric voltage in response to a mechanical stimulation. The inverse piezoelectric effect, strain induced due to applied voltage, forms the principle of actuation of piezoelectric actuators. Piezoelectric effect is generated due to the reorientation or "poling" of randomly orientated neutral-state dipoles in the material to form an anisotropic structure as well as a dipole moment. When the

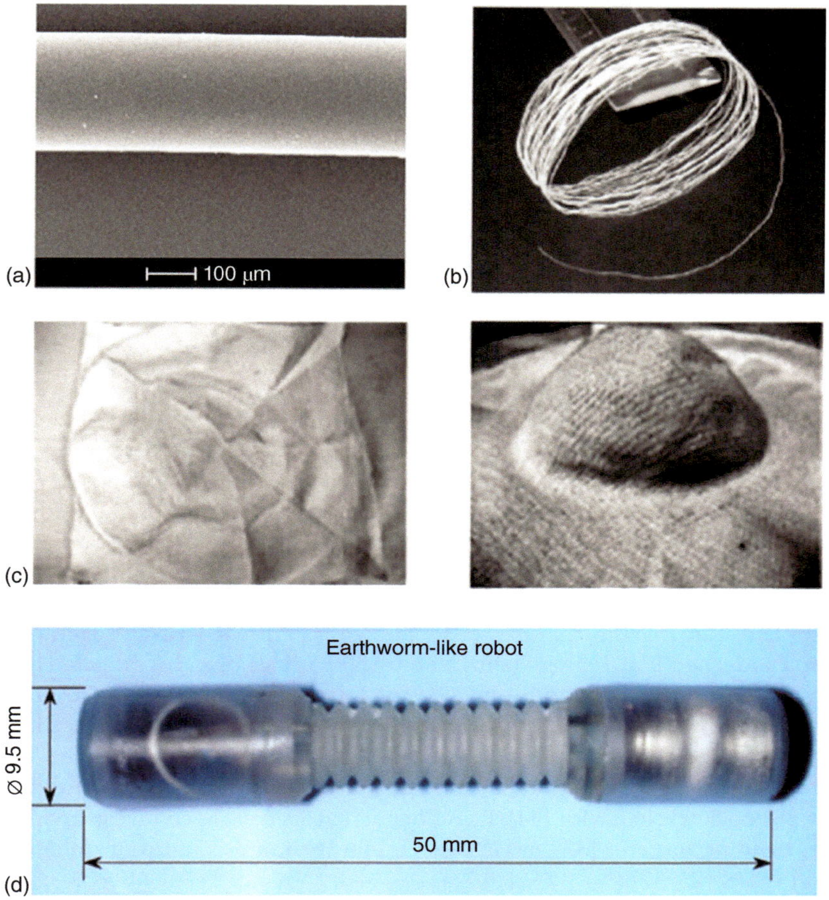

Figure 25.2 Shape memory material-based fiber actuators. (a) Scanning electron microscopy (SEM) image of fiber actuator based on polyurethane multiwalled carbon nanotube composites. Source: Meng and Hu 2008 [13]. Reproduced with permission of Elsevier. (b) Roll of triblock liquid-crystal polymer-based actuator. Source: Ahir et al. 2006 [19]. Reproduced with permission of John Wiley & Sons. (c) Creased shape and recovered smooth knitted SMP fabrics. Source: Hu and Chen 2010 [15]. Reproduced with permission of Royal Society of Chemistry. (d) SMA fiber muscles used in worm-like robot. Source: Kim et al. 2006 [21]. Reproduced with permission of Elsevier.

poled material is subjected to an electric voltage, it may show either contractional or elongational deformation due to reconfiguration of its dipoles [22]. Lead zirconate titanate (PZT) polycrystalline ceramic and polyvinylidene difluoride (PVDF) polymers are the most commonly used piezoelectric materials [22]. Piezoelectric actuators offer many benefits such as compact structures (usually multilayered), precision response, short response time, vacuum and clean room processabilities, and the possibility of operation at cryogenic temperatures [23]. However, they have fragile mechanical properties and very limited actuation strains especially for the ceramic actuators.

Piezoelectric fiber-based actuators (refer Figure 25.3) have been extensively used in various applications such as optical device control, reinforcements in composites, miniaturized rotary motors, controls for medical devices, and artificial muscles, see Table 25.2. Most piezoelectric fiber diameters are in the range of 100–200 μm, with some extremely wide fibers that can go up to 1000 μm. Typical driving electric voltages for actuation are in the order of hundreds of volts. Interestingly, piezoelectric fibers show extremely small linear actuation strains – substantially less than 1%. Piezoelectric fiber composites actuators have been developed [27, 41, 42] for applications in tunable Bragg gratings, helicopter motor blades, tail fins for fighter aircrafts, and inflatable spacecrafts.

Compared with their inorganic counterparts, piezoelectric polymers are lighter, less expensive, easiest to process, and offer better flexibility and deformability [43]. PVDF and its copolymers are well-known piezoelectric polymers with a high-energy conversion factor between its electrical and mechanical domains. Similar to piezoceramics, a poling process is required during fabrication of PVDF to achieve piezoelectricity. Electrospinning has been used to spin and pole PVDF fibers simultaneously to produce fibers of diameters in the range of hundreds of nanometers. Conventional electrospinning process normally produces randomly distributed fiber webs, and thus their piezoelectric response is not significant. Near-field electrospinning (NFES) has been developed to spin fiber in more ordered and controllable manner (also known as "direct-writing" method). Single or patterned PVDF fibers fabricated by NFES have been investigated and explored in actuator applications [26, 39]. The L-isomer of polylactic acid (PLA) and poly-L-lactic acid (PLLA) molecule with its asymmetric structure is a chiral polymer that requires no polling like PVDF to exhibit its piezoelectric behavior. Due to its chirality, deformation by shear in the drawing process is required to display piezoelectric behavior [44]. PLA fiber actuators have been investigated to work as bending actuator [37] and in tweezers for thrombosis sample exfoliation in blood vessel [25, 37].

25.3.3 Electrically Conducting Polymers

Electrically conducting polymers (CPs) can be used as actuators due to dimensional changes resulting from electromechanical ion insertion and removal. CP actuators offer many advantages, including being lightweight and low cost and having high-output stresses and relatively low operating voltages [45]. Among a wide range of available CPs, polypyrrole (PPy) and polyaniline (PANI) have been extensively investigated for their actuation properties. The strain generated by PPy and PANI actuators is usually small, and their performance is limited by a relatively slow response speed [46] and very low energy conversion efficiencies. Most CP actuators require a surrounding aqueous electrolyte, and hence solid or gel electrolytes have been developed to broaden the utility of CP actuators. Fabrication of CP filaments is usually via wet spinning wherein CP-based solution is extruded through the spinneret and the resultant as-spun fibers are kept in a liquid bath to remove the solvent [47–49]. Electrospinning has also been employed to spin CP fibers [50]. The obvious advantage of CPs over electronic EAPs is their low operating voltage.

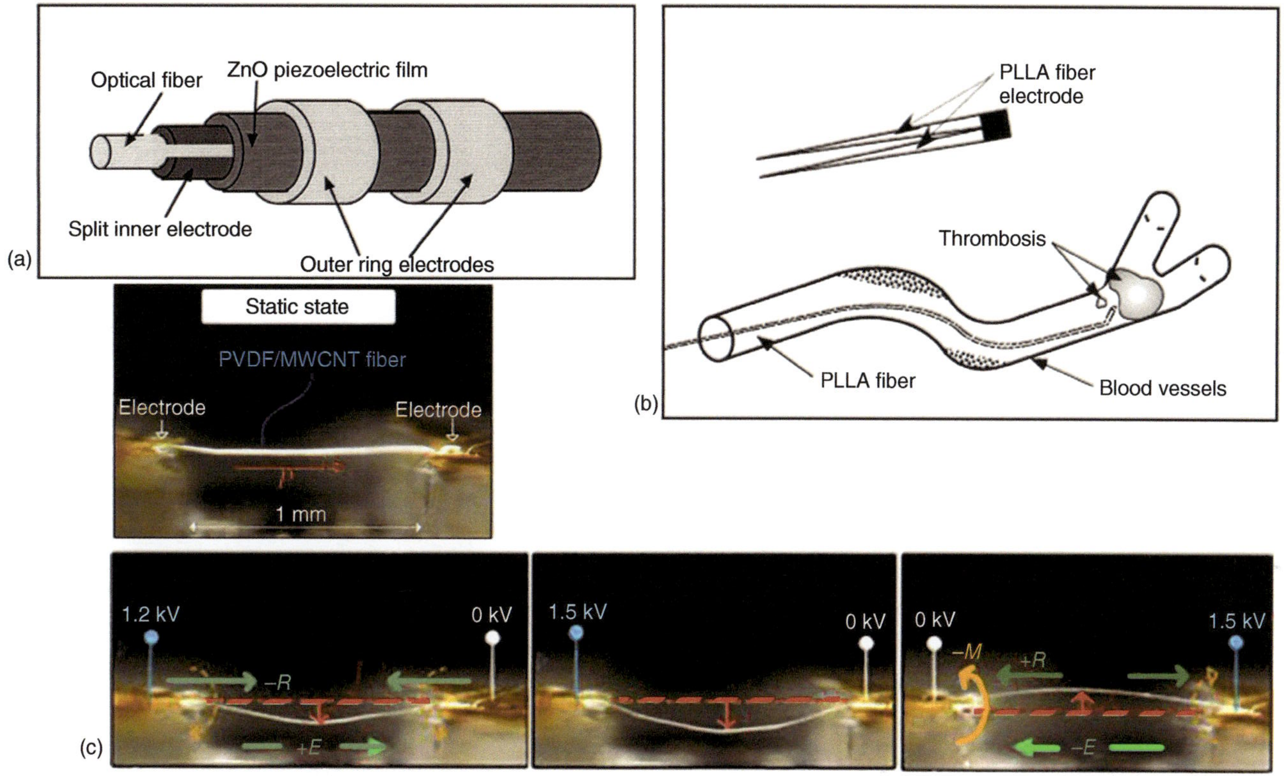

Figure 25.3 Piezoelectric material-based fiber actuators. (a) Hollow piezoelectric fibers actuator with optical fiber core. Source: Trolier-McKinstry et al. 1999 [24]. Reproduced with permission of Elsevier. (b) Piezoelectric poly(L-lactic acid) (PLLA) fiber tweezer and catheter. Source: Tajitsu 2006 [25]. Reproduced with permission of John Wiley & Sons. (c) PVDF/MWCNT composite fiber bending actuator. Source: Liu et al. 2013 [26]. Reproduced with permission of Elsevier.

Table 25.2 Summary of piezoelectric material-based fiber actuators.

Materials and dimension	Actuation stimuli	Maximum linear strain	Applications
PZT, PMN-PT[a)] single-crystal fibers Diameter: 120–250 μm [27–31]	100 V–3 kV	1.5×10^{-4}% to 1.25×10^{-3}%	• Piezoelectric fiber reinforced composite actuator
Hollow ZnO fiber Diameter: ∼140 μm [24]	AC: 0.5–5 V	4.5×10^{-6}%	• Control optical fiber located in the core of the hollow piezoelectric fiber actuator
PZT fiber Diameter: 1000 μm [32–36]	AC: −500 to 500 V	Torsion angle: 1.7°	• As rotary motor • To control miniature orthogonal optical scanning mirror
Chiral poly(L-lactic acid) (PLLA) polymer fiber Diameter: 20–100 μm [25, 37, 38]	AC/DC: −300 to 800 V	2.2×10^{-5}%	• To control catheter and tweezers for thrombosis sample exfoliation in blood vessels
PVDF or PVDF/CNT composite fibers Diameter: 2.6–35 μm [26, 39, 40]	DC: 1200–1500 V	1.9×10^{-3}% to 9.7×10^{-2}%	• Artificial muscles and switches

a) Lead magnesium niobate-lead titanate.

CP fiber actuators (refer Figure 25.4) reported in the literature are summarized in Table 25.3. Their main applications are in biomedical devices, such as microcatheters, and in robotic devices, such as artificial muscles. Their diameters range from 0.9 to 2000 μm. Fibers with very larger diameters are referred to as "dry" state actuators, which are assembled with a gel or solid electrolyte as the layer or core.

25.3.4 Dielectric Electroactive Polymers

Dielectric electroactive polymers (D-EAPs) constitute a class of electronic EAPs, which offer a tremendous potential as flexible actuators. They are lightweight and low cost and can generate large strains at high frequencies. In a typical actuator configuration, a layer of soft D-EAP is sandwiched between two layers of compliant electrodes. When an electric potential is applied across the electrodes, the attractive force between the opposite charges and the repulsive force between the like charges in the compliant electrode generates a compressive force on the D-EAP sandwich layer. This compressive force normalized over the area of the D-EAP is known as "Maxwell stress" [3]. Silicones [56, 57], acrylics [58, 59], and PU [60, 61] are the most commonly used D-EAP materials. Fiber-like D-EAP actuators were first developed by SRI International (SRI) [62], and subsequently

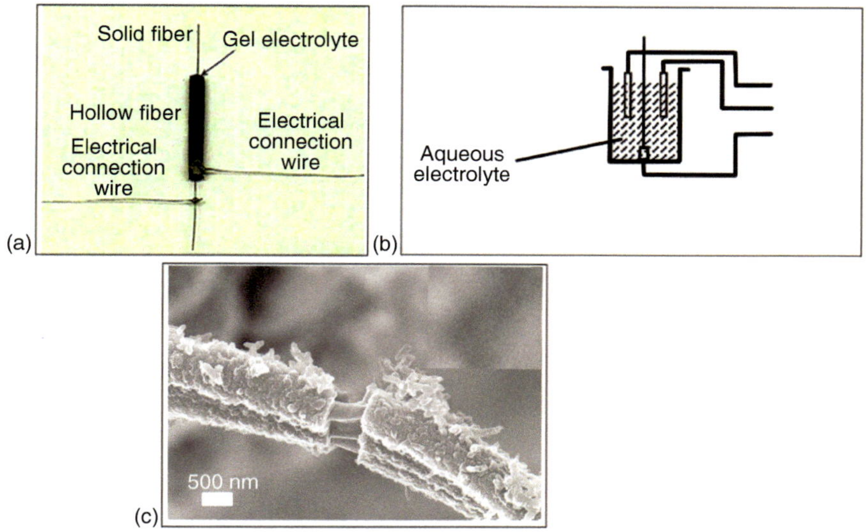

Figure 25.4 Electrically conductive polymer-based fiber actuators. (a) Solid-in-hollow fiber all polymer linear actuator. Source: Lu et al. 2004 [45]. Reproduced with permission of American Chemical Society. (b) Linear fiber actuator working in an aqueous electrolyte. Source: Qi et al. 2004 [47]. Reproduced with permission of American Chemical Society. (c) SEM image of electrospun PU/PANI hybrid fiber to assemble into bundles as artificial myofibril. Source: Gu et al. 2009 [50]. Reproduced with permission of American Chemical Society.

their many modifications have been demonstrated by Ghosh and coworkers [63], Carpi and De Rossi [64], and Cameron et al. [65]. D-EAP fiber actuators normally consist of an electrically conductive inner core, a middle elastomer layer, and an outer conductive layer. Technologies used to fabricate D-EAP fiber actuators include addition of conductive materials on the inside and outside walls of hollow tubes [64], repeatedly dipcoating a cylindrical core alternately into solutions of insulating and conducting precursors [66], and various co-extrusion methods [65].

D-EAP fiber actuators (refer Figure 25.5) reported in the literature generally have relatively larger diameters (~1000 μm) and moderate to high actuation strains in the range of 2–8%, see Table 25.4, although this strain range is promising for many applications and can be further improved by enhancing the material's dielectric properties or mechanical softness. However, the D-EAP actuator is limited by its requirement of high actuating voltages, usually of the magnitude of several kilovolts [67]. To some extent, this can be combatted by decreasing the thickness of the dielectric elastomer or by incorporating a voltage amplifier in the circuit. Miniaturization of D-EAP fiber actuators is another exciting, research-intensive challenge.

25.3.5 Carbon Nanotubes

Carbon nanotubes (CNTs) can be considered as graphene sheets rolled into cylinders with multiple layers and a nanoscale diameter (dubbed single-walled carbon nanotube [SWCNT] and multiwalled carbon nanotube [MWCNT] based on the number of layers) [68]. The actuation principle behind these

Table 25.3 Summary of electrical conductive polymer-based fiber actuators.

Materials and dimension	Actuation stimuli (V)	Maximum linear strain	Applications
PPy tube or PANI fiber Diameter: 10–400 μm [51–53]	AC: −0.5 to 0.8	0.2–0.95%	• To control a steerable microcatheter • Smart wearable garments
PPy tube infiltrated with liquid electrolyte Diameter: 1300–2200 μm [46]	AC: 3.6	Torsion angle: 0.034–0.017°	• Biomedical and biological devices
PANI doped with 2-acrylamido-2-methyl-1-propanesulfonic acid (PANI·AMPSA) Diameter: ~1250 μm [4, 45, 54]	AC: −0.8 to 0.3	0.91%	• Lifelike robots • Artificial limbs • Biomimetic devices
PANI fiber working in aqueous electrolyte Diameter: ~90 μm [47]	AC: −0.2 to 0.6	~1.2%	• Linear actuator
20 PANI fibers twisted yarn, trade name PANION™ Filament diameter: ~60 μm Yarn diameter: ~300 μm [55]	AC: −1 to 1.75	0.82%	• To drive a cantilever device
PU/PANI hybrid fiber bundle working in aqueous methane sulfonic acid Filament diameter: 0.9 μm [50]	AC: −0.2 to 0.8	1.65%	• To mimic myofibrils

carbonaceous materials is complex. Many different independent mechanisms have been proposed, such as double-layer charge injection [69–71], electrostatic force, photothermal phenomenon, and "pneumatic mechanism" [72].

A CNT-based actuator that can generate large strains by melting and solidifying a "guest" or filler component such as paraffin has been reported. For example, CNT yarns infiltrated by paraffin wax form a hybrid yarn that undergoes torsional actuation upon a change in temperature. This has been attributed to the phase transition and thermal expansion of the paraffin wax. Other filler materials activated by pulsed light heating have also been used [73]. Another twisted CNT yarn actuator driven by internal pressure associated with ion insertion of CNTs was developed, behaving in a manner similar to the deformation finger trap stent [74]. Recently, a novel actuation mechanism involving an electromagnetic driving force applied onto spun CNT fibers has been introduced [75, 76]. Besides

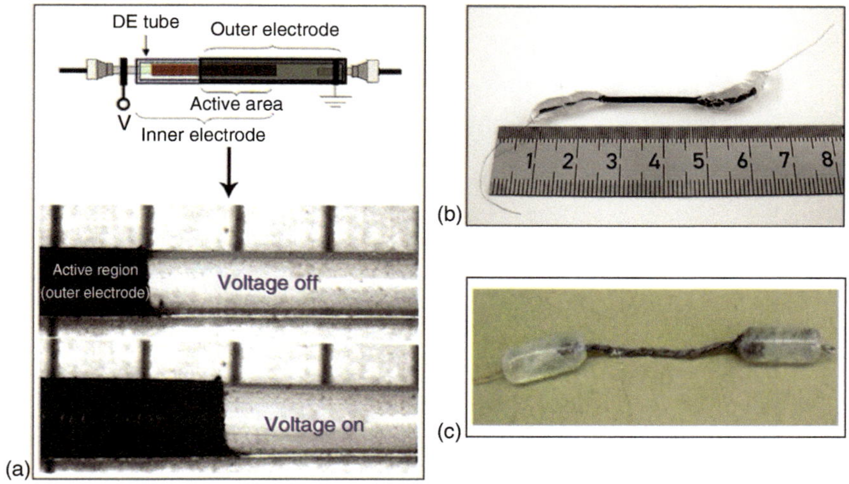

Figure 25.5 Dielectric elastomer (DE)-based fiber actuators. (a) Prototype of D-EAP fiber actuator made of silicone and PU tube. Source: Arora et al. 2007 [63]. Reproduced with permission of Elsevier. (b) Styrene-ethylene-butylene-styrene (SEBS) fiber actuator. Source: Kofod et al. 2010 [66]. Reproduced with permission of Springer. (c) Thermoplastic PU fiber actuator. Source: Cameron et al. 2008 [65]. Reproduced with permission of Elsevier.

Table 25.4 Summary of dielectric elastomer-based fiber actuators.

Materials and dimension	Actuation stimuli (kV)	Maximum linear strain (%)	Applications
Silicone and PU Diameter: 900 μm [63]	DC: ~3	7	• Fiber actuator used in conventional textiles or active, smart structures
Poly-(styrene-ethylene-butylene-styrene) (SEBS) Diameter: 1100 μm [66]	DC: 10	8	• Highly compact actuators
Thermoplastic polyurethane elastomer Diameter: ~1000 μm [65]	DC: 3.5	~2	• Linear actuator

electrical stimulation, solvent or vapor has also been explored as a viable actuation stimulus for CNT-based actuators [77, 78]. Table 25.5 lists various CNT or graphene fiber actuators.

Though the CNT-/graphene-based actuators (refer Figure 25.6) have diverse and complex actuation mechanisms, CNT fibers twisted into continuous strands of yarns offer the potential of forming into other textile structures.

Table 25.5 Summary of CNT-/graphene-based fiber actuators.

Material and dimension	Actuation stimuli	Maximum linear strain	Applications
Twisted CNT yarns working with guest matrix Yarn diameter: 10–200 μm; CNT diameter: 0.009 μm [73, 79]	Heat: 25–200 °C DC: ~280 V Optical stimulus: 100 W lamp	Torsion: 11 500 revolutions/min Tensile: 3% contraction	• Carbon nanotube yarn muscles • To hurl a projectile
Electrolyte filled twisted CNT yarns Yarn diameter: 3.8 μm; CNT diameter: 0.012 μm [74]	AC: −2 to 2 V	Torsion: 590 revolution/min Tensile: 1% contraction	• Torsional carbon nanotube artificial muscles
Over-twisted CNT yarns Diameter: 280–530 μm [75, 80]	DC: <40 V	14%	• To create crawling motion in an electric walker robot
CNT yarn Diameter: 70–270 μm [78]	Ethanol, acetone, toluene, and dichloromethane	15%	• Hierarchically arranged helical fiber actuators • Smart textile woven lifting robot
Graphene fiber coated with electrolyte containing PPy Diameter: ~33 μm [81]	AC: −0.8 to 0.8 V	2.5%	• To fabricate multi-armed tweezers and net actuators
Graphene/graphene oxide fibers Diameter: ~30 μm [77]	Water vapor	5%	• Moisture-triggered actuators and application in smart textile robot and walking robot

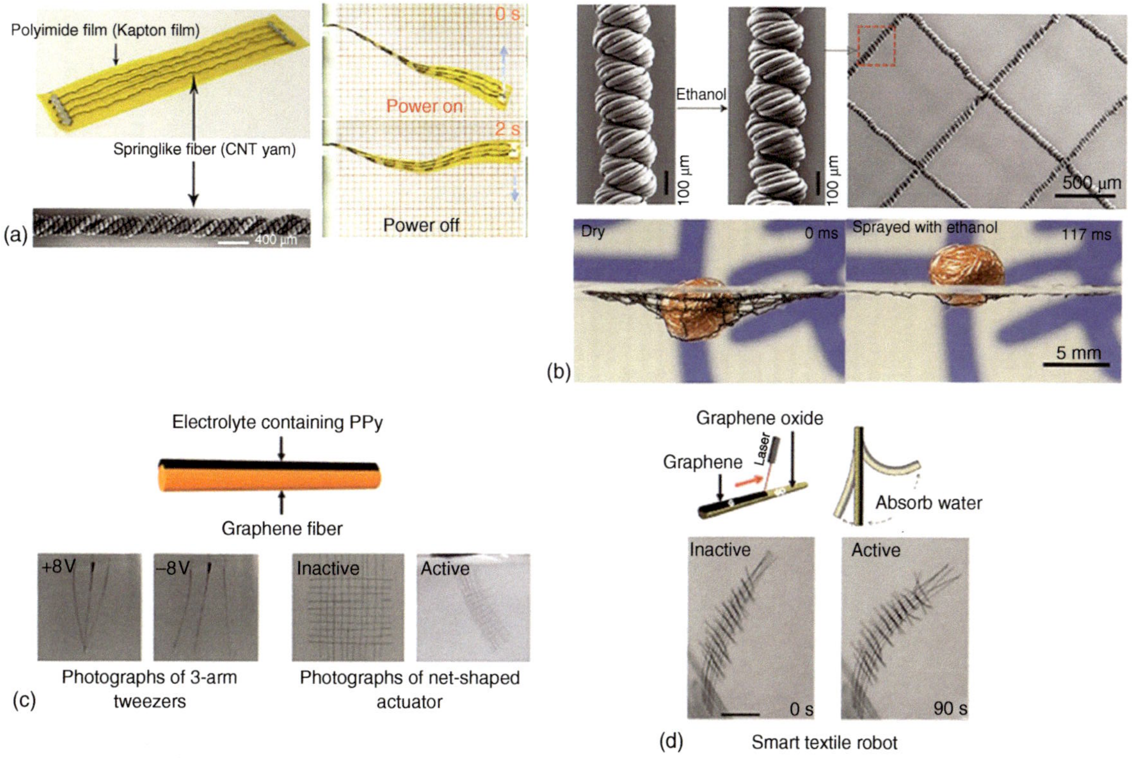

Figure 25.6 CNT-/graphene-based fiber actuators. (a) Kapton fabric ribbon woven from springlike CNT yarn actuators. Source: Chen et al. 2015 [80]. Reproduced with permission of John Wiley & Sons. (b) Woven textile-based lifting robot driven by CNT yarn muscles. Source: Chen et al. 2015 [78]. Reproduced with permission of Springer. (c) Multi-armed tweezers and net actuators made of graphene fiber actuators. Source: Wang et al. 2013 [81]. Reproduced with permission of Elsevier. (d) Moisture-triggered graphene fiber actuators and their application in smart textile robot and walking robot. Source: Cheng et al. 2013 [77]. Reproduced with permission of John Wiley & Sons.

25.3.6 Twisted/Coiled Fibers

Use of twist (or untwist) to produce linear and/or rotational actuation was proposed in the form of a strand-muscle actuator. The simple principle involved the use of length contraction (or extension) upon twisting (or untwisting) of two strands of polymeric fibers [82]. Thermally induced dimensional change in highly twisted and/or coiled polymeric fibers as a means of actuation forms this relatively new but innovative class fiber-based of actuators. Termed as fishing line and sewing thread muscles, these were first reported in 2014 [83]. Interestingly, these actuators made of commonly used polymeric (nylon, polyethylene, etc.) fibers with generally high levels of twist can generate fast, scalable, non-hysteretic tensile, and torsional actuations upon electrothermal or hydrothermal stimulation. Yarns with extreme twist can reach up to 49% contractile deformation and surprisingly lift loads over 100 times heavier than what a human muscle of the same dimension. The simple mechanism behind this actuator involves thermal expansion of the polymer filaments.

Various polymers such as nylon (nylon 6 and nylon 6,6), polyethylene, Kevlar, PVDF, polypropylene, and polyester have been used to fabricate twisted yarn muscles. In terms of applications, fishing line and sewing thread actuators not only work as muscles individually – for example, to electrothermally drive a window shutter or as a muscle to drive a robotic finger [84] – but also possess the capability to be woven or braided into novel textile structures as comfort-adjustable clothing with thermally or electrothermally driven porosity changes.

25.3.7 Other Types

In addition to the often-investigated fiber-based actuators described thus far, there are other types of fiber actuators, such as, pneumatic, electromagnetic, and photothermal. Some of these actuators offer potential future opportunities for development and have been summarized in Table 25.6.

25.4 Fundamental Principle and Types of Fiber Sensors

Fiber-based sensors integrated in a fabric are able to provide a measurable and quantifiable signal when there is a change in the input – referred to as the measurand. These sensors can detect changes in strain, pressure, humidity, temperature, or chemical composition among others and exhibit these changes in a variety of manners. Figure 25.7 explains the operating mechanism of sensors in a manner that makes their principle more apparent.

The information in this section is organized in terms of common measurand types and the transduction principle utilized to quantify the measurand.

Table 25.6 Summary of other types of fiber actuators.

Actuator type	Actuation mechanism	Fiber fabrication	Applications
Polymer gel-based fiber actuator [85–90]	Change in concentration of hydrogen ion or flow of solvated charges induced an asymmetric pressure	Electrospinning, wet spinning, and coagulation spinning	Analogous to muscles in living creatures
Liquid crystal elastomer (LCE) fiber actuator [89–91]	Electroclinic effect or the tilting angle of mesogens changes upon actuation or nematic–isotropic (NI) phase transition	Technology to fabricate LCE fibers has not yet been well developed	Artificial muscles
Magnetic microfibers used in textile actuators [92, 93]	Electromagnetic actuators	By introducing magnetic/ferromagnetic nanoparticle powders into the fiber matter during fiber production	Textile actuators
Photothermal actuator composed of optical fibers [94, 95]	Photothermal effect at the end of optical fiber causes stretching vibration at the end	One end of an optical fiber is cut as a bevel and painted black, and the incident light is along the length of the fiber	Novel optical robots
Pneumatic fiber actuator [96–99]	Pressurized air inflating or deflating a pneumatic bladder	An internal bladder surrounded by a braided structural shell made of threads	Artificial muscles

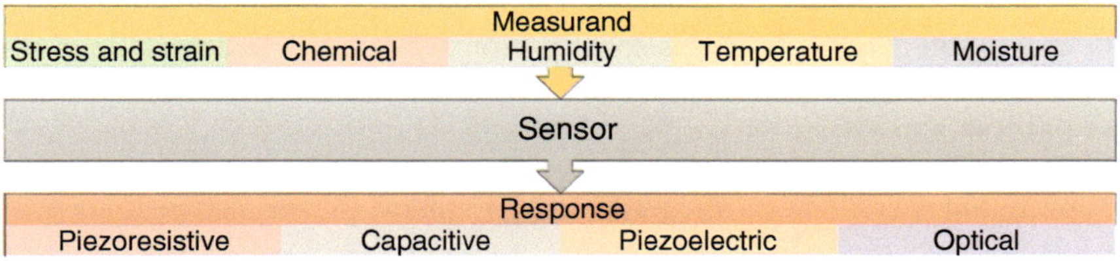

Figure 25.7 Basic working principle of a sensor with the various measurands and the sensor's sensing responses that will be discussed in this chapter.

25.4.1 Strain and Pressure Sensors

Most of the fiber-based sensors reported in the literature are designed to measure displacement or motion. Strain sensors are designed to measure relative displacement or motion and produce a measurable signal. The signal in turn is processed to calculate one or more of the very useful parameters such as strain, pressure (or stress), weight, and posture. This signal can be resistive, capacitive, optical, or piezoelectric. Pressure sensing involves the characterization of the force applied over a unit area of the component via generation of an electrical signal. Since application of pressure in many cases causes a relative displacement in the material, the characterization of such signals can be done again via a sensor capable of generating a resistive, capacitive, optical, or piezoelectric response.

25.4.1.1 Piezoresistive Type

The principle of piezoresistance is based on the simple concept that an electrical resistor may change its resistance upon experiencing a deformation. A *strain gauge* is one such piezoresistive sensor wherein the mechanical motion experienced by the sensor is directly translated to an electronic signal. The dependence of electrical resistance on the shape and material properties is straightforward. For a prismatic bar of length L and cross-sectional area A, it can be expressed as

$$R = \frac{\rho L}{A}$$

The electrical resistivity (ρ) is an intrinsic material property for homogeneous materials and is an invariant; therefore the piezoresistive behavior for homogeneous materials is derived from the changes in the shape parameters (A and L). However, in the case of a biphasic system, such as polymer particle composites, the change in electrical resistance cones from shape parameters as well as change in resistivity. At an atomic level, piezoresistance is due to a change in the energy gap between valence and conduction bands resulting in a change in the number of charge carriers, subsequently changing the resistance [2]. The most common morphology of a piezoresistive fiber includes a polymeric fiber with conductive particles dispersed throughout its volume or on the surface at or around its percolation threshold. Upon deformation, the relative movement of the conducting particles causes measurable change in its electrical resistance through breakage and reformation of conductive pathways. In the case of fiber assemblies, application of pressure on a network of conducting fibers (e.g. yarns/fabrics) results in a change in the contact resistance within the network, and this change can provide information about the location of the pressure being applied on the grid [100, 101]. Table 25.7 summarizes some piezoresistive sensors described in literature and their applications. Figure 25.8 illustrates the fabrication methods for some of these sensors.

Obviously, the research thrust has been toward integrating fiber-based sensors into continuous health monitoring systems. In general, piezoresistive fibers provide a large range of strain measurements in various configurations of bending, twisting, and stretching, enabling a strain sensitivities ranging from the measurement of heartbeats to giant deflections of up to 900% [105]. However, problems like unpredictable response, large hysteresis, and material choice limitations in

Table 25.7 Piezoresistive fiber-based sensors and their applications.

Sensor material [References]	Fabrication method	Sensing mechanism	Sensor performance	Applications
rGO[a)]- and AgNW[b)]-coated cotton fiber film [102, 103]	Dipcoating	Dynamic bridging effect of AgNW[b)] with rGO[a)] under external load	• 5.8 kPa^{-1} pressure sensitivity • 29.5 ms response time • 0.125 Pa detection limit • >10 000 cycles stability	• Measuring wrist hand pulses • Monitor motion of human body compression, bending, twisting sensor
rGO[a)]-coated nylon fabric [104]	Dipcoating	Piezoresistivity of rGO[a)] due to changes in its interlayers under pressure, fiber–fiber interactions, nylon's elastic strain	• 24 kPa^{-1} pressure sensitivity • 3 mm bend radius results in 213 ± 12 kΩ resistance change • At least 6000 cycles stability	• Low-cost forced touch trackpads for laptops • Wearable devices for health monitoring
CNT[c)] wrapped on SEBS[d)] fiber [105]	Benchtop melt drawing	Decrease in fiber diameter decreases CNT[c)] sheath mass per unit length, increasing resistance with strain	• 14.1–65.1 Q_R depending on core diameter • 31.8–44.5 Q_R depending on CNT[c)] layers • Up to 900% strain range	• Amperometric glucose sensing • Weaveable sensors in clothing
PU fiber coated with carbon ink and CNTs[c)] [106]	Dipcoating	Piezoresistive strain response of CNTs[c)] because of their higher deformability due to their fibrous structure	• Up to 100% strain measurement • Gauge factor ~ 3	• In microfluidic channels for delivering analytes • Multimodal physiological sensing
Hybrid Spandex-CNT[c)] yarns [107]	CNT wrapped on outer surface of Spandex via interlocking knitting machine	Increase in resistance with increasing strain due to deformation of percolated networks	• Repeatable strain sensing from 0% to 100% strain for minimum 1000 cycles with low hysteresis • Gauge factor from 0.12 to 0.4	• Adjustable smart clothing, robotics, and medical devices • Smart actuating textiles

(Continued)

Table 25.7 (Continued)

Sensor material [References]	Fabrication method	Sensing mechanism	Sensor performance	Applications
CCF[e] integrated with polyester fiber and Lycra® [108]	Polyester and Lycra form composite core yarn with CCF[e] wrapped around it	CCF[e] yarn shows piezoresistive strain response due to elongational strain with increased resistance	• Can sustain 0.3–0.4 kg force under 23% strain compared with single CCF yarn, which only sustains 0.09 kg force • Measures bpm[f] accurately for humans (10–20 bpm)	• Sensitive strain sensor • Wearable respiration monitoring systems
Stainless steel threads and AgNW embedded into PDMS [109]	Patterning AgNW on PDMS surface, gluing it to cloth substrate, with stainless steel threads as electrodes	Transconductance causes pressure-sensitive current flow between the two electrodes at a fixed potential difference results in resistive pressure sensing	• 4/16 ms response time • 0.6 Pa detection limit • Stable for >5000 cycles • Works in <1 kPa pressure regime	• Ultrasensitive electronic skin • Flexible solar cells and optoelectronic devices • Disposable skin sensor
PPy[g] on SWCNT[h]-coated cellulose yarns [110]	Cellulose yarn coated with SWCNT[h] via bobbin winding, PPy[g] electrodeposited on yarns	Conductive yarns act as piezoresistive strain sensors with differential strain sensing compared with no strain current and 100% strain current	• Up to 100% strain response stable after 4400 cycles	• Capable of energy storage and strain sensing • Self-powered sensory wearables • Possible large scale production

a) Reduced graphene oxide.
b) Silver nanowire.
c) Carbon nanotube(s).
d) Styrene-ethylene-butylene-styrene block copolymer.
e) Carbon-coated fibers.
f) Beats per minute.
g) Polypyrrole.
h) Single-walled carbon nanotube.

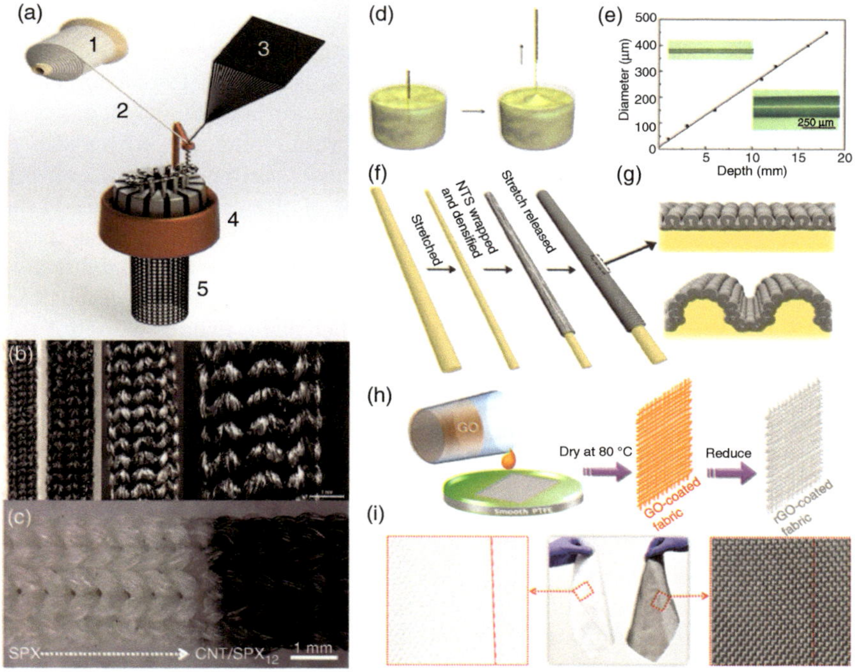

Figure 25.8 Fabrication methods of some piezoresistive sensors: (a) schematic for producing a knitted CNT/Spandex (SPX) textile. The different components are (1) a spool of SPX fibers, (2) an *n*-fiber SPX yarn, (3) a CNT forest, (4) a circular knitting machine, and (5) a knitted CNT/SPX textile. (b) Photograph of knitted CNT/SPX structures, containing 1, 4, 8, and 12 SPX fibers from left to right, respectively. (c) The transition between SPX_{12} yarns and CNT-wrapped SPX_{12} yarns in the knitted textile. Source: (Panels a–c) Foroughi et al. 2016 [107]. Reproduced with permission of American Chemical Society. (d) Schematic illustration of melt drawing used for fibers in Ref. [105]. (e) Influence of immersion depth of wooden rod on rubber fiber diameter with inset showing optical microscope images of rubber core fiber. (f) Detailed steps in fabrication. (g) Differences in buckling of the fibers depending on mass of CNT sheath to that of rubber core. Source: (Panels d–g) Wang et al. 2016 [105]. Reproduced with permission of John Wiley & Sons. (h) Coating nylon fabric with GO and subsequent reduction to rGO. Panel (i) shows the pristine nylon fabric (white) compared with coated nylon fabric. Source: (Panels h and i) Abdul Samad et al. 2017 [104]. Reproduced with permission of Elsevier.

the case of piezoresistive sensing belie the need for a better strain and pressure sensing mechanism explored in Sections 25.3.1.2–25.3.1.4.

25.4.1.2 Capacitive Type

Capacitive sensors convert a mechanical motion to an electrical signal via a change in their capacitance. A capacitor consists of a pair of electrodes separated by a dielectric medium, and a change in either the distance (d) between the electrodes, the area (A) of the electrodes, or the dielectric constant (ε) of the medium can cause a change in the capacitance (C_0), for parallel plate capacitors:

$$C_0 = \varepsilon \frac{A}{d}$$

For designing a highly sensitive capacitive pressure or strain sensor, it would be ideal to have a capacitor design with a large plate area and a narrow gap between the plates, without compromising the performance accuracy and fabrication reproducibility of the sensor [2]. Capacitive sensors provide the advantage of a linear relation between the mechanical deformation and the capacitance in certain sensor configurations, with lower hysteresis compared with piezoresistive sensors. More importantly, capacitive sensors can be formed in fibrous structures relatively easily. A summary of some of the capacitive fiber-based sensors proposed in the literature is provided in Table 25.8. Some capacitive fiber sensors may combine strain sensing mechanisms with capabilities of performing as energy storage or energy harvesting devices as supercapacitors [111, 112]. Various capacitor geometries (parallel plate, cylindrical, etc.) can be implemented in capacitive sensors making multimodal flexible sensors with the ability to sense pressure, axial strain, bending, and twisting. A large variety of materials can be used for capacitive fiber sensors, and the low hysteresis response of such sensors enables more accuracy in the measuring parameters. Figure 25.9 also summarizes the fabrication methods of some of the capacitive fiber-based sensors discussed in Table 25.8.

25.4.1.3 Piezoelectric Type

Piezoelectric materials exhibit an electric polarization when subjected to mechanical stresses, resulting in measurable potential difference. Conversely, they can act as actuators to generate displacement on application of an electric field. Therefore, these materials can be potentially used as self-sensing actuators [115, 116]. In the case of piezoelectric sensors, the mechanical stress applied on the sensor causes a differential potential difference between the two sides or faces of the sensor, and the resulting potential difference is measured as the sensor response. This differs from piezoresistive sensors in the manner that piezoelectric sensors generate a potential difference rather than a resistance change. It is important to remember that piezoelectric strain sensors generate an electrostatic charge only when force is applied to or removed from them and are generally not suitable for static force measurements because of charge leakage. They can be used effectively for transient or dynamic force measurements.

Polyvinylidene fluoride (PVDF) and its copolymers are the most widely used materials for flexible sensors including those in the form of fibers. PVDF consists of long chains of repeating monomer ($-CH_2-CF_2-$). The hydrogen atoms contain a net positive charge with respect to the carbon atom, and hence the monomer unit has an intrinsic dipole moment, causing it to show piezoelectric behavior. PVDF is pyroelectric, and hence its piezoelectric performance shows a strong dependence on temperature [117]. Table 25.9 summarizes some piezoelectric fiber sensors and their applications. Some of these piezoelectric sensors perform pressure sensing in an acoustic medium whereby acoustic waves are converted into electrical signals.

Table 25.8 Capacitive sensors and their applications.

Sensor material [References]	Fabrication method	Sensing mechanism	Sensor performance	Applications
CNT[a] buckled around silicone rubber core fibers [111]	CNT[a] aerogel sheets wrapped around silicone core fiber axially stretched by 300%	Capacitor structure made via buckled CNT[a]/dielectric rubber/buckled CNT[a] structure shows capacitive change in response to strain due to Poisson's ratio of rubber core ~ 0.49	• Strain sensing via 115.7% and 26% capacitive changes during stretching (200%) and twisting (1700 rad/m), respectively • 36.8 kPa pressure sensing range	• Highly deformable batteries, solar cells, wearable sensors, and energy harvesters
Silicone elastomer dielectric and encapsulating layers with ionically conductive ink as electrode [112]	Multicore–shell fiber printing process	Sensor arranged as a cylindrical resistor, a cylindrical capacitor, and a cylindrical ring resistor in series, which shows a resistive and capacitive response to strain	• Good agreement with theoretical model for capacitive and resistive response up to 250% strain • Gauge factor of 0.348 ± 0.11	• Can be integrated with wearable textiles for exo-suits, soft robotics, and health monitoring data
Primaloft® yarn (non-resistive) with carbon fiber yarn (sensing element) [113]	Weft knitting machine used to form knitted structure of carbon fiber trace	Measures differential voltage output and capacitance changes due to charging and discharging cause by finger	• Deviation RMS[b] error of 0.225 between simulation and experimental data for touch location via capacitance	• For robotics and smart garments • Gesture recognition
Pt deposited on cotton fibers and coated with PDMS[c] [114]	Thermal atomic layer deposition of metallic Pt	PDMS[c] acts as dielectric with conductive fiber as electrodes, showing capacitive response with pressure due to change in PDMS[c] thickness	• Stable capacitive response during loading and unloading of 0.4 N load for over 10 000 cycles • Capacitive response mapping for 56-pixel pressure sensors	• For advanced occupant classification system for vehicles • Wearable electronic sensors

a) Carbon nanotube.
b) Root mean square.
c) Polydimethylsiloxane.

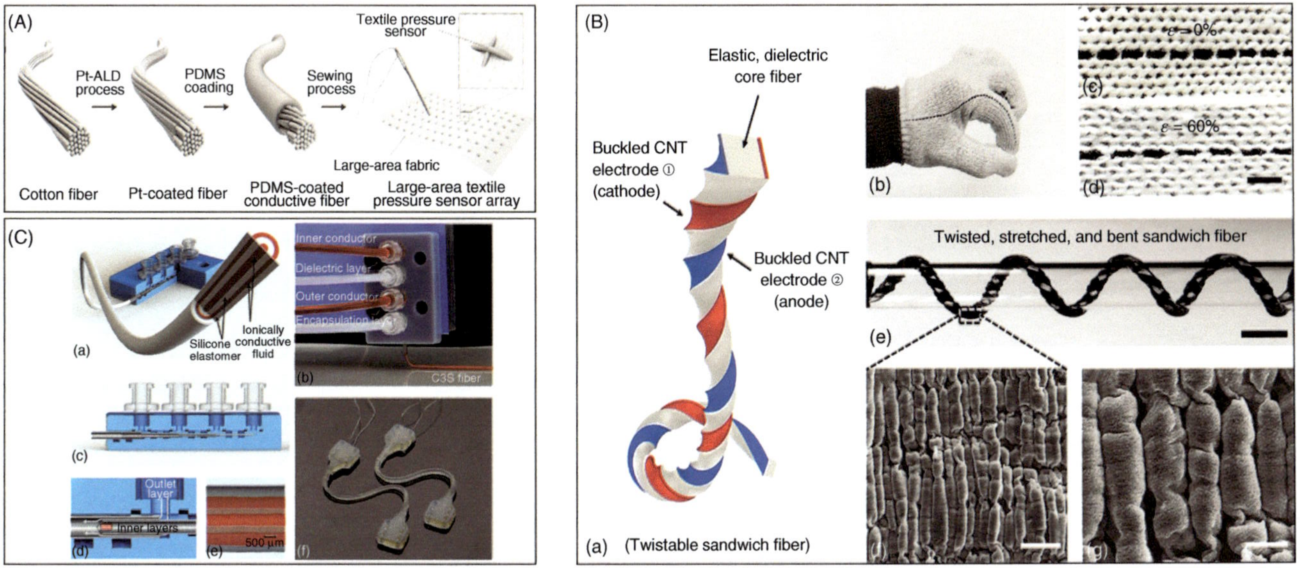

Figure 25.9 Fabrication processes of various capacitive fiber sensors. (A) Schematic illustration showing coating the cotton fiber with Pt via low-temperature atomic layer deposition (ALD) and a polydimethylsiloxane (PDMS) coating on the surface of the conductive fiber and subsequent integration of the pressure sensors into a fabric by sewing. Source: Lee et al. 2016 [114]. Reproduced with permission of Springer. (B) (a) Schematic illustration of twisted sandwich fibers with Ecoflex™ rubber core and two symmetric, buckled CNT electrodes and (b) 20 cm long fiber woven into a commercial glove and its magnified image in (c) before and (d) after applying 60% tensile strain with 8 mm scale bar. (e) Photograph of fiber wound around a glass tube (scale bar: 50 mm). SEM images of buckles formed in CNT due to straining and relaxation: (f) low (scale bar: 50 μm) and (g) high (scale bar: 20 μm) magnification. Source: Choi et al. 2016 [111]. Reproduced with permission of American Chemical Society. (C) (a) Schematic illustration and (b) image of multicore–shell printing process of fibers in Ref. [112]. Source: Frutiger et al. 2015 [112]. Reproduced with permission of John Wiley & Sons. (c, d) Schematic illustrations of the print-head design used for the process and higher magnification view of the printing outlet. (e) Optical image of a printed multicore–shell capacitive soft sensor filaments. (f) Two sensing filaments of different lengths. Source: Frutiger et al. 2015 [112]. Reproduced with permission of John Wiley & Sons.

Table 25.9 Piezoelectric sensors and their applications.

Sensor material [References]	Fabrication method	Sensing mechanism	Sensor performance	Applications
PVDF sheath with an HDPE[a)]/CB[b)] compound electrically conductive core [118]	Melt spinning bicomponent fibers	Piezoelectric response of the fiber with linear increase in output voltage with increasing strain	• 57 V/% voltage strain ratio • 0.313 (V/m)/(N m^2) piezoelectric coefficient	• Powering low power electronics • Heart rate and respiratory signal monitoring
CPC and PVDF-TrFE[c)] with indium filaments and PC[d)] cladding [119]	Thermal drawing of preform	Piezoelectric response via acoustic signal generation of the fibers on receiving pulsed excitations from an acoustic transducer	• Fibers generated audible sound between 7 and 15 kHz at 5 V driving voltage	• Active catheters for pressure and flow measurements in blood vessels • Acoustic microscopy
Aligned PVDF-TrFE[c)] nanofibers [120]	Electrospinning	Generates a voltage across the effective contact length of the nanofibers in response to pressure Shows a positive and negative peak for current in response to bending cycles	• In 0.4–2 kPa pressure range shows sensitivity of 1.1 V/kPa • Current and voltage outputs for bending sensor were 6–40 nA and 0.5–1.5 V, respectively	• Skin conformal pressure sensor in small pressure regime • Accelerometers, vibrometers, orientational sensors
PVDF nanofiber web [121]	Electrospinning	PVDF sandwiched between gold-coated PET[d)] films (electrodes), sound waves exert pressure on film to generate electric signals	• 266 mV/Pa sensitivity • Sensitivity to sound pressure level above 100 dB	• Acoustic sensor for detecting noise in a system

a) High-density polyethylene.
b) Carbon black.
c) Poly(vinylidene-fluoride-trifluoroethylene) copolymer.
d) Polyethylene terephthalate.

25.4.1.4 Optical Type

The use of optical fibers for sensing applications was first implemented in the early 1970s. An optical fiber is a high aspect ratio material that works on the principle of optical total internal reflection (TIR) for guiding light waves [122].

Optical fiber sensors can be described as a sensor wherein physical, chemical, or biological stimuli interact with light guided through an optical fiber to produce a quantifiable optical signal, which contains information about the stimulus being measured. Optical fiber sensors are classified as extrinsic types where the optical fiber acts as a carrier for transporting light to an external sensing system or intrinsic types wherein the entire sensing process is carried out within the optical fiber. In terms of materials, glass optical fibers (GOFs) have been traditionally used for optical fiber sensors, but they present problems in terms of being rigid and brittle. For something that is conformable and flexible, polymeric optical fibers (POFs) are preferred. POFs also have lower manufacturing costs, are lighter due to lower density, and have large core diameters, which make them less prone to breakage during handling. POFs also have a higher sensitivity to strain and temperature compared with GOFs. Some commonly used POF materials include polymethyl methacrylate (PMMA), polystyrene (PS), and amorphous fluorinated polymer (CYTOP) [123].

While it is not within the scope of this chapter to discuss all the different types of optical fiber sensors, in terms of strain and pressure sensing, optical fiber sensors can have various modes of operation – namely, fiber Bragg gratings (FBGs), interferometric sensing, and fiber bend sensing. Fiber-optic sensors used for strain and pressure sensing can have various modes of operation – namely, FBGs, interferometric sensing, and fiber bend sensing. In the case of FBGs, application of any local strain on the optical fiber sensor results in a change in the effective refractive index of the core in the optical fiber, resulting in a change in the wavelength of the reflected light. In interferometric sensing, the input light is split into a reference fiber and a sensory fiber (manipulated by the stimulus). On exposure to an external perturbation, such as longitudinal strain or pressure, a phase shift occurs. When these two recombine, they display an interference pattern at the output. Interferometric sensors for strain and pressure sensing provide high sensitivity and are used for applications such as monitoring structural health [124–127]. In microbend (or macro) sensing, when a fiber is subjected to small deformations, light rays in the core of the fiber exceed the critical angle causing small amounts of light escape the core. This results in a redistribution of energy in the core and the cladding. As the severity of microbending increases, the light intensity in the core decreases. The amount of received light is related to the amplitude of strain or deformation. Microbend sensors are useful for contact mode mechanical sensing applications such as tactile systems and strain detection. The difference between macro- and microbend sensors lies in the bending diameters and in the power loss mechanism; macrobend sensors have power loss due to the large bend causing changes in phase velocity of the light and transition loss due to light coupling in the leaky mode of the optical fiber [123] near the macrobend. Micro- and macrobend sensors are used for many shape and deformation sensing systems

[128–130]. Table 25.10 summarizes some POFs that operate on one of the abovementioned principles of sensing.

Optical fiber sensors offer the advantage of resilience to electromagnetic interference, a wide range of operating temperatures and conditions (such as high voltage or explosive environments), and chemical inertness. These have been under development for many years, and the technology is rather mature compared with all others. Commercial fiber-optic devices include gyroscopes, sensors for temperature, pressure, and vibration measurements [123, 135].

25.4.2 Humidity and Temperature Sensors

The principle of operation of humidity sensors can be same as that of pressure and strain sensors, i.e. capacitive, piezoresistive, optical [136], and mechanical [137] (swelling of the sensor due to moisture). Capacitive humidity sensors respond to changes in humidity levels via a change in the dielectric behavior of their dielectric medium. Resistive humidity sensors show a response via change in their resistance due to the presence of moisture, while optical humidity sensors respond via interferometric sensing wherein change in the relative humidity causes a change in the reflected optical power. Mechanical humidity sensors can detect changes in shape due to swelling in response to humidity changes. In terms of fiber-based sensors, interferometric, capacitive, and resistive responses to humidity changes have been extensively explored with CNTs being the most commonly used sensor material. The change in resistance of CNTs in response to humidity is due to reversible hygroscopic swelling [138]. Other materials such as polyvinyl alcohol (PVA), polyacrylic acid (PAA), hydroxyethyl cellulose (HEC), carbon black (CB), PANI have also been used as resistive humidity sensors. Such sensors are fabricated via simple dipcoating [139], electrospinning [140], and wet spinning [141]. Traditionally, metallic wires such as W, Ni, Cu, and Pt have been used as temperature sensors due to their intrinsic change in resistivity in response to temperature gradients. These metals can also be incorporated into fabric structures via knitting to form flexible temperature sensors [142, 143]. However, metallic fiber sensors in textiles add undesirable bulk and discomfort. Hence, polymeric temperature sensors with small amounts of temperature-sensitive additives such as CNTs, CB, carbon fibers, and metallic particles have been incorporated through fabrication methods such as dipcoating and wet and melt spinning, in an effort toward developing temperature sensing fibers [144–146]. Table 25.11 summarizes some humidity and temperature sensing fibers and their applications.

25.4.3 Chemical Sensors

When it comes to safety in hazardous environments, sensors play a vital role in saving human lives and protecting critical assets. Conformable chemical sensors in the form of fibers not only will be able to detect threats but also will be relatively comfortable to wear as a protective layer. Gas sensors in fibrous forms can provide enough porosity for the gas to penetrate the sensor system and change a key parameter of the sensor.

Table 25.10 Polymeric optical fiber sensors and their applications.

Sensor material [References]	Fabrication method	Sensing mechanism	Sensor performance	Applications
TOPAS COC[a] grade 5013 [131]	Drill and draw method: TOPAS drilled with hexagonal holes and then drawn to fiber	FBG mechanism causes shift in Bragg wavelength of sensor	• Sensitivity of 0.76, 0.80, 1.0 pm/$\mu\varepsilon$[b] at 50, 80, 110 °C, respectively, to small strains up to 0.3% • Sensitivity of 0.75 pm/$\mu\varepsilon$ at 98 °C for large strain up to 3%	• FBG POFs that can operate at temperatures up to 110 °C • In vivo biosensing applications
PMMA mPOF[c] [132]	Drill and draw technique	FBG mechanism wherein resonance wavelength depends linearly on applied force and strain up to 1%	• Strain sensitivity of 0.73 ± 0.02 pm/$\mu\varepsilon$ up to 1.53% strain • Temperature sensitivity of -77 ± 7 pm/°C • Linear resonance tunability of 7 nm	• In vivo biomedical applications for strain and temperature sensing
PMMA doped with <3% polystyrene [133]	N/A	Interferometric sensing causes change in phase of the light in response to elongation	• Elongation sensitivity of $131 \pm 3 \times 10^5$ rad/m • Temperature sensitivity of -212 ± 26 (rad/m)/K	• Structural health monitoring of bridges, highways, aircraft wings, and engineering structures
PMMA embedded into PDMS [134]	PMMA fabricated via preform-drawing method	FBG sensing mechanism enabling shear and pressure sensing due to change in Bragg wavelength shifts	• Pressure sensitivity of 0.8 pm/Pa in the range of 2.4 kPa • Shear stress sensitivity of 1.3 pm/Pa for a range of 0.06 kPa	• Measuring contact stress sensing for the human skin and tissues

a) Cyclic olefin copolymer.
b) Strain.
c) Microstructured polymer optical fiber.

Table 25.11 Humidity and temperature sensing fibers and their applications.

Sensor	Sensor material [References]	Fabrication method	Sensing mechanism	Sensor performance	Applications
Humidity	Cotton thread/CNTs dissolved in nafion or PSS water solutions [139]	Dipcoating	Resistance increases with humidity due to hygroscopic swelling of nafion and cotton, disrupting electron transport between CNTs	• Instantaneous and reversible change in resistance even at low humidity (20%)	• Human physiological sensing
	GO–silicon bilayer structure [137]	Spin coating GO thin films on silicon microbridge	Humidity-induced deformation of GO causes increased interlayer distance of GO sheets resulting in bending of silicon microbridge	• Sensitivity of 28.02 μV/%RH[a)] • Maximum humidity hysteresis ~3%	• Integration into MEMS[b)]/NEMS[c)] for multimodal sensing
	PVA/MWCNT [141]	CVD for preparing MWCNT and subsequent drawing in PVA solution	Sudden increase in electrical resistance of MWCNT/PVA due to swelling PVA matrix causes dispersed particles and interfaces in the composite matrix	• Sensitivity up to 1.89 at 100% RH[a)] • Adjustable switch point between 75 and 84% RH[a)] • High residual sensitivity of 0.63	• Monitoring real-time RH[a)] in electronic devices
Temperature	Copper, nickel, and tungsten wires embedded in knitted polyester fabric [142]	Flatbed knitting	Intrinsic resistance of the wire wrapped around an insulator changes with temperature	• Linear trend of temperature–resistance curve with coefficient of determination over 99.9%	• Garments to monitor skin temperature
	Nylon-6 functionalized with MWCNT/PPy [145]	Electrospun nylon-6 with vacuum-deposited MWCNT and polymerized PPy	Resistance linearly varies with temperature depending on temperature coefficient of resistance of the sensor	• Reliable detection in 25–45 °C range • Average temperature coefficient of resistance −0.204 ± 0.008%/°C	• Smart clothing and physiological monitoring in prosthetic sockets
	Double-helix SWCNT yarns [146]	SWCNT film twisted into double-helix structure	Double-helix structure acts as a thermocouple with the two CNT yarns as the electrodes Linear increase of resistance with temperature	• Sensitivity of 0.08–0.14 Ω/°C • Linear decrease of resistance upon cooling	• Lightweight and flexible thermocouple

a) Relative humidity.
b) Microelectromechanical systems.
c) Nanoelectromechanical systems.

Table 25.12 Chemical sensors and their applications.

Sensor material [References]	Fabrication method	Sensing mechanism	Detected gas	Detection limit
SWCNT[a)]/cotton yarn/PEI[b)] [149]	Dipcoating/immersion	NH_3 increases resistance of sensor due to charge transfer of electrons from NH_3 to SWCNT,[a)] which increases distance between conductance and valence band in SWCNT[a)]	NH_3	N/A
(PVC,[c)] cumene-PSMA,[d)] PSE,[e)] and PVP[f)]) Functionalized SWCNT on cotton fiber [150]	Dipcoating followed by embroidering	Percolation of gas molecules into the porous film, increasing resistance of sensor fiber	Body odors	50 ppm
PANI–nylon 6 composite fabric [153]	Immersion of nylon fabric in aniline and subsequent polymerization	PANI behaves as a p-type material, and NH_3 acts as a strong reducing agent to decrease conductivity	NH_3	1000 ppm
HCSA[g)] doped-PANI nanofibers [154]	Directly via chemical synthesis or indirectly via electrospinning	Change in sensor resistance on exposure to gas is due to hygroscopic nature of the material and size of alkyl group of the alcohol, which affects the penetration of alcohol molecules into the polymer	Methanol, ethanol, 2-propanol	N/A

a) Single-walled CNT.
b) Polyethyleneimine.
c) Polyvinyl chloride.
d) Polystyrene-*co*-maleic anhydride.
e) Partial isobutyl/methyl mixed ester.
f) Polyvinylpyrrolidone.
g) Camphorsulfonic acid

Source: Seesaard et al. 2015 [150]. https://www.mdpi.com/1424-8220/15/1/1885.

Chemical sensors in fiber form can be made of TiO_2- and CNT-incorporated porous carbon fibers [147], PANI-coated optical fibers [148], CNT coated onto polyethylenimine (PEI)/cotton yarns [149], functionalized CNT on cotton fibers [150], and pitch-based carbon fibers [151, 152].

Summary of some of the relevant chemical sensor concepts and their applications proposed in the literature is provided in Table 25.12. In all of these cases, the sensing response comes from relative change in the electrical resistance of the sensing fiber. Development of chemical sensors in the fiber form is an active area of scientific research.

25.5 Conclusions and Outlook

Flexible electronics is emerging as an important technology, critical in many applications including wearables, human–machine interfaces, healthcare, microrobotics, and electronic textiles. One of the earliest development in flexible electronics was thin silicon wafer solar cells assembled on polymer substrates for flexibility [155]. Many of the advances over the years have resulted in a wealth of demonstrators. The academic and commercial interest is driven primarily by the need for reduced cost, large area, roll to roll, and flexible systems that require conformal, distributed, and integrated functionality, which are unavailable from more traditional brittle material and device platforms.

Fiber-based sensing and actuation are a relatively new area of research and development within flexible electronics and mostly relevant to electronic textiles. In most cases, initial applications are being explored. The tremendous commercial interest in these technologies come from their potential to benefit many current applications as well as new applications that are beyond the abilities of current semiconductor-based sensors and actuators. In the case of actuators, many of the polymer-based materials and principles being employed feature a combination of high work density, large strain, and high stress, comparable or more than that of the natural muscles. Some of the electric field-driven EAP technologies demonstrate particularly high performance, but the requirement of high voltages may prove to be an obstacle. On the other hand, ionic EAPs often suffer from slow response and low electromechanical coupling. In all cases, life cycle issues such as creep deformation and fatigue life need further investigation.

Soft, polymer-based sensors have been under development for a long time and had resulted in the commercial production of various devices. Research and development in fiber sensors is relatively new with the exception of optical fiber sensors. Fiber-based sensors can be employed as part of clothing as a diagnostic and monitoring tool. For example, heat, moisture, salinity, or pressure sensor arrays can be used to monitor a patient in real time. The body heat and pressure distribution, sweat content, and posture can all reveal vital information on pathological symptoms. Textile-embedded electronics is becoming a reality, and textile-embedded sensors are likely to become ubiquitous. Electronic devices that can be seamlessly integrated to textiles require not only dimensional but also mechanical compatibility, such as flexibility, conformability, and tolerance to stretch (10–20% strain). Fibrous structures offer not only high porosity and permeability but also large surface area for the measurement of biopotentials. Fiber-based devices open up a vast area of new possibilities for deployment of sensors, energy harvesting, etc. Current technologies to precisely manipulate fibers into useful structures may help in developing sensors that can combine multimodal functionalities into one and provide versatile sensing regimes. Another aspect in the development of fiber-based sensors that has not been touched upon here is the development of self-powered fiber sensors. Research in this area focuses on developing fiber sensors that act as supercapacitors for energy storage and conversion or thermoelectric materials (such as poly-(3,4-ethylenedioxythiophene):polystyrene sulfonate [PEDOT:PSS], PEI, Bi_2Te_3) that can harvest body heat into usable forms of energy. In this way, fiber-based sensors can transcend to form a complete system that can be self-powered and sensitive to various parameters at the same time.

The opportunities also come with challenges, such as dealing with material and geometric nonlinearities associated with large strains. Moreover, sensors that are typically designed to sense a specific measurand are most likely to respond to other stimuli that are not of interest. For example, a piezoresistive strain measurement sensor may respond to temperature. In the realm of fabrication or manufacturing of fiber-based devices, some of the common methods of fabricating sensors such as dipcoating might not be the most robust when it comes to multiple use and wash cycles. To overcome this, sensing capabilities must be integrated into the more fundamental fiber form – which can act as a building block for the entire textile.

Other problems such as biocompatibility of certain sensing materials also should be addressed especially when used in intimate skin applications. Another aspect that has a growing interest in the research community is the routing of signals from these fiber-based sensors to processors and transducers. Methods to interconnect the sensor and actuators to other electronic elements of a circuit need to be examined to allow development of flexible circuits that can transfer data in a facile, fast, and reliable manner. Finally, scaling up of fiber-based sensors in terms of a low-cost, high-output manufacture is an area that still needs significant research. Adapting existing fiber manufacturing techniques such as melt spinning so that fiber sensors can be manufactured on a large scale seems to be one way in which this problem can be tackled. Further work by the material community including fiber scientists should help make these technologies widely applicable in a range of applications where fiber-based electronic systems are desirable.

References

1 Trung, T.Q. and Lee, N. (2016). Flexible and stretchable physical sensor integrated platforms for wearable human-activity monitoring and personal healthcare. *Advanced Materials* 28: 4338–4372.

2 Kloeck, B. and De Rooij, N.F. (1994). Mechanical sensors. In: *Semiconductor Sensors* (ed. S.M. Sze), 153–204. New York: Wiley.

3 Bar-Cohen, Y. (2004). *Electroactive Polymer (EAP) Actuators as Artificial Muscles: Reality, Potential, and Challenges*, 2e. Bellingham, WA: SPIE.

4 Baughman, R.H. (2005). Playing nature's game with artificial muscles. *Science* 308: 63–65.

5 Gregorio, C.C., Granzier, H., Sorimachi, H., and Labeit, S. (1999). Muscle assembly: a titanic achievement? *Current Opinion in Cell Biology* 11: 18–25.

6 Fink, J.K. (2012). Sensor types and polymers. In: *Polymeric Sensors and Actuators* (ed. Pilla, Srikanth), 1–42. Wiley.

7 Peng, H., Sun, X., Weng, W., and Fang, X. *Polymer Materials for Energy and Electronic Applications*. London, United Kingdom: Academic Press.

8 Seelecke, S. and Muller, I. (2004). Shape memory alloy actuators in smart structures: modeling and simulation. *Applied Mechanics Reviews* 57: 23–46.

9 Lam Po Tang, S. and Stylios, G.K. (2006). An overview of smart technologies for clothing design and engineering. *International Journal of Clothing Science and Technology* 18: 108–128.

10 Hu, J. (2007). Development of shape memory polyurethane (SMPU) fiber. In: *Shape Memory Polymers and Textiles* (ed. Hu, Jinlian). pp. 305–337. Boca Raton, FL, USA: Woodhead Publishing.

11 Meng, Q., Hu, J., Zhu, Y. et al. (2007). Morphology, phase separation, thermal and mechanical property differences of shape memory fibres prepared by different spinning methods. *Smart Materials and Structures* 16: 1192.

12 Meng, Q., Hu, J., Zhu, Y. et al. (2007). Polycaprolactone-based shape memory segmented polyurethane fiber. *Journal of Applied Polymer Science* 106: 2515–2523.

13 Meng, Q. and Hu, J. (2008). Self-organizing alignment of carbon nanotube in shape memory segmented fiber prepared by in situ polymerization and melt spinning. *Composites Part A: Applied Science and Manufacturing* 39: 314–321.

14 Meng, Q. and Hu, J. (2008). Study on poly(Îμ-caprolactone)-based shape memory copolymer fiber prepared by bulk polymerization and melt spinning. *Polymers for Advanced Technologies* 19: 131–136.

15 Hu, J. and Chen, S. (2010). A review of actively moving polymers in textile applications. *Journal of Materials Chemistry* 20: 3346.

16 Chan Vili, Y.Y.F. (2007). Investigating smart textiles based on shape memory materials. *Textile Research Journal* 77: 290–300.

17 Gok, M.O., Bilir, M.Z., and Gurcum, B.H. (2015). Shape-memory applications in textile design. *Procedia Social and Behavioral Sciences* 195: 2160–2169.

18 Stylios, G.K. and Wan, T. (2007). Shape memory training for smart fabrics. *Transactions of the Institute of Measurement and Control* 29: 321–336.

19 Ahir, S.V., Tajbakhsh, A.R., and Terentjev, E.M. (2006). Self-assembled shape-memory fibers of triblock liquid-crystal polymers. *Advanced Functional Materials* 16: 556–560.

20 Zhuo, H., Hu, J., and Chen, S. (2008). Electrospun polyurethane nanofibres having shape memory effect. *Materials Letters* 62: 2074–2076.

21 Kim, B., Lee, M.G., Lee, Y.P. et al. (2006). An earthworm-like micro robot using shape memory alloy actuator. *Sensors and Actuators A: Physical* 125: 429–437.

22 Choi, S. and Han, Y. (2010). *Piezoelectric Actuators: Control Applications of Smart Materials (1)*. Boca Raton, FL: CRC Press.

23 Ballas, R.G. (2007). Piezoelectric materials. In: *Piezoelectric Multilayer Beam Bending Actuators*, 17–30. Berlin Heidelberg: Springer-Verlag.

24 Trolier-McKinstry, S., Fox, G.R., Kholkin, A. et al. (1999). Optical fibers with patterned ZnO/electrode coatings for flexural actuators. *Sensors and Actuators A: Physical* 73: 267–274.

25 Tajitsu, Y. (2006). Development of electric control catheter and tweezers for thrombosis sample in blood vessels using piezoelectric polymeric fibers. *Polymers for Advanced Technologies* 17: 907–913.

26 Liu, Z.H., Pan, C.T., Lin, L.W., and Lai, H.W. (2013). Piezoelectric properties of PVDF/MWCNT nanofiber using near-field electrospinning. *Sensors and Actuators A: Physical* 193: 13–24.

27 Wilkie, W.K., Inman, D.J., Lloyd, J.M., and High, J.W. (2006). Anisotropic laminar piezocomposite actuator incorporating machined PMN-PT single-crystal fibers. *Journal of Intelligent Material Systems and Structures* 17: 15–28.

28 NASA Technology Transfer Program (2006). Piezoelectric fiber composite actuator portfolio. https://technology.nasa.gov/patent/LAR-TOPS-17 (accessed 22 June 2017).

29 Bent, A.A., Hagood, N.W., and Rodgers, J.P. (1995). Anisotropic actuation with piezoelectric fiber composites. *Journal of Intelligent Material Systems and Structures* 6: 338–349.

30 Williams, R.B., Inman, D.J., Schultz, M.R. et al. (2004). Nonlinear tensile and shear behavior of macro fiber composite actuators. *Journal of Composite Materials* 38: 855–869.

31 Nguyen, C. and Kornmann, X. (2006). A comparison of dynamic piezoactuation of fiber-based actuators and conventional PZT patches. *Journal of Intelligent Material Systems and Structures* 17: 45–55.

32 Pan, C.L., Feng, Z.H., Ma, Y.T., and Liu, Y.B. (2008). Small torsional piezoelectric fiber actuators with helical electrodes. *Applied Physics Letters* 92: 012923.

33 Pan, C.L., Ma, Y.T., Liu, Y.B. et al. (2008). Torsional displacement of piezoelectric fiber actuators with helical electrodes. *Sensors and Actuators A: Physical* 148: 250–258.

34 Han, W.X., Zhang, Q., Ma, Y.T. et al. (2009). An impact rotary motor based on a fiber torsional piezoelectric actuator. *The Review of Scientific Instruments* 80: 014701.

35 Pan, C.L., Feng, Z.H., Ma, Y.T. et al. (2011). Coupled torsional and longitudinal vibrations of piezoelectric fiber actuator with helical electrodes. *IEEE Transactions on Ultrasonics, Ferroelectrics, and Frequency Control* 58: 829–837.

36 Pan, C.L., Ma, Y.T., Yin, J. et al. (2011). Miniature orthogonal optical scanning mirror excited by torsional piezoelectric fiber actuator. *Sensors and Actuators A: Physical* 165: 329–337.

37 Tajitsu, Y. (2008). Piezoelectricity of chiral polymeric fiber and its application in biomedical engineering. *IEEE Transactions on Ultrasonics, Ferroelectrics, and Frequency Control* 55: 1000–1008.

38 Itoh, S., Sawano, M., Hikawa, H. et al. (2010). Fundamental study of piezoelectric motion of chiral polymeric fibers for realizing tubular-type polymeric actuator for large macroscale physical object. *Japanese Journal of Applied Physics* 49: 09MD14.

39 Pu, J., Yan, X., Jiang, Y. et al. (2010). Piezoelectric actuation of direct-write electrospun fibers. *Sensors and Actuators A: Physical* 164: 131–136.

40 Chang, C., Tran, V.H., Wang, J. et al. (2010). Direct-write piezoelectric polymeric nanogenerator with high energy conversion efficiency. *Nano Letters* 10: 726–731.

41 French, J.D., Weitz, G.E., Luke, J.E. et al. (1996). Production of continuous piezoelectric fibers for sensor/actuator applications. In: *Proceedings of the 10th IEEE International Symposium on Applications of Ferroelectrics, 1996. ISAF'96*, vol. 2, 867–870.

42 Williams, R.B., Park, G., Inman, D.J., and Wilkie, W.K. (2002). An overview of composite actuators with piezoceramic fibers. In: *20th International Modal Analysis Conference*, 421–427. Los Angeles, CA.

43 Martins, P., Lopes, A.C., and Lanceros-Mendez, S. (2014). Electroactive phases of poly(vinylidene fluoride): determination, processing and applications. *Progress in Polymer Science* 39: 683–706.

44 Li, Z. and Cheng, Z. (2009). Piezoelectric and electrostrictive polymer actuators: fundamentals. In: *Biomedical Applications of Electroactive Polymer Actuators*, 317–334. Wiley.

45 Lu, W., Smela, E., Adams, P. et al. (2004). Development of solid-in-hollow electrochemical linear actuators using highly conductive polyaniline. *Chemistry of Materials* 16: 1615–1621.

46 Fang, Y., Pence, T.J., and Tan, X. (2011). Fiber-directed conjugated-polymer torsional actuator: nonlinear elasticity modeling and experimental validation. *IEEE/ASME Transactions on Mechatronics* 16: 656–664.

47 Qi, B., Lu, W., and Mattes, B.R. (2004). Strain and energy efficiency of polyaniline fiber electrochemical actuators in aqueous electrolytes. *The Journal of Physical Chemistry B* 108: 6222–6227.

48 Uh, K., Yoon, B., Lee, C.W., and Kim, J. (2016). An electrolyte-free conducting polymer actuator that displays electrothermal bending and flapping wing motions under a magnetic field. *ACS Applied Materials & Interfaces* 8: 1289–1296.

49 Spinks, G.M., Mottaghitalab, V., Bahrami-Samani, M. et al. (2006). Carbon-nanotube-reinforced polyaniline fibers for high-strength artificial muscles. *Advanced Materials* 18: 637–640.

50 Gu, B.K., Ismail, Y.A., Spinks, G.A. et al. (2009). A linear actuation of polymeric nanofibrous bundle for artificial muscles. *Chemistry of Materials* 21: 511–515.

51 De Rossi, D., Lorussi, F., Mazzoldi, A. et al. (2001). Active dressware: wearable proprioceptive systems based on electroactive polymers. In: *Proceedings of the 5th International Symposium on Wearable Computers*, 161–162.
52 Santa, A.D., Mazzoldi, A., and Rossi, D. (1996). Steerable microcatheters actuated by embedded conducting polymer structures. *Journal of Intelligent Material Systems and Structures* 7: 292–300.
53 De Rossi, D. and Mazzoldi, A. (1999). Linear fully dry polymer actuators. In: *Proceedings of SPIE 3669, Smart Structures and Materials 1999: Electroactive Polymer Actuators and Devices*, vol. 3669, 35–44.
54 Ding, J., Zhou, D., Spinks, G. et al. (2003). Use of ionic liquids as electrolytes in electromechanical actuator systems based on inherently conducting polymers. *Chemistry of Materials* 15: 2392–2398.
55 Lu, W., Norris, I.D., and Mattes, B.R. (2005). Electrochemical actuator devices based on polyaniline yarns and ionic liquid electrolytes. *Australian Journal of Chemistry* 58: 263–269.
56 Zhang, X., Wissler, M., Jaehne, B. et al. (2004). Effects of crosslinking, prestrain, and dielectric filler on the electromechanical response of a new silicone and comparison with acrylic elastomer. In: *Proceedings of SPIE 5385, Smart Structures and Materials 2004: Electroactive Polymer Actuators and Devices (EAPAD)*, vol. 5385, 78.
57 Kofod, G. and Sommer-Larsen, P. (2005). Silicone dielectric elastomer actuators: finite-elasticity model of actuation. *Sensors and Actuators A: Physical* 122: 273–283.
58 Kofod, G., Sommer-Larsen, P., Kornbluh, R., and Pelrine, R. (2003). Actuation response of polyacrylate dielectric elastomers. *Journal of Intelligent Material Systems and Structures* 14: 787–793.
59 Ma, W. and Cross, L.E. (2004). An experimental investigation of electromechanical response in a dielectric acrylic elastomer. *Applied Physics A* 78: 1201–1204.
60 Anonymous (1997). An experimental investigation of electromechanical responses in a polyurethane elastomer. *Journal of Applied Physics* 81: 2770–2776.
61 Su, J., Zhang, Q.M., Kim, C.H. et al. (1997). Effects of transitional phenomena on the electric field induced strain–electrostrictive response of a segmented polyurethane elastomer. *Journal of Applied Polymer Science* 65: 1363–1370.
62 Pelrine, R.E., Kornbluh, R.D., and Joseph, J.P. (1998). Electrostriction of polymer dielectrics with compliant electrodes as a means of actuation. *Sensors and Actuators A: Physical* 64: 77–85.
63 Arora, S., Ghosh, T., and Muth, J. (2007). Dielectric elastomer based prototype fiber actuators. *Sensors and Actuators A: Physical* 136: 321–328.
64 Carpi, F. and De Rossi, D. (2004). Dielectric elastomer cylindrical actuators: electromechanical modelling and experimental evaluation. *Materials Science and Engineering C* 24: 555–562.
65 Cameron, C.G., Szabo, J.P., Johnstone, S. et al. (2008). Linear actuation in coextruded dielectric elastomer tubes. *Sensors and Actuators A: Physical* 147: 286–291.

66 Kofod, G., Stoyanov, H., and Gerhard, R. (2010). Multilayer coaxial fiber dielectric elastomers for actuation and sensing. *Applied Physics A* 102: 577–581.

67 Ashley, S. (2003). Artificial muscles. *Scientific American* 289: 52–59.

68 Mirfakhrai, T., Madden, J.D.W., and Baughman, R.H. (2007). Polymer artificial muscles. *Materials Today* 10: 30–38.

69 Kosidlo, U., Omastova, M., Micusik, M. et al. (2013). Nanocarbon based ionic actuators: a review. *Smart Materials and Structures* 22: 104022.

70 Baughman, R.H., Cui, C., Zakhidov, A.A. et al. (1999). Carbon nanotube actuators. *Science* 284: 1340–1344.

71 Carpi, F., Kornbluh, R., Sommer-Larsen, P., and Alici, G. (2011). Electroactive polymer actuators as artificial muscles: are they ready for bioinspired applications? *Bioinspiration & Biomimetics* 6: 045006.

72 Spinks, G.M., Wallace, G.G., Fifield, L.S. et al. (2002). Pneumatic carbon nanotube actuators. *Advanced Materials* 14: 1728–1732.

73 Lima, M.D., Li, N., Andrade, M. et al. (2012). Electrically, chemically, and photonically powered torsional and tensile actuation of hybrid carbon nanotube yarn muscles. *Science* 338: 928–932.

74 Foroughi, J., Spinks, G.M., Wallace, G.G. et al. (2011). Torsional carbon nanotube artificial muscles. *Science* 334: 494–497.

75 Guo, W., Liu, C., Zhao, F. et al. (2012). A novel electromechanical actuation mechanism of a carbon nanotube fiber. *Advanced Materials* 24: 5379–5384.

76 Meng, F., Zhang, X., Li, R. et al. (2014). Electro-induced mechanical and thermal responses of carbon nanotube fibers. *Advanced Materials* 26: 2480–2485.

77 Cheng, H., Liu, J., Zhao, Y. et al. (2013). Graphene fibers with predetermined deformation as moisture-triggered actuators and robots. *Angewandte Chemie International Edition* 52: 10482–10486.

78 Chen, P., Xu, Y., He, S. et al. (2015). Hierarchically arranged helical fibre actuators driven by solvents and vapours. *Nature Nanotechnology* 10: 1077–1083.

79 Chun, K., Hyeong Kim, S., Kyoon Shin, M. et al. (2014). Hybrid carbon nanotube yarn artificial muscle inspired by spider dragline silk. *Nature Communications* 5: 3322.

80 Chen, P., He, S., Xu, Y. et al. (2015). Electromechanical actuator ribbons driven by electrically conducting spring-like fibers. *Advanced Materials* 27: 4982–4988.

81 Wang, Y., Bian, K., Hu, C. et al. (2013). Flexible and wearable graphene/polypyrrole fibers towards multifunctional actuator applications. *Electrochemistry Communications* 35: 49–52.

82 Suzuki, M. and Ichikawa, A. (2005). Toward springy robot walk using strand-muscle actuators. In: *Climbing and Walking Robots* (eds. Armada, M. A. and Santos, Pablo de Gonzalez), 479–486. Berlin Heidelberg: Springer-Verlag.

83 Haines, C.S., Lima, M.D., Li, N. et al. (2014). Artificial muscles from fishing line and sewing thread. *Science* 343: 868.

84 Wu, L., Jung, D.A., Rome, R.S. et al. (2015). Nylon-muscle-actuated robotic finger. In: *Proceedings of SPIE 9431, Active and Passive Smart Structures and Integrated Systems 2015*, vol. 9431, 94310I-12.

85 Lee, S.J., Lee, D.Y., Song, Y.S., and Cho, N.I. (2007). Chemically driven polyacrylonitrile fibers as a linear actuator. *Solid State Phenomena* 124–126: 1197–1200.

86 Samatham, R., Park, I.S., Kim, K.J. et al. (2006). Electrospun nanoscale polyacrylonitrile artificial muscle. *Smart Materials and Structures* 15: N152.

87 Shahinpoor, M., Norris, I.D., Mattes, B.R. et al. (2002). Electroactive polyacrylonitrile nanofibers as artificial nanomuscles. In: *Proceedings of SPIE 4695, Smart Structures and Materials 2002: Electroactive Polymer Actuators and Devices (EAPAD)*, vol. 4695, 351–358.

88 Gonzalez, M.A. and Walter, W.W. An investigation of electrochemomechanical actuation of conductive polyacrylonitrile (PAN) nanofiber composites. In: *Proceedings of SPIE 9056, Electroactive Polymer Actuators and Devices (EAPAD) 2014*, vol. 9056, 90563J-9.

89 Yu, L. and Gu, L. (2009). Effects of microstructure, crosslinking density, temperature and exterior load on dynamic pH-response of hydrolyzed polyacrylonitrile-blend-gelatin hydrogel fibers. *European Polymer Journal* 45: 1706–1715.

90 Michardiere, A., Mateo-Mateo, C., Derre, A. et al. (2016). Carbon nanotube microfiber actuators with reduced stress relaxation. *Journal of Physical Chemistry C* 120: 6851–6858.

91 Naciri, J., Srinivasan, A., Jeon, H. et al. (2003). Nematic elastomer fiber actuator. *Macromolecules* 36: 8499–8505.

92 Wiak, S., Firych-Nowacka, A., and Smolka, K. (2008). Computer model of 3-D magnetic micro fibres used in textile actuators. In: *2008 18th International Conference on Electrical Machines*, 1–6.

93 Rubacha, M. and Zieba, J. (2006). Magnetic textile elements. *Fibres & Textiles in Eastern Europe* 14: 49–53.

94 Otani, Y., Matsuba, Y., and Yoshizawa, T. (2002). Two-dimensional movement of photothermal actuator composed of optical fibers. In: *Proceedings of SPIE 4902, Optomechatronic Systems III*, vol. 4902, 78–82.

95 Otani, Y., Matsuba, Y., and Yoshizawa, T. (2001). Photothermal actuator composed of optical fibers. In: *Proceedings of SPIE 4564, Optomechatronic Systems II*, vol. 4564, 216–219.

96 Chou, C.-P. and Hannaford, B. (1994). Static and dynamic characteristics of McKibben pneumatic artificial muscles. In: *Proceedings of the 1994 IEEE International Conference on Robotics and Automation*, vol. 1, 281–286.

97 Yeh, T., Wu, M., Lu, T. et al. (2010). Control of McKibben pneumatic muscles for a power-assist, lower-limb orthosis. *Mechatronics* 20: 686–697.

98 Connolly, F., Polygerinos, P., Walsh, C.J., and Bertoldi, K. (2015). Mechanical programming of soft actuators by varying fiber angle. *Soft Robotics* 2: 26–32.

99 Tondu, B. (2012). Modelling of the McKibben artificial muscle: a review. *Journal of Intelligent Material Systems and Structures* 23: 225–253.

100 Shimojo, M., Namiki, A., Ishikawa, M. et al. (2004). A tactile sensor sheet using pressure conductive rubber with electrical-wires stitched method. *IEEE Sensors Journal* 4: 589–596.

101 Hui, Z., Ming, T.X., Xi, Y.T., and Sheng, L.X. (2006). Pressure sensing fabric. *Proceedings MRS* 001, 920.

102 Wei, Y., Chen, S., Dong, X. et al. (2017). Flexible piezoresistive sensors based on aerodynamic bridging effect of silver nanowires toward graphene. *Carbon* 113: 395–403.

103 Wei, Y., Chen, S., Lin, Y. et al. (2016). Silver nanowires coated on cotton for flexible pressure sensors. *Journal of Materials Chemistry C* 4: 935–943.

104 Abdul Samad, Y., Komatsu, K., Yamashita, D. et al. (2017). From sewing thread to sensor: Nylon® fiber strain and pressure sensors. *Sensors and Actuators B: Chemical* 240: 1083–1090.

105 Wang, H., Liu, Z., Ding, J. et al. (2016). Downsized sheath–core conducting fibers for weavable superelastic wires, biosensors, supercapacitors, and strain sensors. *Advanced Materials* 28: 4998–5007.

106 Mostafalu, P., Akbari, M., Alberti, K.A. et al. (2016). A toolkit of thread-based microfluidics, sensors, and electronics for 3D tissue embedding for medical diagnostics. *Microsystems & Nanoengineering* 2: 16039.

107 Foroughi, J., Spinks, G.M., Aziz, S. et al. (2016). Knitted carbon-nanotube-sheath/spandex-core elastomeric yarns for artificial muscles and strain sensing. *ACS Nano* 10: 9129–9135.

108 Huang, C., Shen, C., Tang, C., and Chang, S. (2008). A wearable yarn-based piezo-resistive sensor. *Sensors and Actuators A: Physical* 141: 396–403.

109 Lai, Y., Ye, B., Lu, C. et al. (2016). Extraordinarily sensitive and low-voltage operational cloth-based electronic skin for wearable sensing and multifunctional integration uses: a tactile-induced insulating-to-conducting transition. *Advanced Functional Materials* 26: 1286–1295.

110 Huang, Y., Kershaw, S.V., Wang, Z. et al. (2016). Highly integrated supercapacitor-sensor systems via material and geometry design. *Small* 12: 3393–3399.

111 Choi, C., Lee, J.M., Kim, S.H. et al. (2016). Twistable and stretchable sandwich structured fiber for wearable sensors and supercapacitors. *Nano Letters* 16: 7677–7684.

112 Frutiger, A., Muth, J.T., Vogt, D.M. et al. (2015). Capacitive soft strain sensors via multicore–shell fiber printing. *Advanced Materials* 27: 2440–2446.

113 Vallett, R., Young, R., Knittel, C. et al. (2016). Development of a carbon fiber knitted capacitive touch sensor. *MRS Advances* 1: 2641–2651.

114 Lee, J., Yoon, J., Kim, H.G. et al. (2016). Highly conductive and flexible fiber for textile electronics obtained by extremely low-temperature atomic layer deposition of Pt. *NPG Asia Materials* 8: e331.

115 Ikeda, H. and Morita, T. (2011). High-precision positioning using a self-sensing piezoelectric actuator control with a differential detection method. *Sensors and Actuators A: Physical* 170: 147–155.

116 Rakotondrabe, M., Ivan, I.A., Khadraoui, S. et al. (2014). Simultaneous displacement/force self-sensing in piezoelectric actuators and applications to robust control. *IEEE/ASME Transactions on Mechatronics* 20: 519–531.

117 Sirohi, J. and Chopra, I. (2000). Fundamental understanding of piezoelectric strain sensors. *Journal of Intelligent Material Systems and Structures* 11: 246–257.

118 Nilsson, E., Lund, A., Jonasson, C. et al. (2013). Poling and characterization of piezoelectric polymer fibers for use in textile sensors. *Sensors and Actuators A: Physical* 201: 477–486.

119 Egusa, S., Wang, Z., Chocat, N. et al. (2010). Multimaterial piezoelectric fibres. *Nature Materials* 9: 643–648.

120 Persano, L., Dagdeviren, C., Su, Y. et al. (2013). High performance piezoelectric devices based on aligned arrays of nanofibers of poly(vinylidenefluoride-co-trifluoroethylene). *Nature Communications* 4: 1633.

121 Lang, C., Fang, J., Shao, H. et al. (2016). High-sensitivity acoustic sensors from nanofibre webs. *Nature Communications* 7: 11108.

122 Chin, K., Fang, Z., and Qu, R. (2012). *Fundamentals of Optical Fiber Sensors (1)*, Wiley Series in Microwave and Optical Engineering. Somerset, NJ: Wiley.

123 Rajan, G. (2014). *Optical Fiber Sensors : Advanced Techniques and Applications*. Boca Raton, FL: CRC Press.

124 Her, S.C., and Yang, C.M. (2012). Dynamic strain measured by Mach-Zehnder interferometric optical fiber sensors. *Sensors* 12 (3): 3314–3326.

125 Xu, F., Ren, D., Shi, X. et al. (2012). High-sensitivity Fabry-Perot interferometric pressure sensor based on a nanothick silver diaphragm. *Optics Letters* 37: 133–135.

126 Zhang, Y., Yuan, L., Lan, X. et al. (2013). High-temperature fiber-optic Fabry-Perot interferometric pressure sensor fabricated by femtosecond laser. *Optics Letters* 38: 4609–4612.

127 Tian, Z. and Yam, S.S.H. (2009). In-Line abrupt taper optical fiber Mach-Zehnder interferometric strain sensor. *IEEE Photonics Technology Letters* 21: 161–163.

128 Chen, Y. and Liu, C. (2016). Radius and orientation measurement for cylindrical objects by a light section sensor. *Sensors* 16: 1981.

129 Bianchi, M., Haschke, R., Büscher, G., Ciotti, S., Carbonaro, N. and Tognetti, A., 2016. A multi-modal sensing glove for human manual-interaction studies. *Electronics* 5 (3): 42.

130 Noh, Y., Liu, H., Sareh, S. et al. (2016). Image-based optical miniaturized three-axis force sensor for cardiac catheterization. *IEEE Sensors Journal* 16: 7924–7932.

131 Markos, C., Stefani, A., Nielsen, K. et al. (2013). High-Tg TOPAS microstructured polymer optical fiber for fiber Bragg grating strain sensing at 110 degrees. *Optics Express* 21: 4758–4765.

132 Yuan, W., Stefani, A., and Bang, O. (2011). Tunable polymer fiber Bragg grating (FBG) inscription: fabrication of dual-FBG temperature compensated polymer optical fiber strain sensors. *IEEE Photonics Technology Letters* 24: 401–403.

133 Silva-Lopez, M., Fender, A., MacPherson, W.N. et al. (2005). Strain and temperature sensitivity of a single-mode polymer optical fiber. *Optics Letters* 30: 3129–3131.

134 Zhang, Z.F., Tao, X.M., Zhang, H.P., and Zhu, B. (2013). Soft fiber optic sensors for precision measurement of shear stress and pressure. *IEEE Sensors Journal* 13: 1478–1482.

135 Jha, A.R. (2007). *Fiber Optic Technology - Applications to Commercial, Industrial, Military, and Space Optical Systems* Raleigh, NC, USA: SciTech Publishing.

136 Arregui, F.J., Liu, Y., Matias, I.R., and Claus, R.O. (1999). Optical fiber humidity sensor using a nano Fabry-Perot cavity formed by the ionic self-assembly method. *Sensors and Actuators B: Chemical* 59: 54–59.

137 Yao, Y., Chen, X., Guo, H. et al. (2012). Humidity sensing behaviors of graphene oxide-silicon bi-layer flexible structure. *Sensors and Actuators B: Chemical* 161: 1053–1058.

138 Na, P.S. and Kim, H. (2005). Investigation of the humidity effect on the electrical properties of single-walled carbon nanotube transistors. *Applied Physics Letters* 87: 093101.

139 Shim, B.S., Chen, W., Doty, C. et al. (2008). Smart electronic yarns and wearable fabrics for human biomonitoring made by carbon nanotube coating with polyelectrolytes. *Nano Letters* 8: 4151–4157.

140 Lin, Q., Li, Y., and Yang, M. (2012). Investigations on the sensing mechanism of humidity sensors based on electrospun polymer nanofibers. *Sensors and Actuators B: Chemical* 171–172: 309–314.

141 Li, W., Xu, F., Sun, L. et al. (2016). A novel flexible humidity switch material based on multi-walled carbon nanotube/polyvinyl alcohol composite yarn. *Sensors and Actuators B: Chemical* 230: 528–535.

142 Husain, M.D., Kennon, R., and Dias, T. (2014). Design and fabrication of temperature sensing fabric. *Journal of Industrial Textiles* 44: 398–417.

143 Husain, M.D. and Kennon, R. (2013). Preliminary investigations into the development of textile based temperature sensor for healthcare applications. *Fibers* 1: 2–10.

144 Zhang, C., Ma, C., Wang, P., and Sumita, M. (2005). Temperature dependence of electrical resistivity for carbon black filled ultra-high molecular weight polyethylene composites prepared by hot compaction. *Carbon* 43: 2544–2553.

145 Blasdel, N.J., Wujcik, E.K., Carletta, J.E. et al. (2015). Nanocomposite resistance temperature detector. *IEEE Sensors Journal* 15: 300–306.

146 Shang, Y., Li, Y., He, X. et al. (2013). Highly twisted double-helix carbon nanotube yarns. *ACS Nano* 7: 1446–1453.

147 Yun, J., Kim, H., and Lee, Y. (2013). A hybrid gas-sensing material based on porous carbon fibers and a TiO_2 photocatalyst. *Journal of Materials Science* 48: 8320–8328.

148 Shao, S., Huang, Y., and Tao Simultaneous, S. (2014). Monitoring of ammonia and moisture using a single fiber optoelectrode as a transducer. *IEEE Sensors Journal* 14: 847–852.

149 Zhang, W., Tan, Y.Y., Wu, C., and Silva, S.R.P. (2012). Self-assembly of single walled carbon nanotubes onto cotton to make conductive yarn. *Particuology* 10: 517–521.
150 Seesaard, T., Lorwongtragool, P., and Kerdcharoen, T. (2015). Development of fabric-based chemical gas sensors for use as wearable electronic noses. *Sensors* 15 (1): 1885–1902.
151 Lee, S.K., Im, J.S., Kang, S.C. et al. (2012). Effects of improved porosity and electrical conductivity on pitch-based carbon nanofibers for high-performance gas sensors. *Journal of Porous Materials* 19: 989–994.
152 Kim, J., Lee, S.H., Park, S., and Lee, Y. (2014). Preparation and gas-sensing properties of pitch-based carbon fiber prepared using a melt-electrospinning method. *Research on Chemical Intermediates* 40: 2571–2581.
153 Hong, K.H., Oh, K.W., and Kang, T.J. (2004). Polyaniline–nylon 6 composite fabric for ammonia gas sensor. *Journal of Applied Polymer Science* 92: 37–42.
154 Pinto, N.J., Ramos, I., Rojas, R. et al. (2008). Electric response of isolated electrospun polyaniline nanofibers to vapors of aliphatic alcohols. *Sensors and Actuators B: Chemical* 129: 621–627.
155 Crabb, R.L. and Treble, F.C. (1967). Thin silicon solar cells for large flexible arrays. *Nature* 213: 1223–1224.

26

Textile-Based Electronics: Polymer-Assisted Metal Deposition (PAMD)

Casey Yan and Zijian Zheng

Laboratory for Advanced Interfacial Materials and Devices, Institute of Textiles and Clothing, The Hong Kong Polytechnic University, Hung Hom, Kowloon, Hong Kong SAR, China

26.1 Introduction

26.1.1 The Rise of Textile-Based Electronics

With many silicon-based technology breakthroughs demonstrated in the past few decades, lightweight electronic gadgets such as cell phones, GPS navigation, and MP3 had emerged and successfully dominated consumer's market. Today, semiconductor electronics get even more advanced, faster in clock speed, and physically smaller in size. With these technological advances and huge demand for mobile devices, recently, electronics evolutionarily transform from "portable" to even "wearable" with many thought-provoking worn-on-the-body applications targeting daily communication, entertainment, fitness tracking, etc. Well-known examples are Google Glass, Apple Watch, and other various fitness trackers developed by Jawbone, Nike, Fitbit, Garmin, etc. However, most of these wearable electronics are accessory-based devices that are still rigid and bulky. When it comes to what people define as "wearables" that can effortlessly accommodate the shape of human body and at the same time can provide electronic functions, flexibility and stretching ability of the device are the concerns. Apparently, conventional rigid silicon-based electronics cannot do this job well. Still, a flexible and stretchable platform for wearable electronics is highly demanded [1].

Among the many flexible and stretchable material substrates, the most familiar one is textile. Textile consisting of trillion thousands of fibers creates a versatile platform for the implementation of electronics because of not only its functional aspect, such as the large surface area of one fiber that make them desirable to be incorporated with plenty of electronic stuffs at micro- to nanoscale, but also their unique attributes such as commodity, comfort, wearing ability, and ultrahigh flexibility intrinsically inherited from the nature [2]. As textile is the most in dispensable necessity that we cannot live without, the implementation of electronics onto textile injects new research direction for the future development of electronics. In fact, such an idea to integrate electronics in textiles had already been demonstrated by Park and coworkers, who presented the first prototype of textile-based

Handbook of Fibrous Materials, First Edition. Edited by Jinlian Hu, Bipin Kumar, and Jing Lu.

electronics, Georgia Tech Wearable Motherboard (GTWM), by directly attaching devices and computer chips onto textiles for monitoring human vital signs and wireless communications [3–5]. Recently, because of the increasing market demands especially from the medical and sports sectors, direct attachment of electronics onto textiles becomes technically viable. Some have already been well branded by many renowned sports companies, such as Adidas miCoach training shirt that detects one's heart rate during physical training and the North Face Hustle Audio Jacket that features an external joystick panel on sleeves to control iPod. Recently, Google and Samsung have also engaged in developing electronic garment prototypes, such as Project Jacquard and Smart Suit. Other than sports and entertainment, wearable electronic garments directly attached with small electronic components are also readily recognized in both medical and military sectors [6, 7].

To date, most of the wearable electronic garments still rely heavily on direct attachment or embedment of conventional off-the-shelf electronic devices and other bulky components. The wearing ability of these wearable electronics, however, has never been compromised due to the rigidness of the electronic devices that are conventionally constructed on the rigid silicon wafers. Besides, wearable electronic devices as tangible extras adding on clothing are bulky and bothersome to the wearers. Therefore, to truly develop textile-based electronics, the ultimate goal is to fully get rid of all the off-the-shell, rigid, and bulky silicon-based electronic devices and to construct all electronic devices from scratch on flexible and stretchable textile substrates such as fibers, yarns, and fabrics that can soon be seamlessly integrated into wearable forms via different textile technologies. The recent rise of textile-based electronics attracts many research efforts with many promising textile-based electronic devices demonstrated such as sensors [8], solar cells [9], energy storage devices [10–12], etc.

26.1.2 The Essential: High-Performance Conductive Textiles

As we all know, conductive traces are inevitable components in constructing electronic circuits to enable movements of electrons for driving all the electronic devices. There is no exception for textile-based electronics where high-performance flexible and stretchable textile conductors as interconnects, contacts, and electrodes are highly demanded. Considering the criteria for conductive textiles qualifying for electronic purposes, conductivity of the textile conductor is the first concern that should be taken into serious account. In particular, textile conductors should possess a high conductivity to ensure the performance of the device, and the conductivity should retain stable without significant performance drop throughout the devices' lifespan. Other than conductivity, concerns regarding the mechanical attributes and robustness of the textile conductors are also important. First, in order to allow electronics fully integrate into the textile structure, conductive textiles must be flexible enough in such an extent that a certain degree of stretching/bending is allowed. Second, a stable performance of the conductive textile must be maintained even under stretching/bending, as textiles are meant to be worn on the body and have to endure repeated cycles of bending, stretching, and release of these actions because of daily human motions. Third, it is the washing durability of the conductive textiles. No wonder daily uses of textiles require hand to mild machine washing. Therefore, it is of particular importance whether the textile-based electronics can withstand such a vigorous mechanical abrasion underwater. Fourth, the fabrication process of these textile conductors must be compatible with the industry, so as to allow scale production of textile-based electronics.

Synthesizing high-performance, stable, and washable conductive textile is one biggest challenge that materials scientists have ever encountered. Addressing the needs of conductive textiles, many fabrication methods for high-performance conductive textiles have been reported in the past decades, such as yarns and fabrics directly drawn or incorporated with metal or metal oxide [13, 14], coated with intrinsically conducting polymers (ICPs) [15–17], spun using carbon nanotubes (CNTs) [18–20], etc. However, most of them fail in at least one or two abovementioned criteria to meet the needs. For example, textiles directly drawn or incorporated with metal or metal oxide impart extra rigidness on the final textile garments. Further, textiles coated with ICPs suffer majorly from chemical instability as conductivity of ICPs depends on different doping levels, which can be decreased dramatically during washing, so as the conductivity of the ICPs [16]. Even though CNTs can provide superior mechanical properties and electrical conductivity, the cost of the material is one of the major obstacles. Moreover, the potential hazard of CNTs to human body is not yet confirmed. Therefore, among all the conductive materials, only metals can beat to provide the best conductivity and stability at low cost for electronic purposes.

26.1.3 Fabrication of Metallic Textiles and the Challenges

Fabrication of metallic textiles once was a hot topic in the field of both academic research and industry. The simplest strategy is to directly deposit a layer of metal on the textile surface via electroplating [21]. Indeed, this strategy demands external electricity supply, which imparts high cost to the fabrication. Moreover, metallization by electroplating requires conductive substrates. As all textile fibers are intrinsically nonconductive, electroplating on textiles requires surface coating of conductive materials, which impart extra steps to the fabrication. Apparently electroplating is not a promising approach to yield metallic textiles. Other work involved to achieve metallic textiles is to incorporate metal wires into textiles [22, 23]. Even though high conductivity can be achieved, low flexibility and bending ability of these textiles remain big issues due to the rigidness of the metal wires. In addition, heavyweight brought by the metal wires is one major trouble during transport, not to mention the adoption in any worn-on-the-body applications.

Metal particles are later directly deposited onto textile surface by means of physical and chemical deposition techniques [24–29]. These techniques provide a great upscale potential in the textile industry to fabricate metallic textiles in bulk. However, a number of limitations still exit. First, these deposition techniques require high vacuum condition, which are still cost-ineffective. Second, the adhesion of the as-deposited metal particles on the textile surface is so poor that metal

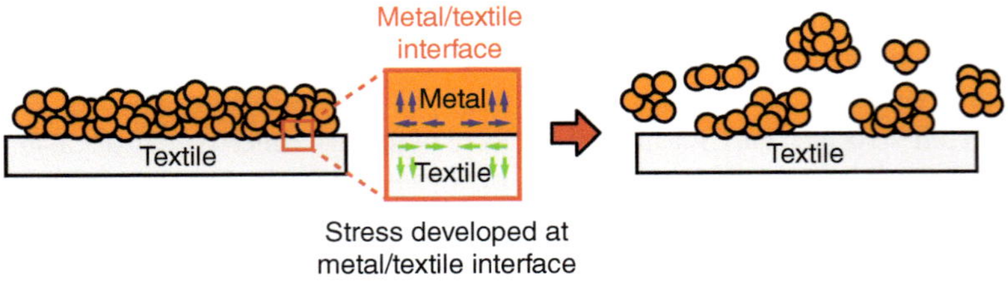

Figure 26.1 Schematic illustration showing stress developed at metal/textile interface when metal particles are barely attached onto the textile surface. Illustration is not drawn to scale.

particles are barely attached on the textile surface by weak physical interactions. Due to the mismatch between materials' physical properties, large amount of stress developed at the metal/textile interface can result in falling off of these metal particles from the textile surface (Figure 26.1). Third, the metal-deposited textile by these deposition techniques suffers majorly from the shadow effect. Those textile fibers on top shield the fibers underneath and block the passage of the sputtered or deposited metal particles. As a result, only outermost textile fibers can be coated with metal particles. With such a poor fiber contact, metallic textiles prepared by sputtering usually suffer from poor conductivity, which is not suitable for high-performance electronic applications.

In later development, electroless metal deposition (ELD) stands out as an alternative approach to depositing conformal metal particles on flexible polymeric substrates. Conventional ELD is an autocatalytic redox reaction, which involves the reduction of metal ions in the solution on the surface-catalyzed/activated substrate to form a conformal metal layer. Compared with other metal deposition techniques aforementioned, ELD is particularly attractive to achieve metal-deposited polymeric surface since ELD is an aqueous- and air-compatible process where external electricity, conductive substrates (required in electroplating), and vacuum condition (required in both physical and chemical vapor depositions) are eliminated. Thus, ELD is especially suitable for most of the intrinsically nonconductive textile materials that cannot withstand high temperatures and harsh environment. Demonstrations of ELD on textiles have been widely reported [30–34]. Notwithstanding, as the catalysts immobilized on the textile surface are physically attached via weak physisorption on the textile surface, adhesion of the as-deposited metal thin films on the polymeric substrates is still not assured owing to the metal/textile interfacial instability. Metal thin films can be easily delaminated from the textile surface.

26.1.4 Polymer Brushes Tackle Metal/Textile Interfacial Challenge

Metal/textile interfacial mismatch remains one significant challenge in the fabrication of robust and high-performance metallic textiles. To address this challenge, the metal/textile interface must be engineered in order to accommodate a stable metal/textile interfacial structure. To allow such a structure, one material must be present in between metal and textile for the establishment of a stable interface to ease material's mechanical mismatch. Recently, surface

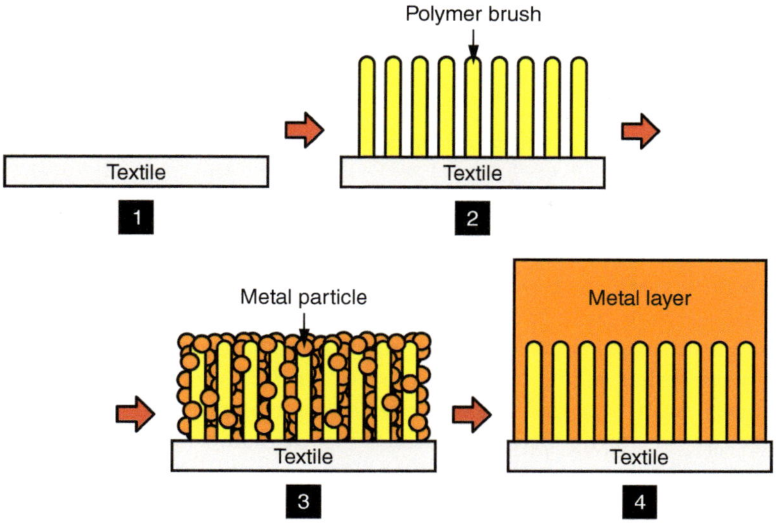

Figure 26.2 Polymer brush as an interfacial layer in between textiles and the deposited metal enables outstanding adhesion of metal on the textiles surface. Illustration is not drawn to scale.

modification of materials using polymer brushes emerges as one promising strategy to control surface architecture by changing the entire surface chemistry and topographical structures. In addition, polymer brushes possess large amount of functional groups along the polymer chains, which can be further utilized in subsequent chemical reactions. Taking both the advantages of polymer brushes as well as the aqueous- and air-compatible ELD process, an all-solution strategy, namely, PAMD, has been developed to fabricate highly durable, flexible, and stretchable metal conductors of various substrates including plastics, elastomers, papers, and textiles [35].

Literally, PAMD approach is a combined strategy of polymer brush grafting and ELD, involving (i) grafting of polymer brushes forming an interfacial layer on the textile surface and (ii) subsequent ELD of metals. The interfacial instability between the mechanically mismatched textiles and metal thin films can be tackled (Figure 26.2). Besides, outstanding adhesion of metal on the textiles surface is exhibited with robust metal coating to withstand repeated cycles of bending, stretching, and even washing. More importantly, unlike other conventional metal deposition strategies, PAMD is an all-solution process and is potentially favorable for scale production of flexible metallic textile conductors of both natural and synthetic substrates including cotton, wool, silk, polyester, nylon, Kevlar, spandex, etc. to suit different end purposes.

In this chapter, we specifically look into the PAMD process and briefly explore the underlying mechanism of PAMD. Also, we shall briefly explain the reaction chemistry behind PAMD, as well as the selection of polymer brush to enable PAMD. In Sections 26.3 and 26.4, we particularly focus on how PAMD realizes textile metal conductors and their applications in textile-based electronics.

Last but not least, the challenges of PAMD in realizing soft metal conductors in the foreseeable future will be discussed at the end of this chapter.

26.2 Polymer-Assisted Metal Deposition (PAMD)

26.2.1 What Are Polymer Brushes?

PAMD utilizes polymer brushes to assist ELD for strongly adhered metal thin films on substrate surface. The core part of the PAMD strategy is to combine polymer brush grafting and ELD to construct a thin layer of polymer brush in between the substrate and the as-deposited metal layer. Some basic understandings of polymer brushes will be introduced to readers in this section.

Polymer brushes are regarded as assemblies of macromolecular chains covalently tethered one end on a substrate surface. Once with sufficiently high grafting density, polymer chains are forced to induce significant stretch away from the tethered sites on the substrate surface to form a brushlike configuration [36, 37]. Polymer brushes grafted on a surface is commonly used to modify surface architecture of a substrate. Not only the surface chemistry is tailored by introducing new functional groups, but also the amount of functional groups can be dramatically increased for being further utilized in subsequent chemical reactions. Compared with self-assembled monolayer (SAM) [38], long-chain polymer brushes in a "brush-protruding" structure can accommodate many more functional groups along the polymer chains of repeated functional monomer units (Figure 26.3). With such a significant increase in the amount of functional groups on the surface, surface behavior of a brush-modified substrate can be controlled by the interaction of polymer brushes with the surrounding environment. The most well-known applications of polymer brushes are their switchable self-responsive behaviors to the surrounding solvent, ions, pH, temperature, etc. [39].

Grafting of polymer chains on one surface can be accomplished by either two brush grafting strategies. One is called the "grafting-to" strategy while another is called the "grafting-from" strategy. The "grafting-to" strategy involves the grafting of pre-synthesized long polymer chains on a substrate surface (Figure 26.4). These pre-synthesized polymer chains have functional binding sites for binding on the complementary sites on the substrate. "Grafting to" relies heavily on the diffusion of large polymer chains from the polymer solution to the substrate surface. Thus, it is very difficult to obtain polymer brushes with high grafting density, particularly due to steric hindrance of these long polymer chains that eventually prevent the surrounding polymer chains from approaching to the substrate surface and binding to the neighboring complementary sites. As the binding process proceeds with time, the substrate surface is getting more crowded and packed with bonded polymer chains. Those remaining polymer chains have to diffuse against the concentration gradient in order to reach the substrate surface, which is highly unfavorable for further binding. Thus, low grafting density will be resulted, leaving polymer chains in either "mushroom" or "pancake" configuration.

In order to achieve a "brush-protruding" configuration, materials scientists start manipulating the polymerization process in situ, and it is where the

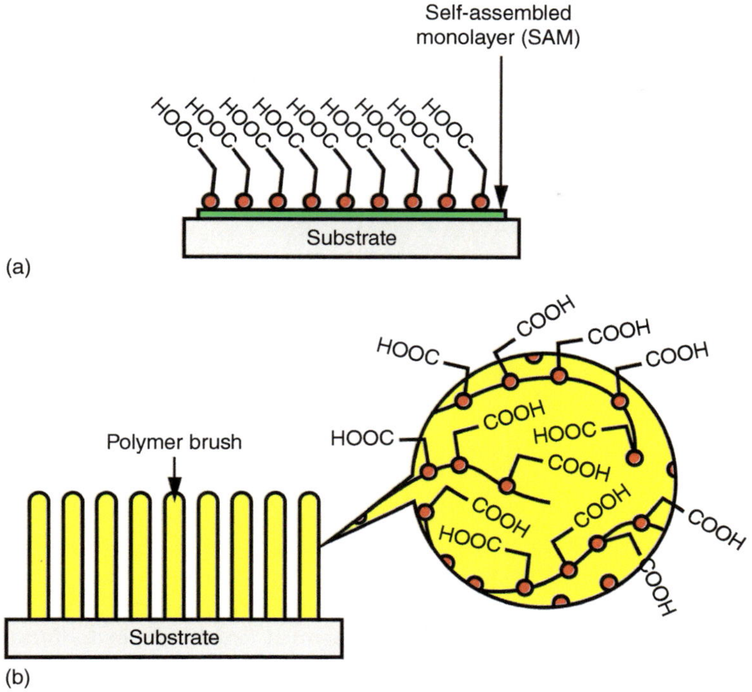

Figure 26.3 Schematic illustration comparing the amount of functional groups (–COOH as an example) on (a) SAM-modified and (b) polymer brush-modified substrate surfaces. Illustration is not drawn to scale.

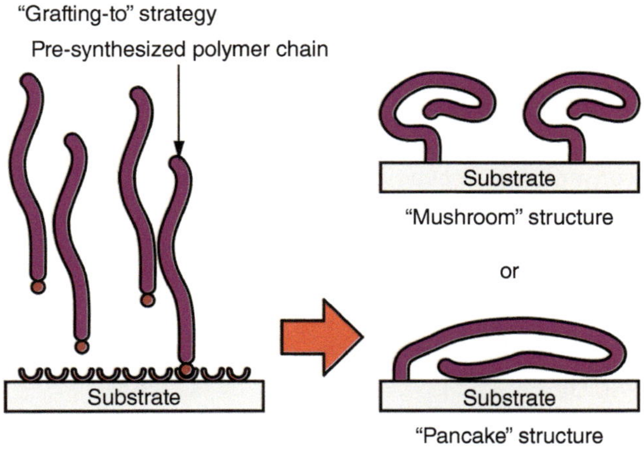

Figure 26.4 "Grafting-to" strategy only allows low brush grafting density, thus resulting in "mushroom" or "pancake" structure on the substrate surface, particularly due to the steric hindrance of the polymer chains as well as unfavorable diffusion against the concentration gradient. Illustration is not drawn to scale.

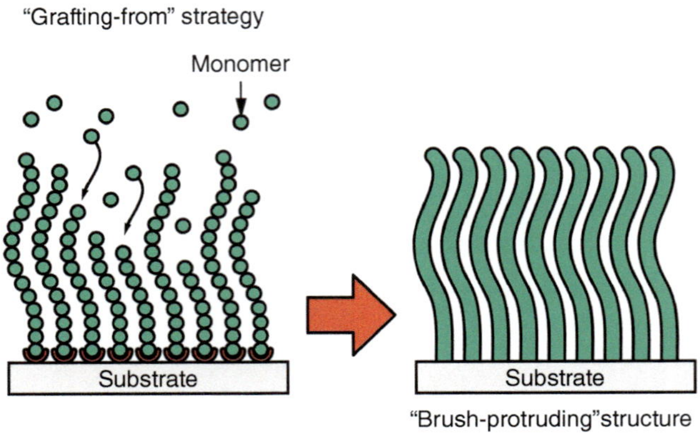

Figure 26.5 "Grafting-from" strategy allows high brush grafting density for a "brush-protruding" structure so as to maximize the amount of functional groups on the substrate surface for further chemical reaction. Illustration is not drawn to scale.

"grafting-from" strategy originates from. The "grafting-from" strategy involves in situ polymerization, i.e. polymerization of a polymer chain synthesis is initiated from a surface-immobilized initiator (Figure 26.5). To make in situ polymerization form a surface, a layer of initiator must first be deposited onto the substrate for the initiation of a polymerization process.

In PAMD, the "grafting-from" strategy is usually adopted to ensure a high grafting density of polymer brushes. There are mainly two polymerization strategies for the growth of polymer chains from the surface. First, it is the surface-initiated atom transfer radical polymerization (SI-ATRP) [40] that is the most commonly used strategy in PAMD. One supreme advantage of adopting SI-ATRP is that it is capable of preparing well-defined polymer brushes of desired molecular weight. However, it usually takes longer time to obtain a desired brush length. In addition, nitrogen protection is required throughout the entire polymerization process. Therefore, concerning the feasibility in upscaling the whole PAMD process, atom transfer radical polymerization (ATRP) is always not the first choice. On the contrary, conventional free radical polymerization (FRP) is always preferred in scale production as the whole process can be done under ambient conditions. However, as FRP is a kind of polymerization strategy that cannot be controlled, no well-defined polymer brushes of desired molecular weight can be obtained.

26.2.2 Mechanism of PAMD

PAMD is a universal strategy that takes both the advantages of polymer brushes and ELD to achieve conformal metal layer under aqueous- and air-compatible conditions [35]. Here, we particularly focus on the PAMD process on textiles, which consists of four steps: (i) surface modification of textiles with a functional organosilane layer, (ii) grafting of polymer brushes on the organosilane-modified textiles, (iii) immobilization of catalytic moieties onto the polymer brushes, and (iv) ELD of metal on the catalyst-loaded areas (Figure 26.6).

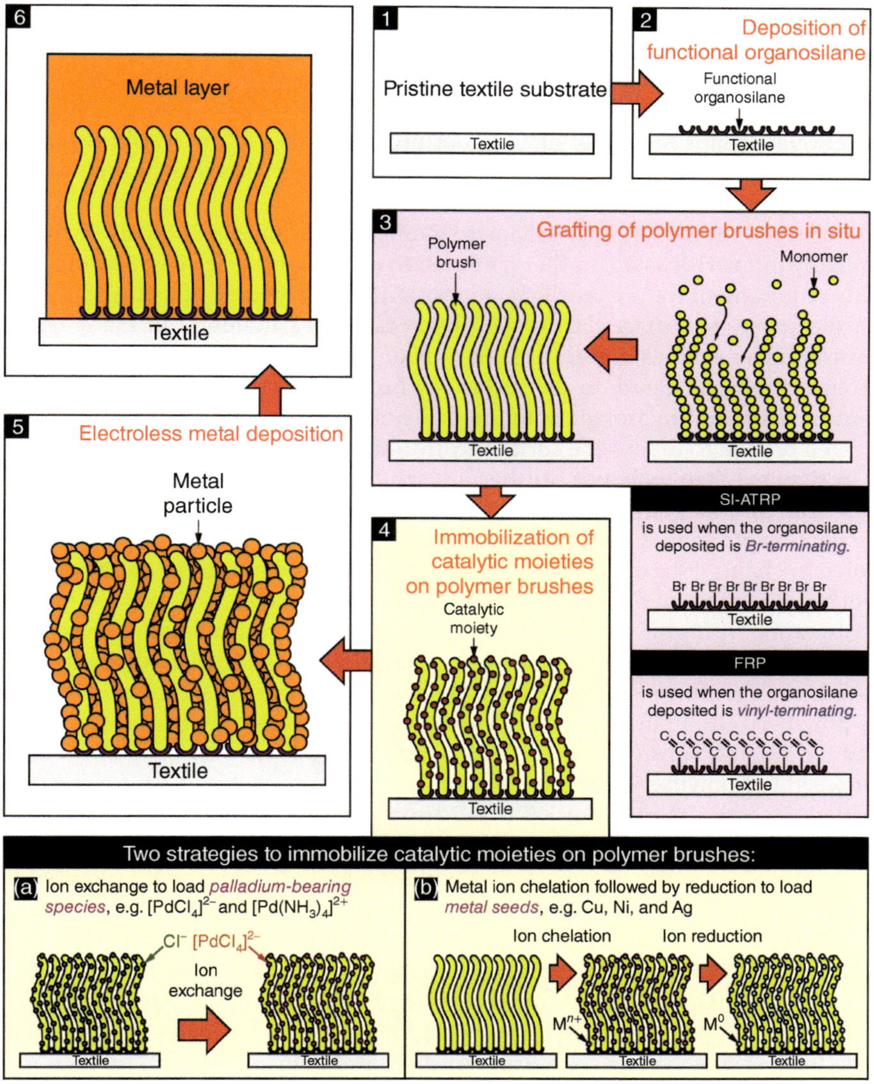

Figure 26.6 Typical PAMD process on a textile substrate. (1) A pristine textile substrate is first deposited with a layer of functional organosilane as shown in (2). (3) Surface modification of textiles with a functional organosilane can initiate grafting of polymer brushes in situ from the organosilane-modified textile surface. (4) Immobilization of catalytic moieties onto the polymer brushes either by (a) ion exchange to load palladium-bearing species or by (b) metal ion chelation and reduction to load metal seeds. (5) Electroless metal deposition (ELD) of metal on the catalyst-loaded areas. Panel (6) represents the typical metal layer on the brush-grafted textile substrate. Illustration is not drawn to scale.

In brief, one textiles surface intended to be coated with metal is first modified with a layer of functional organosilane. In general, any kinds of textile materials can be used as the substrates in PAMD. The function of such an organosilane layer is to provide initiation sites for the subsequent in situ grafting of polymer brushes.

In a typical silanization process on textile materials, organosilane is hydrolyzed under suitable pH where the hydrolyzed organosilanes are then condensed on the textile surface by self-assembly. This procedure is particularly critical for textile materials to proceed to subsequent brush grafting as most of the textiles lack functional groups for initiating polymer brush grafting. If the textile surface is not modified by organosilane, there will be no polymer brushes grafted on the textile surface.

Moreover, the use of organosilane to modify textile surface is highly favorable to many industries as it is rather common in materials surface engineering. The only requirement for organosilane deposition is the presence of hydroxyl groups on the substrate surface. Hydroxyl groups can be easily found in many textile natural fibers such as cotton. For man-made fibers such as polyester, hydroxyl groups can be rendered on the surface by just immersing the textiles into concentrated sodium hydroxide solution or by exposing the textile substrates with oxygen plasma to render the surface hydrophilic.

In the next step, polymer brushes are synthesized in situ from the organosilane initiators. Depending on the types of organosilanes deposited, suitable polymerization strategy is adopted. For example, SI-ATRP will be used to grow the polymer chains if the organosilane previously deposited is Br terminating; or FRP will be used to grow the polymer chains if the organosilane is deposited is vinyl terminating. It is also worth mentioning that polymer brush length varies upon the adoption of different polymerization strategies. For ATRP, well-defined brush length can vary from ~20 to ~50 nm, depending on the types of polymers as well as polymerization parameters such as ligand to transition metal ratio, types of counterion and solvent, etc. [40]. Conversely, in FRP, no well-defined molecular weight of polymer brush can be achieved. Instead, the brush length can only achieve in a broad range of ~5 to ~20 nm, depending on the polymerization parameters such as monomer concentration and polymerization temperature. Even so, FRP strategy eliminates the undesirable inert atmosphere required in ATRP, which is more suitable in scale production.

After grafting of polymer brushes on the textile surface, catalytic moieties are immobilized onto the polymer brushes for initiating subsequent metal ELD. In PAMD, the most commonly used catalysts to initiate metal ELD are either (i) palladium-bearing species or (ii) metal seeds. Their respective loading mechanisms on the brushes are quite different. To load the palladium-bearing species on the brushes, it is done via an ion exchange process. Examples of palladium-bearing catalyst for ELD include tetrachloropalladate(II) anion ($[PdCl_4]^{2-}$) and tetraamminepalladium(II) cation ($[Pd(NH_3)_4]^{2+}$). It is worth mentioning that in order to allow successful ion exchange, the selection of palladium-bearing species is highly specific in which the grafted polymer brushes must bear suitable functional groups for capturing these palladium-bearing species. For instance, $[PdCl_4]^{2-}$ is one of the most reported catalytic moieties to couple with poly[2-(methacryloyloxy)ethyl]trimethylammonium chloride (PMETAC) brushes, due to the high affinity of the quaternary ammonium groups (QA^+) in the PMETAC brushes toward $[PdCl_4]^{2-}$, while another catalytic moiety $[Pd(NH_3)_4]^{2+}$ can only couple with poly(methacryloyl ethyl phosphate) (PMEP) brushes, due to the high affinity of the phosphate groups in the PMEP brushes toward $[Pd(NH_3)_4]^{2+}$.

Another method to immobilize catalytic moieties onto the brushes is the absorption of metal seeds on the brushes as catalysts for the initiation and growth of metals. To do so, chelation of metal cations onto the grafted polymer brushes is first required, followed by subsequent reduction of these metal ions to form a thin layer of metal seeds on the brushes. Such an extra step of metal ion reduction is required that is usually accomplished by sodium borohydride solution. Chelation of metal ions on the brushes is specially suitable for those brushes that possess carboxylate groups, for instance, poly(acrylic acid) (PAA). Details will be provided in Section 26.2.3.2.

In the final step of PAMD, the catalytic moieties loaded substrates are immersed into the ELD plating bath for metal deposition, where the areas immobilized with catalytic moieties act as effective catalytic sites for metal ELD. It should be noted that the loading amount of catalytic moieties plays an important role in the ELD process, which affects the amount of metal deposited on the brushes. Azzaroni et al. evidenced the effect of the metal deposited by varying the amount of the catalytic moieties loaded on the brushes [41]. Scarce loading of the catalytic moieties can only result in thin metal deposited on the substrate. Other than the amount of catalytic moieties, thickness of the metal deposited is also affected by brush length, ELD time, and concentration of the ELD bath [42].

26.2.3 Brush Selection for ELD

The selection of polymer brush is highly critical and selective in PAMD and was systematically reviewed by Liu et al. [43]. It should be noted that only those polymer brushes that possess specific functional groups for the immobilization of catalytic moieties to initiate subsequent ELD are only applicable in PAMD. To help further understand the system, some polymer brushes used in PAMD for ELD will be introduced as follows according to their intrinsic charges that the brushes are carrying, i.e. cationic, anionic, and nonionic.

26.2.3.1 Cationic Polymer Brushes

Cationic polymer brushes, as its name suggests, are positively charged polyelectrolytes in an aqueous solution. Examples of cationic polymer brushes are the earlier mentioned PMETAC and poly[1,1′-bis(4-vinylbenzyl)-4,4′-bipyridinium dinitrate] (PVBVN), with chemical structures illustrated in Figure 26.7.

The mostly reported cationic polymer brush in PAMD is the PMETAC brush that possesses QA^+ groups for strong coupling with anionic catalytic moiety $[PdCl_4]^{2-}$ to initiate ELD. Azzaroni et al. first demonstrated the utilization

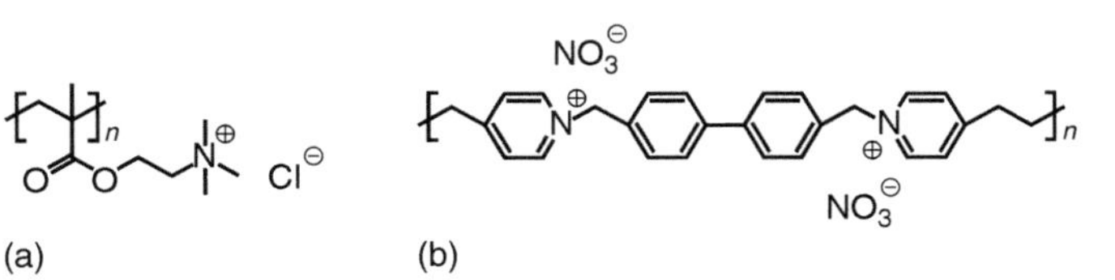

Figure 26.7 Chemical structures of two cationic polymer brushes: (a) PMETAC and (b) PVBVN that can be used in PAMD.

of PMETAC brushes that were grafted on wafer surface using SI-ATRP [41]. PMETAC brushes captured anionic catalytic moieties $[PdCl_4]^{2-}$ for subsequent ELD where a robust metal layer was selectively deposited with outstanding adhesion properties.

In 2009, Cui et al. demonstrated the fabrication of silver-coated polyimide (PI) film via ultraviolet (UV)-induced surface graft copolymerization of 1,1′-bis(4-vinylbenzyl)-4,4′-bipyridinium dinitrate (VBVN) to yield PVBVN bonded to the PI film surface [44]. Even though PVBVN is cationic, its chemical route for subsequent ELD is quite different from that of PMETAC as previously discussed. As mentioned, PMETAC requires the coupling of a catalytic moiety for subsequent ELD. On the contrary, PVBVN consists of viologen derivatives. Upon UV radiation, viologen derivatives on the brushes undergo redox reaction in which the metal ions in the solution can be readily reduced on the surface of the viologen-grafted substrate to form a very thin layer of metals. In their experiment, Cui et al. successfully grafted PVBVN brushes on the PI surface via a UV-induced polymerization, followed by exposure to UV irradiation in which silver nanoparticles were deposited on the surface of the PVBVN brushes through photo-induced reduction of silver ions in the silver salt solutions to obtain conformal and uniform silver layer on the PI surface.

26.2.3.2 Anionic Polymer Brushes

Anionic polymer brushes are negatively charged polyelectrolytes in an aqueous solution. Examples of anionic polymer brushes include PAA and PMEP with chemical structures respectively illustrated in Figure 26.8.

The chemical route for PAA enabling ELD is quite selective, thanks to the carboxylate groups present in the PAA brushes that allow chelation of metal cations. The mechanism is to first chelate metal cations on carboxylate groups in the PAA brushes, followed by a reduction of metal cations to obtain a thin layer of metal seeds for subsequent metal ELD. Garcia et al. reported the ELD of copper on different polymeric substrates by grafting PAA via a GraftFast technology [45, 46]. Taking the advantage of PAA brush that can complex with copper(II) ions, after grafting of PAA brushes, the PAA-grafted sample was immersed into copper(II) ion source to induce an chelation process, followed by a reduction of the copper(II) ions using sodium borohydride as a reducing agent. As a result, a thin layer of metal seeds is adsorbed on the bushes, which act as efficient catalysts for the subsequent ELD. Finally, the sample was put into a copper ELD bath for the subsequent growth of copper metals. This strategy can eliminate the use of costly palladium-bearing catalyst.

For another anionic brush PMEP, the mechanism in fact is quite similar to the cationic PMETAC brushes, which is capable of coupling with specific counterions as catalytic moiety due to the presence of specific functional groups. PMEP brush possesses phosphate groups for strong coupling with cationic catalytic moiety

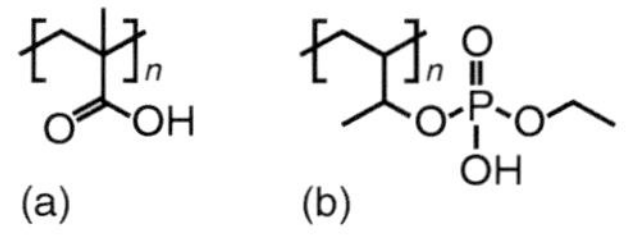

Figure 26.8 Chemical structures of two anionic polymer brushes: (a) PAA and (b) PMEP that can be used in PAMD.

Figure 26.9 Chemical structures of two nonionic polymer brushes: (a) P4VP and (b) PAN that can be used in PAMD.

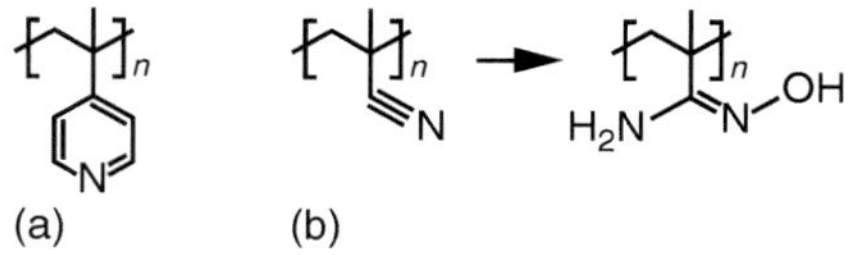

$[Pd(NH_3)_4]^{2+}$ to initiate ELD. Grafting of PMEP was demonstrated by Liu et al. [47]. They fabricated two oppositely charges polymer brushes that were cationic PMETAC and anionic PMEP via SI-ATRP on a substrate. Respective immobilization of catalytic anionic moieties $[PdCl_4]^{2-}$ for PMETAC and cationic moieties $[Pd(NH_3)_4]^{2+}$ for PMEP was then carried out. Subsequent copper and nickel ELD yielded bimetallic patterns on the substrate surface.

26.2.3.3 Nonionic Polymer Brushes

Nonionic polymer brushes are neutrally charged polyelectrolytes in an aqueous solution. Even though they are non-charge-bearing, they possess special functional groups with binding mechanism for the attachment of catalytic moieties for subsequent ELD. Examples of nonionic polymer brushes are poly(4-vinylpyridine) (P4VP) and polyacrylonitrile (PAN) with chemical structures illustrated in Figure 26.9.

Yang et al. first demonstrated the grafting of P4VP on poly(tetrafluoroethylene) (PTFE) and poly(vinylidene fluoride) (PVDF) films via UV-induced grafting [48]. After grafting of P4VP brushes, direct immobilization of the palladium catalyst was carried out for the subsequent ELD of nickel. Due to the presence of nitrogen-containing pyridine ring on the P4VP brushes, the palladium species complexed with the nitrogen moiety in the pyridine ring, therefore enabling a strong chemisorption of the palladium species on the grafted polymer brushes to initiate subsequent ELD.

Similarly, grafting of PAN brushes was demonstrated by Yu et al. via a UV-induced grafting on poly(tetrafluoroethylene-*co*-hexafluoropropylene) (FEP) films [49]. However, due to the lack of any functional groups on PAN brushes for the immobilization of the catalytic moieties for ELD, cyano groups on PAN brushes were firstly amidoximated using hydroxylamine hydrochloride to introduce amidoxime groups on the PAN brushes. Amidoxime groups are capable of complexing with palladium species. The amidoximated PAN brushes were then subjected to sensitization in the $SnCl_2$ solution, followed by activation in the $PdCl_2$ solution. Subsequent ELD was then carried out to yield a conformal copper layer on the substrate surface.

26.2.3.4 Advantages of Using Polymer Brushes as ELD Platform

The major innovation in PAMD strategy is the use of polymer brushes that act as a bridging layer in between the substrate and the as-deposited metal so as to ease the stress developed at the substrate/metal interface. As a result, outstanding metal adhesion on the substrates can be ensured. Theoretically, metal deposition can be assisted by adopting a layer of SAM once it possesses suitable functional groups for metal deposition [50, 51]. However, when comparing polymer brushes to SAM in assisting ELD, polymer brush has two major advantages that cannot be accomplished by SAM. First, polymer brush can numerously increase the number of immobilization sites for catalytic moieties or metal seeds attachment for metal ELD. Polymer brushes can provide a "three-dimensional" functional layer, which accommodates much more functional binding sites or accommodation sites for the attachment of catalytic moieties. With an increased amount of catalysts or metal seeds loaded onto the polymer brushes, the rate of ELD can be significantly increased under the same ELD condition, due to the increased number of reactive sites for the growth of metal. Conversely, catalyst uptake in SAM is rather poor than in polymer brushes as SAM only have "one-dimensional" functional layer, which limits the number of surface binding sites for the attachment of catalytic moieties for metal ELD. Second, polymer brush allows an interpenetrating matrix structure to be established that enables a remarkable adhesion between the as-deposited metal and the substrate. In this context, polymer brush as an interfacial layer can be regarded as a layer of adhesive that bridges the deposited metal and the substrate surface. Outstanding robustness of metal layer against repeated cycles of mechanical actions is exhibited, and remarkable adhesion of metal on the substrate surface is therefore enabled.

26.2.3.5 Fabrication of Metallic Textiles via PAMD

Textiles inheritably of great flexibility and wearability are suitable platforms for electronics. One advantage of using textiles as the substrate for PAMD is that the metal-coated textiles can be flexibly integrated into many fabrics and even garments as conductive interconnects, contacts, and electrodes for many electronic purposes. In 2010, Liu et al. first reported on the preparation of highly durable conductive metal-coated cotton yarns via PAMD approach [52]. They first modified the cotton fiber surface with Br-terminating silane for the grafting of PMETAC brushes by SI-ATRP. Subsequent ELD of copper on the cotton fiber surface resulted in copper-coated cotton yarns with outstanding mechanical durability and electrical stability upon repeated cycles of bending, stretching, and washing (Figure 26.10a). Application of these metallic cotton textiles was also demonstrated as interconnects in textile electronics by connecting a 9 V battery and a light emitting diode (LED), showing great potential to be used as electrical wires in wearable and flexible electronic devices (Figure 26.10b,c).

However, concerning the feasibility in scale production, such demonstration by Liu and coworkers is not practical as ATRP requires long polymerization time as well as protection under inert nitrogen atmosphere. Therefore, further modification has been made in 2014 by Wang et al. for a more practical and scalable approach to yield metallic textiles, in which the polymerization of PMETAC was carried out via in situ FRP instead of SI-ATRP [42]. In the fabrication, vinyl-terminating silane was deposited on the cotton fibers to allow FRP to graft PMETAC brushes on the fiber surface. The polymerization time can be greatly reduced to ~30 minutes with comparable sheet resistance achieved under the same experimental conditions. Since the whole fabrication is aqueous and air compatible, the fabrication can be scaled up and incorporated with pad-dry-cure technology – a textile finishing technique commonly used in dyeing and finishing with special treatment on fabrics. Importantly, PAMD can be applied on wide varieties of natural and synthetic fibers such as nylon, polyester, Kevlar, and spandex (Figure 26.10d–g), showing a great versatility of

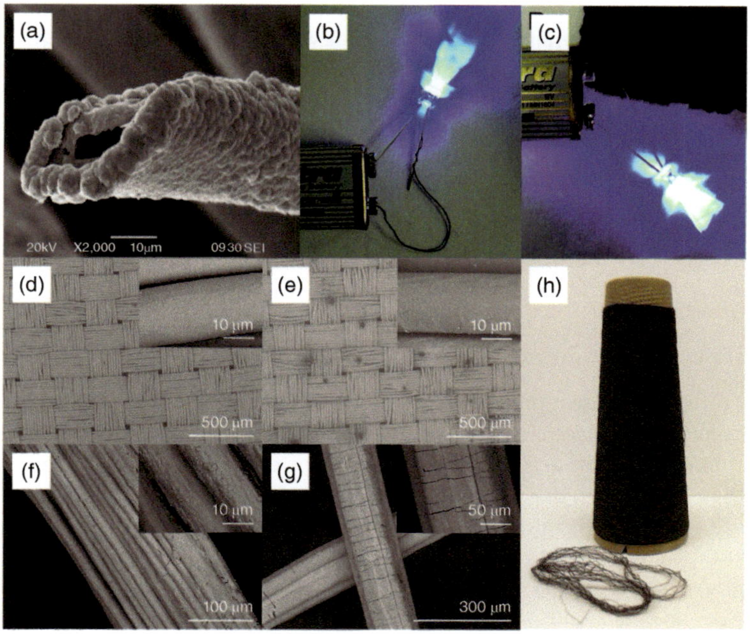

Figure 26.10 (a) Scanning electron microscopy (SEM) images of the cross-sectional morphologies of PMETAC-modified cotton fiber that was immersed in copper ELD plating bath for 30 minutes. (b) Conductive yarn or (c) fabric was used as electrical wire for powering a blue LED. The yarn or fabric was placed in contact with the battery and LED without using extra conductive glue or paste. Source: (Panel c) Reproduced with permission from Liu et al. [52]. Copyright 2010, American Chemical Society. SEM images of Cu-coated (d) nylon fabric, (e) polyester fabric, (f) Kevlar yarns, and (g) spandex monofilaments. Insets are SEM images of fibers in high magnification. Source: Reproduced with permission from Wang et al. [42]. Copyright 2014, John Wiley & Sons. (h) Digital image of a 500 m long nickel-coated cotton yarn wound on a spinning cone. Source: Reproduced with permission from Liu et al. [53]. Copyright 2015, Macmillan Publishers Limited.

PAMD on different textile substrates that well suit different end applications. With such a potential in scale-up fabrication, Liu et al. first demonstrated the feasibility of scale fabrication of 500 m long nickel-coated cotton yarn prepared on spinning cone in one batch (Figure 26.10h) [53]. These nickel-coated cotton yarns still possess both textile-like flexibility and an outstanding conductivity ($\sim$1.3 Ω/cm), which are highly suitable for flexible electronic applications.

As the performance of the as-fabricated metallic textiles is crucial to the fabrication of any electronic devices, Wang et al. analyzed the effect of different experimental parameters in PAMD on the electrical performance of the metallic textiles. It is found that critical parameters determining the performance of metallic textiles in PAMD are majorly threefold, including polymerization time, polymerization temperature, and ELD time [42]. It is also worth mentioning that the conductivity of the as-fabricated metallic textiles highly depends on the fibrous structure of the textile materials. For example, metal-coated spun yarns provide relatively lower conductivity when compared with those using filament yarns as substrates due to more fiber-to-fiber contacts. Thus, the contact resistance will be higher when compared with metal-coated filament yarns. Even so, natural staple fibers are of more advantages in holding up more active materials in some device fabrication, thanks to the larger surface areas that the fibers can provide with, as well as the intrinsic ability to absorb foreign materials essential for boosting device performance.

26.3 Strategy to Fabricate Patterned Metallic Traces in PAMD

26.3.1 Why Patterning Is Required?

Theoretically, infinitely large metal-coated textiles can be prepared by PAMD for scale applications in large-area electromagnetic shielding, electrostatic discharge, etc. However, metal patterns are always demanded for the constructions of electronic circuitries. To pattern precise metal traces on textile surface, conventional resist-based lithographic techniques long been developed in the semiconductor industries should be borrowed, which involve multiple steps such as masking, resist patterning, etching, metal deposition, lift-off, annealing, etc. Nonetheless, these techniques are all incompatible with soft polymeric substrates especially textiles, particularly due to the harsh chemical environment where most textiles cannot withstand, as well as the high temperature required in the annealing step (~900 °C) that can almost damage all textile substrates. Concerning the incompatibility in transitional patterning methods with textile substrates, rather than adopting the old technologies, recently, the continuous roll-to-roll (R2R) printing to realize flexible electronics comes into the spotlights. The continuous manner and the high throughput enabled in the R2R printing technology have always been impressive to allow greater production capacity because it is ultrafast, accurate, and less facility-demanded (only common printing facilities are required). In the textile industry, the printing sectors are in fact very well developed, especially the fabrication of various prints and patterns using screen printing and ink-jet printing. PAMD is also applicable to the fabrication of patterned metal-coated textiles by printing strategies, which can be realized by either two strategies: (i) catalytic moiety ink patterning or (ii) copolymer ink patterning.

26.3.2 Catalytic Moiety Ink Patterning

The method of patterning metallic trace in PAMD on textile is to pattern the catalytic moieties that are responsible for the initiation of metal ELD. The schematic illustration of this approach is shown in Figure 26.11. Only the areas immobilized with catalytic moieties can result in metal deposition to yield patterned metal coatings on the textile surface. In 2013, Guo et al. first demonstrated the synthesis of well-defined metal traces on textile substrates via matrix-assisted catalytic printing (MACP) [54]. Such a printing strategy is in fact a modified PAMD process, where the catalytic moieties were first prepared as printable ink with suitable viscosity tuned for different printing techniques such

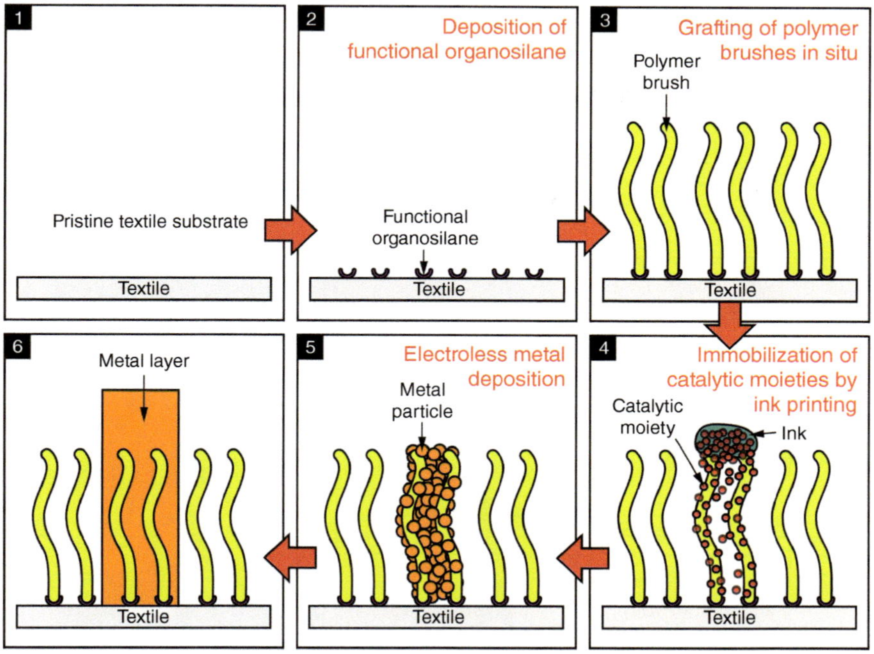

Figure 26.11 Schematic illustration showing the procedures to obtain metal patterns on textile surface by catalytic moiety ink patterning. (1) Pristine textile substrate is first deposited with a layer of functional organosilane as shown in (2). (3) Grafting of polymer brushes in situ from the organosilane initiators. (4) Printing of ink containing catalytic moieties with a delivering matrix polymer, poly(ethylene glycol) (PEG) as the moiety carrier and thickening agent. Catalytic moieties then diffuse from the ink to the polymer brushes. After diffusion, localized brush area is loaded with catalytic moieties. (5) Electroless metal deposition (ELD) to obtain patterned metal trace on the textile surface to obtain metal layer on textile surface in (6). Illustration is not drawn to scale.

as screen printing and ink-jet printing. To do so, the viscosity of the catalytic moiety solution must first be fine-tuned. Guo and coworkers used a delivering matrix polymer, poly(ethylene glycol) (PEG) serving as both the carrier of the catalytic moieties as well as the thickening agent for tuning the ink viscosity.

As a proof of concept, a layer of PMETAC brushes was first grafted on papers and fabrics via FRP. Subsequently, catalytic $[PdCl_4]^{2-}$ as ink was printed on the desired areas intended for metal deposition via screen printing or ink-jet printing. The catalytic moieties as a result diffused from the delivering matrix polymer PEG to the receiving matrix polymers, i.e. the grafted PMETAC brush layer. As the area immobilized with $[PdCl_4]^{2-}$ determines the area for the subsequent ELD of metal, patterned metal-coated paper and fabric can be resulted with feature size of the metal patterns ranging from micro- to millimeter scales. Importantly, such a patterning process is compatible with industrial printing facilities such as screen printing and ink-jet printing in the textile industry, which is of great importance in making patterned electronic circuits for textile-based electronic devices.

26.3.3 Copolymer Ink Patterning

Recently, PAMD strategy was further modified for an ultrafast fabrication of metallic patterned textiles. A first printable copolymer ink used for the rapid fabrication of metallic textiles in PAMD was proposed by Yu et al. [55]. The schematic illustration of this approach is shown in Figure 26.12. To achieve this,

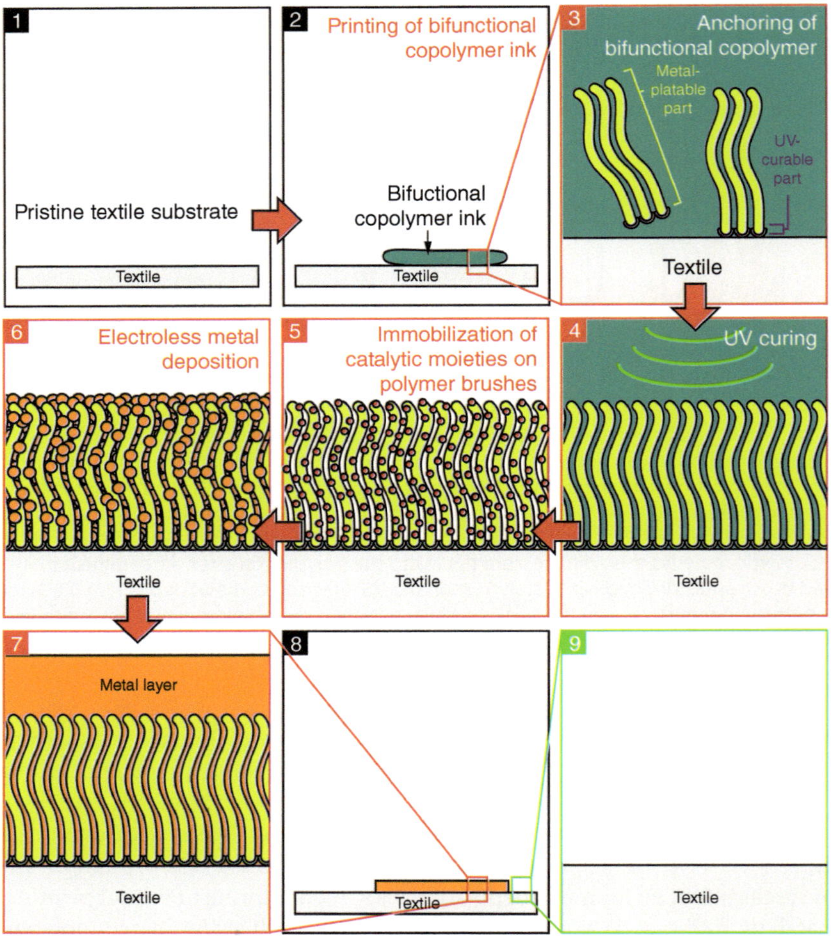

Figure 26.12 Schematic illustration showing the procedures to obtain metal patterns on textile surface via copolymer ink patterning. (1) Pristine textile substrate is first deposited with a layer of bifunctional copolymer ink as shown in (2) via screen printing or ink-jet printing. The bifunctional copolymer consists of two major components that are a metal-platable part and a UV-curable part as indicated in (3). (4) UV radiation simultaneously triggers the self-cross-linking and surface grafting of the copolymer under ambient conditions. (5) Localized brush area is loaded with catalytic moieties. (6) Electroless metal deposition (ELD) to obtain patterned metal trace on the textile surface as shown in (7) and (8). Areas without ink printing result in none of the metal deposited as shown in (8) and (9). Illustration is not drawn to scale.

a bifunctional UV-curable and metal-platable copolymer poly(4-methacryloyl benzophenone-*co*-2-methacryloyloxy ethyltrimethylammonium chloride) [P(MBP-*co*-METAC)] was first synthesized as ink. Instead of controlling the areas where the catalytic moieties immobilized onto like catalytic moiety ink patterning, this strategy takes a step forward to pattern the brushes on to the area where metal particle is intended to deposit onto. After printing of this copolymer on the textile surface via screen printing or ink-jet printing, a one-step curing procedure of the copolymer ink using UV radiation can yield metal-platable copolymers securely cross-linked on the textile surface. Subsequent loading of catalytic moieties and ELD can readily yield the textile surface with strongly surface-deposited metallic patterns. Importantly, the copolymer ink can be fabricated easily under ambient condition and is readily compatible with commercialized screen printing and ink-jet printing facilities. Such a fabrication can greatly reduce the fabrication time when compared with the conventional PAMD process.

26.4 Applications in Textile-Based Electronics

26.4.1 Supercapacitor

In 2015, Liu et al. first demonstrated a flexible and wearable yarn-based supercapacitor (SC) utilizing nickel-coated cotton yarns synthesized by PAMD [53]. The nickel-coated cotton yarns served as current collectors in the as-prepared SC, which processed high conductivity and robustness in repeated bending with only 2% increase in electrical resistance after 5000 bending cycles. With subsequent electrochemical deposition of reduced graphene oxide (RGO) on the nickel-coated cotton fiber surface and the deposition of electrolyte gel, a solid-state yarn SC was obtained (Figure 26.13a). Owing to the robustness of the nickel metal and also the hierarchical structures offered by the nickel-coated cotton fibers, the solid-state yarn SC not only achieved remarkable performance with efficient ionic and electronic transport (Figure 26.13b,c) but also maintained good cyclic stability (82%) even after 10 000 cycles of charging and discharging (Figure 26.13d). Moreover, the capacitance of the SC yarn increased proportionally in a first order as the device length increased. As indicated in Figure 26.13e, the capacitance of the SC yarn increased from 0.28 to 1.52 F with the length increased from 3 to 17 cm, showing a 20-fold increased performance when compared with other work reported in the literature [56, 57].

Later in 2016, Huang et al. successfully fabricated fabric-based SC using nickel-coated cotton fabrics (Ni cotton) synthesized by PAMD [58]. They eliminated any harsh processing conditions required in the conventional fabrication of fabric-based SC and reported a cost-effective and continuous approach for high-performance SC fabric via a one-step electrospinning of carbon nanowebs (C-web) on Ni cotton to first form C-web@Ni-cotton electrode. Then, two pieces of C-web@Ni-cotton electrodes were assembled together with a piece of pristine cotton fabric as the separator in between to form a symmetrical SC fabric device (Figure 26.14a). Fabric samples after deposition of nickel and

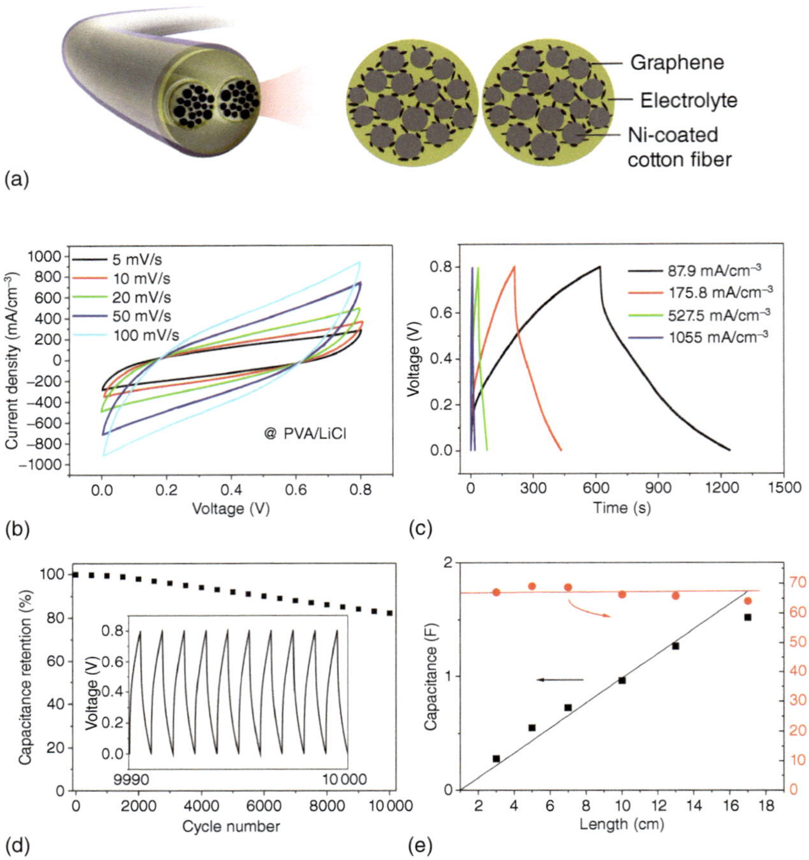

Figure 26.13 Performance of solid-state supercapacitor (SC) yarns. (a) Schematic illustration of the structure of one SC yarn. (b) Cyclic voltammetry (CV) curves of the device at scan rates ranging from 5 to 100 mV/s. (c) Galvanostatic charge/discharge (GCD) curves of the device at different current densities. (d) Cycle life of the device at a current density of 439.6 mA/cm^3. The inset is the GCD curve from the 9990th to the 10 000th cycle. (e) Device capacitance as a function of the device length. Source: Reproduced with permission from Liu et al. [53]. Copyright 2015, Macmillan Publishers Limited.

subsequently with C-web are shown in Figure 26.14b. Remarkably, Ni cotton retained its good flexibility (Figure 26.14c) even though nickel particles were densely deposited on the cotton fibers (Figure 26.14d,e). After the electrospinning process, a three-dimensional nanoweb was formed on the surface of Ni cotton (Figure 26.14f). To demonstrate the practical application of such C-web@Ni-cotton fabric electrode, solid-state SC fabric devices using a polyvinyl alcohol (PVA)/LiCl gel electrolyte was fabricated. As shown in Figure 26.14g,h, the solid-state device exhibits a high areal capacitance of 275.8 mF/cm^2 at 1 mA/cm^2 with a low internal resistance. Importantly, such SC fabrics can be further integrated into commercial textiles with any desirable forms, indicating application potentials in wearable electronics.

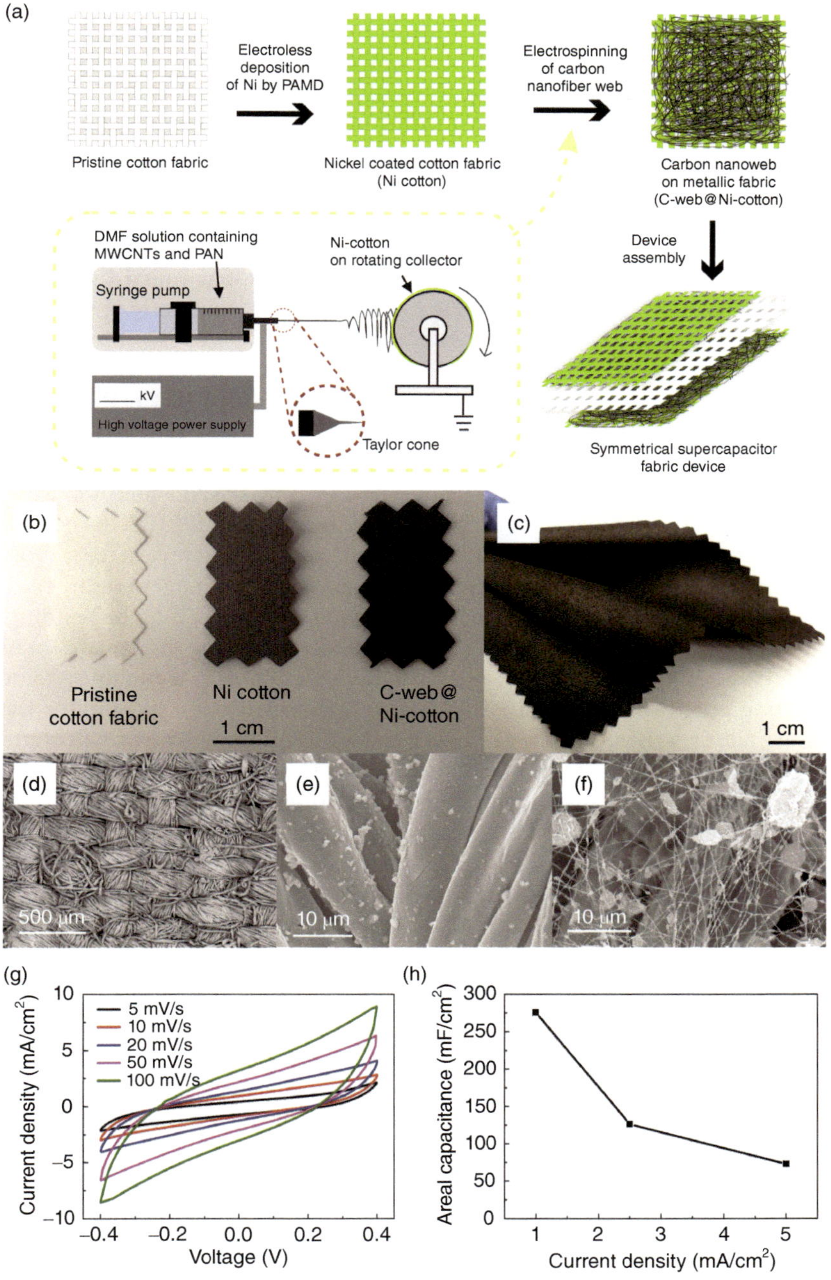

Figure 26.14 (a) Schematic illustration of the fabrication procedures of supercapacitor fabrics. The inset shows the details of the one-step electrospinning setup. (b) Photographs of the pristine cotton fabric, nickel-coated cotton fabric (Ni cotton), and carbon nanofiber web-coated Ni-cotton fabric (C-web@Ni-cotton). (c) Photograph of the Ni-cotton fabric with large dimension, showing good flexibility. (d, e) SEM images showing the surface morphology of the Ni-cotton fabric. (f) SEM image of the C-web@Ni-cotton fabric electrode. (g) CV curves of solid-state C-web@Ni-cotton supercapacitor fabric. (h) Summary of the areal capacitance of the supercapacitor at different current densities. Source: Reproduced with permission from Huang et al. [58]. Copyright 2016, Royal Society of Chemistry.

26.4.2 Triboelectric Nanogenerator

Zhao et al. fabricated the first machine-washable textile-based triboelectric nanogenerator (t-TENG) by adopting the metallic textile fabricated by PAMD [59]. Such a textile-based sensor was successfully demonstrated as an effective human respiratory monitor to detect human respiratory patterns. The whole fabrication of the device was based on plain weaving of copper-coated polyethylene terephthalate (PET) yarns (Cu-PET) as warp and PI-coated Cu-yarns (PI-Cu-PET) as weft to form a flexible fabric (Figure 26.15a). In the fabrication, 1-ply PET yarn was first coated with a thin layer of copper metal by PAMD technique to obtain Cu-PET yarn (Figure 26.15b). Subsequently, two 1-ply Cu-PET yarns were twisted together to obtain a 2-ply Cu-PET yarn to further decrease the linear resistance of the yarn (Figure 26.15c). After coating PI on the PI-Cu-PET yarn (Figure 26.15d,e), 2-ply Cu-PET yarns and PI-Cu-PET yarns were woven to form a flexible fabric. As a proof of concept, the metallic yarns were wound onto a cone for industrial automatic weaving (Figure 26.15f,g). The as-prepared woven fabric as t-TENG (Figure 26.15h) comprised 9 pieces of 2-ply Cu-PET as warp and 60 pieces of PI-Cu-PET as weft (Figure 26.15i). As copper and PI can serve as one effective triboelectric couple, every yarn crisscross intersection in the device is regarded as a triboelectric nanogenerator. As a result, upon even very subtle deformation of the device by tapping or bending, the contact area at each yarn crisscross intersection changes thus leads to an effective generation of triboelectric charges. Thus, such device is highly suitable for detecting human subtle motions such as human respiration to monitor the respiratory patterns. Besides, the device is remarkably durable that can withstand washing cycles in the washing machines, thanks to the robust metallic yarns fabricated by PAMD for the outstanding metal adhesion on the yarn surface. Upon machine washing up to 20 cycles in the standard washing machine, 2-ply Cu-PET yarns can maintain a linear resistance $<0.6\,\Omega/\text{cm}$ (Figure 26.15j), with copper particles still strongly adhered on the yarn surface (Figure 26.15k–o).

26.4.3 Solar Cell

In 2017, Zhen et al. successfully fabricated foldable polymer solar cells (PSCs) on woven fabrics by a freestanding wet transfer method [60]. Since the conventional method to fabricate solar cells by direct spin coating is not applicable to common textile fabrics due to their intrinsically nonplanar properties and rugged surface morphologies, a novel new fabrication concept is therefore proposed by Zhen and coworkers, in which a wet transfer method is utilized to laminate the PSC onto the woven fabric electrode. The fabrication procedure of the device is illustrated in Figure 26.16a. In this work, they first utilized PAMD technique to obtain patterned silver-coated PET (fabric A) and nylon (fabric B). PSC device fabrication was then carried out on the flat silicon substrates by successive spin coating a water-soluble sacrificial layer PSSNa, active layer, and doped PH1000 as the top electrode. The resulting silicon/PSSNa/active layer/PH1000 layer was then put into the water. As PSSNa layer is water soluble, the PSC can be easily detached from the silicon substrate and transferred onto patterned silver-coated

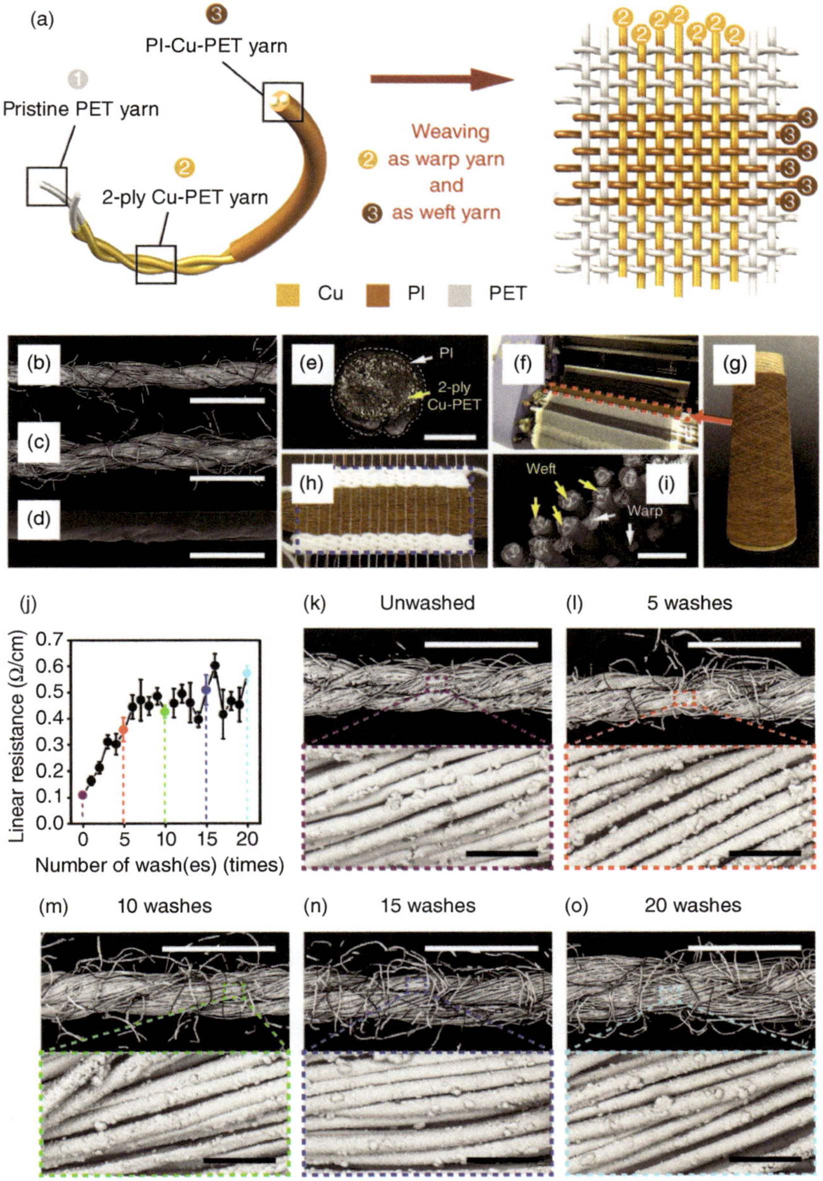

Figure 26.15 Fabrication and structure of the t-TENG. (a) Schematic illustration of the preparation of warp and weft yarns and structure of the woven t-TENG. SEM images of (b) 1-ply Cu-PET yarn, (c) 2-ply Cu-PET yarn, (d) PI-Cu-PET yarn (scale bar: 1 mm), and (e) cross section of PI-Cu-PET yarn (scale bar: 250 μm). Yellow dashed line outlines the cross section of the 2-ply Cu-PET yarn while the white dashed line outlines the cross section of PI wrapping around the 2-ply Cu-PET yarn. Digital image of (f) sample weaving loom carrying out automatic weaving of (g) metallic yarns wound on a cone. (h) As-woven t-TENG on the loom with a device area of ≈6 cm × 4 cm (blue dashed box). (i) SEM image of the cross section of the woven structure of the t-TENG, showing the up and down interlacement of the weft and warp yarns (scale bar: 1 mm). Washing durability tests. (j) Linear resistance and (k–o) SEM images of the unwashed 2-ply Cu-PET yarn and 2-ply Cu-PET yarn washed up to 20 cycles of standard washing (white scale bar, 1 mm; black scale bar, 50 μm). Source: Reproduced with permission from Zhao et al. [59]. Copyright 2016, John Wiley & Sons, Inc.

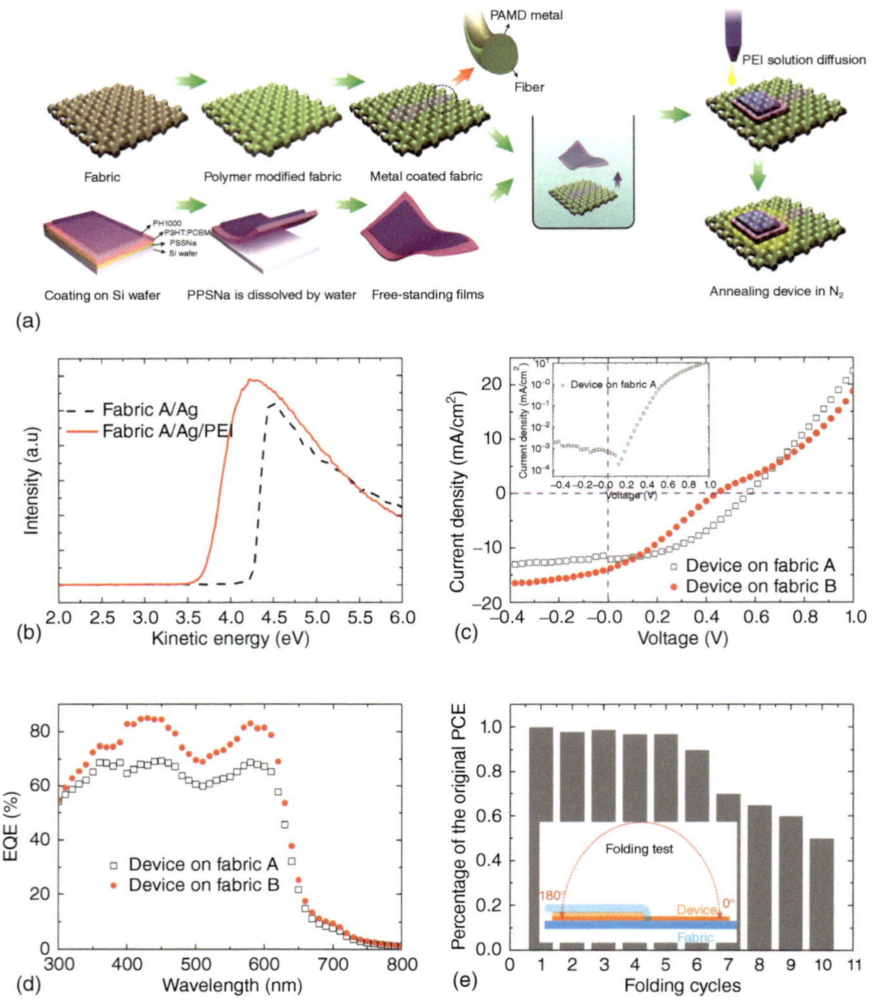

Figure 26.16 (a) Schematic illustration of the fabrication process of the PSCs on a woven fabric. (b) UPS spectra of a woven fabric electrode with/without PEI. (c) *J–V* characteristics of the PSCs on a woven fabric under AM 1.5 illumination (inset: dark *J–V* characteristics of the PSCs on fabric A). (d) EQE spectra of the PSCs on a woven fabric. (e) Performance degradation of the PSCs on fabric A in the folding test. Source: Reproduced with permission from Zhen et al. [60]. Copyright 2017, Royal Society of Chemistry.

fabric substrates. Finally, interface modification layer (IML) between the woven fabric electrode and active layer was obtained by the diffusion of the dilute solution of polyethylenimine (PEI) as the interface modification material. Such a PEI thin film is highly important in achieving high power conversion efficiency (PCE) performance by decreasing the work function of the fabric silver electrode to 3.6 ± 0.2 eV (Figure 26.16b). The performance of the PSC on fabric A exhibits a power conversion efficiency of 2.90% with a V_{OC} of 0.58 V, an fill factor (FF) of

0.41, and a J_{SC} of 12.10 mA/cm^2 (Figure 26.16c), which is close to the photovoltaic performance compared with those devices using the same active materials, while the performance on fabric B exhibits a power conversion efficiency of 1.85% with a V_{OC} of 0.46 V, an FF of 0.29, and a J_{SC} of 13.58 mA/cm^2. The external quantum efficiency (EQE) spectra of the PSCs based on fabric electrodes and also the performance degradation of the PSC sample up cycles of folding tests are, respectively, shown in Figure 26.16d,e. Still, the stability of the device upon folding requires further improvement in the future by proper encapsulation method.

26.5 Conclusion, Future Outlook, and Challenges

Wearable electronics are one of the most exciting technologies that attract considerable attention after the smart phone era. Meanwhile, advanced portable technology motivates and promotes further development of many textile-based electronic devices. With an urge to realize electronic devices based on textile fibers, yarns, and fabrics as the substrate materials, high-performance, durable, and washable textile metal conductors acting as interconnects, contacts, and electrodes are inevitably essential. In this chapter, a universal strategy, namely, PAMD, aiming to tackle the interfacial instability of metal/textile interface has been discussed. Remarkably, PAMD qualifies the as-deposited metal on textiles for excellent durability to withstand repeated cycles of bending and stretching without significant conductivity failure. In addition, PAMD is compatible with common printing technologies and textile dyeing and finishing technologies, providing a promising fabrication of high-performance, durable textile metal conductors for electronic purposes.

Although PAMD can solve critical challenges in the fabrication of metallic textiles such as metal adhesion on the textiles surface, durability, electrical conductivity, and compatibility in scale production, there are some limitations in the PAMD process. First, PAMD adopts ELD as the final step to deposit metal. Constrained by the ELD chemistry, limited choices of metal are only allowed such as copper, nickel, silver, and gold. Second, PAMD involves quite a lot of steps to go through. Its experimental design is based on a bottom-up approach, which means that the fabrication process consists of repetitive step-by-step immersion and drying procedures that must be done after one another. In the production point of view, the fabrication process is still tedious. Even though recently a printing approach has been proposed to reduce the steps taken in the PAMD process, printing can result in lateral diffusion in textile materials if the printing parameters are not well controlled.

Sooner or later, it is expected that more textile-based electronics such as flexible and wearable displays, energy harvesting devices, and robotic skins will be demonstrated by using the textile metal conductors fabricated by PAMD. Particularly, more focuses and efforts will be put in improving device's performance stability, reliability, washing durability, and compatibility of the manufacturing process with the current textile manufacturing industry. It is anticipated that different encapsulation methods of the devices will be demonstrated and optimized in order to fully incorporate textile-based devices into garments. However, the integration of these textile-based electronic devices into garment wearables is still not an easy feat. First, it still takes a lot redesign and experiment works that require technology transfer of PAMD from a lab-scale fabrication to the bulk production in the industry. Second, fabrication of textile-based electronic devices is still tough and challenging as it involves all precise synthesis, construction, and alignment of electronic devices on flexible form of materials, which demands revolutionary development of all the compatible machineries. Third, the construction of microelectronics on textiles still requires many research efforts as it involves the incorporation of multiple-level electronic components such as conductive traces, construction of integrated circuits, packaging, cable connections, and the construction of the whole wearable systems that are still not yet established at such an early development stage. Most importantly, the safety issues still require the incorporation of multidisciplinary expertise to pave the way for mass production of washable textile-based electronics in the near future.

References

1 Service, R.F. (2003). *Science* 301: 909.
2 Park, S. and Jayaraman, S. (2003). *MRS Bulletin* 28: 585.
3 Park, S., Gopalsamy, C., Rajamanickam, R., and Jayaraman, S. (1999). *Studies in Health Technology and Informatics* 62: 252.
4 Firoozbakhsh, B., Jayant, N., Park, S., and Jayaraman, S. (2000). *Proceedings IEEE International Conference on Multimedia and Expo*, 1253. New York.
5 Park, S., Mackenzie, K., and Jayaraman, S. (2002). *Proceedings of the ACM/IEEE 39th Design Automation Conference*, 170. New Orleans.
6 Tao, X. (2005). *Wearable Electronics and Photonics*. Cambridge: Woodhead Publishing Ltd.
7 Langenhove, L.V. (2007). *Smart Textiles for Medicine and Healthcare: Materials, Systems and Applications*. Cambridge: Woodhead Publishing Ltd.
8 Lee, J., Kwon, H., Seo, J. et al. (2015). *Advanced Materials* 27: 2433.
9 Zhang, Z.T., Yang, Z.B., Wu, Z.W. et al. (2014). *Advanced Energy Materials* 4: 1301750.
10 Lee, Y.H., Kim, J.S., Noh, J. et al. (2013). *Nano Letters* 13: 5753.
11 Aboutalebi, S.H., Jalili, R., Esrafilzadeh, D. et al. (2014). *ACS Nano* 8: 2456.
12 Huang, Q.Y., Wang, D.R., and Zheng, Z.J. (2016). *Advanced Energy Materials* 6: 1600783.
13 Drew, C., Liu, X., Ziegler, D. et al. (2003). *Nano Letters* 3: 143.
14 Lotus, A.F., Bender, E.T., Evans, E.A. et al. (2008). *Journal of Applied Physics* 103: 024910.
15 Dall'Acqua, L., Tonin, C., Peila, R. et al. (2004). *Synthetic Metals* 146: 213.
16 Bhat, N.V., Seshadri, D.T., and Radhakrishnan, S. (2004). *Textile Research Journal* 74: 155.
17 Kim, B., Koncar, V., and Dufour, C. (2006). *Journal of Applied Polymer Science* 101: 1252.
18 Jiang, K.L., Li, Q.Q., and Fan, S.S. (2002). *Nature* 419: 801.

19 Zhang, M., Atkinson, K.R., and Baughman, R.H. (2004). *Science* 306: 1358.
20 Lima, M.D., Fang, S.L., Lepro, X. et al. (2011). *Science* 331: 51.
21 Little, B.K., Li, Y.F., Cammarata, V. et al. (2011). *ACS Applied Materials & Interfaces* 3: 1965.
22 Lou, C.W. (2005). *Textile Research Journal* 75: 466.
23 Chen, H.C., Lee, K.C., Lin, J.H., and Koch, A. (2007). *Journal of Materials Processing Technology* 184: 124.
24 Lai, K., Sun, R.J., Chen, M.Y. et al. (2007). *Textile Research Journal* 77: 242.
25 Wei, Q.F., Yu, L.Y., Wu, N., and Hong, S.J. (2008). *Journal of Industrial Textiles* 37: 275.
26 Depla, D., Segers, S., Leroy, W. et al. (2011). *Textile Research Journal* 81: 1808.
27 Jur, J.S., Sweet, W.J., Oldham, C.J., and Parsons, G.N. (2011). *Advanced Functional Materials* 21: 1993.
28 Kalanyan, B., Oldham, C.J., Sweet, W.J., and Parsons, G.N. (2013). *ACS Applied Materials & Interfaces* 5: 5253.
29 Sweet, W.J., Oldham, C.J., and Parsons, G.N. (2014). *ACS Applied Materials & Interfaces* 6: 9280.
30 Lu, Y.X., Liang, Q., and Xue, L.L. (2012). *Applied Surface Science* 258: 4782.
31 Zhang, H.R., Zou, X.G., Liang, J.J. et al. (2012). *Journal of Applied Polymer Science* 124: 3363.
32 Lee, H.M., Choi, S.Y., Jung, A., and Ko, S.H. (2013). *Angewandte Chemie International Edition* 52: 7718.
33 Montazer, M. and Allahyarzadeh, V. (2013). *Industrial and Engineering Chemistry Research* 52: 8436.
34 Wang, W.C., Li, R.Y., Tian, M. et al. (2013). *ACS Applied Materials & Interfaces* 5: 2062.
35 Yu, Y., Yan, C., and Zheng, Z.J. (2014). *Advanced Materials* 26: 5508.
36 Milner, S.T. (1991). *Science* 251: 905.
37 Brittain, W.J. and Minko, S. (2007). *Journal of Polymer Science Part A: Polymer Chemistry* 45: 3505.
38 Ulman, A. (1996). *Chemical Reviews* 96: 1533.
39 Azzaroni, O., Brown, A.A., and Huck, W.T.S. (2007). *Advanced Materials* 19: 151.
40 Barbey, R., Lavanant, L., Paripovic, D. et al. (2009). *Chemical Reviews* 109: 5437.
41 Azzaroni, O., Zheng, Z.J., Yang, Z.Q., and Huck, W.T.S. (2006). *Langmuir* 22: 6730.
42 Wang, X.L., Yan, C., Hu, H. et al. (2014). *Chemistry - An Asian Journal* 9: 2170.
43 Liu, X.Q., Zhou, X.C., Li, Y., and Zheng, Z.J. (2012). *Chemistry - An Asian Journal* 7: 862.
44 Cui, W.S., Wu, D.Z., Wang, W.C. et al. (2009). *Surface and Coatings Technology* 203: 1885.
45 Garcia, A., Berthelot, T., Viel, P. et al. (2010). *ACS Applied Materials & Interfaces* 2: 3043.
46 Garcia, A., Polesel-Maris, J., Viel, P. et al. (2011). *Advanced Functional Materials* 21: 2096.

47 Liu, Z.L., Hu, H.Y., Yu, B. et al. (2009). *Electrochemistry Communications* 11: 492.

48 Yang, G.H., Lim, C., Tan, Y.P. et al. (2002). *European Polymer Journal* 38: 2153.

49 Yu, Z.J., Kang, E.T., and Neoh, K.G. (2002). *Langmuir* 18: 10221.

50 Guo, R.H., Li, Y.N., Lan, J.W. et al. (2013). *Journal of Applied Polymer Science* 130: 3862.

51 Guo, R.H., Jiang, S.X., Zheng, Y.D., and Lan, J.W. (2013). *Journal of Applied Polymer Science* 127: 4186.

52 Liu, X.Q., Chang, H.X., Li, Y. et al. (2010). *ACS Applied Materials & Interfaces* 2: 529.

53 Liu, L.B., Yu, Y., Yan, C. et al. (2015). *Nature Communications* 6: 7260.

54 Guo, R.S., Yu, Y., Xie, Z. et al. (2013). *Advanced Materials* 25: 3343.

55 Yu, Y., Xiao, X., Zhang, Y.K. et al. (2016). *Advanced Materials* 28: 4926.

56 Le, V.T., Kim, H., Ghosh, A. et al. (2013). *ACS Nano* 7: 5940.

57 Kou, L., Huang, T.Q., Zheng, B.N. et al. (2014). *Nature Communications* 5: 3754.

58 Huang, Q.Y., Liu, L.B., Wang, D.R. et al. (2016). *Journal of Materials Chemistry A* 4: 6802.

59 Zhao, Z., Yan, C., Liu, Z. et al. (2016). *Advanced Materials* 28: 10267.

60 Zhen, H., Li, K., Chen, C. et al. (2017). *Journal of Materials Chemistry A* 5: 782.

27

Fibers for Medical Compression

Bipin Kumar[1], *Harishkumar Narayana*[2], *and Jinlian Hu*[2]

[1] *Indian Institute of Technology Delhi, Department of Textile Technology, TX135, Hauz Khas, New Delhi 110016, India*
[2] *The Hong Kong Polytechnic University, Institute of Textiles and Clothing, Hung Hom, Kowloon, Hong Kong, China*

27.1 Introduction

It has been estimated that 2% of the general population in the world (age group: 18–64) is suffering from chronic venous disorders such as venous ulcers, varicose veins, edema, deep vein thrombosis, etc. [1]. This rate is further increased to 4% in people over the age of 65. This has significant socioeconomic impact, costing 1% of total healthcare budgets in developed countries, for example, in the United States, this costs US$2.5 billion to treat 6 million patients every year [1]. This number is bound to increase with a growing aging population and changing lifestyle (sitting working culture with limited limb movements) that increases the risk of poor blood circulation [2, 3]. Thus there is a huge market in compression products for improving life quality of world population.

As an efficient and long-term medical intervention for chronic venous disorders, compression therapy is preferred so as to minimize the post-thrombotic syndrome, prevent ulcer recurrence, and alleviate the related symptoms, such as leg pain and swelling caused by damaged veins [4–6]. Herein, a textile bandage or stocking is used to apply certain level of pressure around the affected site on the limb. The objectives of compression therapy are to reduce the venous hypertension in the affected area, to reduce the swollen limb to minimum the size, and also to maintain a uniform pressure gradient in the leg from toe to knee to improve the venous return to the heart [7]. External compression is given to the affected leg portion during compression therapy using compression materials like bandage, stockings, or pressure garments [8]. The efficacy of the treatment is undoubtedly dependent on the pressure generated at the interface between the textiles (used for compression) and the skin [9]. This pressure is called interface pressure or sub-bandage pressure. The interface pressure exerted here depends on several factors like limb shape and size, application technique, physical and structural properties of the bandage, physical activities taken by the patient, etc. [10, 11]. It has been observed by several researchers that the interface pressure decreases

Handbook of Fibrous Materials, First Edition. Edited by Jinlian Hu, Bipin Kumar, and Jing Lu.

over a period of time and hence reduces the effectiveness of the treatment of the compression system [5, 12–15]. Undoubtedly, the pressure drop or the success of the treatment entirely depends on the fiber and fabric construction used in the product; therefore, more understanding of the underlying interaction and performance of different fibrous materials and structures is needed. This could be useful for the academicians, R&D managers, product manufacturers, processing industries, doctors, health practitioners, researchers, nurses, and all others related to medical compression.

In this chapter, it is aimed to review several aspects of textile fibers used for compression therapy. Although the above subject demands the understanding of the fundamental nature of compression therapy, compression or pressure itself requires multidisciplinary approaches involving various concepts of physics, biological science, biomaterials, fabric engineering, structural dynamics, material science, technical textiles, and instrumentation to better deal the subject from different perspective. For simplification and meeting the desired requirements, this chapter introduces several multidisciplinary topics and describes the complex subject with very basic and simple discussions and gradually proceeds to list some of the recent developments in designing smart textiles for compression.

27.2 Compression Therapy

27.2.1 Pathophysiology and Implications of Chronic Venous Disorders

The cardiovascular system is a closed circulatory system in which blood flows in one direction only. It is organized around a central organ called the heart and is made up of three different types of blood vessels (arteries, veins, and capillaries; Figure 27.1). The arterial system is responsible to circulate blood from the heart to tissue and organs, and the venous system carries blood back to the heart. The venous system helps in the general circulation to send oxygen depleted blood rich in cell metabolism waste to the heart. The venous return toward the heart depends on the venous pressure that is developed in the veins. For a proper venous return, it is the combined action of the foot pump and calf muscle during movement and the efficacy of the venous valve system that ensures proper functioning and maintains the balance of blood flow in the body.

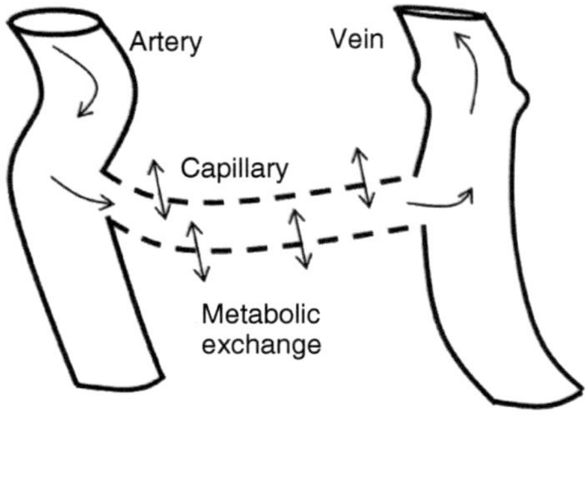

Figure 27.1 Schematic of the fluid transport of the human in the cardiovascular system.

Due to aging or limited movement, there exists high risk of imbalance in the cardiovascular function of the human body, and this results in loss of cutaneous substance, which might progresses over an indeterminate time period. This results in the development of chronic wounds, e.g. venous ulcer, leg ulcer, arterial ulcer, mixed ulcer, pressure ulcers, etc. Among these wounds, majority of cases are reported for venous hypertension (~80%), arterial insufficiency (~10%), or a combination of both [16]. One should be aware that chronic wounds may occur from venous or arterial disease, diabetes, arthritis, or malignancy. The basic characteristics of the venous and arterial ulcers are reviewed in Ref. [17]. The treatment for these etiologies will be different for each case. It has been advised that a comprehensive assessment of the patient, limb, vascular status, limb, and ulcer should be done to determine the underlying etiology and to formulate an appropriate compression management plan.

The level of chronic venous diseases in older patient is considerably higher in percentage, and their recurrence rate is also higher. Increase in venous pressure is obtained due to damage to the veins or venous valves. Also the action of foot and calf muscle movement does not function properly for older people due to their limited movement. The poor pump mechanism and the incompetent venous valve lead to venous hypertension due to increase in venous pressure. If the venous valves do not perform well or there is not enough pressure in the veins to push back the blood toward the heart, the pooling of blood occurs, and this causes edema (swelling of leg). If not managed, venous hypertension causes excessive pooling of fluid in the surrounding muscle that generates high pressure and lack of availability of oxygen and food. This causes skin deterioration and eventually the venous leg ulcers occur (Figure 27.2). Patients suffering from this experience a decreased quality of life because of continuous pain, discomfort, clinical depression, anxiety, social isolation, high cost for their treatment, lay beliefs, and also professional/patient conflict. Majority of patients (~68%)

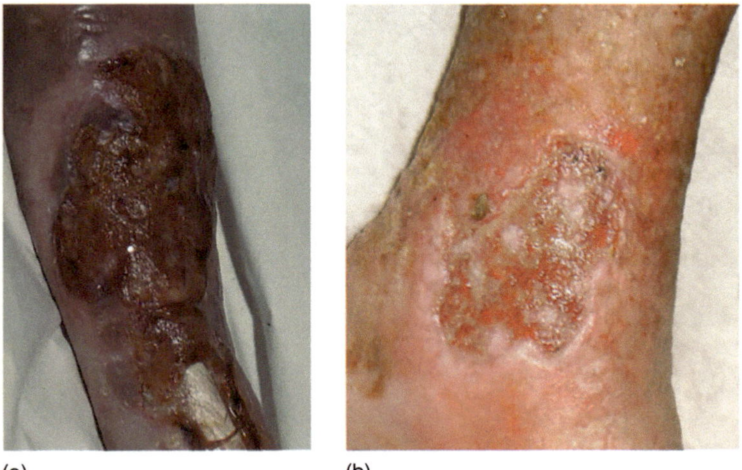

(a) (b)

Figure 27.2 (a) Chronic venous deficiency (CVD). (b) Venous leg ulcer.

experience depression, anger, fear, depression, etc. These wounds have a high reoccurrence rate and are considered a major healthcare problem. It is also a difficult task for the healthcare practitioner to deal with these diseases because of their underlying implications associated with patient's life.

27.2.2 Need for External Pressure and Its Physiopathology

Compression therapy has been shown to improve the lives of patients by significantly reducing pain, reducing the swollen limb to minimum size, maintaining that size, increasing mobility, and allowing the patient to participate for the treatment of his limb whenever possible. The pressure is exerted to the affected limb part by some external means like medical compression bandage, stockings, pressure garments, etc. There are two pressure mechanisms acting within venous system: the hydrostatic pressure and the oncotic pressure [18]. The hydrostatic pressure is equal to the great saphenous vein pressure, while the oncotic pressure is the osmotic pressure created by protein colloid in plasma. The oncotic pressure difference causes reabsorption of fluid, while the hydrostatic pressure causes filtration. The amount of lymph formed depends upon the permeability of the capillary wall and the gradient of hydrostatic pressure and oncotic pressure between blood and tissue. Increasing venous or hydrostatic pressure in the vein leads to more depositions of venous fluid from the veins to surrounding tissues because of increasing filtration through venous tubes. Compression therapy aims to decrease the pressure difference across the venous tubes and increases the action of the venous pumps in order to restore venous blood flow. It is based on application of external pressure equal to the excess venous pressure in order to restore normal transmural pressure. The main effects of applying external pressure are as follows:

(i) It reduces the venous diameter and increases the interstitial pressure in the surrounding and hence leads to an increase in blood flow in the deep veins, a reduction in pathologic reflux, and a reduction of the hydrostatic pressure (Figure 27.3a).

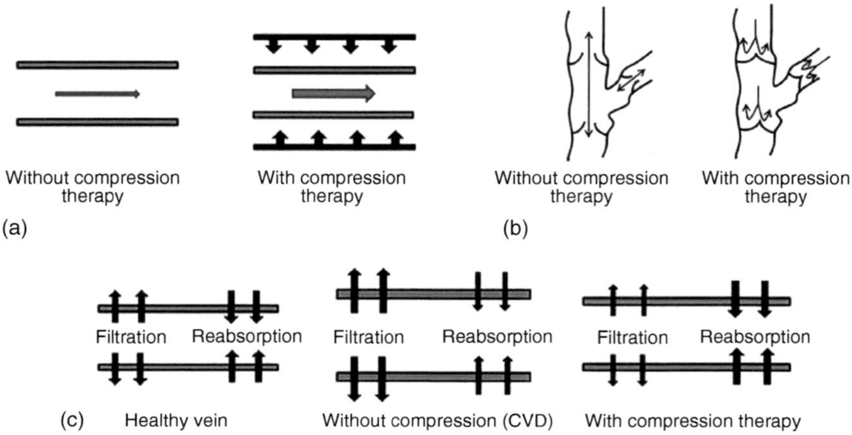

Figure 27.3 Effect of compression. (a) Increase in venous flow. (b) Restoration of venous valve function. (c) Increase in reabsorption of interstitial liquid.

(ii) It restores the valve function by bringing the walls of the veins closer together (Figure 27.3b).
(iii) It reduces blood pressure in the superficial venous system.
(iv) It reduces the pressure differences between the capillaries and the tissue to prevent back flow.
(v) It increases the cutaneous microcirculation, favors white cell detachment from the endothelium, and prevents further adhesion.
(vi) It reintegrates the interstitial liquids into the vessels (Figure 27.3c).

27.3 Role of Fibers in Compression Therapy

27.3.1 Fiber-Based Compression Modalities

Several compression methods and combinations are available for treatment of venous disorders. The fiber-based compression products can be categorized based on the elasticity, pressure, material, and structure (Figure 27.4). The pressure or compression is the ley to the treatment success, and the primary methods for applying compression are bandages, stockings, and intermittent pneumatic compression (IPC) [5, 16]. Bandages are preferred for the patients who require frequent dressing changes and recommended during the therapy phase of treatment for venous ulceration and also to control edema. Bandages are more practical for those incapable of applying compression stockings or for patients with fragile skin. Compression bandages are often more useful than the topical dressings for managing peripheral edema. The disadvantages of compression by bandages are the variability of pressure achieved even when applied by experienced professionals and patient compliance because of discomfort. Medical compression stockings are used to provide compression, which

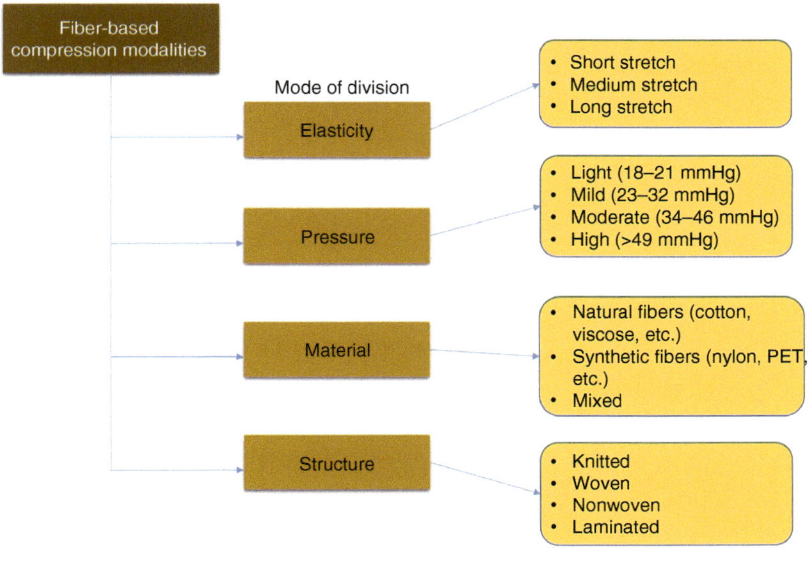

Figure 27.4 Fiber-based compression modalities.

is maximum at the ankle and gradual reduction in compression as the limb circumference increases. Unlike compression bandaging, the pressure generated with stockings is less dependent on the person applying it. It is imperative that the appropriate size and compression rating be prescribed for the condition and the patient being treated. IPC is useful for those who are not able to tolerate compression bandaging, who have reduced calf muscle function or limited ankle mobility, or who have peripheral arterial disease where other forms of compression are contradicted. In this technique, the limb is sequentially inflated and deflated, so as to simulate normal circulatory action and venous foot and calf muscle action. The disadvantages are that it is costly, can be bulky and noisy, and requires power supply.

27.3.2 Compression Requirements

Success of the compression therapy requires proper management plan designed to handle the underlying pathology and contributing factors. Several important factors need to be present for a health practitioner to use compression systems, like knowing how to use different compression systems, knowing the best available compression systems available for the treatment, being able to identify the etiology of the ulcer, and the willingness of the patient to agree to the commencement of compression treatment and for this to be sustained. Undoubtedly, the first step is to achieve the right amount of compression. A general guide to the amount of compression recommended for various indications is given in Table 27.1. It is likely that calf muscle function and variable ankle mobility may account for much of the variability in the success of compression therapy.

Although, the compression redistributes blood from the venous system into the central parts of the body in the case of the patients with poor blood flow. However, the product should also maintain a pressure gradient, which is paramount for such treatment. Figure 27.5a shows that the ankle pressure is different when the subject is lying, rising, standing, and walking for a normal and diseased person. So, when a normal person stands still, the pressure of venous blood in the leg increases from knee to foot because of gravity. Compression

Table 27.1 Recommended compression.

Degree of compression (kPa)	Indication
<2.6	Prevention of deep vein thrombosis (graduated compression stocking), mild edema, aching legs (occupational leg symptoms)
2.6–4	Mild varicose veins, mild to moderate edema, long-haul flights, varicose veins during and after pregnancy
4–5.3	Venous ulcers (including healed ulcers), deep vein thrombosis, superficial thrombophlebitis, varicose veins with severe edema, post-thrombotic syndrome, mild lymphedema
>5.3	Severe lymphedema, severe chronic venous insufficiency

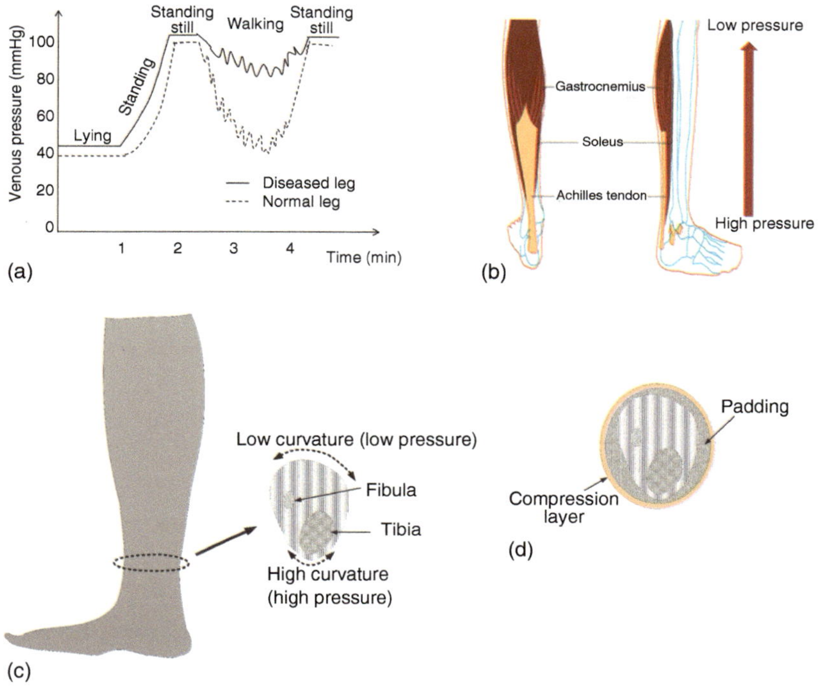

Figure 27.5 (a) Changes of ankle pressure. (b) Pressure gradient due to compression. (c) Irregular cross section of the leg causing high and low pressure around the (d) fibers as padding in pressure redistribution.

product should produce gradient pressure from foot to knee to propel blood toward heart (Figure 27.5b). Apart from the gradient requirement, the pressure should also be uniform around the circumference. In general, the cross section of the leg is not circular due to which the pressure around the circumference could vary significantly. The radius of curvature is small over bony prominences such as tibia or fibula due to which they can experience high pressure (Figure 27.5c). This all indicates the need for pressure management where it is aimed to prevent high pressure at critical regions such as the tibia or fibula and to ensure uniform distribution of pressure equally around the circumference at a particular height. In order to distribute pressure evenly around the limb, it is essential that high pressures created at the tibia and fibula regions should be absorbed. Fibers are therefore used as padding layer to reshape legs (Figure 27.5d), which are not narrower at the ankle than the calf. It helps to reshape the limb more like a cone shape so that the pressure gradient can be achieved with more pressure at the ankle and less at the calf.

27.3.3 Practical Challenges

Current compression products have the following shortcomings: (i) The application of a bandage needs trained personal to wrap the bandage on the patient's leg with constant force or extension, hence raising the treatment cost to hospitals and patients [19]. For maximum treatment efficacy, a required pressure gradient from "toe to knee" is critical yet arduous to achieve during bandage mounting due to reasons including different leg attributes (shape or size) among patients and difference in bandage materials/designs. Even after proper training to a staff, it is difficult to ensure the uniform bandage extension on the limb of the patient. (ii) Maintenance of the initially set pressure over time is another major challenge, due to the time and temperature dependence of the material behaviors [20–22], leading to a diminished treatment efficacy. (iii) For some patients, a dynamic (massaging) compression is recommended, but the required equipment like the IPC is costly, noisy, bulky, and, once attached, severely constraining the patients [23]. (iv) Also, all of the existing devices are incapable of offering heating therapy, preferred especially for non-active patients to improve blood flow.

27.4 Theoretical Insights into Pressure Prediction

Prediction of interface pressure applied by a bandage to a wounded leg segment could facilitate the efficiency of compression products. The Laplace's law is used to predict the interface pressure generated by the bandage immediately after the application that is a function of the tension in the fabric, the number of layers wrapped, the bandage width, and the circumference of the limb. The derivation of Laplace's law is attributed to the French scientist Pierre Simon de Laplace (1749–1827). The Laplace's law states that the force in the walls of the container is dependent on both the pressure of the container's content and its radius. The derivation of Laplace's law can be easily described by simplistic models like spheres or cylinders that can serve as models for many organs (Figure 27.6). In the case of air-filled cylinder removed from a longer tube, the forces in the wall of the cylinder must counterbalance the outward tension of the trapped air. In other words, $F_{wall} = F_{air}$. The total downward force applied by the wall to contain the air for a half cylinder is expressed as

$$F_{wall} = 2 \times T \times (L \times w) \tag{27.1}$$

where T is the tension in the wall per unit area and L and w are the length and wall thickness of the cylindrical vessel, respectively.

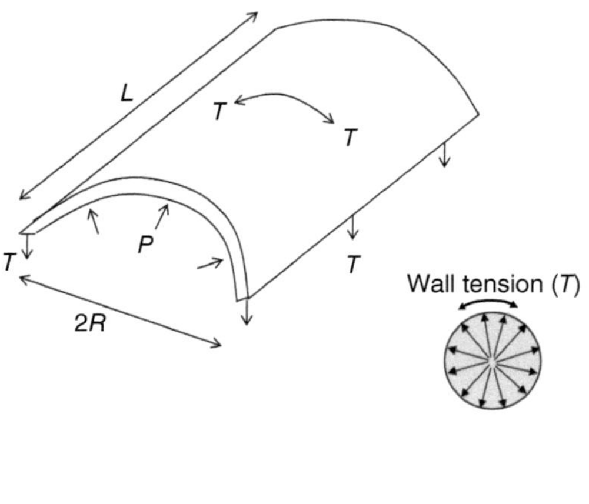

Figure 27.6 Derivation of Laplace's law for a cylinder.

The air inside the cylinder exerts pressure on the walls of the cylindrical vessel. The force exerted by the air pressure is equal to

$$F_{\text{air}} = P \times 2 \times R \times L \tag{27.2}$$

where P is the pressure difference across the wall and R is the radius of vessel. As the cylindrical container is neither expanding nor contracting, hence the tensions applied by the walls should be equal to those generated by its compressed contents. These simplify to

$$F_{\text{air}} = P \times 2 \times (R \times L) = F_{\text{wall}} = T \times 2 \times (L \times w)$$

$$\text{or } P = \frac{T \times w}{R} \tag{27.3}$$

This law is frequently used in medical science with an improved understanding of the functioning of the human body, the pathology of vascular wall, varicose vein, bladder rupture, the physics of respiratory physiology, and shape of the ventricles of the heart. This law is also used in compression therapy to predict interface pressure applied by a bandage on limb surface. To use Laplace's law in clinical practice, Thomas [24] modified Laplace's law to calculate pressure of compression system, which is described as

$$P = \frac{F \times n}{R \times W} \tag{27.4}$$

where P is the interface pressure (N/m^2), F is the longitudinal force applied to the bandage while wrapping (N), n is the number of layers of the bandage wrapped, R is the radius of the limb (m), and W is the bandage width (m).

The use of Laplace's law to determine interface pressure remains a controversial issue. Medical compression bandages are mostly applied in the form of multiple or overlapping layers, which results in the increase of the overall thickness of bandage fabric over the limb. Bandage application in spiral with 50% overlap results in two layers of bandage, while spiral with 33% overlap results in three layers of bandage. Four layers of bandage result from figure-of-eight bandage application with 50% overlap. Also multilayer multicomponent bandaging system involves the use of multiple layers of different medical bandages like padding, crepe bandage, compression bandage, adhesive bandage, etc. The modified equation of Laplace's law, described by Thomas [24], does not consider the increase in radius because of the preceding layers of the bandage. The reported model is based on the assumption that the bandage thickness has a negligible effect on the pressure. This may result in erred prediction of interface pressure for multiple layers. Schuren and Mohr [25] found that the interface pressure based on theoretical mathematical equations is not supported by experimental results. When compared between single layer and four-layer bandaging systems, it has been observed that the final pressure achieved by a multilayer bandaging system is not equal to the sum of the pressure exerted by each individual layer as predicted by Laplace's law [26, 27]. So, it is mandatory to use Laplace's law carefully to predict interface pressure as it could influence the selection of a particular bandage system according to a patient's condition and circumstances. The effect might be negligible or small for a few layers of bandage.

It is important to consider the bandage thickness to estimate interface pressure exerted by multilayer multicomponent bandaging system in order to avoid overestimating the pressure. The error can be 19% or even more at the ankle (circumference 25 cm) in the case of four-component multilayer bandaging systems. Al Khaburi et al. [28] have proposed a model based on thick wall cylinder theory to predict interface pressure by multiple layers of medical compression bandages:

$$P = \sum_{i=1}^{n} \frac{T[d + t + 2t(i-1)]}{((1/2)w[d + 2t(i-1)]^2) + (wt[d + t + 2t(i-1)])} \tag{27.5}$$

where P is the total pressure due to n layers of bandage (N/m^2), T is the tension in the bandage (N), d is the limb diameter (m), w is the bandage width when it is extended (m), t is the bandage thickness when it is extended (m), and n is the number of layers wrapped.

Interface pressure generated by the bandage varies significantly in the active patients as they walk due to the variation in limb size or shape as a result of calf muscle activity and the associated dynamic variation in the leg dimensions. The amount of variation in the interface pressure is dependent on the stiffness of the bandaging system. Al Khaburi et al. [29] have also developed mathematical models to predict the pressure change under dynamic conditions using Chord modulus, which can be calculated using force–elongation curve of bandage fabric. Chord modulus is defined as the slope of the straight line drawn between any two points on the force–elongation curve. Chord modulus indicates the change in force in dynamic conditions. The change in interface pressure due to change in limb shape is described as

$$\Delta P = \frac{2 \times \Delta d \times E \times w}{d^2} \quad \text{(using thin wall cylinder theory)} \tag{27.6a}$$

$$\Delta P = \frac{\Delta d \times E \times w}{2 \times r \times (r + w)} \quad \text{(using thick wall cylinder theory)} \tag{27.6b}$$

where ΔP is the change in the interface pressure that is caused by change in leg diameter (N/m^2), E is the chord modulus (N/m^2), w is the bandage thickness, d is the original leg diameter (m), Δd is the change in the leg diameter (m), and r is the leg radius (m).

27.5 Fibrous Material and Construction Used in Compression and Their Performance

Textile material, structure, and fabrication technology determines the product properties and performance and so decides the treatment outcomes and success. A textile fabric is defined as an assembly of fibers, yarns, or combinations of these. Different fabric construction has been frequently used for the preparation of compression product to apply pressure to the diseased limb. There are many ways to manufacture a textile fabric. Each manufacturing method is able to produce a wide range of fabric that depends on the raw materials used, the equipment, the machinery employed, and the manufacturing processes involved. The commonly used structures in the medical compression are woven, knitted, nonwoven, and spacer fabrics. Woven construction consists of two sets of perpendicular yarns (also known as warp and weft yarns) that are interlaced together with the help of weaving process. Knitted construction consists of intersecting loops that are produced by a knitting process. It is characterized by courses and wales. Nonwovens are also used commonly for the preparation of padding material for the multilayer compression bandaging system. Nonwovens are defined as the manufactured sheet, web, or batt of directionally or randomly oriented fibers that are bonded by friction, cohesion, or adhesion. Spacer fabric is three-dimensional knitted fabrics that consist of two separate knitted substrates. These two separate substrates are kept apart by spacer yarn. Spacer fabrics are much like a sandwich and feature with two complementary slabs of fabrics with a third layer tucked in between.

Textile structures used in compression bandages or stockings are made from fibers or yarns. Fiber is the first building block of any textile structure, and the fiber has to be converted into yarn to finally weave or knit any structure. Yarn, used for compression, is a heterogeneous structure (consist two or more fiber type) that can be produced by different covering methods (e.g. single or double covering, core spinning, stitch covering, core twisting, and air jet covering) [11]. Core–sheath yarn is more popular as it contains the elastomeric filament covered by staple cotton, PET, or nylon fibers. Among them polyamide-based conventional covered spandex elastomeric yarns are more frequently used in making compression stockings or other pressure garments. Sometimes, the elastomeric yarn is directly used in the fabric structure in the form of inlay fabric structure with the help of tuck, float, and stiches in the construction (Figure 27.7). The construction (tuck, float, stitch, thread density) and material properties (fiber extensibility, denier, twist, type of covering) need to be optimized to achieve desired thickness, tension, and stiffness from a compression product.

During the compression treatment, it is expected from a textile to sustain a uniform interface pressure gradient over the limb for faster recovery. The efficiency of different textiles to provide sustained pressure varies because of the

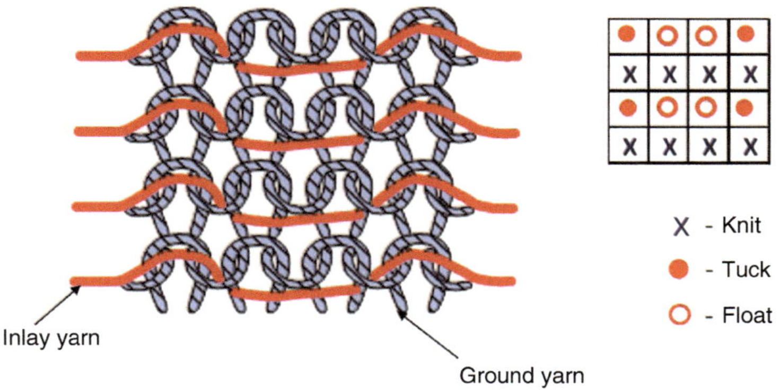

Figure 27.7 Inlay structure used for medical stocking for compression.

differences in their structure and constituent material type. A textile applies pressure over the limb because of the internal stress developed in the structure during its application over limb by applying external force. The capacity of a product to sustain pressure is greatly dependent on its ability to maintain this internal stress developed in the bandage under wrapped position. All textile fibers or structures are made from polymers that are viscoelastic in nature. Because of the viscoelasticity, the stress developed in the textile structure under constant extension decreases over time. This process is called stress relaxation that refers to the material behavior to relax the internal stress over time under a fixed level of elongation (Figure 27.8a). This can be experimentally measured by applying a constant elongation to the specimen and measuring the internal stress required to maintain that elongation as a function of time. Figure 27.8b,c shows the general relaxation behavior of textile fibers at different strains and temperatures, respectively.

The capacity of a bandage to sustain pressure is greatly dependent on its ability to maintain the internal stress developed in the bandage under wrapped position. Because of viscoelastic behavior of fiber or yarn, the stress developed in the bandage structure decreases over time. The reduction of the internal stress in the bandage over time could be an important factor for the interface pressure drop during course of compression treatment and hence could be a deciding factor to evaluate bandage performance [12, 13, 30]. Kumar et al. [14, 22] showed the pressure drop by compression bandages due to relaxation of stress in the fibers

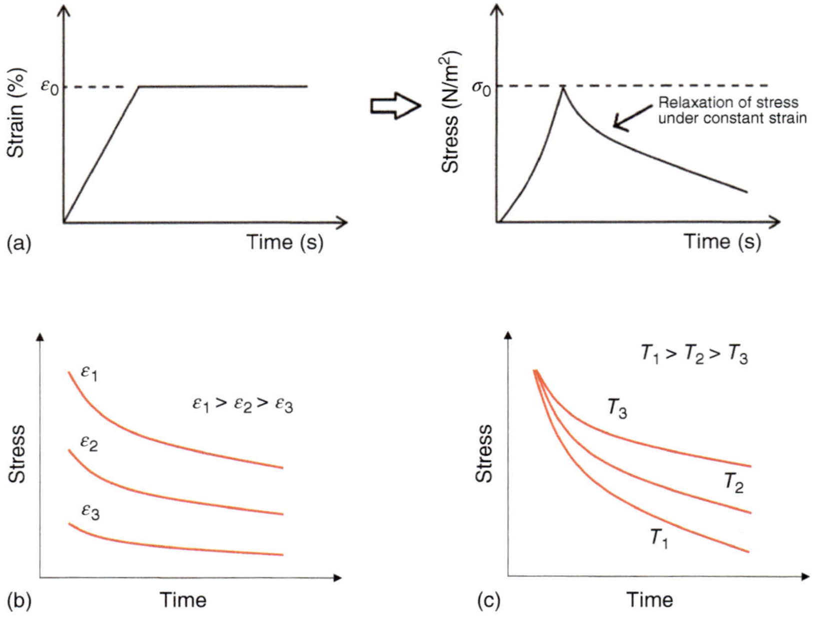

Figure 27.8 (a) Stress relaxation behavior of the material. (b) Generalized relaxation behavior of textile behaviors at different extensions. (c) Generalized relaxation behavior of textile behaviors at different temperatures.

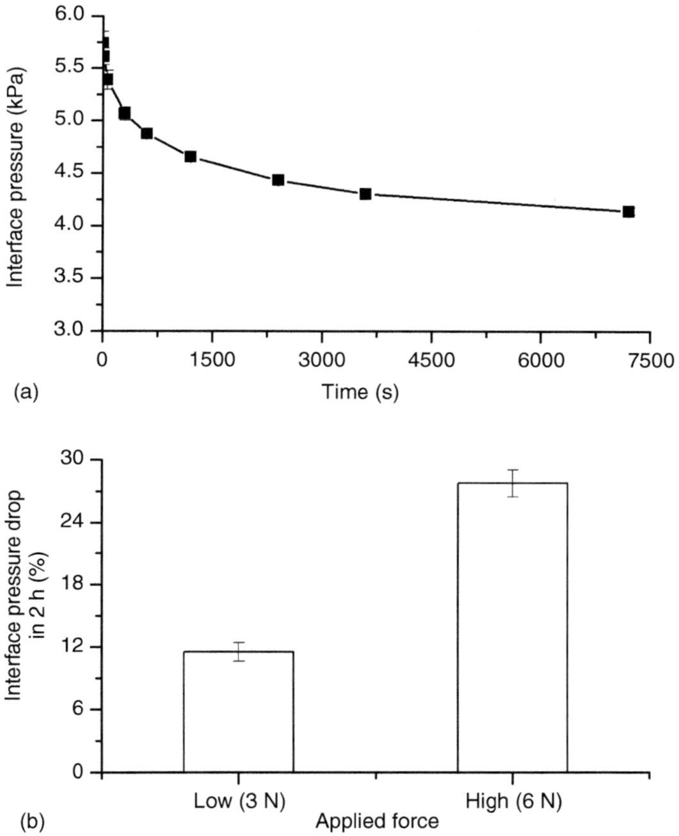

Figure 27.9 (a) Pressure profiles over time for a cotton bandage. (b) Pressure drop at different applied forces.

(Figure 27.9). Understanding viscoelastic behaviors of different fibrous materials present in the compression product is therefore critical and helps to design and evaluate long-term product performance.

Kumar et al. [12–14, 21, 22, 30–35] also conducted extensive study on different materials and other factors and obtain critical information on the product performance. Pressure relaxation by a bandage depends significantly on the relaxation behavior of fibers or yarns used in it. Relaxation time is commonly used to describe stress relaxation behavior of a viscoelastic material, which indicates the time required to reach from unrelaxed state to new relaxed state of the material. It has been found that cotton and viscose fibers have the lower relaxation time due to which the relaxation of stress is faster. Synthetic fibers (e.g. nylon, PET, elastase) have low stress relaxation as compared with cotton fiber because of their higher relaxation time. This could be the reason for different pressure profiles obtained for bandages developed from different fibers (Figure 27.10). Higher and faster pressure drop for 100% cotton and viscose samples were obtained because of their poor ability to sustain internal stress in the structure. Elastomeric yarns

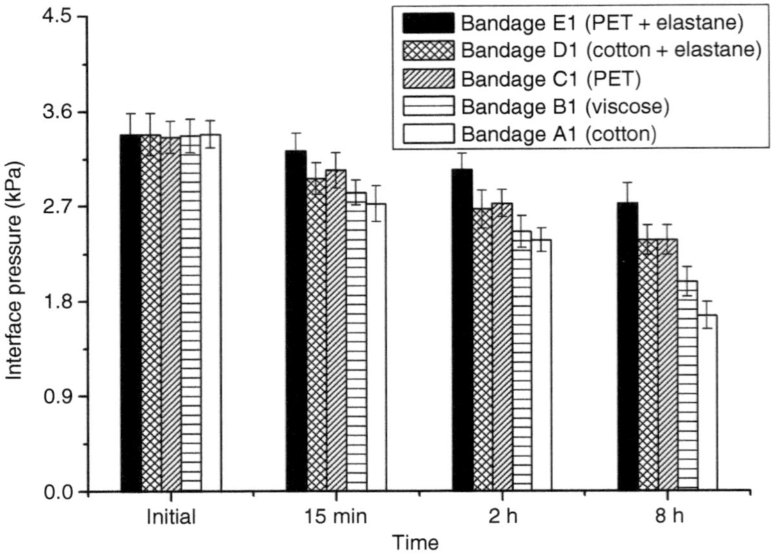

Figure 27.10 Effect of material type on interface pressure profile generated over time for different bandages.

have excellent elastic and rheological properties. They can maintain the internal stress for longer time because of their good elastic properties [36]. It has been also observed that fabrics containing elastomeric yarn tend to be tighter and have more number of threads per unit area. So, a compression product containing elastomeric yarn is best suited to maintain the pressure. Kumar et al. [14, 22] also observed that the pressure drop can be minimized by increasing the tightness of the structure (Figure 27.11). It has been observed that pressure drop is less for a structure having higher thread density compared with the structure having lower thread density. More pressure drop is expected at higher applied force in the bandage.

27.6 Innovation in Compression Products

A number of innovations have been noticed in the field of compression to use smart fabrics or structure with added benefits or functions for the treatment. Schafer and Jung [37] proposed the suitability of wide bandage fabric with more transverse stability and easy tearing capacity. Herein, the tearing capacity is obtained by utilizing the spun crape threads with *S* and *Z* twisting directions. In other innovation, the leno structure is used to provide high frictional resistance between warp and weft yarns particularly to improve the durability of the bandage, where frequent washing is necessary [38]. Application of bandage at a constant extension is always a challenge for unexperienced nurses or practitioners, and to ease the application process, an innovative design with visual pattern is used to assure uniform wrapping. In this, the

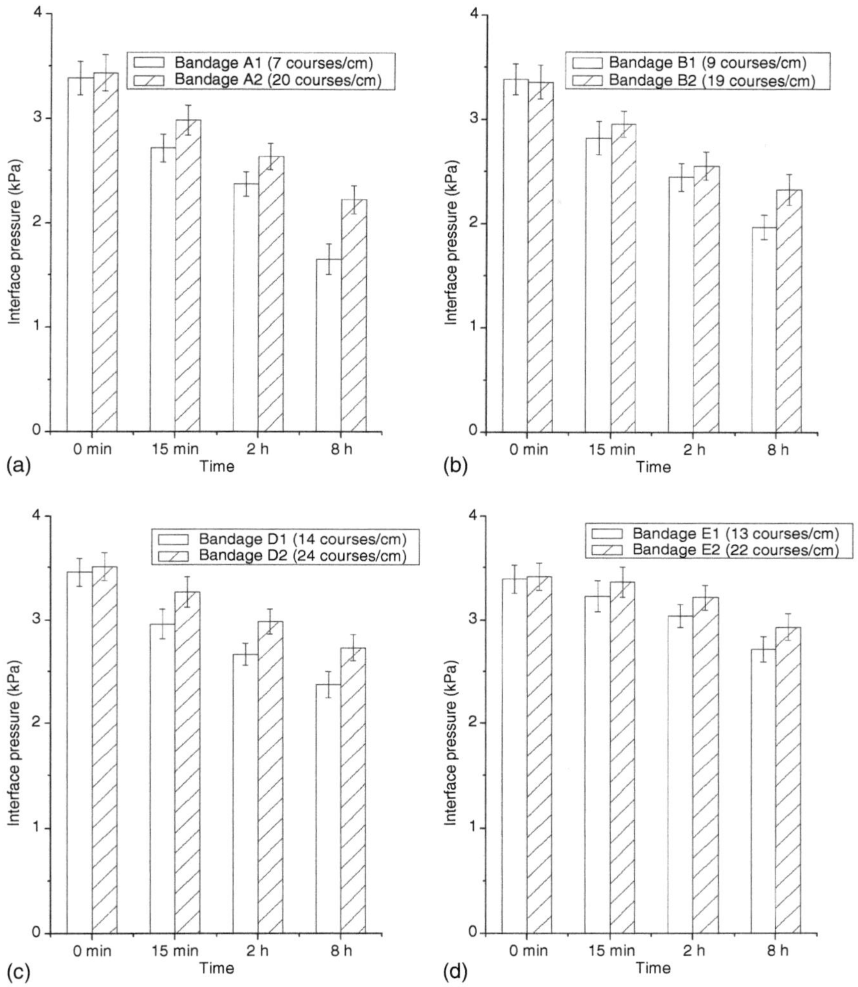

Figure 27.11 Effect of thread density on the pressure profile of different bandages. (a) Cotton, (b) viscose, (c) cotton + elastane, and (d) PET + elastane.

continuous pattern is formed across the length of bandage, so that the shape of each geometric shape is changed when tension is applied to the bandage and shape of the deformed pattern is indicative of compression force applied (Figure 27.12a). Another approach to maintain constant force is to use yarns made from synthetic elastomeric polymer containing triblock A-B-C copolymer (styrene–butadiene–styrene) [39]. In general, if a normal bandage is stretched too much during the application, the compressive force over the bandage may be overly great and cause damage, for example, by restriction of blood supply. But the bandage manufactured by block A-B-C copolymer achieves an effective compressive force at an extension of between 20% and 60% after which any further extension of the bandage is accompanied by only a small increase in compressive

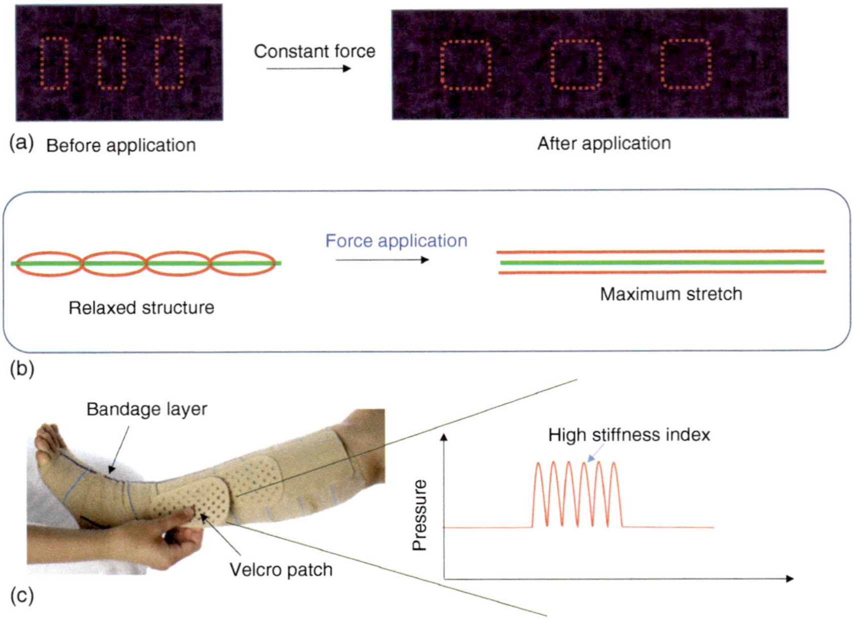

Figure 27.12 (a) Mechanism to obtain constant stretch. (b) Thonic bandage to apply at constant extension. (c) Velcro patch to control bandage stiffness.

force. This way it eliminates the risk of high force due to over stretching. More recently, the use of Thonic bandage (Figure 27.12b) is becoming popular, which helps the medical practitioners to apply bandage at a constant extension [40]. It suggests the use of leno structure for the cotton yarn to control the stretchability of the bandage. In addition to application at constant extension, the stiffness of compression product is also critical. The functioning of calf muscle pump helps in returning the blood to the heart; and the function of calf pump gets enhanced during compression therapy when an inelastic bandage (short-stretch bandage) is applied to the limb. More recently, an innovative method is proposed to provide a low, sustained resting pressure but high, efficient working pressure [41]. Herein, the Velcro patches (Figure 27.12c) are used above bandage layer to increase the stiffness of the underlying material and enhance the working pressure.

Multilayer compression bandaging system comprises crepe bandage, padding bandage, compression bandage, and adhesive bandage. The whole system is bulky and provides discomfort to the patient. With a view to replace both the compression and padding bandages, 3D knitted single-layer bandages have been introduced, which is produced by spacer technology [42]. The fabric sample contains two different layers of fabric that are combined with an inner spacer yarns. The construction allows using different materials for its layers and spacer yarns and hence could be engineered for a wide range of applications. It provides softness, good resilience, and a cushioning effect to the body. The need for health monitoring is becoming popular in the twenty-first century to get continuous feedback of the pressure results and access the efficacy of product remotely. In such product, a pressure sensor can be integrated for providing an indication of the compression applied by a bandage to a human or animal body. Such pressure sensing device can be directly designed from the conductive yarns integrated in the fabric structure [43].

27.7 Shape Memory Fibers for Compression

There exist several limitations of conventional compression treatments in the current clinical practice. The level of pressure is difficult to achieve due to several influencing factors including the difference in limb size or shape, material properties, and application procedure [14]. Additionally, the pressure gradient achieved by the bandages or stockings varies over times due to changes in limb shape (i.e. reduction of edema over time), and there exists no external means to adjust the pressure gradient. Inevitably pressures drop during use, as for all compression bandages in the market. And there exists no current solution to control the pressure and its distribution by the bandage locally or as a whole. Clearly, controlling or readjusting pressure on the leg would serve many advantages to health practitioners and would make the treatment even more effective.

The aforementioned challenges in compression products (i.e. easy application at constant extension and external pressure control) could be resolved if there is a possibility of shape and stress control in the bandage materials. The stimulus-responsive polymers, such known as memory polymers (MPs), could be useful here as they demonstrate shape memory ability that permits programming or fixing of different deformed shapes from an original (permanent) shape (Figure 27.13a) and simultaneously allows the shape recovery to the original shape by external stimulus such as heat [44, 45]. Many applications of MPs have been demonstrated in aerospace, biomedical, transport, construction, electronic, textile, and consumer products [46–48]. This shape control ability of MP could be utilized in compression bandage to provide easy and alternative method of uniform wrapping; thus it will eliminate the need for a trained personal for bandage application. In the textile, the shape memory effect can be achieved by incorporating MP filaments in the fabric structure (Figure 27.13b). To achieve constant extension, the deformed shape (constant strain) of an MP-based bandage could be programmed first using shape fixity potential of the MP, and then the bandage could be wrapped over the limb with no further extension. The bandage could be then activated directly on the limb to initiate the recovery process that would result in the uniform shrinkage throughout the wrapping area, hence achieving nearly constant extension in the entire length. This process could be easily achieved using external stimulus such as heat.

More recently, another new phenomenon in MP, i.e. stress memory, is also discovered [49]. Similar to shape memory, stress memory is a phenomenon whereby the stress in a polymer can be programmed, stored at temperature below its thermal transition temperature T_{trans}, and then retrieved reversibly, when needed, with an external thermal stimulus at $T > T_{trans}$ (Figure 27.13c,d), thus to realize the in situ material stress control. In other words, the integration of MP fibers into bandage fabric structures could allow in situ stress control [50–52]. As the bandage interface pressure depends on the internal stress in the fabric, so the possibility of stress alteration could help in achieving recommended pressure level in the wrapped state of the bandage. Also, this will allow the bandage to provide

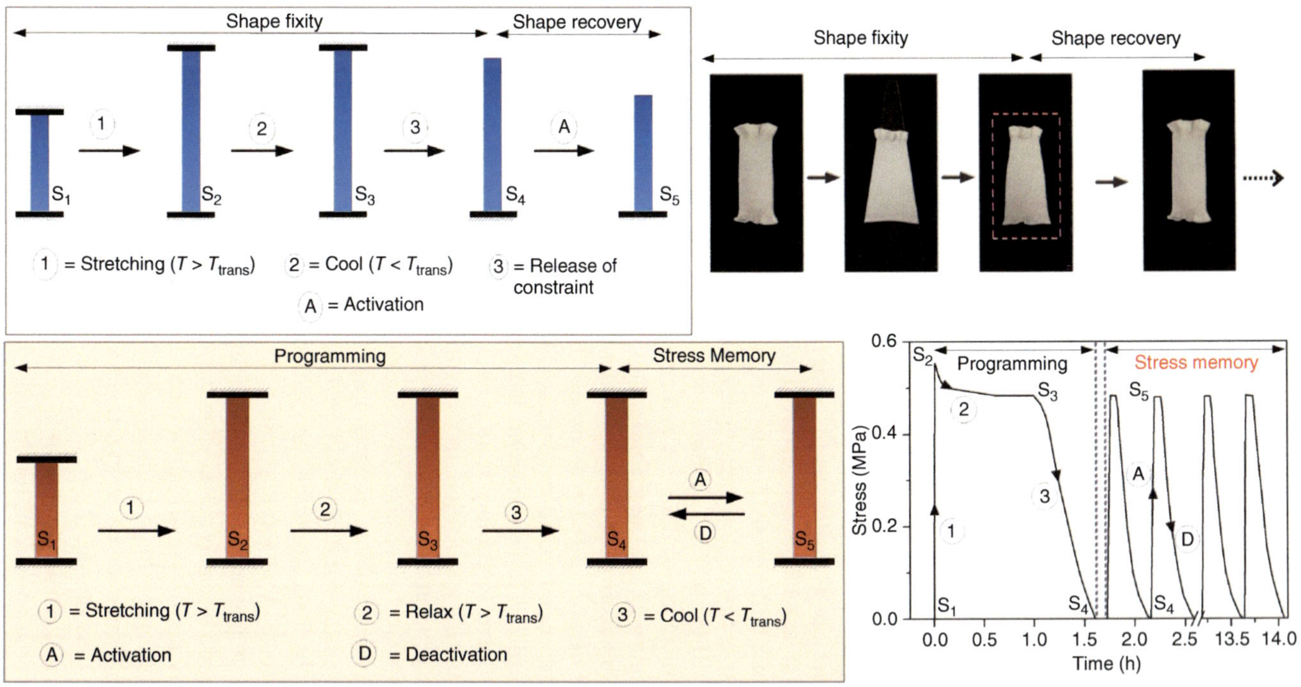

Figure 27.13 (a) Thermomechanical process to observe shape memory (fixity and recovery) in a memory polymer (MP). (b) Demonstration of shape memory using a memory fabric. (c) Stress memory programming in the MP. (d) Results of stress memory response of an MP.

dynamic massage effect via alternating stress modulation and so would eliminate the need for IPC.

More recently, Hu and coworkers [49–53] have designed and proposed stimulus responsive memory fibers for the smart compression management. The fibers were prepared from poly(1,6-hexanediol adipate) (PHA) (UBE Industries, Japan) having molecular weight of 3000 g/mol as a reversible phase (soft segment), 4,4′-diphenylmethanediisocyanate (MDI) (Aldrich Chemical Company, USA), and chain extender 1,4-butanediol (BDO) (Acros Organics, USA) as a fixed phase (hard segment). These filaments could be incorporated in the stocking fabric for external pressure control functions. The pressure testing of memory filaments integrative compression stocking allows to provide extra pressure (Figure 27.14a) and massage effect (Figure 27.14b) in the same deformation constraint just by change in an external heat stimulus to trigger the stocking. The

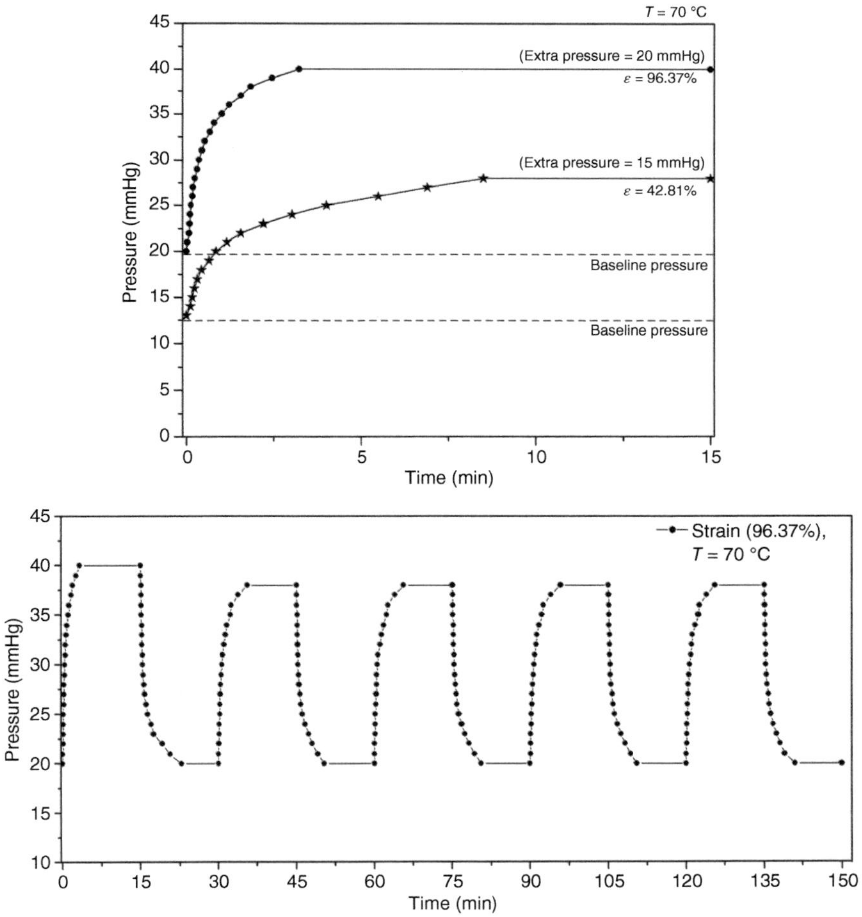

Figure 27.14 Experimental results of memory filaments integrative stocking fabric. (a) Pressure–time profile at different strains. (b) Pressure–time profile showing massage effect.

novel memory phenomenon could be further implemented scientifically with applied technology in multidisciplinary arenas such as massage devices, artificial muscles, and smart fabrics, where the stimuli-responsive stress is needed.

27.8 Conclusions

The present chapter introduces the principles of compassion therapy and the role of different textile material and structures. More emphasis is given on the interface pressure exerted by a compression product on the limb. The relaxation behavior of the fibers as well as related structures plays a critical role in maintaining the pressure over time. Elastomeric fiber in such product is a must if one has to maintain elasticity and eliminate the problem of pressure drop over time. At the end, the future potential of memory filaments in the field of compression is briefly introduced. The memory fabric allows to change the internal stress in its structure by external heating. It is possible to control or manage the pressure in wrapped position. This smart compression system could offer several advantages in compression management. First, it could give more freedom to nurses to control or readjust pressure to achieve an appropriate pressure whenever needed. Second, the advantage of memory compression system over other conventional compression products is that when the pressure drops below a targeted level, then it would be possible to readjust pressure level and therefore could provide sustained compression. Future work needs to be focused on designing and optimization of such filament and fabrics for smart compression management.

References

1 Pascarella, L. and Shortell, C.K. (2015). Medical management of venous ulcers. *Seminars in Vascular Surgery* 28 (1): 21–28.
2 Ruckley, C. (1997). Socioeconomic impact of chronic venous insufficiency and leg ulcers. *Angiology* 48 (1): 67–69.
3 Bergan, J.J., Schmid-Schönbein, G.W., Smith, P.D.C. et al. (2006). Chronic venous disease. *New England Journal of Medicine* 355 (5): 488–498.
4 Partsch, H. (2013). Compression therapy in leg ulcers. *Reviews in Vascular Medicine* 1 (1): 9–14.
5 Partsch, H. (2014). Compression for the management of venous leg ulcers: which material do we have? *Phlebology* 29 (1 Suppl): 140–145.
6 Mosti, G., Picerni, P., and Partsch, H. (2012). Compression stockings with moderate pressure are able to reduce chronic leg oedema. *Phlebology* 27 (6): 289–296.
7 Cullum, N.A., Nelson, E.A., Fletcher, A., and Sheldon, T. (2001). *Compression for Venous Leg Ulcers*. The Cochrane Library, 2.
8 Kumar, B., Das, A., and Alagirusamy, R. (2014). *Science of Compression Bandages*. WPI Publishing.

9 Partsch, H., Clark, M., Bassez, S., et al. (2006). Measurement of lower leg compression in vivo: recommendations for the performance of measurements of interface pressure and stiffness. *Dermatologic Surgery* 32 (2): 224–233.

10 Hirai, M. (1998). Changes in interface pressure under elastic and short-stretch bandages during posture changes and exercise. *Phlebology* 13 (1): 25–28.

11 Liu, R., Guo, X., Lao, T.T., and Little, T. (2016). A critical review on compression textiles for compression therapy: textile-based compression interventions for chronic venous insufficiency. *Textile Research Journal* 87 (9): 1121–1141.

12 Kumar, B., Das, A., and Alagirusamy, R. (2012). Prediction of internal pressure profile of compression bandages using stress relaxation parameters. *Biorheology* 49 (1): 1–13.

13 Kumar, B., Das, A., and Alagirusamy, R. (2013). An approach to examine dynamic behavior of medical compression bandage. *Journal of the Textile Institute* 104 (5): 521–529.

14 Kumar, B., Das, A., and Alagirusamy, R. (2013). Effect of material and structure of compression bandage on interface pressure variation over time. *Phlebology* 29 (6): 376–385.

15 Mosti, G. and Partsch, H. (2010). Inelastic bandages maintain their hemodynamic effectiveness over time despite significant pressure loss. *Journal of Vascular Surgery* 52 (4): 925–931.

16 Farah, R.S. and Davis, M.D. (2010). Venous leg ulcerations: a treatment update. *Current Treatment Options in Cardiovascular Medicine* 12 (2): 101–116.

17 Schmidt, K., Debus, E.S., St, J., et al. (2000). Bacterial population of chronic crural ulcers: is there a difference between the diabetic, the venous, and the arterial ulcer? *VASA, Zeitschrift fur Gefasskrankheiten* 29 (1): 62–70.

18 Partsch, H. (2003). Understanding the pathophysiological effects of compression. In: *EWMA Position Document. Understanding Compression Therapy* (ed. Suzie Calne), 2–4. London: Medical Education Partnership Ltd.

19 Partsch, H. (2014). Compression therapy for deep vein thrombosis. *VASA* 43 (5): 305–307.

20 Kumar, B., Das, A., and Alagirusamy, R. (2014). Effect of material and structure of compression bandage on interface pressure variation over time. *Phlebology* 29 (6): 376–385.

21 Kumar, B., Das, A., and Alagirusamy, R. (2012). An approach to determine pressure profile generated by compression bandage using quasi-linear viscoelastic model. *Journal of Biomechanical Engineering-Transactions of the ASME* 134 (9).

22 Kumar, B., Das, A., and Alagirusamy, R. (2013). Study of the effect of composition and construction of material on sub-bandage pressure during dynamic loading of a limb in vitro. *Biorheology* 50 (1–2): 83–94.

23 Feldman, J.F., Stout, N.L., Wanchai, A., et al. (2012). Intermittent pneumatic compression therapy: a systematic review. *Lymphology* 45 (1): 13–25.

24 Thomas, S. (2003). The use of the Laplace equation in the calculation of sub-bandage pressure. *EWMA Journal* 3 (1): 21–23.

25 Schuren, J. and Mohr, K. (2008). The efficacy of Laplace's equation in calculating bandage pressure in venous leg ulcers. *Wounds Uk* 4 (2): 38.

26 Dale, J.J., Ruckley, C.V., Gibson, B., et al. (2004). Multi-layer compression: comparison of four different four-layer bandage systems applied to the leg. *European Journal of Vascular and Endovascular Surgery* 27 (1): 94–99.
27 Ruckley, C.V., Dale, J.J., Gibson, B., et al. (2003). Multi-layer compression: comparison of four different four-layer bandage systems applied to the leg. *Phlebology* 18 (3): 123–129.
28 Al Khaburi, J., Nelson, E.A., Hutchinson, J., and Dehghani-Sanij, A.A. (2011). Impact of multilayered compression bandages on sub-bandage interface pressure: a model. *Phlebology* 26 (2): 75–83.
29 Al Khaburi, J., Nelson, E.A., Hutchinson, J., and Dehghani-Sanij, A.A. (2011). Impact of variation in limb shape on sub-bandage interface pressure. *Phlebology* 26 (1): 20–28.
30 Das, A., Kumar, B., Mittal, T., et al. (2012). Pressure profiling of compression bandages by a computerized instrument. *Indian Journal of Fibre and Textile Research* 37 (2): 114–119.
31 Kumar, B. and Das, A. (2014). Design and development of a Computerized Wicking Tester for longitudinal wicking in fibrous assemblies. *Journal of the Textile Institute* 105 (8): 850–859.
32 Kumar, B., Das, A., and Alagirusamy, R. (2012). Analysis of sub-bandage pressure of compression bandages during exercise. *Journal of Tissue Viability* 21 (4): 115–124.
33 Kumar, B., Das, A., and Alagirusamy, R. (2012). Analysis of factors governing dynamic stiffness index of medical compression bandages. *Biorheology* 49 (5-6): 375–384.
34 Kumar, B., Das, A., and Alagirusamy, R. (2013). Study on interface pressure generated by a bandage using in vitro pressure measurement system. *Journal of the Textile Institute* 104 (12): 1374–1383.
35 Kumar, B., Das, A., and Alagirusamy, R. (2013). *Science of Compression Bandages*. Woodhead Publishing India Pvt. Ltd. ISBN: 9781782422686.
36 Senthilkumar, M., Anbumani, N., and Hayavadana, J. (2011). *Elastane Fabrics–A Tool for Stretch Applications in Sports*. India: NISCAIR-CSIR.
37 Schafer, E. and Jung, H. (1984). Wide bandage fabric. Google Patents.
38 Hampton, R. and Hanes, P.F. Jr., (1980). Woven elastic compression bandage. Google Patents.
39 Miller, N.D. (1998). Woven or knitted elastic bandages having controlled compressive forces. Google Patents.
40 Gonon, P. (2014). Elastic bandage for bandaging and methods for manufacturing such a bandage. Google Patents.
41 Damm, J., Lundh, T., Partsch, H., and Mosti, G. (2017). An innovative compression system providing low, sustained resting pressure and high, efficient working pressure. *Veins and Lymphatics* 6 (1): 10–11.
42 Lee, G., Rajendran, S., and Anand, S. (2009). New single-layer compression bandage system for chronic venous leg ulcers. *British Journal of Nursing* 18 (15): S4–S18.
43 Seitz, P. (2013). Textile pressure sensor. Google Patents.
44 Leng, J., Lan, X., Liu, Y., and Du, S. (2011). Shape-memory polymers and their composites: stimulus methods and applications. *Progress in Materials Science* 56 (7): 1077–1135.

45 Xie, T. (2010). Tunable polymer multi-shape memory effect. *Nature* 464 (7286): 267–270.
46 Mather, P.T., Luo, X., and Rousseau, I.A. (2009). Shape memory polymer research. *Annual Review of Materials Research* 39: 445–471.
47 Hu, J., Zhu, Y., Huang, H., and Lu, J. (2012). Recent advances in shape-memory polymers: structure, mechanism, functionality, modeling and applications. *Progress in Polymer Science* 37 (12): 1720–1763.
48 Zhao, Q., Qi, H.J., and Xie, T. (2015). Recent progress in shape memory polymer: new behavior, enabling materials, and mechanistic understanding. *Progress in Polymer Science* 49–50: 79–120.
49 Hu, J.L., Kumar, B., and Narayan, H.K. (2015). Stress memory polymers. *Journal of Polymer Science Part B: Polymer Physics* 53 (13): 893–898.
50 Kumar, B., Hu, J.L., and Pan, N. (2016). Smart medical stocking using memory polymer for chronic venous disorders. *Biomaterials* 75: 174–181.
51 Kumar, B., Hu, J., Pan, N., and Narayana, H. (2016). A smart orthopedic compression device based on a polymeric stress memory actuator. *Materials and Design* 97: 222–229.
52 Kumar, B., Hu, J., and Pan, N. (2016). Memory bandage for functional compression management for venous ulcers. *Fibers* 4 (1): 1–10.
53 Narayana, H., Hu, J., Kumar, B., and Shang, S. (2016). Constituent analysis of stress memory in semicrystalline polyurethane. *Journal of Polymer Science Part B: Polymer Physics* https://doi.org/10.1002/polb.24000.

28

Electrospun Nanofibers for Environmental Protection: Water Purification

Hongyang Ma[1,2], *Christian Burger*[2], *Benjamin Chu*[2], *and Benjamin S. Hsiao*[2]

[1] *Beijing University of Chemical Technology, State Key Laboratory of Organic-Inorganic Composites, College of Materials Science and Engineering, 15 North Third Ring Road, Beijing 100029, China*
[2] *Stony Brook University, Department of Chemistry, 100 Nicolls Road, Stony Brook, NY 11794-3400, USA*

28.1 Introduction

The water crisis in the twenty-first century not only brings pressing challenges to humanity but also creates tremendous opportunities for the membrane community to tackle these challenges. Great efforts have already been made for research and development of membranes with new structures and properties using new nanomaterials for water purification [1–3]. In this endeavor, nanofibrous composite membranes are a novel class of new membrane, consisting of multiple fibrous layers with different fiber diameters, serving as support layer, barrier layer, or additive to the barrier layer [4–7]. The unique features of these membranes include high porosity and functionalizability, which can lead to high permeability, high rejection, low fouling, low energy consumption, and low environmental impact for water purification applications.

Nanofibers with diameters in the nanoscale range can be fabricated from a variety of natural or synthetic, inorganic, organic, or polymeric materials by using a variety of technologies, including template synthesis, self-assembly, electrospinning, and chemical/mechanical separation [8–12]. Although many nanofiber systems produced by different methods have been demonstrated, only a few are really practical to use in water purification when considering factors of safety, stability, and cost. Among these fabrication technologies, electrospinning is one of the scalable methods that can produce useful membranes and meet the considerations [13–15].

The principle of electrospinning is simple but the description of the process is complex. Electrospinning is based on the interplay between the surface tension of a polymer solution and an external electrostatic field that pulls this solution [16–20]. A droplet of polymer solution can be formed at the tip of a spinneret because of the surface tension. But when this droplet is charged and placed in an external electrical field, the electrospinning process can take place. In specific, with increasing applied voltage, the surface tension of polymer solution can be overcome by the electrical force. At that point, the tip of the droplet will

Handbook of Fibrous Materials, First Edition. Edited by Jinlian Hu, Bipin Kumar, and Jing Lu.

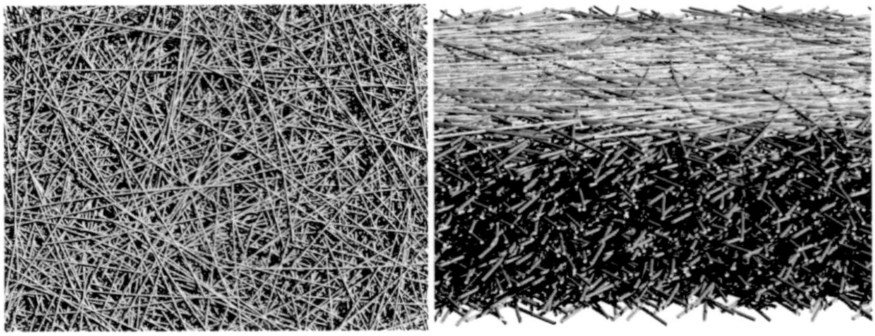

Figure 28.1 Simulations of electrospun nanofibrous composite membrane with top view (a) and cross-sectional view (b). Source: Adapted from Ma et al. 2013 [21]. Reproduced with permission from Bentham Science Publishers.

deform into a Taylor cone and eject a fine jet of polymer solution. While the jet stream is traveling between the spinneret and the collector, the solvent evaporates and the traveling jet becomes unstable due to the combined effects of increasing viscoelasticity and enhanced charge repulsion as a result of increasing polymer concentration. This process leads to significant stretching of the jet with final velocity near a fraction of sound speed and results in a fine fiber with diameters of submicrons on the surface of the collector. As the nanofiber is created without guided direction, the resulting scaffold has a nonwoven structure with high porosity (Figure 28.1) [21].

The fabrication and morphology of electrospun nanofiber scaffold can be affected by several parameters of the polymer solutions, such as concentration, viscosity, conductivity, polymer nature, and solvent, as well as by processing parameters, such as applied voltage, flow rate, jet traveling distance, temperature, and humidity. The morphology and properties of electrospun scaffolds such as fiber uniformity, fiber diameter, pore size, pore size distribution, porosity, and mechanical strength can be controlled by adjusting these parameters [13]. With increasing interests in commercialization of the electrospinning process [22], more and more researchers have engaged in research and development of scalable methods, resulting in flourishing activities not only in fabrication but also in application. The primary attractive feature of electrospun nanofiber scaffolds is the high surface-to-volume ratio and high porosity as a result of the very small fiber diameters (submicron to nanometer). Combined with their functionality, electrospun fibrous scaffolds can be used in a broad range of applications, including tissue engineering [23, 24], drug delivery [24–26], air and water purification [2, 27–29], gas storage [30], sensors, and electrodes [31–34]. A few comprehensive reviews and book chapters regarding these applications have already been published [35].

In this chapter, we will focus on the application of electrospun nanofiber scaffolds for water purification, where these nanofiber scaffolds can function either as a barrier layer or as a supporting layer in the assembly of composite fibrous membranes having unique properties such as high porosity, high surface area, and interconnected water channels created by the fibrous network. The hierarchical structure of these nanofibrous composite membranes are highly permeable; thus it will provide high flux, high rejection, and low fouling properties, which is essential to meet the increasing performance demand of the water industry today [2, 6, 7].

28.2 Characters of Electrospun Nanofiber Scaffold

The filtration performance of nanofibrous composite membranes, including flux, rejection, and serviceable life span are highly affected by the electrospun nanofiber component, either as a barrier or as a support [7, 36]. Therefore, the understanding of the nanofiber scaffold properties will be essential for the design and fabrication of nanofibrous composite membranes with high filtration efficiency.

28.2.1 Porosity

The permeation flux of a membrane intimately depends on the porosity of the scaffold, if used as a barrier layer [37–39]. In general, the higher the porosity, the higher the permeation flux. Depending on the fabricate technique, its pore distribution can be inhomogeneous, resulting in a surface porosity that differs from the bulk porosity with both of them having an effect on the permeation flux. The surface porosity can be estimated based on scanning electron microscopy (SEM) measurements and software analysis, while the bulky porosity, defined as the volume ratio of cavity to mat, can usually be estimated by the following equation [40]:

$$\varepsilon = 1 - \frac{V_b}{V_f} = 1 - \frac{m_f}{A_f \times z_f \times \rho_b} \tag{28.1}$$

where ε is the bulky porosity and V_b, V_f, and ρ_b are the bulk volume, fiber mat volume, and bulk density, respectively. By measuring the mass (m_f), area (A_f), and thickness (z_f) of the fiber mat, the porosity of the electrospun nanofiber scaffold can be estimated. It is worthwhile to note that the thickness of a fibrous mat can simply be measured with a micrometer [41], but the compaction of the membrane under pressure during filtration can create significant deviation. For this purpose, a solvent gravimetric method has been reported that can improve the accuracy of the measurement [42].

The typical membrane produced by the phase inversion method usually has an anisotropic pore structure [43]. As a result, the surface porosity and the bulk porosity are quite different in the membrane produced by phase inversion, with the surface porosity being lower. The surface porosity controlled by the solvent extraction route is typically 10–30 vol%, while the bulk porosity is 50–80 vol% [44–46]. It should be noted that some commercially available membranes made by this method using cellulose nitrate/ester can also have a relatively similar porosity (60–80 vol%) on the surface as well as in bulk [47]. Electrospun nanofiber scaffold, however, always display a similar porosity for both surface and bulk, which is often above 80 vol% and could be up to 95 vol% [48, 49].

The porosity of the barrier layer directly affects the permeation flux of the filtration membrane. Empirically, the permeation flux is proportional to the porosity of the media in a linear manner [50]. For a filtration media with tubular pores, the relationship between the flux and porosity can be expressed by the Hagen–Poiseuille equation [49, 51]:

$$J = \frac{r_p^2 \Delta P \varepsilon}{8 \mu z} \tag{28.2}$$

where J is the permeation flux (m/s), r_p represents the effective pore radius (m), $\Delta P/z$ is the differential pressure per unit thickness (Pa/m), μ is the solution viscosity (Pa s), and ε is the porosity of the filtration media, respectively.

It is clear that the flux increases with increasing porosity when the pore geometry is kept constant. For an electrospun nanofiber scaffold with tortuous and interconnected pores, the tortuosity (τ) needs to be considered. The correlation of tortuosity and porosity (ε) is empirically suggested by Eq. (28.3) [52]:

$$\tau = \frac{(2 - \varepsilon)^2}{\varepsilon} \tag{28.3}$$

Thus, the relationship between the flux and porosity becomes more complex when the tortuosity is considered. Nevertheless, Eq. (28.2) provides a simple correlation of the permeation flux and porosity, which implies that an electrospun nanofiber scaffold with its high porosity of 80–95 vol% will consistently provide a high permeation flux [44, 53, 54]. This property can be visually understood by a simulation. Figure 28.2 represents the morphology of nonwoven nanofiber scaffolds with different porosities. All fiber diameters remain the same, the fibers are

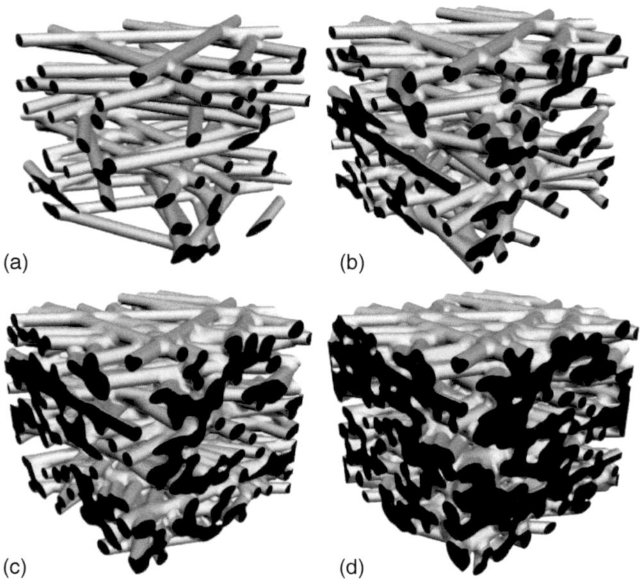

Figure 28.2 Representative electrospun nanofiber scaffolds with porosity of (a) 80 vol%, (b) 60 vol%, (c) 40 vol%, and (d) 20 vol%, respectively.

randomly deposited together, and fiber-crossing points can be "soldered" to form pseudo-nanotrusses [36]. The pore space-to-material volume ratio varied from 4 : 1 to 1 : 4.

It is evident that high porosity can reduce hydraulic resistance and increase the volume of the water paths and therefore can contribute a high permeation flux to the membrane.

28.2.2 Pore Size

The pore size and pore size distribution of a nanofibrous scaffold can also affect the permeation flux and rejection capability [44, 55]. Larger pore sizes will increase the flux but lower the rejection, when the porosity is kept constant. From Eq. (28.2), it is clear that the permeation flux (J) is proportional to the square of the effective pore radius (r_p), as shown in Eq. (28.4):

$$J \sim r_p^2 \tag{28.4}$$

For a nonwoven electrospun nanofiber scaffold, when the fibers were deposited and collected randomly and the porosity of the fibrous mat was kept constant at 80 vol%, the pore geometry and size distribution can be correlated with the fiber diameter. This relationship is illustrated in Figure 28.3 [7].

Empirically, the maximum and average pore size was found to scale linearly with the fiber diameter at the optimal flux performance [56]. The mean pore size (d_{avp}) was found to be about 3 ± 1 times the average fiber diameter (d_{avf}), and the maximum pore size (d_{mxp}) was about 10 ± 2 times the mean fiber diameter, as shown in Eqs. (28.5) and (28.6):

$$d_{\text{avp}} = (3 \pm 1)d_{\text{avf}} \tag{28.5}$$

$$d_{\text{mxp}} = (10 \pm 2)d_{\text{avf}} \tag{28.6}$$

Thus, the pore sizes can be adjusted by controlling the fiber diameter of the electrospun membrane, allowing them to satisfy the requirements for different applications [56, 57]. It is clear that the large pore size decreases the hydraulic resistance in a filtration process and therefore increases the permeation flux, but this may reduce the rejection capability (or selectivity).

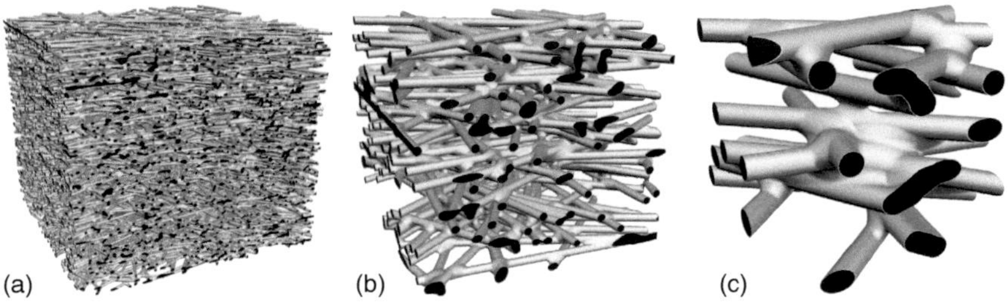

Figure 28.3 Illustration of the correlations between pore size and fiber diameter at a constant porosity of 80 vol% in a fixed volume. The relative fiber diameter ratio of panels (a), (b), and (c) is 1 : 3 : 10. Source: Ma et al. 2011 [7]. Reproduced with permission of Royal Society of Chemistry.

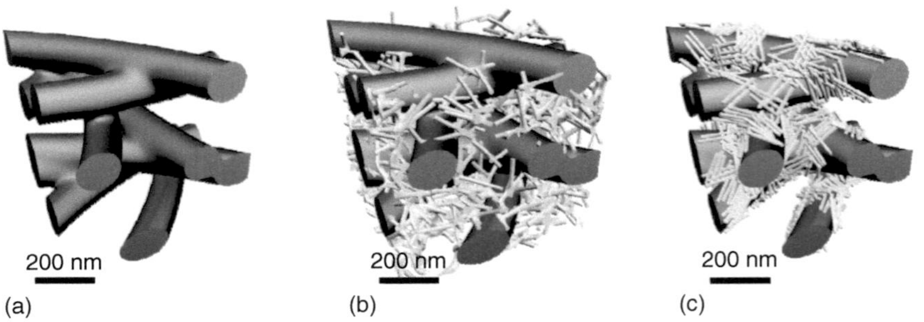

Figure 28.4 Schematic representations of electrospun nanofiber scaffolds: (a) scaffold before nanowhisker infusion, (b) infused nanowhiskers forming loose cross-linked mesh, and (c) nanowhiskers collapsed onto the scaffold, forming bundles. Source: Ma et al. 2012 [60]. Reproduced with permission of American Chemical Society.

The pore size uniformity in the barrier layer can greatly affect the selectivity of the membranes for separation [58, 59]. In general, the narrower pore size distribution (more uniform pore sizes) will lead to higher rejection and better selectivity [39, 60]. Membranes with narrow pore size distribution often need to be fabricated by special methods, such as templating [61, 62], self-assembly [63], track etching [64], and microfabrication [65]. The membranes prepared by these methods can have high permeation flux and rejection capability but often suffer from fouling [66] and cost issues. Nonwoven nanofiber scaffolds prepared by electrospinning typically have a relatively broad pore size distribution in 1–2 orders of magnitude range [40]. It is worthy to mention that the impregnation of ultrafine nanofibers (e.g. cellulose nanowhiskers or cellulose nanofibers [CNs]) into the electrospun nanofiber scaffolds can decrease the pore size and narrow down the pore size distribution, as shown in Figure 28.4 [60].

The above behavior is probably due to the "leveling effect." That is, the ultrafine nanofibers can penetrate into the large pores in the electrospun scaffold and subdivide them into smaller pores while leaving the small pores empty. It should be noted that for a membrane produced by the phase inversion method, the pores mentioned above are referring to the surface pores, while bulk pores located in the interior of the membrane affect the rejection to a lesser degree [67].

28.2.3 Surface Area

In a fibrous scaffold, the specific surface area is inversely proportional to the radius of the fiber. Therefore, the thinner the fiber diameter (d_f, m), the higher the specific surface area (S, m^2/g), where the two have the following relationship [68]:

$$S = \frac{4}{d_f \times 10^6} \tag{28.7}$$

Based on this equation, the diameter of electrospun nanofiber is in submicrons, so the specific surface area is typically 10–40 m^2/g [69]. The high surface area of nanofibrous composite membranes offers high retention of contaminant through adsorption (or depth filtration). A high adsorption capacity can be achieved when a large number of functional adsorption sites located on the surface of the scaffold are reachable by the targeted contaminants [70, 71]. Common fibrous scaffolds [72], such as polyester nonwoven substrate, glass fiber, or cellulose paper, have larger fiber diameters of a few microns to a few tens of microns. Based on Eq. (28.7), their surface area will be 10–100 times lower than that of an electrospun nanofiber scaffold of the same mass, i.e. 10–100 times lower retention capability in an adsorption process. Thus, reducing the diameter of the electrospun nanofiber scaffold can efficiently increase the adsorption capacity in water purification [73]. However, the reduction of the fiber diameter can also lead to a decrease in pore size and increase the pressure drop across the membrane. This situation becomes more critical when the targeted contaminants are adsorbed on the surface of the fiber, thereby further decreasing both porosity and average pore size, reducing the permeation flux, and increasing the pressure drop [74].

The increase in surface roughness of the electrospun nanofiber can also enhance the surface area [75–77]. There are numerous strategies, mostly using the phase separation behavior during the formation of nanofiber in electrospinning, to create a rough surface (Figure 28.5). These strategies including the incorporation of nanoparticles (e.g. polyhedral oligomeric silsesquioxane (POSS), silica/titanium dioxide particles) into the spinning solution [78, 79], spinning of a bicomponents system followed by leaching out one component [80], or using a special volatile solvent to induce porous nanofibers [81]. In addition, some posttreatment approaches, such as post-coating [82] or etching [83] of electrospun nanofiber scaffolds, have also been demonstrated to create a rough fiber surface to increase the surface area.

28.2.4 Pore Geometry

The pore geometry in the barrier layer of a membrane can have an either interconnected (e.g. tubular or tortuous paths with interconnectivity) or dead-end nature (e.g. closed voids) [47, 84–86]. In general, tubular or tortuous interconnected pores, regarded as fluid paths, are beneficial to achieve a high permeation flux. This is because the interconnected pores can reduce the hydraulic resistance when the fluid is traveling in the membrane [87]. In addition, the interconnected pores can reduce the fouling problem of the membrane in the filtration process [85]. However, the typical tubular and gyroidal pores, as shown in Figure 28.6a,b, are difficult to achieve with conventional methods, where a special fabrication process is necessary [39].

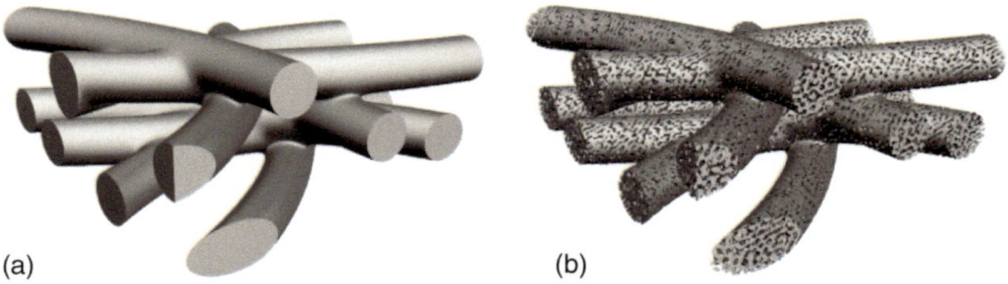

(a) (b)

Figure 28.5 Illustration of the nonwoven electrospun nanofibers (a) and porous nanofibers (b).

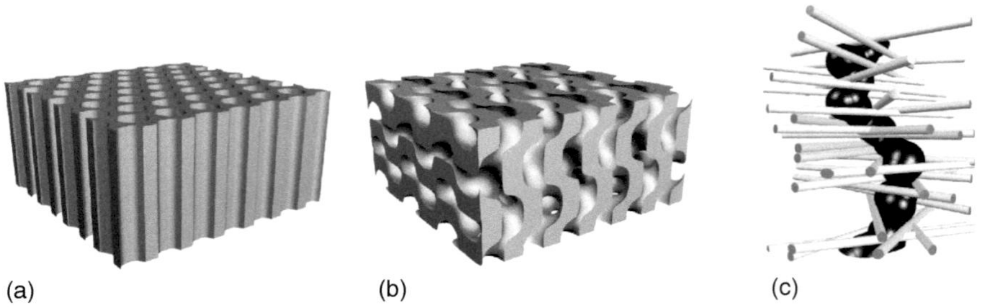

Figure 28.6 Schematic representative of (a) straight tubular pores, (b) tortuous gyroidal pores, and (c) tortuous interconnected pores in an electrospun nanofiber scaffold, where water can go through the water path easily.

It is interesting to note that the pores in an electrospun scaffold are interconnected and spontaneously created in the spinning process, as shown in Figure 28.6c. This indicates that the nature of the electrospun nanofiber scaffolds will lead to higher flux and lower fouling performance than conventional membranes. These advantages have been verified experimentally. The tortuosity in the electrospun nanofiber scaffold has been proven to be relatively low when compared with those of the conventional membranes containing some dead-end cavities [49, 89]. As a result, membranes containing electrospun nanofibers as the barrier layer always have a higher permeation flux than membranes made by the phase inversion method.

Furthermore, the interconnected pores in electrospun nanofiber scaffolds that can reduce the fouling tendency have also been seen experimentally [47]. Although the fouling problem of the membrane is closely related to the chemical and physical interactions between the membrane and the feed solution, the pore geometry, especially the interconnections among the pores, also plays an important role to affect the membrane fouling property. There are two fouling mechanisms that can be reduced by using an electrospun nanofibrous composite membrane. One is the internal concentration polarization, for example, in the forward osmosis (FO) process. The concern of internal concentration polarization may be reduced significantly due to the rapid equilibrium of the permeate water and draw solution in the interconnected pores of the electrospun scaffold as a result of the higher water flux [90]. Another fouling improvement is due to the following consideration. When the foulants in the feed solution enter the membrane and adhere to the surface of the pores, the irreversible fouling occurs. As a result, the pore size will decrease as the pores are blocked [88]. If the pores are of the dead-end type, as in conventional membranes, the tendency of blocking the pores by the foulants is high. In contrast, electrospun nanofiber scaffolds contain interconnected pores, which are more difficult to block completely and are relatively easier to be cleaned by flushing or back-flushing than conventional membranes [91]. Therefore, the usable lifetime of the membrane based on electrospun nanofiber scaffolds could be longer, and a higher recovery of the flux could be achieved.

28.2.5 Mechanical Properties

The mechanical properties of electrospun nanofiber scaffolds in filtration are mainly associated with the capability to resist compression or deformation [92, 93]. The compaction of the membrane is inevitable in a pressure-driving filtration process, which can significantly reduce the permeation flux [94–96]. The irreversible deformation of the pores in the barrier layer can also lead to notable reduction of permeation flux as a result of the decrease in porosity, pore size, and pore geometry change. Therefore, a membrane with a good compaction resistance will exhibit high durability, which means that the permeation flux can be retained at the high initial level. Membranes produced by the phase inversion method usually have good mechanical properties as some cavities are disconnected and the solid portion forms a strong support. However, after a long-term operation under pressure, the structure of the phase inversion membrane can also change and be stabilized at a certain status [97]. For electrospun nanofiber scaffolds with very high porosity (e.g. 80–95%), adhesion between the nanofibers will be relatively weak [49]. Thus, an electrospun nanofiber scaffold can easily be compacted when compared with a phase inversion membrane. The effective porosity of the electrospun nanofiber scaffold does not depend on the initial porosity of the membrane, but is a function of the applied pressure [50, 98]. Figure 28.7 shows the compaction effect on a nonwoven nanofiber scaffold, where the porosity is changed from 80 to 5 vol%, respectively. It is seen that as the scaffold becomes denser, pore size, pore size distribution, and pore geometry all change correspondingly. In most cases, the deformation of the nanofiber scaffold is irreversible, whereas the permeation flux could only be partially recovered after the pressure release. This implies that the electrospun nanofiber scaffold is more suitable for low-pressure filtration, where a high energy-saving benefit can be obtained. However, if the electrospun nanofiber layer does not account for the major volume component in the filtration membrane, as to be described in the design of thin-film nanofibrous composite (TFNC) membranes, the system can stand high-pressure operations and still yield significant energy-saving benefits.

By adjusting the electrospinning parameters, such as a special solvent, suitable humidity and temperature, or post-heating treatment, the adhesion between the nanofibers could be significantly improved [99, 100]. For example, when polyethersulfone (PES) electrospun nanofiber scaffolds are fabricated from a mixed solvent of *N*,*N*-dimethylformamide (DMF)/*N*-methylpyrrolidone (NMP)

Figure 28.7 Compaction effect of electrospun nanofiber scaffold from 80 to 55 vol% and 30 to 5 vol%.

at different ratios, the mechanical properties can be greatly enhanced [101]. With increasing NMP content (from 0 to 50 wt%) at the fixed PES concentration of 26 wt%, the diameter of electrospun nanofibers was found to increase from 550 to 760 nm, while the corresponding modulus and ultimate tensile strength were found to improve by 570% and 360%, respectively. The increase in mechanical properties was not just due to the increased fiber diameter, but was mostly due to the enhanced junction points between the interconnected nanofibers by partial fusion as a result of the residual solvent in nanofibers. Similarly, chemical cross-linking at the crossover points between the nanofibers is an efficient method to improve the mechanical properties [74]. It is interesting to note that cross-linked polyvinyl alcohol (PVA) nanofiber scaffolds using glutaraldehyde (GA) exhibited a consistent flow rate without the sign of compaction. This is because the cross-linking reaction resulted in the formation of a rigid fibrous network structure, where the crossover points between the fibers were "soldered" together [36, 57]. The concept of cross-linking the nanofiber scaffold to overcome the compaction problem has led us to design the TFNC membranes, which will later be further elaborated on. In this new membrane design, a coating of dense barrier layer (e.g. cellulose nanofibers) is applied on the surface of electrospun nanofibrous composite membrane (e.g. polyacrylonitrile [PAN]), where the barrier matrix can induce a nanocomposite layer below the barrier layer. The nanocomposite layer containing electrospun nanofiber scaffold and the matrix component of barrier layer significantly increased the Young modulus of the final composite membrane, where no yielding point could be observed during the tensile stretching process [60]. Similarly, bi-vinyl and tri-vinyl monomers were polymerized on the surface of the PAN electrospun nanofibers, which significantly improved the mechanical properties of the membranes due to the formation of a nanocomposite structure. The resulting membranes showed an excellent ultrafiltration (UF) performance that could simultaneously remove bacteria and viruses from contaminated water but still retain their high permeation flux [74, 92].

28.2.6 Materials

The electrospun nanofibers can be fabricated from a variety of natural or synthetic, inorganic, organic, or polymeric materials. Typically, polymer materials such as PAN, PVA, polysulfone (PSU), PES, and polyvinylidene fluoride (PVDF) are employed to fabricate electrospun nanofibers for water purification [2, 13]. Common solvents used to dissolve abovementioned polymers in electrospinning are DMF, dimethyl sulfone (DMSO), and NMP, which has a relative high boiling point. It is clearly that the surface properties of the electrospun nanofibers depend on the chosen materials, where PVDF, PSU, PES, and polystyrene (PS) nanofibrous membranes exhibit hydrophobic property, while PVA and PAN membranes possess relative hydrophilic property. The filtration performance of electrospun nanofibers, functioning either as a barrier layer or as a substrate, is intimately dependent on the material properties, aside from the structural characteristics.

28.3 Applications of Electrospun Nanofibrous Composite Membranes

Highly permeable nanofibrous composite membranes with multilayered hierarchical structures, where an electrospun nanofiber scaffold acts as a barrier layer and/or a support, can be obtained [2, 6, 7, 38]. This new membrane structure offers high permeation flux and high rejection rate. One-, two-, three-, and even four-layered structures can be constructed, depending on the demands of rejection vs. mechanical stability of the membranes. Some examples of these membranes are as follows.

28.3.1 As a Barrier Layer

Typically, microfiltration membranes have a pore size in the range of 0.1–10 μm, which is ideal to remove micron- or submicron-size contaminants, such as waterborne bacteria, for drinking water application. The purification process is through size exclusion separation. However, if smaller charged contaminants such as dyes, viruses, proteins, and heavy metal ions are involved, the size exclusion mechanism will not work. To remove these contaminants, the adsorption process has to be utilized [102–105]. For this purpose, electrospun nanofiber scaffolds, having pore sizes in the range of 0.1–10 μm, are good candidates for a barrier layer for both filtration and adsorption purposes. This indicates that we can design a microfiltration membrane that also behaves as a good adsorbent medium.

28.3.1.1 Size Exclusion

Waterborne bacteria typically have diameters above 0.2 μm. For example, the dimension of *Escherichia coli* is 0.5 × 2.0 μm and that of *Brevundimonas diminuta* is 0.3 × 0.9 μm [7, 106, 107]. Therefore, a membrane designed with a mean pore size of 0.2 μm can be expected to successfully remove waterborne bacteria from contaminated water (Figure 28.8) [2].

For microfiltration, the electrospun nanofibrous composite membrane can have either one layer (i.e. self-standing nanofiber scaffold) [49] or a two-layered structure involving an electrospun nanofiber scaffold and a nonwoven substrate (e.g. polyethylene terephthalate [PET] nonwoven) [2]. The nanofiber scaffold is

Figure 28.8 Microfiltration membrane for removal of bacteria from contaminated water. Source: Ma et al. 2012 [2]. Reproduced with permission of Springer.

used as the barrier or selective barrier, where the nonwoven substrate functions as the mechanical support. In the case of self-standing electrospun nanofibrous microfiltration membranes, both PVDF and PSU have been demonstrated to remove micron-size particles from water [91, 98]. In specific, the electrospun nanofibrous membranes having mean fiber diameters ranging from 380 to 470 nm exhibited a fairly high flux (>100 l/(m^2 h)/psi) and good rejection (>91%) of 1.0 μm particles. However, these membranes failed to retain 0.5 and 0.1 μm particles through size exclusion (the rejection ratio was only 47% and 14%, respectively). However, when the mean fiber diameter was reduced to about 100 nm, the two-layered electrospun PAN nanofibrous composite membranes and cross-linked PVA nanofibrous composite membranes (both use a PET nonwoven substrate) could retain 0.2 μm particles with a very high rejection (>93%). When challenged with 10^6 colony-forming unit (cfu)/ml of *E. coli* aqueous suspension, these membranes showed an impressive rejection ratio of 99.9999%. The typical SEM cross-sectional images of electrospun nanofibrous composite membranes after filtration of 0.2 μm particles and *E. coli* are shown in Figure 28.9, respectively [57, 58]. The unique characteristics of these membranes were that they exhibited two to three times higher flux than that of commercially available microfiltration membranes (e.g. GS9035).

Alternatively, the electrospun PAN nanofiber scaffold can be modified with nano-absorbents, such as cellulose nanofibers (fibers with 5 nm diameter [108, 109]) that can be derived from natural cellulose resources, forming a nanoweb within the electrospun nanofiber scaffold. The nanoweb structure not only improved the mechanical properties of electrospun nanofiber scaffolds but also minimized the pore size and narrowed the pore size distribution [60, 110]. As a result, the rejection of 0.2 μm particles was larger than 98%, and the rejection of both of *E. coli* and *B. diminuta* were over 99.9999%. Figure 28.10 represents the structure of cellulose nanofibers infused into electrospun PAN nanofibrous composite membranes.

Similarly, the surface of PAN electrospun nanofibers could be modified with a polymer network formed by in situ polymerization of bi-vinyl and tri-vinyl monomers [92]. These membranes also exhibited 99.9999% rejection to *E. coli* while retaining a high permeation flux. The surface modification of an electrospun nanofiber scaffold could also introduce adsorption functionality into the

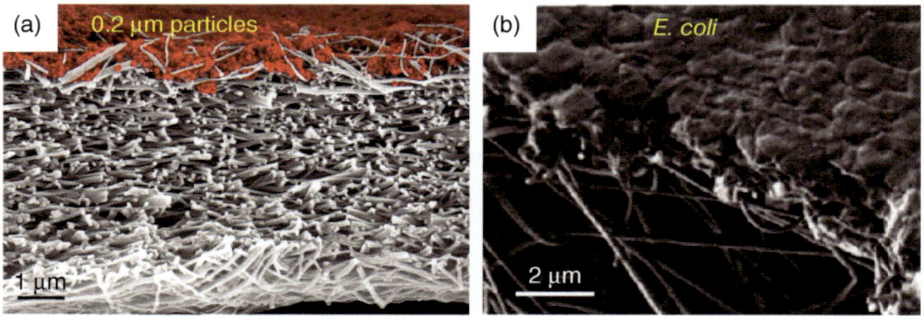

Figure 28.9 Electrospun nanofibrous composite membranes after filtration of 0.2 μm particles (a) and *E. coli* (b). Source: Adapted from Liu et al. 2013 [57] and Wang et al. 2012 [58].

Figure 28.10 Depiction of a two-layered fibrous structure with a top electrospun nanofibrous layer infused with cellulose nanofibers [73].

membrane and remove charged species, such as viruses, dyes, proteins, and heavy metal ions [74]. This can be elaborated as follows.

28.3.1.2 Adsorption

The separation mechanism through adsorption, also known as depth filtration, is based on electrostatic interactions, hydrophobic interactions, or ligand coordination between the functional groups on the surface of nanofibers and the targeted contaminants [21, 111–113]. Therefore, the adsorption capacity mainly depends on the surface area of the membrane and the density of functional groups. It should be noted that the adsorption capacity is limited by the separation mechanism, where the theoretical maximum rejection (or retention) ratio depends on the amount of functional groups on the surface of the scaffold in depth filtration. There are trade-offs among the flux, retention, pressure drop, and thickness of the membrane associated with the membrane usage. Nonwoven fibrous membranes with large fiber diameters usually led to large pore sizes when the porosity was fixed [7]. These membranes could exhibit a high permeation flux and low pressure drop. In this scenario, the larger the fiber diameter, the higher the permeation flux. However, the retention of the contaminants is associated with the total amount of the functional groups, which depend on the surface-to-volume ratio of the scaffold. In this case, the membranes with smaller fiber diameters will provide a larger surface-to-volume ratio and, therefore, the higher retention capability. Obviously, the permeation flux decreases when the average pore size of the membrane defined by the fiber diameter becomes small, and the pressure drop will increase correspondingly. Moreover, during the actual filtration process, as the contaminants are adsorbed on the fiber surface, the pore sizes will be further decreased, therefore reducing the flux and increasing the pressure drop [74]. The service life span of the membrane is also related to the thickness of the membrane. The thicker the membrane, the higher the usage life. However, the increase of the membrane thickness will lead to a decline in permeation flux and an increase in pressure drop, which has to be taken into account. Furthermore, the surface chemistry of the fibrous membrane is another key factor in the adsorption process. Generally, the adsorption equilibrium should be achieved as fast as possible in order to capture the contaminant molecules under the given pressure or flow rate [21]. Based on electrostatic interactions, commercially available nonwoven microfibrous mats, such as PET, polypropylene (PP), or glass fiber mats, have been used for the adsorption of Metanil Yellow (dye), *Klebsiella terrigena*, *E. coli*, *B. diminuta* (bacteria), and *MS2* (bacteriophage) after surface modification with amino-based chemicals [72, 114]. With the proper treatment, high flux and high adsorption capacity could be achieved. However, the surface area of these fibrous mats was limited due to the large fiber diameters (typically a few tens of microns). Nanofibrous composite membranes could be used to significantly increase the adsorption capacity since typical diameters of electrospun nanofibers are in the range of a 100 nm [73, 74]. In this case, an electrospun nanofibrous composite membrane with a properly functionalized surface could also be used as an adsorption medium [21].

Some example surface modification schemes of electrospun nanofiber scaffolds are as follows. A regenerated cellulose nanofibrous membrane has been prepared by hydrolysis of electrospun cellulose acetate (CA) nanofibers. After surface grafting with special ligands, the membranes could be used either for adsorption of proteins (e.g. bovine serum albumin [BSA]) or for filtration purification of antibody IgG [115, 116]. In another study, the electrospun PSU nanofibrous membrane covalently bonded with ligand could be used for BSA adsorption [117]. In yet another study, a facile method involving the surface wrapping of electrospun PAN nanofibers with a positively charged polymer through in situ polymerization of bi-vinyl monomers has been demonstrated [92]. This membrane system exhibited very high retention of BSA in static adsorption as well as dynamic adsorption separation (or depth filtration) with high permeation flux. This positively charged membrane can work as an effective microfiltration membrane but also simultaneously removed 99.99% of bacteriophage, *MS2*, a virus model.

The fiber diameter of an electrospun nanofibrous composite membrane is typically higher than 100 nm, whereas a further decrease in fiber diameter is usually limited by practical electrospinning conditions [118]. To further increase the active surface area, a novel hierarchical structure of a nanofibrous composite membrane has been demonstrated. In this system, the electrospun nanofiber scaffold is used as a network, where much smaller nanofibers (diameters in the range of a few nanometers) are infused into the gaps between the electrospun fibers. As a result, the pore size can be decreased, offering both high rejection of large contaminants due to size exclusion and high retention of small contaminants due to the increase in surface area for adsorption. For example, two strategies to form such a membrane structure have been demonstrated. One is through the infusion of cellulose nanofibers (having ~5 nm fiber diameter) into electrospun PAN mats having diameters in the range of 100–500 nm, forming a very fine nanoweb immersed in the electrospun nanofiber scaffold (Figure 28.11a–d) [60, 119].

This membrane system exhibits 16 times higher adsorption capacity against the positively charged crystal violet dye molecules than that of a commercially available membrane. Meanwhile, in dynamic adsorption of *MS2* (the virus model), the retention ratio could reach 99%, which is also better than that of the commercial system. This is because commercially available microfiltration membranes are

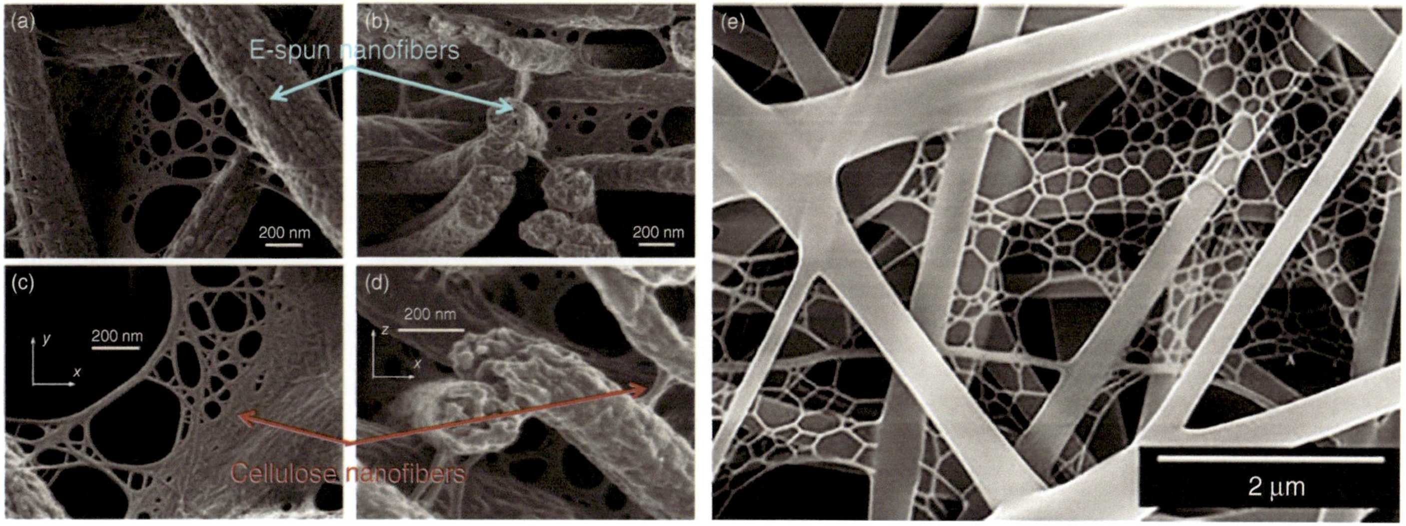

Figure 28.11 Top view (a, b) and cross-sectional view (c, d) of SEM image of the nanostructure of the electrospun PAN scaffold infused with cellulose nanofibers. Source: Ma et al. 2012 [60]. Reproduced with permission of American Chemical Society. (e) Electrospun nylon 6 nanofibers containing nanowebs. Source: Ding et al. 2006 [119]. Reproduced with permission of Elsevier.

often produced by the phase inversion method, where the resulting porous structure has a lower porosity (60–80%) [43]. For example, the commercially available porous membranes, made of nitrocellulose-based cellulose ester, has exhibited good adsorption and is commonly used for protein binding in the field of microbiology [120]. However, this membrane is not as good as the composite nanofibrous microfiltration membranes containing both electrospun PAN nanofiber scaffolds and infused webs of ultrathin cellulose nanofibers. The composite nanofibrous microfiltration membranes can be made by techniques, thus possessing different configurations. For example, they can be fabricated by the "electronetting" method. With this approach, both scaffolds containing electrospun nanofibers with a mean diameter of 100–500 nm and a web of finer nanofibers with a mean diameter of 10–40 nm could be simultaneously produced in a co-mingled manner. The average pore size of the nanoweb could be about 16 nm, as shown in Figure 28.11e [32, 119, 121]. The formation of nanowebs (or nanonets) has been attributed to the phase separation of charged droplets generated during electrospinning. Such a membrane system showed excellent retention to *E. coli* and viruses, such as *H1N1* (diameter: 80–120 nm) and *SARS* (diameter: ~100 nm) in air filtration, but they have not been tested in water filtration.

It should be noted that direct electrospinning of polymers with suitable functional groups, such as chitosan [122, 123] or PVA/polyethylenimine (PEI) polymer blend [124], could produce positively charged nanofiber scaffolds to adsorb negatively charged macromolecules or microorganisms such as viruses. These scaffolds do not require further surface functionalization. However, the mechanical strengths of these electrospun scaffolds are usually weak, which should be supported on a strong substrate. In this case, a post-cross-linking step is usually necessary in order to immobilize the functional nanofibrous scaffold in the composite membrane.

Recently, a composite nanofibrous membrane system containing carbon nanotubes (CNTs) has been demonstrated for bacteria/virus removal [125, 126]. In this system, single-walled or multiwalled CNTs were deposited onto a commercially available microfiltration membrane, where the resulting membranes could remove bacteria (e.g. *E. coli*) by screen filtration with very high retention ratio. In addition, these membranes could also remove viruses (e.g. *MS2*) by depth filtration (adsorption) with 99.9% retention ratio. Unfortunately, this membrane system exhibited a relatively short lifetime, high cost, and the need to be stored in the wet state to preserve permeation flux. However, even though one can overcome these hurdles, the intrinsic risk of using potentially toxic CNTs for drinking water applications may outweigh any benefits in performance improvement. We thus believe the electrospinning technique is a more versatile and safer technique to produce nanofibrous membranes for drinking water applications.

28.3.2 As a Support Layer

Electrospun nanofiber scaffolds deposited onto nonwoven microfibrous substrates have also been used as a mid-layer to support another denser barrier layer, thus forming high-performance membranes for varying filtration applications including ultrafiltration (UF), nanofiltration (NF), reverse osmosis (RO), forward osmosis (FO), pervaporation, and membrane distillation (MD) [2]. In this section, we will review the structure and properties relations of these membranes containing electrospun nanofiber scaffolds as a support layer for selective applications (i.e. UF, NF/RO, and pervaporation).

28.3.2.1 Ultrafiltration

Ultrafiltration (UF) membranes with typical pore sizes between 2 and 100 nm are typically used for separation of proteins, emulsified oily droplets, macromolecules, and suspended solids (in that size range) with water [127]. Ultrafiltration is not fundamentally different from microfiltration, except that the sizes of the targeted contaminant particles are smaller. The major problems in ultrafiltration are the relatively low filtration efficiency and membrane fouling [128]. For example, conventional ultrafiltration membranes are typically prepared by the phase inversion method, resulting in a low surface porosity (10–30 vol%) that would severely limit the membrane permeation flux [129]. The low porosity, dead-end pore structure, compounding with the use of relatively hydrophobic materials, such as PSU, PES, PAN, and PVDF, can drastically increase the fouling tendency, even though the membranes usually exhibit excellent chemical and mechanical stability [130]. It turns out that the new structure in TFNC membrane (as shown in Figure 28.12) [93], containing the electrospun nanofiber scaffold as a mid-layer support, can overcome these challenges.

In TFNC membranes, the bottom layer is a nonwoven substrate (e.g. PET nonwoven) containing microfibers with a typical thickness of ~100 μm, providing mechanical stability for the system. The mid-layer is an electrospun nanofiber scaffold (thickness ~ 50 μm) with high porosity (above 80 vol%) and interconnected pores, which replaces the conventional porous support produced by the phase inversion method. The materials used in the electrospun nanofiber middle layer can be the same as conventional filter materials, such as PAN

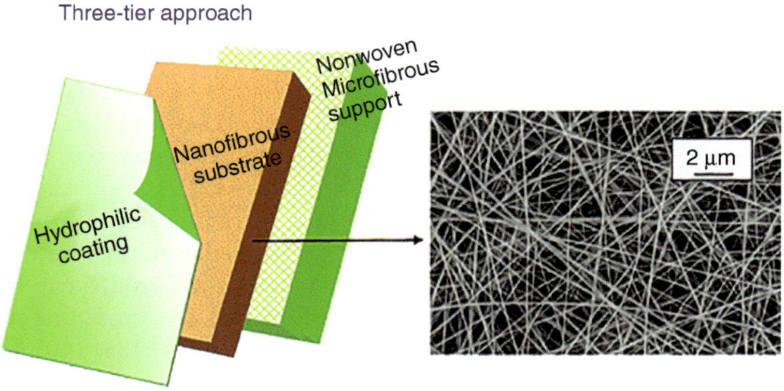

Figure 28.12 Schematic representation of a TFNC membrane with the three-tier structure: hydrophilic thin coating top layer, electrospun nanofiber scaffold mid-layer, and nonwoven microfibrous substrate bottom layer. Source: Wang et al. 2005 [93]. Reproduced with permission of American Chemical Society.

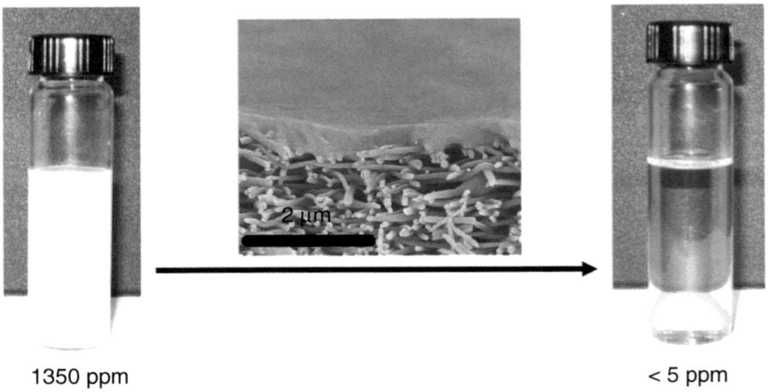

Figure 28.13 Purification of emulsified oil/water mixture with PVA-CN (1.25 wt%)-TFNC ultrafiltration membrane. Source: Yoon et al. 2009 [48]. Reproduced with permission of Elsevier.

[50, 131], PVA, or PVDF [132, 133]. The electrospun mid-layer can support a thin barrier layer (thickness of typically from 50 to 1000 nm). To reduce the fouling tendency, hydrophilic but water-resistant materials are usually considered, such as cross-linked PVA [48, 132], cross-linked PVA/cellulose nanofibers [134], cross-linked PEG/cellulose nanofibers [135], cross-linked polyamide [136], chitosan [50, 133], regenerated cellulose/chitin [137, 138], polysaccharide nanofibers [96, 131, 139], and even CNTs [93].

The performance of the three-layered TFNC membrane was found to be significantly better than conventional UF membranes. For example, in the system containing cross-linked PVA as the barrier layer supported by an electrospun PAN nanofiber scaffold, the permeation flux in separating emulsified oil droplets with diameter of 70–1000 nm and water (Figure 28.13) was 2–10 times higher than those of commercial membranes while being able to maintain a 99.0% rejection ratio [48, 124]. The system of emulsified oil droplets and water was used to simulate bilge water from ships or oily wastewater from oil fields.

Further improvement of the permeation flux could be achieved by incorporating nanofillers (e.g. cellulose nanofibers or CNTs) into the barrier layer of cross-linked PVA. This is because the naturally occurring interfacial gaps between the nanofibers and the polymer matrix in the barrier layer can serve as "water channels" to guide the water transport through the interconnected nanofiber scaffold, thus increasing the permeation flux [93, 134, 135]. Unfortunately, with GA cross-linked PVA as the membrane barrier layer, the coating process was ill-controlled due to the high viscosity, and the gel-like barrier has to be kept in the wet state before use. Alternately, bi-vinyl groups can be introduced into the PVA or PEG-based barrier layer, and the cross-linking reaction can be carried out by photoinitiation or heat initiation [135, 140]. In one study, chitosan without cross-linking was used as the barrier layer of a TFNC membrane [50]. The resulting membranes exhibited a good rejection ratio but relatively low permeation flux for the separation of oil emulsion and water, probably due to a relatively large barrier layer thickness. Cellulose and chitin regenerated from ionic liquids have also been used as the barrier layer

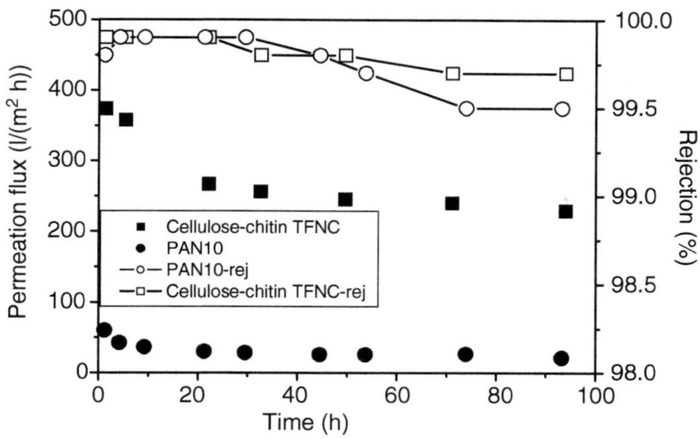

Figure 28.14 Cross-flow ultrafiltration of oil/water emulsion using TFNC membrane containing cellulose–chitin blend barrier layer and PAN10 membrane up to 100 hours. The applied pressure was 30 psi and temperature was 35 ± 2 °C. Source: Ma et al. 2010 [137]. Reproduced with permission of Elsevier.

to fabricate TFNC membranes. In this system, as the barrier thickness was optimized, 2–10 times higher flux was obtained in the TFNC membrane than the commercial counterparts while the membrane also maintained a good rejection ratio (>99.5%, seen in Figure 28.14) [137, 138].

The performance of the TFNC membranes has been reviewed recently [2]. For ultrafiltration applications, polysaccharide nanofibers were found to be very good candidates to fabricate the barrier layer, which led to good balance of filtration efficiency, i.e. high permeation flux, high rejection, and good antifouling tendency (due to the hydrophilicity) [131]. Furthermore, it is worthwhile to note that cellulose nanofiber-based TFNC membranes exhibited excellent chemical resistance, bacteria resistance, hypochlorite resistance, and applicability in a broad pH range [96]. The hierarchical structure of the cellulose nanofiber (CN) based TFNC membrane is illustrated in Figure 28.15, where a nonwoven microfibrous substrate (with fiber diameters in tens of microns) is used as the bottom support,

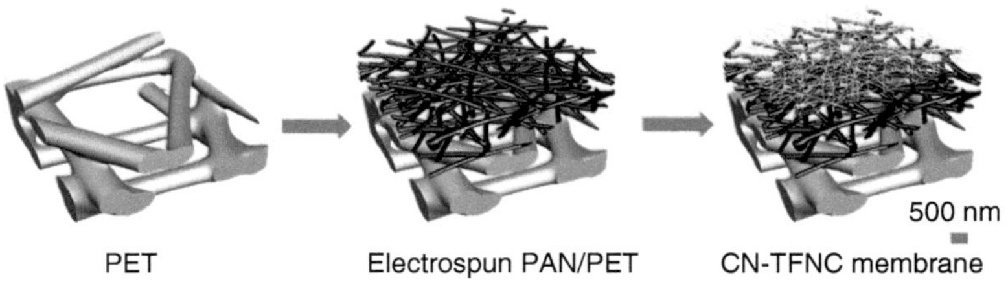

Figure 28.15 Schematic representation of the formation of a cellulose nanofiber (CN)-based TFNC membrane: nonwoven microfibrous substrate (with fiber diameters in the tens of micron size range) as the bottom support, electrospun nanofiber scaffold (with fiber diameters in the 100 nm to micron size range) as the mid-layer support, and cross-linked ultrafine cellulose nanofibers (with nominal fiber diameters in the five to tens of nanometer size range) as the top barrier layer.

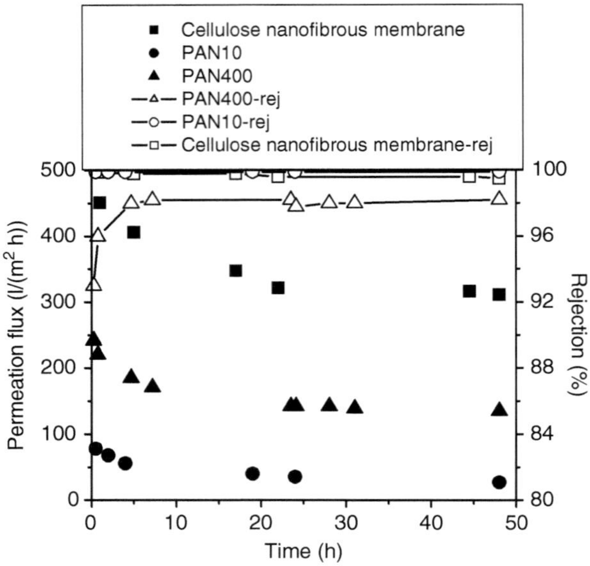

Figure 28.16 Permeation flux and rejection ratio of TFNC membrane containing cellulose nanofiber barrier layer as a function of time; measurement was carried out at a constant pressure of 30 psi and temperature of 37 °C in ultrafiltration of oil/water emulsions. The ultrafiltration performance of commercial PAN10 and PAN400 membranes was also included for comparison. Source: Ma et al. 2011 [131]. Reproduced with permission of American Chemical Society.

an electrospun nanofiber scaffold (with fiber diameters between 100 nm and few microns) as the mid-layer, and cross-linked ultrafine cellulose nanofibers (with fiber diameters between five and tens of nanometers) as the barrier layer.

The CN-TFNC membranes showed 3–11 times higher permeation flux compared with commercial ultrafiltration membranes, such as PAN10 and PAN400 (Sepro), and a rejection ratio higher than 99.5% (Figure 28.16), where the oil concentration in the permeate could exceed the environmental standard for wastewater disposal (i.e. less than 10 ppm) [131]. The high permeation flux could be attributed to the thinness and hydrophilicity of the CN barrier layer, leading to the reduction of hydraulic resistance and oil fouling tendency, but it was also due to the high porosity and interconnected pores in the electrospun nanofiber support.

In TFNC membranes, there is some integration between the barrier layer and the electrospun nanofibrous support, that is, the top part of the electrospun scaffold is partially impregnated in the nanofiber barrier layer [131, 137]. As evidence, the contour of electrospun nanofibers could be seen from the top view of the TFNC membrane (Figure 28.17). Such integration could increase the mechanical stability of the barrier layer and decrease the thickness of the barrier layer. The last effect would further increase the permeation flux [38]. In addition, the nanofiber scaffold embedded in the barrier layer can also serve as a way to introduce "water channels" to guide water transportation in the barrier layer, which is

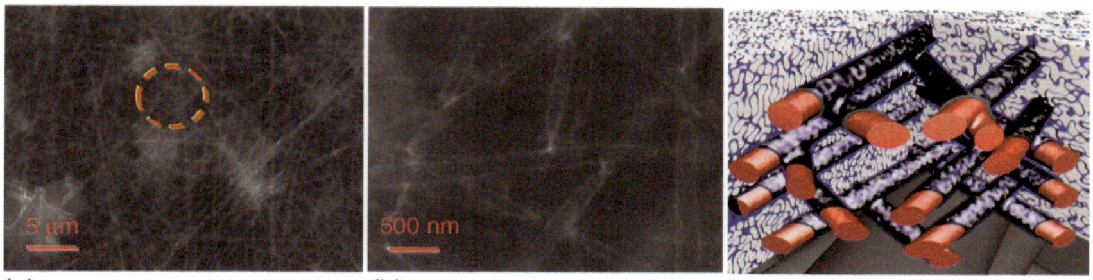

(a) (b) (c)

Figure 28.17 (a) SEM image of electrospun nanofibers embedded in the polymer matrix barrier layer of a TFNC membrane. (b) SEM image of magnification of the circle area in (a). (c) Representative of a fibrous network formed by electrospun nanofibers distributed in polymer matrix. Source: Ma et al. 2011 [131], 2010 [137]. Reproduced with permission of American Chemical Society.

also beneficial to increase the permeation flux and will be further discussed later [141].

As discussed earlier, in a pressure-driving filtration process, porosity of the porous membrane usually changes as a result of compaction, especially after a long-term and high-pressure operation. Thus, the permeation flux of the membrane often decreases significantly over time (e.g. 10–65% of the initial value) [97]. In general, a membrane with a very high porosity can be significantly compacted during the filtration process, independent of whether its porous structure is from the phase inversion process or from the electrospinning process. As a result, membranes with high porosity, such as electrospun nanofiber scaffolds (>80 vol%), are often more suitable to be used in low-pressure applications.

In TFNC membranes, electrospun PAN nanofiber scaffolds have been used as a mid-layer to support the cellulose nanofiber barrier layer. In this membrane system, the permeation flux was found to reduce to 42% after 21 hours of compaction using pure water at 30 psi. This was similar to the commercial membrane (PAN10), which was reduced to 48% under the same operating conditions. The compacted TFNC membrane, however, showed five times higher permeation flux than the compacted PAN10 membrane for removal of microparticle (diameter 100 nm) and a similar rejection ratio [96]. The electrospun PVDF membrane was also used as a mid-layer to support the barrier layer made of cross-linked chitosan. This TFNC membrane system also exhibited a significantly higher permeation flux and a lower fouling tendency than the commercial UF membranes for filtration of a BSA solution [133]. The low fouling tendency could be attributed to the hydrophilicity of the barrier layer. Finally, electrospun post-cross-linked PVA nanofiber scaffolds with GA were used as a support layer to support a relatively thick cross-linked PVA barrier layer, where the permeation flux of the resulting TFNC membrane, was quite stable and was over two times higher than that of the corresponding commercial membrane [93]. These studies clearly demonstrated that the high porosity and interconnected pore structure of the nanofiber scaffold played an essential role in improving the permeation flux of the membrane, even under the intense compaction conditions. The filtration performance of a variety of TFNC membranes is summarized in Table 28.1.

Table 28.1 Filtration performance of TFNC UF membranes.

Membranes	Top layer (thickness: nm)	Middle layer	Bottom layer	MWCO (kDa)	Permeability (l/(m^2 h)/psi)	Rejection ratio (%)	Applied pressure (psi)	References
PVA-TFNC	PVA (1800)	PVA e-spun	PET	—	1.30	>99.5	100	[132]
PVA-TFNC	PVA (500)	PAN e-spun	PET	40	2.56	99.0	90	[48]
PVA-TFNC	PVA (600)	PAN e-spun	PET	10	6.85	99.7	30	[134]
PVA-TFNC	PVA-MWCNT (1800)	PAN e-spun	PET	—	3.30	99.8	100	[93]
PVA-TFNC	PVA-UFCN (850)	PAN e-spun	PET	17.5	11.67	99.5	30	[134]
PVA-TFNC	PVA-UV (ND)	PVA e-spun	PET	—	1.83	99.5	30	[140]
PVA-TFNC	PVA (400)	PAN e-spun	N/A	—	4.67	99.5	45	[142]
Cellulose-TFNC	Cellulose (300)	PAN e-spun	PET	2000	9.30	99.8	30	[137]
Cellulose-TFNC	Microcrystal cellulose (600)	PAN e-spun	PET	5000	5.50	99.8	30	[138]
Chitin-TFNC	Chitin (600)	PAN e-spun	PET	5000	8.00	99.6	30	[138]
Cellulose/chitin-TFNC	Cellulose/chitin (600)	PAN e-spun	PET	5000	7.70	99.7	30	[138]
Chitosan-TFNC	Chitosan (1000)	PAN e-spun	PET	—	1.00	99.9	130	[50]
UFCN-TFNC	Cellulose nanofiber (100)	PAN e-spun	PET	2000	10.40	99.6	30	[131]
UFCN-TFNC	Cellulose nanowhisker (200)	PAN e-spun	PET	2000	9.10	99.7	30	[131]
UFCN-TFNC	Chitin nanofiber (450)	PAN e-spun	PET	2000	7.20	99.9	30	[131]
PEBAX-TFNC	PEBAX (800)	PVA e-spun	PET	—	0.57	>99.5	100	[132]
PEBAX-TFNC	PEBAX-MWCNT (800)	PVA e-spun	PET	—	1.61	99.8	100	[93]
PAN10	PAN (N/A)	N/A	PET	70	0.87	99.9	30	[137]
PAN400	PAN (N/A)	N/A	PET	2000	4.50	98.2	30	[137]

MWCNT, multiwalled carbon nanotube.

28.3.2.2 Nanofiltration and Reverse Osmosis

The pore sizes of nanofiltration membranes are usually less than 5 nm. When the pore size is less than 2 nm, the membrane can be used to remove divalent ionic compounds, such as $MgSO_4$, $MgCl_2$, Na_2SO_4, and dye molecules [128, 143]. The mechanism that normally describes the nanofiltration process is based on the combined effect of size exclusion and ionic repulsion interactions, or the so-called Gibbs–Donnan effect [144]. Typical commercially available nanofiltration membranes are fabricated with cellulose acetate using the phase inversion method [145] or with cross-linked polyamide by interfacial polymerization [146] on an ultrafiltration support.

The format of TFNC membranes based on electrospun nanofiber scaffolds has offered a new pathway to improve the filtration efficiency of nanofiltration [141]. In this format, electrospun nanofiber scaffolds (e.g. PAN and PES) could be used directly as the layer, instead of a UF support by the phase inversion method, to support the top barrier layer made by interfacial polymerization. For example, interfacial polymerization between trimesoyl chloride (TMC) in hexane and piperazine (PIP)/bipiperazine (BP) mixtures in water was carried out directly on the electrospun scaffold, where ionic liquids were also used as the inert additives in the water phase [141, 147]. The best performing TFNC nanofiltration membrane of this system exhibited 36% higher permeation flux than NF270, which was a top performing commercial membrane, for the removal of 2000 ppm of $MgSO_4$ (both membranes showed a similar rejection ratio of 97%). The higher permeation flux of the TFNC membrane was attributed to the combined effect of "water channels" due to the formation of nanocomposite structure and the high porosity of the electrospun nanofiber scaffold as the support layer. To reduce the compaction effect, a hot-pressed electrospun PAN nanofiber scaffold has been used as a support layer for fabrication of TFNC membranes, which exhibited a stable permeation flux even after a long operation [100]. In this system, high permeation flux (102 l/(m^2 h)) has been achieved for nanofiltration of $MgSO_4$ with 2000 ppm, which is about two times higher than for NF270. However, this system also exhibited a relatively low rejection ratio (86%). In another study, electrospun PVDF nanofiber scaffold was used as a support to carry an interfacially polymerized TMC and *p*-phenylenediamine (PPD) barrier layer, which also showed a relatively low rejection of 80.7% against to 2000 ppm of $MgSO_4$ [148].

In our opinion, the diameter of electrospun nanofibers (>100 nm) is relatively too large to support a thin barrier with thickness less than 100 nm and form a uniform nanocomposite structure. To overcome this problem, a three-layered TFNC membrane, containing a top layer of ultrafine cellulose nanofibers (fiber diameter ~ 5 nm), was used to host the interfacial polymerization process. There are several benefits to fabricating TFNC nanofiltration membranes with this approach [149]. The first benefit is that the cellulose nanofiber scaffold can retain the advantage of high porosity as in electrospun nanofiber scaffolds. Second, the effective pore size of the cellulose nanofiber scaffold is in the range of 20 nm, which is 1–2 orders of magnitude smaller than that of electrospun nanofiber scaffolds [150]. Third, as the average diameter of cellulose nanofibers is quite small (~5 nm), they can form an interconnected network in the barrier layer. The

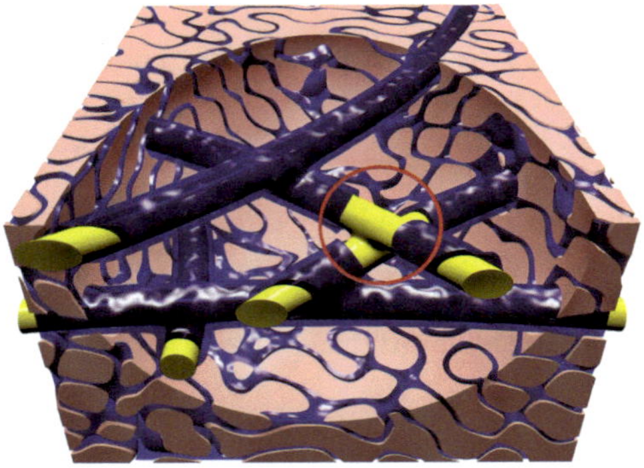

Figure 28.18 Schematic representation of the nature of water channels in the nanocomposite barrier layer. A skeleton of overlapped cellulose nanofibers (yellow) guides a continuously connected system of directed water channels (blue) formed by the connected hollow cylindrical gaps between the nanofibers and the polymer matrix (pink). The cutout in the red circle sketches the cross-linked nature of the nanofiber interconnects. The nanofiber skeleton is anchored at occasional direct contacts with the polymer matrix (not shown). Non-directed molecular cavities in the polymer matrix also contribute to the overall water flow through the barrier layer. Source: Ma et al. 2012 [151]. Reproduced with permission of American Chemical Society.

nanocomposite structure in the barrier layer thus would lead to interconnected water channels, created by the naturally occurring gap between the cellulose nanofiber and the matrix polyamide (Figure 28.18) [151]. The presence of these narrow water channels can greatly enhance the transportation of water molecules without sacrificing the selectivity.

The difference between using the regular interfacial polymerization pathway (i.e. the aqueous phase resides in the nanofiber scaffold) and the reversed interfacial polymerization pathway (i.e. the organic phase resides in the nanofiber scaffold) was investigated in the fabrication of TFNC membranes for NF and RO applications [149]. In this study, the initial scaffold possessed a three-layered structure, containing a top cellulose nanofiber layer and electrospun PAN nanofiber mid-layer. With the regular interfacial polymerization pathway (using TMC and PIP monomers), the loose part of the barrier layer was formed on top of the cellulose nanofiber scaffold, while the dense part was generated below the loose layer (Figure 28.19). The resulting TFNC membrane, exhibiting a high flux of about 40 l/(m^2 h) and 98% rejection ratio against 2000 ppm of $MgSO_4$, was better than that of the NF270 commercial nanofiltration membrane (having 72% rejection ratio under the same conditions). However, with the reversed interfacial polymerization pathway, smaller pore sizes in the barrier layer were obtained because the dense part of the barrier layer was left alone on top of the cellulose nanofiber scaffold. To transition from NF membranes to RO membranes, the barrier layer produced by the reversed interfacial polymerization pathway was further reacted with an excess amount of secondary amino groups (i.e. 1% TMC in hexane), which notably increased the NaCl rejection ratio from 74% to 91% (using 500 ppm NaCl feed solution under 100 psi applied pressure).

Figure 28.19 Schematic representative of TFNC-based nanofiltration membrane with nanocomposite barrier layer, made by typical interfacial polymerization process with the aqueous phase in the scaffold (the green region represents the barrier layer with loose structure, the yellow region represents the barrier layer with dense structure, the red fibers represent cellulose nanofibers, and the blue fibers represent electrospun nanofibers). Source: Wang et al. 2014 [149]. Reproduced with permission of Royal Society of Chemistry.

Recently, another method was demonstrated to fabricate TFNC RO membranes. In this approach, an electrospun cellulose acetate nanofiber scaffold was used to support multilayered polyelectrolytes of opposite charges in a layer-by-layer (LBL) manner [152]. Unfortunately the performance of this membrane system was not sufficient for practical considerations. For example, the rejection ratio of this composite membrane system for the removal of 2000 ppm NaCl was only in the range of 6–15%, while the corresponding water permeability of these membranes was about 40 l/(m^2 h).

28.3.2.3 Other Applications

TFNC membranes appear to be very suitable for forward osmosis (FO) applications due to the high permeability. However, the subject of this research is rarely reported in the literature [2, 153] and will surely be a subject of growing interest. Recently, two studies on the use of TFNC membranes for forward osmosis have been demonstrated [90, 154]. In these studies, PSU and PAN/cellulose acetate electrospun nanofiber scaffolds were used to support interfacially polymerized TMC-MPD polyamide barrier layers. The resulting TFNC membranes offered permeation fluxes two to five times higher than the commercial HTI osmotic membranes and high salt rejection for forward osmosis applications.

Electrospun PAN nanofiber scaffolds have also been used as the support layer for the fabrication of TFNC pervaporation membranes, where the barrier layer was either cross-linked PVA or graphene oxide (Figure 28.20) [155, 156]. These TFNC membranes were used to dehydrate an ethanol–water mixture, where higher flux and higher separation factors were obtained for both systems when compared with commercially available pervaporation membranes. The superior property of the TFNC format is due to the synergy of a highly porous substrate

Figure 28.20 The schematic diagram indicating the morphological arrangement of GO sheets (not shown in the correct scale, the red circle area represents a packing of GO sheets) and the supporting TFNC membrane scaffold. Source: Yeh et al. 2013 [156]. Reproduced with permission of Royal Society of Chemistry.

and a dense but thin barrier layer. It can be expected that a TFNC pervaporation membrane designed for the purification of volatile organics-contaminated water would also exhibit high flux and high separation factors.

Electrospun PAN nanofiber scaffolds also can play an important role for the membrane bioreactor (MBR) application. MBR is a facial water purification process, having a suspended growth bioreactor, typically used in municipal and industrial wastewater treatments. In this operation, the membrane is the most essential component in the system, where the issue of membrane fouling has been noted as the most pressing technology challenge. Typically, the MBR process can purify wastewater with sufficient quality to be discharged in the coastal areas or be used for urban irrigation [157, 158]. The membrane for MBR application, regarding nominal pore size, should be in the range of 0.05–0.20 μm, which falls into the categories between microfiltration and ultrafiltration. Therefore, a TFNC membrane with high porous structure, high pore interconnectivity, and appropriately designed pore size can be expected to be very useful in MBR, which would lead to high filtration efficiency. The use of suitable hydrophilic materials for electrospinning will be a critical component to produce membranes with good antifouling capability. Unfortunately, not much research has been carried out for this application.

Finally, electrospun nanofiber scaffolds have been proven very useful for MD applications. MD is a thermally driven approach for water purification, especially useful when there are substantial impurities with different nonvolatile contaminants (e.g. inorganics, organics, biological and polymeric materials) and different amount of contaminants in water [52, 159, 160] (e.g. the fracturing water from shale gas exploitation [161–163]) and when the pressure driven system is undesirable. The MD process is driven by a temperature gradient involving two phase transitions, i.e. liquid/vapor and vapor/liquid, subsequently taking place on the membrane surface. In this process, water in the feed solution passes through the membrane in the vapor state, while the impurities are left behind. Compared with conventional distillation and pressure-driving membrane processes, MD shows the advantages in contaminant rejection (i.e. near 100% rejection for nonvolatile components) and lower operating pressures. However, the disadvantages of this operation also include low permeation flux and high scaling or fouling problems (as the operation is taken place at higher temperatures, it promotes the mineralization growth). For MD operations, the ideal electrospun nanofibrous membranes must have a superhydrophobic surface and an average pore size around 200 nm as conventional MD membranes. This was confirmed by several recent studies, indicating that hydrophobic electrospun scaffolds are very suitable for desalination of seawater by using the MD approach [164, 165], where high flux, high rejection, and low energy consumption have been achieved.

28.4 Conclusions

TFNC membranes, using electrospun nanofibers as a critical component, are a new generation of membranes with merits of high flux, high rejection, low fouling, and low energy consumption for water filtration. Electrospun nanofibers, either as a barrier component or as a supporting component, play a key role in improving the membrane performance. In specific, electrospun nanofiber scaffolds offer high porosity, high interconnectivity, adjustable pore size and distribution, high surface-to-volume ratio, and good mechanical properties. With the rapid development of scalable electrospinning processes and the ease to modify the surface property, electrospun nanofiber scaffolds can be expected to be implemented in new membrane systems for many different types of water purification.

28.5 Future Prospects

In spite of having many unique advantages over other filtration materials for water purification, electrospun nanofibrous membranes still face some challenges for implementation: (i) the quality and characteristics of the electrospun membranes need to be precisely controlled in large-scale production, (ii) the solvent (especially volatile organic solvent) used in industrial electrospinning process can be recycled and reused, and (iii) the surface modification of the nanofibrous membranes can be implemented in a green and economic manner. We believe all these challenges will be overcome in the near future where electrospun nanofibrous membranes will truly be implemented in many different types of water applications.

Acknowledgments

This work was supported by a SusChEM award from the National Science Foundation in the United States (DMR-1409507), the National Natural Science

Foundation of China (51673011), the State Key Laboratory of Organic-Inorganic Composites at Beijing University of Chemical Technology (oic-201503004), and the Fundamental Research Funds for the Central Universities (buctrc201501).

References

1 Shannon, M.A., Bohn, P.W., Elimelech, M. et al. (2008). *Nature* 452: 301–310.
2 Ma, H.Y., Hsiao, B.S., and Chu, B. (2012). *Functional Nanofibers and Applications*, Chapter 15 (ed. Q. Wei), 331–370. Cambridge, UK: Wood Publishing Limited.
3 Ramakrishna, S., Ma, Z., and Matsuura, T. (2011). *Polymer Membranes in Biotechnology: Preparation, Functionalization and Application*, 1–254. London: Imperial College Press.
4 Burger, C., Hsiao, B.S., and Chu, B. (2006). *Annual Review of Materials Research* 36: 333–368.
5 Thavasi, V., Singh, G., and Ramakrishna, S. (2008). *Energy & Environmental Science* 1: 205–221.
6 Yoon, K., Hsiao, B.S., and Chu, B. (2008). *Journal of Materials Chemistry* 18: 5326–5334.
7 Ma, H.Y., Burger, C., Hsiao, B.S., and Chu, B. (2011). *Journal of Materials Chemistry* 21: 7507–7510.
8 Meyyappan, M. and Sunkara, M.K. (2009). *Inorganic Nanowires: Applications, Properties, and Characterization*, 1–453. Florida: CRC Press, Inc.
9 Chakarvarti, S.K. and Vetter, J. (1998). *Radiation Measurements* 29: 149–159.
10 Lee, C.C., Grenier, C., Meijer, E.W., and Schenning, A.P.H.J. (2009). *Chemical Society Reviews* 38: 671–683.
11 Teo, W.E. and Ramakrishna, S. (2006). *Nanotechnology* 17: R89–R106.
12 Isogai, A., Saito, T., and Fukuzumi, H. (2011). *Nanoscale* 3: 71–85.
13 Huang, Z.M., Zhang, Y.Z., Kotaki, M., and Ramakrishna, S. (2003). *Composites Science and Technology* 63: 2223–2253.
14 Zong, X., Kim, K., Fang, D. et al. (2002). *Polymer* 43: 4403–4412.
15 Greiner, A. and Wendorff, J.H. (2008). *Advances in Polymer Science* 219: 107–171.
16 Reneker, D.H. and Chun, I. (1996). *Nanotechnology* 7: 216–223.
17 Greiner, A. and Wendorff, J.H. (2007). *Angewandte Chemie International Edition* 46: 5670–5703.
18 Reneker, D.H., Yarin, A.L., Fong, H., and Koombhongse, S. (2000). *Journal of Applied Physics* 87: 4531–4547.
19 Li, D. and Xia, Y. (2004). *Advanced Materials* 16: 1151–1170.
20 Shin, Y.M., Hohman, M.M., Brenner, M.P., and Rutledge, G.C. (2001). *Applied Physics Letters* 78: 1149–1151.
21 Ma, H.Y., Hsiao, B.S., and Chu, B. (2013). *Current Organic Chemistry* 17: 1361–1370.

22 There are numerous companies to produce electrospinning apparatus and products, e.g. http://www.elmarco.com/; http://www.donaldson.com/index.html; http://www.nabond.com/; http://holmarc.com/nano_fiber_electrospinning_station.html, etc.
23 Pham, Q.P., Sharma, U., and Mikos, A.G. (2006). *Tissue Engineering* 12: 1197–1211.
24 Sill, T.J. and von Recum, H.A. (2008). *Biomaterials* 29: 1989–2006.
25 Zeng, J., Xu, X., Chen, X. et al. (2003). *Journal of Controlled Release* 92: 227–231.
26 Liang, D., Hsiao, B.S., and Chu, B. (2007). *Advanced Drug Delivery Reviews* 59: 1392–1412.
27 Chase, G.G., Varabhas, J.S., and Reneker, D.H. (2011). *Journal of Engineered Fibers and Fabrics* 6: 32–38.
28 Park, H.S. and Park, Y.O. (2005). *Korean Journal of Chemical Engineering* 22: 165–172.
29 Qin, X. and Wang, S. (2006). *Journal of Applied Polymer Science* 102: 1285–1290.
30 Jo, S.M. (2012). *Hydrogen Storage*, Chapter 8 (ed. J. Liu), 181–210. Shanghai: InTech.
31 Miao, J., Miyauchi, M., Simmons, T.J. et al. (2010). *Journal of Nanoscience and Nanotechnology* 10: 5507–5519.
32 Ding, B., Wang, M., Wang, X. et al. (2010). *Materials Today* 13: 16–27.
33 Ding, B., Wang, M., Yu, J., and Sun, G. (2009). *Sensors* 9: 1609–1624.
34 Liu, Y., Ma, H.Y., Hsiao, B.S. et al. (2016). *Polymer* 107: 163–169.
35 Botes, M. and Cloete, T.E. (2010). *Critical Reviews in Microbiology* 36: 68–81.
36 Chu, B., Hsiao, B.S., and Yoon, K. (2008). *AATCC Review* 8: 31–33.
37 Graton, L.C. and Fraser, H.J. (1935). *Journal of Geology* 43: 785–909.
38 Chu, B. and Hsiao, B.S. (2009). *Journal of Polymer Science Part B: Polymer Physics* 47: 2431–2435.
39 Ulbricht, M. (2006). *Polymer* 47: 2217–2262.
40 Barhate, R.S., Loong, C.K., and Ramakrishna, S. (2006). *Journal of Membrane Science* 283: 209–218.
41 Zong, X., Ran, S., Kim, K. et al. (2003). *Biomacromolecules* 4: 416–423.
42 Wang, R., Shi, L., Tang, C.Y. et al. (2010). *Journal of Membrane Science* 355: 158–167.
43 Broens, L., Altena, F.W., Smolders, C.A., and Koenhen, D.M. (1980). *Desalination* 32: 33–45.
44 Cheryan, M. (1998). *Ultrafiltration and Microfiltration Handbook*, Chapter 3, 71–95. Pennsylvania, PA: Technomic Publishing Company, Inc.
45 Fane, A.G., Fell, C.J.D., and Waters, A.G. (1981). *Journal of Membrane Science* 9: 245–262.
46 Deniz, S. (2006). *Desalination* 200: 42–43.
47 Ho, C.C. and Zydney, A.L. (1999). *Journal of Membrane Science* 155: 261–275.
48 Yoon, K., Hsiao, B.S., and Chu, B. (2009). *Journal of Membrane Science* 338: 145–152.

49 Barhate, R.S., Loong, C.K., and Ramakrishna, S. (2007). *Journal of Membrane Science* 296: 1–8.
50 Yoon, K., Kim, K., Wang, X. et al. (2006). *Polymer* 47: 2434–2441.
51 Andrade, J.S. Jr., Costa, U.M.S., Almeida, M.P. et al. (1999). *Physical Review Letters* 82: 5249–5252.
52 Alkhudhiri, A., Darwish, N., and Hilal, N. (2012). *Desalination* 287: 2–18.
53 Fane, A.G., Fell, C.J.D., and Waters, A.G. (1983). *Journal of Membrane Science* 16: 211–224.
54 Genne, I., Kuypers, S., and Leysen, R. (1996). *Journal of Membrane Science* 113: 343–350.
55 Boure, T. and Vanholder, R. (2004). *Nephrology Dialysis Transplantation* 19: 293–296.
56 Fridrikh, S.V., Yu, J.H., Brenner, M.P., and Rutledge, G.C. (2003). *Physical Review Letters* 90: 144502.
57 Liu, Y., Wang, R., Ma, H.Y. et al. (2013). *Polymer* 54: 548–556.
58 Wang, R., Liu, Y., Li, B. et al. (2012). *Journal of Membrane Science* 392–393: 167–174.
59 Yang, S.Y., Ryu, I., Kim, H.Y. et al. (2006). *Advanced Materials* 18: 709–712.
60 Ma, H.Y., Burger, C., Hsiao, B.S., and Chu, B. (2012). *Biomacromolecules* 13: 180–186.
61 Zalusky, A.S., Olayo-Valles, R., Wolf, J.H., and Hillmyer, M.A. (2002). *Journal of the American Chemical Society* 124: 12761–12773.
62 Yan, F. and Goedel, W.A. (2004). *Advanced Materials* 16: 911–915.
63 Park, C., Yoon, J., and Thomas, E.L. (2003). *Polymer* 44: 6725–6760.
64 Apel, P. (2001). *Radiation Measurements* 34: 559–566.
65 Kuiper, S., van Rijn, C.J.M., Nijdam, W., and Elwenspoek, M.C. (1998). *Journal of Membrane Science* 150: 1–8.
66 Tracey, E.M. and Davis, R.H. (1994). *Journal of Colloid and Interface Science* 167: 104–116.
67 Saljoughi, E., Sadrzadeh, M., and Mohammadi, T. (2009). *Journal of Membrane Science* 326: 627–634.
68 Boland, E.D., Coleman, B.D., Barnes, C.P. et al. (2005). *Acta Biomaterialia* 1: 115–123.
69 Gibson, P., Schreuder-Gibson, H., and Rivin, D. (2001). *Colloids and Surfaces A: Physicochemical and Engineering Aspects* 187–188: 469–481.
70 Li, J., Sculley, J., and Zhou, H. (2012). *Chemical Reviews* 112: 869–932.
71 Yang, K. and Xing, B. (2010). *Chemical Reviews* 110: 5989–6008.
72 Yeh, E., Ostreicher, E., Sale, R. et al. (2002). Method for charge-modifying polyester. US Patent 2002/0155225; Pignot, C. and Artaud, L. Device for the diffusion of a volatile product and preparation process. US Patent 2002/0037385; Hou, K.C., Bretl, D.S., and Hembree, R.D. (2003). Microorganism filter and method for removing microorganism from water. US Patent 6, 565, 749; Ostreicher, E. (1992). Use of cationic charge modified filter media. US Patent 5, 085, 780.
73 Chu, B., Hsiao, B.S., and Ma, H.Y. (2011). High flux high efficiency nanofiber membranes and methods of production thereof. US Patent 2011/0198282;

Chu, B., Hsiao, B.S., Ma, H.Y., et al. (2012). High flux microfiltration membranes with virus and metal ion adsorption capability for liquid purification. WO Patent 2012/027242.
74 Ma, H.Y., Chu, B., and Hsiao, B.S. (2012). Functionalization of nanofibrous microfiltration membranes for water purification. WO Patent 2012/094407.
75 Barthlott, W. and Neinhuis, C. (1997). *Planta* 202: 1–8.
76 Sun, T., Feng, L., Gao, X., and Jiang, L. (2005). *Accounts of Chemical Research* 38: 644–652.
77 Feng, L., Li, S., Li, Y. et al. (2002). *Advanced Materials* 14: 1857–1860.
78 Tuteja, A., Choi, W., Ma, M. et al. (2007). *Science* 318: 1618–1622.
79 Wang, Z., Ma, H.Y., Chu, B., and Hsiao, B.S. (2017). *Separation Science and Technology* 52: 221–227.
80 Zhang, Y.Z., Su, B., Venugopal, J. et al. (2007). *International Journal of Nanomedicine* 2: 623–638.
81 Lin, J., Ding, B., Yu, J., and Hsien, Y. (2010). *ACS Applied Materials & Interfaces* 2: 521–528.
82 Ding, B., Kim, J., Kimura, E., and Shiratori, S. (2004). *Nanotechnology* 15: 913–917.
83 Min, M., Wang, X., Yang, Y. et al. (2011). *Journal of Nanoscience and Nanotechnology* 11: 6919–6925.
84 Martin, C.R. (1994). *Science* 266: 1961–1966.
85 Cheremisinoff, P.N. (1995). *Handbook of Water and Wastewater Treatment Technology*, Chapter 13, 506–510. New York: Marcel Dekker, Inc.
86 Witte, D., Dijkstra, P.J., Berg, J.W.A., and Feijen, J. (1996). *Journal of Membrane Science* 117: 1–31.
87 Golden, A.P. and Tien, J. (2007). *Lab on a Chip* 7: 720–725.
88 Ho, C.C. and Zydney, A.L. (2001). *Industrial and Engineering Chemistry Research* 40: 1412–1421.
89 Gao, K., Hu, X., Dai, C., and Yi, T. (2006). *Materials Science and Engineering B* 131: 100–105.
90 Bui, N.N., Lind, M.L., Hoek, E.M.V., and McCutcheon, J.R. (2011). *Journal of Membrane Science* 385–386: 10–19.
91 Gopal, R., Kaur, S., Ma, Z. et al. (2006). *Journal of Membrane Science* 281: 581–586.
92 Ma, H.Y., Hsiao, B.S., and Chu, B. (2014). *Journal of Membrane Science* 452: 446–452.
93 Wang, X., Chen, X., Yoon, K. et al. (2005). *Environmental Science and Technology* 39: 7684–7691.
94 Bohonak, D.M. and Zydney, A.L. (2005). *Journal of Membrane Science* 254: 71–79.
95 Park, N., Kwon, B., Kim, I.S., and Cho, J. (2005). *Journal of Membrane Science* 258: 43–54.
96 Ma, H.Y., Hsiao, B.S., and Chu, B. (2014). *Journal of Membrane Science* 454: 272–282.
97 Kallioinen, M., Pekkarinen, M., Manttari, M. et al. (2007). *Journal of Membrane Science* 294: 93–102.

98 Gopal, R., Kaur, S., Feng, C.Y. et al. (2007). *Journal of Membrane Science* 289: 210–219.
99 Yoon, K., Hsiao, B.S., and Chu, B. (2009). *Polymer* 50: 2893–2899.
100 Kaur, S., Barhate, R., Sundarrajan, S. et al. (2011). *Desalination* 279: 201–209.
101 Tang, Z., Qiu, C., McCutcheon, J.R. et al. (2009). *Journal of Polymer Science Part B: Polymer Physics* 47: 2288–2300.
102 Crittenden, J.C., Trussell, R.R., Hand, D.W. et al. (2005). *Water Treatment Principles and Design*, 2e, 1–1901. New Jersey: Wiley.
103 Yang, R., Aubrecht, K.B., Ma, H.Y. et al. (2014). *Polymer* 55: 1167–1176.
104 Yang, R., Su, Y., Burger, C. et al. (2015). *Polymer* 60: 9–17.
105 Liu, Y., Ma, H.Y., Liu, B. et al. (2015). *Film Sheeting* 31: 379–400.
106 facts about E. coli: dimensions, as discussed in "bacteria: diversity of structure of bacteria" in Britannica Online Encyclopedia. http://www.britannica.com/facts/5/463522/E-coli-as-discussed-in-bacteria (acceptable 22 September 2013).
107 Tolliver, D.L. and Schroeder, H.G. (1983). *Microcontamination* 1: 34–43.
108 Su, Y., Burger, C., Ma, H.Y. et al. (2015). *Biomacromolecules* 16: 1201–1209.
109 Su, Y., Burger, C., Ma, H.Y. et al. (2015). *Cellulose* 33: 3127–3135.
110 Sato, A., Wang, R., Ma, H.Y. et al. (2011). *Journal of Electron Microscopy* 60: 201–209.
111 Crini, G. (2006). *Bioresource Technology* 97: 1061–1085.
112 Mohan, D. and Pittman, C.U. (2007). *Journal of Hazardous Materials* 142: 1–53.
113 Busca, G., Berardinelli, S., Resini, C., and Arrighi, L. (2008). *Journal of Hazardous Materials* 160: 265–288.
114 Chanrai, N.G. and Burde, S.G. (2004). Recovery of oil from spent bleached earth. US Patent 6, 780, 321.
115 Ma, Z., Kotaki, M., and Ramakrishna, S. (2005). *Journal of Membrane Science* 265: 115–123.
116 Ma, Z. and Ramakrishna, S. (2008). *Journal of Membrane Science* 319: 23–28.
117 Ma, Z., Masaya, K., and Ramakrishna, S. (2006). *Journal of Membrane Science* 282: 237–244.
118 Ryu, Y.J., Kim, H.Y., Lee, K.H. et al. (2003). *European Polymer Journal* 39: 1883–1889.
119 Ding, B., Li, C., Miyauchi, Y. et al. (2006). *Nanotechnology* 17: 3685–3691.
120 Merck (2013). Microfiltration membranes for filtration and venting applications. http://www.millipore.com/publications.nsf/a73664f9f981af8c852569b9005b4eee/af9dca91de8ae07e852574fd004cf186/$FILE/PF1176EN00_EM.pdf (accessible 22 September 2013).
121 Wang, X., Ding, B., Yu, J. et al. (2010). *Nanotechnology* 21: 055502.
122 Schiffman, J.D. and Schauer, C.L. (2007). *Biomacromolecules* 8: 594–601.
123 Horzum, N., Boyaci, E., Eroglu, A.E. et al. (2010). *Biomacromolecules* 11: 3301–3308.
124 Fang, X., Xiao, S., Shen, M. et al. (2011). *New Journal of Chemistry* 35: 360–368.
125 Brady-Estevez, A.S., Kang, S., and Elimelech, M. (2008). *Small* 4: 481–484.

126 Brady-Estevez, A.S., Schnoor, M.H., Vecitis, C.D. et al. (2010). *Langmuir* 26: 14975–14982.

127 Koros, W.J., Ma, Y.H., and Shimidzu, T. (1996). *Journal of Membrane Science* 120: 149–159.

128 Fane, A.G. and Fell, C.J.D. (1987). *Desalination* 62: 117–136.

129 Rahimpour, A. and Madaeni, S.S. (2007). *Journal of Membrane Science* 305: 299–312.

130 Koops, G.H. (2013). *Encyclopedia of Desalination and Water Resources, Membrane Processes*, vol. II. Isle of Man: Eolss Publishers Co. Ltd. http://www.desware.net/Membrane-Processes.aspx.

131 Ma, H.Y., Burger, C., Hsiao, B.S., and Chu, B. (2011). *Biomacromolecules* 12: 970–976.

132 Wang, X., Fang, D., Yoon, K. et al. (2006). *Journal of Membrane Science* 278: 261–268.

133 Zhao, Z., Zheng, J., Wang, M. et al. (2012). *Journal of Membrane Science* 394–395: 209–217.

134 Ma, H.Y., Yoon, K., Rong, L. et al. (2010). *Industrial and Engineering Chemistry Research* 49: 11978–11984.

135 Wang, Z., Ma, H.Y., Hsiao, B.S., and Chu, B. (2014). *Polymer* 55: 366–372.

136 Zhang, H., Zheng, J., Zhao, Z., and Han, C.C. (2013). *Journal of Membrane Science* 442: 124–130.

137 Ma, H.Y., Yoon, K., Rong, L. et al. (2010). *Journal of Materials Chemistry* 20: 4692–4704.

138 Ma, H.Y., Hsiao, B.S., and Chu, B. (2011). *Polymer* 52: 2594–2599.

139 Wang, Z., Ma, H.Y., Chu, B., and Hsiao, B.S. (2017). *Journal of Applied Polymer Science* 134: 44583.

140 Tang, Z., Wei, J., Yung, L. et al. (2009). *Journal of Membrane Science* 328: 1–5.

141 Yoon, K., Hsiao, B.S., and Chu, B. (2009). *Journal of Membrane Science* 326: 484–492.

142 Wang, X., Zhang, K., Yang, Y. et al. (2010). *Journal of Membrane Science* 356: 110–116.

143 Hilal, N., Al-Zoubi, H., Darwish, N.A. et al. (2004). *Desalination* 170: 281–308.

144 Krasemann, L. and Tieke, B. (2000). *Langmuir* 16: 287–290.

145 Strathmann, H., Scheible, P., and Baker, R.W. (1971). *Journal of Applied Polymer Science* 15: 811–828.

146 Petersen, R.J. (1993). *Journal of Membrane Science* 83: 81–150.

147 Yung, L., Ma, H.Y., Wang, X. et al. (2010). *Journal of Membrane Science* 365: 52–58.

148 Kaur, S., Sundarrajan, S., Gopal, R., and Ramakrishna, S. (2012). *Journal of Applied Polymer Science* 124: E205–E215.

149 Wang, X., Yeh, T.M., Wang, Z. et al. (2014). *Polymer* 55: 1358–1366.

150 Ghosh, A.K. and Hoek, E.M.V. (2009). *Journal of Membrane Science* 336: 140–148.

151 Ma, H.Y., Burger, C., Hsiao, B.S., and Chu, B. (2012). *ACS Macro Letters* 1: 723–726.

152 Ritcharoen, W., Supaphol, P., and Pavasant, P. (2008). P. *European Polymer Journal* 44: 3963–3968.
153 Song, X., Liu, Z., and Sun, D.D. (2011). *Advanced Materials* 23: 3256–3260.
154 Bui, N.N. and McCutcheon, J.R. (2013). *Environmental Science and Technology* 47: 1761–1769.
155 Yeh, T.M., Yang, L., Wang, X. et al. (2010). *Journal of Renewable and Sustainable Energy* 4: 041406.
156 Yeh, T.M., Wang, Z., Mahajan, D. et al. (2013). *Journal of Materials Chemistry A* 1: 12998–13003.
157 Yang, W.B., Cicek, N., and Ilg, J. (2006). *Journal of Membrane Science* 270: 201–211.
158 Meng, F.G., Chae, S.R., Drews, A. et al. (2009). *Water Research* 43: 1489–1512.
159 Souhaimi, M.K. and Matsuura, T. (2011). *Membrane Distillation: Principles and Applications*, 1–460. London, UK: Elsevier.
160 Gryta, M. (2008). *Journal of Membrane Science* 325: 383–394.
161 Kerr, R.A. (2010). *Science* 328: 1624–1626.
162 McGlade, C., Speirs, J., and Sorrell, S. (2013). *Energy* 55: 571–584.
163 Vengosh, A., Warner, N., Jackson, R., and Darrah, T. (2013). *Procedia Earth and Planetary Science* 7: 863–866.
164 Feng, C., Khulbe, K.C., Matsuura, T. et al. (2013). *Separation and Purification Technology* 102: 118–135.
165 Maab, H., Francis, L., Al-saadi, A. et al. (2012). *Journal of Membrane Science* 423–424: 11–19.

29

Fibers for Filtration

Govindharajan Thilagavathi and Siddhan Periyasamy

PSG College of Technology, Department of Textile Technology, Coimbatore, 641004, India

29.1 Introduction

Textiles have been used as one of the important filter media from time as much as textiles were known for clothing. The use of textiles for filtration could be remembered from simple ancient people practices like separation of sand particles and other dirt while taking drinking water from river, ponds, etc. Textile like woven fabrics, by virtue, has the required structure for filtration, having structures with the openness on the orders of as close as to microns. The fabrics have good mechanical properties such as tensile, tear, and impact resistance good enough to withstand the normal course of forces. Hence textile fabrics could be used as an efficient filter media. Further, based on filtration requirements, the openness of the fabrics, gram per square meter (GSM), mechanical properties, etc. can be altered through the control of various governing parameters like type of fiber, yarn count, interlacement patterns, ends per inch (EPI) and picks per inch (PPI), number of layers of fabrics, etc. With the developments in high-performance and functional fibers and their use in technical textiles, filtration with specific requirements like high temperature, chemical, hot chemical, and micro- and nano-filtrations has become possible [1, 2].

The topic of textile filtration is extensive wherein numbers of variables can play its vital role based on the type of filtrations for which they are intended [3]. For example, textile filter media could be used in the form of fibers, rovings, yarns, and fabrics, which are made from yarns like interlaced (woven) fabrics and interloped (warp and weft knitted) fabrics and from fibers like the nonwoven fabrics. Nonwovens in particular can be constructed with varying degrees of porosity, thickness, and density coupled with structural isotropicity as it is made from random alignment of fibers [4]. Hence the nonwovens are promising textile structures to function as effective filter media. Apart from the structural diversity of textiles, considerations of filtration mechanisms too map well with the textile filter media for their effective performance in filtration applications.

Though the performance of the filter media would depend on various structural attributes, material-specific intensive properties are to be considered based on

Handbook of Fibrous Materials, First Edition. Edited by Jinlian Hu, Bipin Kumar, and Jing Lu.

the types of filtration in question. In textiles, these intensive properties emanate from the fibers as they are the foundational building blocks of the textiles [5].

Filtration is one type of technical applications of the fibers. Most of the fibers could be a suitable candidate for developing filter media for any one type of filtration; however, the choice has to be obviously made based on the techno-economical requirements/considerations. Therefore, in this chapter, it is attempted to provide a comprehensive understanding of textile fibers for their suitability in filtration applications.

29.2 Filtration and Filter Media

The term filtration could be perceived as a classification under a broad term separation. Separation, hence, is a process of taking something apart from the mixture of either homogeneous or nonhomogeneous type belonging to solid–solid, solid–liquid, liquid–liquid, solid–gas, liquid–gas, and gas–gas. Based on the type of mixture, various principles could be employed for the separation. In such mixtures, high proportionate compound could be regarded as bulk phase, while the compound(s) in low proportion(s) could be regarded as mixed phase(s). Based on the kind of mixing, the mixtures could be classified as intimate and non-intimate mixing depending on the degree of chemical and physical interactions. If the mixture has spontaneous chemical interaction, such mixing could be classified as intimate and is classified as non-intimate if the chemical interaction between the mixed and bulk phase is nonspontaneous. In nonspontaneous mixtures interactions would be enacted mainly by physical forces coupled with some weak chemical forces. Among these kinds of separations, the term filtration is generally assigned to the separation of solid from liquid in the solid–liquid mixture. Such solid–liquid mixture could be a dispersion (non-intimate but made so by surface active agents) or a solution (intimate mixing) based on which the principle employed to remove the solid would differ. For example, if the solid–liquid mixture is a solution, the principle of separation of the dissolved solid could be by precipitation followed by straining while it could be simple straining or particle agglomeration followed by straining for dispersion type of solid–liquid mixture. Such similar separation principles could be used for other mixtures like liquid–liquid (solution or dispersion), liquid–gas, and solid–gas.

In the Handbook of Filter Media by Elsevier, the definition of filter medium is given as following [6]: *A filter medium is any material that, under the operating conditions of the filter, is permeable to one or more components of a mixture, solution or suspension, and is impermeable to the remaining components.* Hence, the filter media could be broadly regarded as the substrate/object through which the actual separation of the compound in question in the mixture is taking place. In most cases the mixed compound is having a distinct phase from the bulk phases. However, if both the mixed and bulk phases have a single phase like in the case of solutions, as stated above, it could be precipitated and then filtered through the filter media. Such separation of the particles through the filter media could be by simple straining (sieving) mechanism or by other mechanisms such as interception, Brownian diffusion, inertial separation, and electrostatic attraction that are briefly described in Section 29.2.2. Based on the characteristics of the particles, any one of the mechanisms would dominate over the other in the filtration process. Accordingly, the filter media should be designed and selected. Additionally, there are other points such as cake formation aspects and its effect in particular type of filtration, its removal, temperature at which the filtration is carried out, reuse of the filter media, pressure considerations such as flow pressure and pressure drop, etc. based on which the design and selection of filter media would depend.

29.2.1 Types of Filtration

As briefed in Section 29.2, filtration types could be understood based on number of criteria such as based on the type of bulk and mixed phase. If the bulk phase is liquid, it is regarded as liquid filtration, and if it is gas, it is regarded as gas filtration. These two types are encountered in most cases under filtration while the other type of solid to solid comes more appropriately under separation than filtration. Further, based on the temperature of the mixture, the filtration could be regarded as hot/cold filtrations [7]; based on the type of filter media and the mixed phase, it could be regarded as surface or depth filtration [8, 9]. Based on the pressure difference/pressure drop due to filtration, it could be regarded as ambient pressure filtrations and vacuum/pressure filters [10]. Almost all the filtration applications could be brought under these categories [11].

29.2.2 Filtration Theories

Generally, a theory of filtration, by common perception, is considered due to the physical blocking effect based on the particle and filter media pore's average size and distribution that is generally referred to as straining theory/mechanism. However, there are situations wherein the filtration does happen outside the purview of this general theory of filtration. Hence, when the cause for such filtration effects is scientifically studied, distinct mechanisms such as interception, Brownian diffusion, inertial separation, and electrostatic attraction have been discovered in addition to the general straining theory. The schematic of the various mechanisms of particle filtration is shown in Figure 29.1. Interception mechanism is application to large particles that follows the streamline of the fluid carrying the particles, while the inertial mechanism is for the large particles, which travel off the streamline of the fluid.

In both cases the particle is arrested by the filter media when the particle strikes the media grain; however, such striking of particles takes place differently in these two mechanisms, i.e. in the interception mechanism, it happens when the particles flow very close to the media grain, while with the inertial it happens due to the random movement of the large particles due to the inertial force itself, leading to such particle striking and particle capturing. Brownian diffusion is applicable to the particles with diameter having less than 1 μm. Such small particles moving along with the flow of the fluid would exactly not have the straight line path as in the case of a large particle, which is due to the very small size and mass. Hence such particles like micro dusts and viruses would have a random

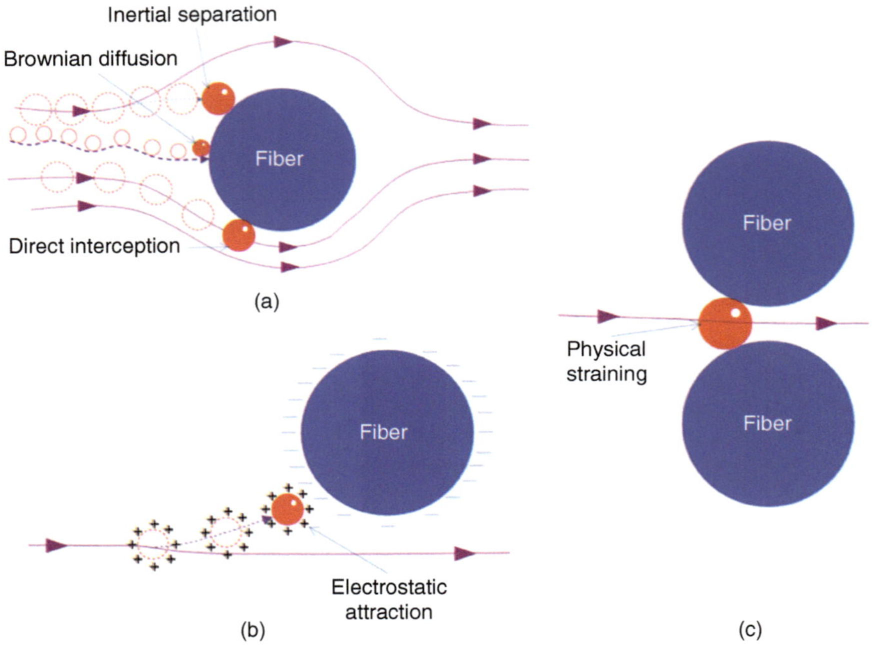

Figure 29.1 Schematic representation of various mechanisms of filtrations: (a) inertial separation, Brownian diffusion, and direct interception, (b) electrostatic attraction, and (c) physical straining.

motion within the flowing liquid that would be governed by the temperature of the fluid. Such randomly moving small particles are captured by the media grain through adherence, agglomeration, and consolidation by the subsequent particle capturing effects. In the case of electrostatic attraction, the particle capturing takes place due to the attractive forces between the filter media and the particles that are oppositely charged. Interception, inertial separation, Brownian diffusion, and straining types of mechanisms depend on the physical aspects of the particles and the filter media, and hence such effects could be the cause in various filtration applications. However, the electrostatic mechanism is possible through introduction of ionizing chemical elements in the filter media for specific filtration applications for which the chemistry of the particle to be filtered of is also known. Similar mechanisms have been reported in the literatures [12, 13].

29.2.3 Requirements of Filter Media

The specific requirements of the filter media for general kinds of fluid filtration could be listed as follows: high filtration efficiency; low pressure drop; no clogging; easy cleaning; easy maintenance; good dimensional stability under the filtration conditions; no change in filtration properties through change in pore size and distribution; versatility; durability; good cake formation; easy removal of cake; high performance to weight ratio; high performance to volume ratio;

good temperature and chemical stability; having good mechanical properties (strength, elastic recovery); good surface properties leading to good adsorption of the particles; good abrasion properties; no or required surface charges as per the filtration requirements; low resistance to fluid flow; no microbial growth or resistance to microbial attack; resistance to environmental factors like redox reactions, hydrolytic degradations, and any other corrosive reactions; resistance to reaction with the particles being filtered, which may be of polycompound belonging to organic and inorganic in nature or in certain cases in which the particles are known to be reactive; and automatic withering off of the filtered particles due to gravitation in appropriate applications [3].

29.3 Fibrous Materials as Filter Media

Textile fiber is unit of matter characterized by fineness, flexibility, and a high ratio of length to width. The fiber must also have sufficiently high temperature stability and a certain minimum strength, extensibility, and elasticity. The uniqueness of the fibrous substance is that it provides large surface area and interstices to the products made of it, due to the fact that the fibers are very small in its cross section, leading to large number of overlaps [14]. Such inherent porosity present in the fibrous substrates can be effectively used as a filter media.

29.3.1 Classification of Fibers

Understanding on classification of fibers would facilitate the fiber selection for developing textile filter media. Meeting the standard definition of a textile fiber, many fibers are naturally available, and man-made fibers using natural and synthetic polymer have been made imitating the natural fibers. Hence, fibers are broadly classified as natural fibers and man-made fibers. Under the man-made fiber categories, there are two major subdivisions, one being the regenerated fibers produced using natural polymers and the other being synthetic/artificial fibers produced using polymers synthesized by man. Cellulosics and polyamides are the two major categories of natural fibers. Examples of the fibers under cellulosic category are cotton, jute, flax, ramie, and hemp, while silk and wool are the major natural fibers under natural category. Regenerated polymer fibers are made first by dissolving the natural fibers in a suitable solvents and then extruding them in a spinneret to form fibers. Examples of the fibers belonging to this category are rayons (viscose, acetate, triacetate, polynosic, cuprammonium), lyocell, Tencel, and casein. Synthetic polymer fibers are produced first by synthesizing the polymer from organic and inorganic monomers and then converting them as fibers. Examples of the fibers belonging to this category are nylons (nylon 66, nylon 6, nylon 610, etc.), polyethylene terephthalate (PET), acrylic, polyethylene, polypropylene, glass, carbon, etc. [15, 16].

29.3.2 Fine and Morphological Structures of Fibers

In order to understand the effective role of the fiber as the building block material of the filter media, it is essential to understand the general structure of fibers.

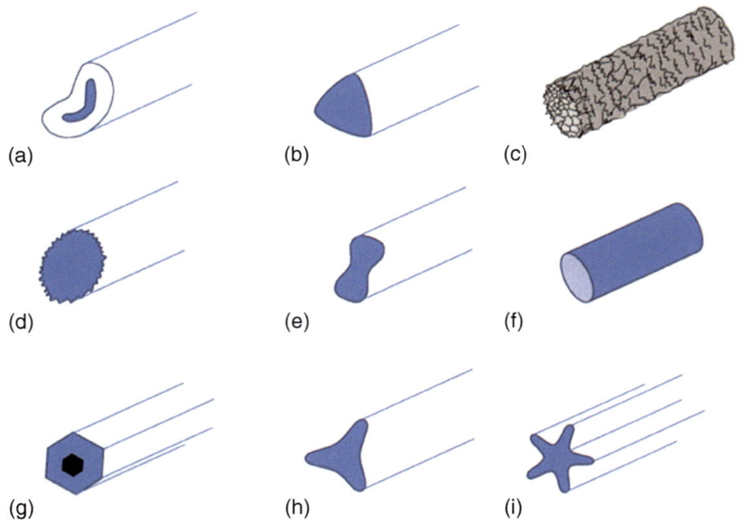

Figure 29.2 Schematic of surface macromorphology of various textile fibers: (a) kidney-shaped (cotton), (b) triangular (silk), (c) oval to round – scales overlapping (wool), (d) circular striated (rayon), (e) dog bone-shaped (acrylic), (f) circular (PET, nylon), (g) polygonal (flax), (h) trilobal (nylon), and (i) multilobular (acetate).

The structure of the fibers, like any materials, could be regarded as the macro surface morphological structure and the fine structure. The macro surface morphological structure is important in number of aspects as it decides the specific surface area of contact for a given mass/volume of materials. Such surface contacts would contribute for any one of the four filtration mechanisms, resulting in the filtration. Surface macromorphological structures of textile fibers are schematically presented in Figure 29.2. Generally, the natural and regenerated fibers have non-round cross section, while the synthetic fibers produced through melt spinning can have round cross section or any other shapes like triangular, trilobal, etc. based on the type of spinneret [17, 18]. Similarly the longitudinal shape of the natural fibers also has generally some irregular rough surface like in the case of wool; it has scaly surface, and with cotton it has twisted convolutions. However, in the case of man-made fibers, the longitudinal surface has generally smooth surface with some striations for noncircular cross-sectional fibers.

One of the important parameters apart from the cross-sectional and longitudinal forms of the fibers is the size of the fiber itself, which normally range about 10–70 μm. In the lower end range, cotton fiber varieties could be located, while on the higher end range, wool fiber varieties could be located. As far as the synthetic fibers are concerned, their diameter could be altered easily; however, producing finer fibers below 1 μm would be challenging in terms of stable structure development [17]. It is a known fact that for a given mass/volume, if the fiber size decreases, the total surface area would increase. Based on this fact, the size of the fibers would greatly influence the performance of the filtration. With the developments in synthetic fiber, the microdenier fibers (<1 denier) have high surface proportionately compared with that of the regular fiber fineness of about

1.5 denier [19]. The filter media developed with microdenier fiber could be made with the average pore size as low as 12–25 μm compared with the pore size of about 35–66 μm for those structures made with regular fibers. Such filter media made with microdenier fibers would yield high filtration efficiency as a result of high specific surface area [20].

In addition to the size/diameter of the fiber, its surface roughness would also have an effect, i.e. the rougher the surface, the more chances for the physical interactions of the particles through the various mechanisms of filtrations. For the purpose, the fibers could be given surface treatments through chemical or physical methods, which are other emerging areas of research [21, 22].

Besides the surface structures, the fine structure of the fibers could also play a significant role in the performance of the filter media. The fine structure of any material and also the fiber refers to the study of molecular arrangements. As it has been discussed in Section 29.3.1, fibers are made of polymers, which are regarded as the macromolecules. Like any other material, for being solids, it has to have crystalline regions (with few exceptions like glass); however, due to the macromolecular nature, it would not have 100% crystallinity but less than anything below, generally ranging about 30–60% for apparel and technical grade fibers and about 60% to <100% for high-performance grade fibers. Though the high degree of crystallinity would be preferred for a good mechanical performance, but it may not be preferred to have very high degree, leading to 100% as it would seriously affect the other unique properties of textile fibers such as the flexibility, extensibility along with elasticity, etc. [5]. Therefore, the semicrystalline nature of the fiber is the inherent advantages of the fibers meeting the functional requirements of a filter media and also for other textile-related applications. So, the partly crystalline and partly amorphous arrangements of the fibers form the fine structural aspects of the fibers.

Based on various parameters, polymer molecular arrangements in the fibers are depicted by two distinct models, namely, fibrillar model and micellar model as shown in Figure 29.3. The major difference in these two types of arrangements of polymer molecules inside the fiber is based on the length of the polymer chain. If the polymer chain is long like in the case of natural fibers and in high-performance fibers, they would run over relatively to a long distance inside the fiber, and hence, they would not form a definite short-range ordered structures but long-range orders in the fiber structures. In contrast, the micellar type of arrangement is for the short polymer chain. As the short chains cannot run over a long length, many chain ends could be seen in the amorphous regions. Moreover,

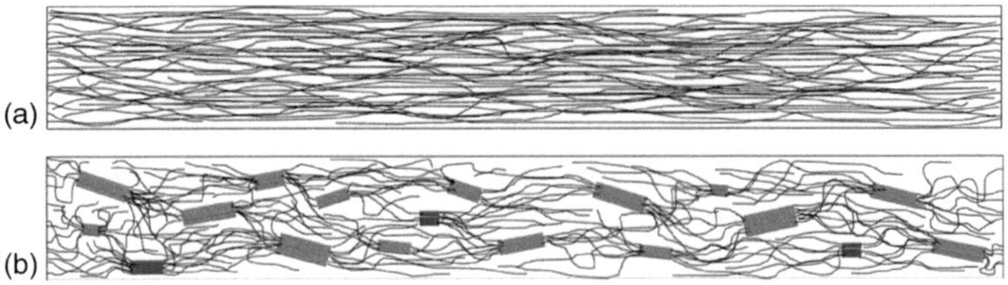

Figure 29.3 Fine structure models of fibers: (a) fibrillar model and (b) micellar model.

due to the relative high degree of freedom and movements, the short chains can adjust themselves to associate with the nearby polymer molecules, leading to a definite short-range order that is perceived as the clear crystalline regions [5].

Fibers belonging to these two models differ distinctly in their properties. Those belonging to fibrillar models exhibit good strength, modulus, and less extension with elasticity. In the case of natural fibers belonging to cellulosics, such arrangement is also one of the causes for increasing wet strength. The fibrillar model could be perceived as amorphous region dispersed in the ordered region, while the micellar model could be perceived as the crystalline regions dispersed in the amorphous regions. This implies that fiber belonging to micellar models would have high amorphous contents, while the fibers belonging to fibrillar model would have high crystalline/ordered regions. Hence the micellar model fibers would have generally low mechanical properties but would have good extensibility, flexibility, and good absorption. In filtration applications, the fine structure of the fiber would influence in terms of various property requirements. For example, if the filtration requires molecular absorption for the removal of the intimately mixed substance, then the choice of fiber would be from those belonging to the micellar model as it would have good absorption. However, if the filtration requirement emphasis is much on the mechanical properties, then the choice would be from those fibers belonging to fibrillar models.

It is evident that fibers remain as the only preferred material choice for various applications because of the aforementioned number of unique properties such as the fineness and flexibility along with the mechanical, thermal, chemical, and environmental properties. Particularly, the various possible fine and ultra/nano-fine sizes and the surface and fine structural aspects of fibers make them as good raw materials in the development of filter media. In Section 29.4, applications of fibers in various filtration applications would be discussed.

29.4 Forms of Fibrous Substrates for Filtration

Various forms of fibrous have been developed over a period of time for technical applications. Some of them are fiber assembly itself stuffed within a perforated cage, roving (it is a bulky strand of fiber twisted together with few twists/inch, an intermediate product prior to yarn formation), yarn, and fabrics. Of these, filter media made of fabrics find more applications as they can be used directly without special preparation as it would be required with rovings and yarns. There are different types of fabrics such as interlaced (woven), interloped (knitted), and entangled (nonwoven) with distinct properties based on the methods of manufacturing. They are briefly discussed below for a comprehensive understanding.

29.4.1 Roving and Yarns

The process of conversion of fiber into yarn is called as spinning process, which involves various stages. The yarn is produced in the final machine called ring

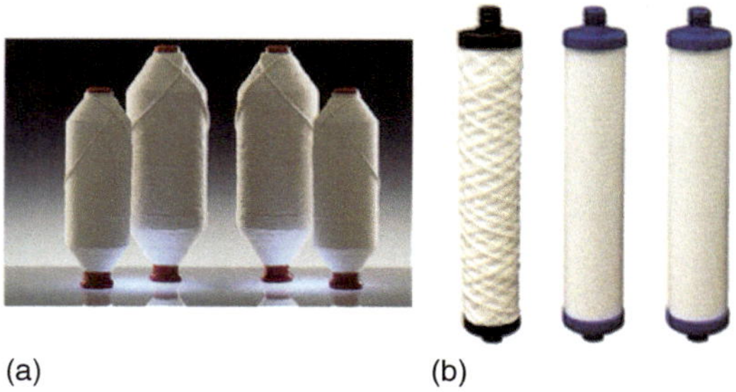

Figure 29.4 Cartridge filter media made of roving. (a) Roving package and (b) cartridge filter made of roving.

frame, which is the most popular way of manufacturing yarn. The feed material to this machine is the roving and is a bulky strand of fiber twisted together with few twists/inch, an intermediate product just prior to yarn formation. Such rovings could be wound in a particular pattern for obtaining the filter media for filtration applications (Figure 29.4). These kinds of filters are used in cartridge filters with the micron range of 0.5–100 μm. The filtration solution could be passed across the either inward to outward or outward to inward so that the mixed phase could be removed during the flow of the fluid through the media. These kinds of structures have its own advantage and disadvantages because of which the other types of fibrous filter media based on fabrics have become the choice. The main advantage of these kinds of filter would be that it would be suitable where the cylindrical type of filter media is preferred, and also on economical view point, these kinds of filter media are produced at a much earlier stages of fabric conversion (except the entanglement type of fabric), and so it may be cheaper. The possible disadvantages of these kinds of filters would be poor filtration performance due to preferential orientation of the fiber in the rove and yarn, and also there are chances for distorted shortcut flow of the fluid through the least resistant path as a result of structural distortion, which is more susceptible in these kinds of filter media.

29.4.2 Interlaced Fabrics

The fabrics made by interlacement of yarn through weaving machine are called woven/interlaced fabrics. Interlacement of yarns involves set of threads, one running along the length of the fabric, called as warp yarns, and the other one running across the fabric, called as weft yarns. Such threads interlace one over the other, and based on the type of interlacement, the fabric basic structures are classified as plain and twill, for which there are further derivatives such as matt, basket weaves as plan derivatives and satin, and sateen as twill derivatives. Apart from the weave structure aspects, the thread spacing that is expressed as EPI for warp yarns and PPI for weft yarns greatly governs the openness. Another important parameter of fabrics is called cover factor, derived from yarn size and thread spacing, also a measure of the openness of the fabric. Simple plain weave union fabrics with same yarn characteristics and thread spacing have the highest thread interaction for a given cover factor of the fabric. Schematics of fabrics are shown in Figure 29.5. Based on requirements of filter media fabric the types of openness required, various fabric parameters such as thread spacings, yarn count, and interlacement

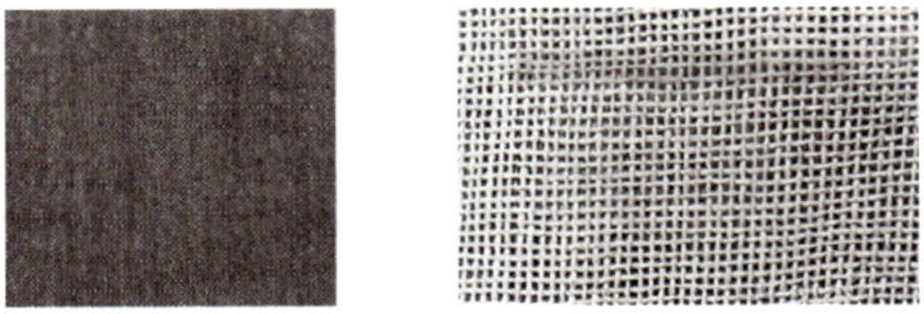

(a) (b)

Figure 29.5 Interlaced fabric structures: (a) high cover factor and (b) low cover factor.

pattern could be altered. This type of filter media would have good mechanical properties and would be more appropriate for surface filtration.

29.4.3 Nonwovens

Nonwovens are a kind of fabrics made directly from fibers without undergoing the various intermediate processes of yarn and fabric manufacture. The process involves preparation of fibrous web through various techniques and then bonding the fibers by any one of the methods like mechanical, chemical, and thermal means based on the type of fiber and possibilities. Various methods of nonwoven preparation machines are available like needle punching machine for mechanical interlocking of fibers, wet laid nonwoven machine for chemical bonding, and spun bonding nonwoven machine for thermal bonding for thermoplastic polymer fibers. Essentially the fabrics of this type greatly differ from the fabrics made from interlacement in terms of mechanical properties, bulkiness, and porosity. For a given GSM of the fabrics, the mechanical strength of the interlaced fabrics would be the strongest as the fibers are well aligned in these structures compared with the nonwovens where the fibers are weakly bound comparatively. Due to the same factor of poor compact binding of fibers in the nonwoven, the nonwoven fabrics are bulkier and thicker for a given GSM, and hence, they would have more free space in its structure that makes it suitable for depth filtration. The mean flow pore ratings (MFPs) of typical nonwoven fabric for filtration media range from 1 to 500 μm. For the MFP of 15 μm and below, the fabric needs calendaring. Typical forms of the bulky nonwovens are shown in Figure 29.6. Because of the large porosity and the thick structures, the nonwoven fabrics could be used as effective filter media for depth filtration involving various types of filtration mechanisms.

29.4.4 Interloped Fabrics

Typical interloping patterns are shown in Figure 29.7. Figure 29.7a shows the interloping by weft knitting machine, and Figure 29.7b shows the microscopic view of the surface of the weft knitted fabrics, which indicate the presence of the

Figure 29.6 Needle-punched nonwoven.

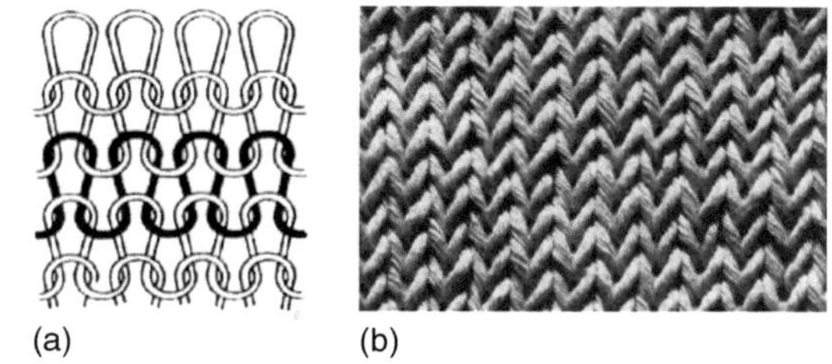

Figure 29.7 Interloped structures: (a) weft knitted structure and (b) microscopic view of weft knitted fabric surface.

looping pattern. As far as filtration is concerned, knitted structures find limited applications where there would be a need to undergo structure expansion of the media in order to take up the pressure fluctuations and protect the media failure and to regulate the pressure drop. However, such expansions could alter the physical dimensions of the interstices, affecting the filtration. Hence, this might be taken care if sufficient cake has formed to an extent that the filtration is actually effectuated by the cake rather than the fabrics. At this stage, the fabric mainly takes up the role of providing the support for the cake and undergoes extension as stated above. These structures are mainly used in air/dust filtrations.

29.4.5 Fibrous Membranes

Membrane is a substance characterized by very fine in thickness of the order of micron, and nanodimensions, flexible, have required mechanical, chemical, and temperature resistance. With the recent advent of technology developments in nanofiber spinning, one of the successful outcomes is the nanofibrous webs, which could be considered as the fibrous membranes with very fine pores for filtration purposes. Many research works have been initiated and reported in these areas that are discussed and presented in Section 29.5. Elsewhere the effect of nanoparticle-assisted nanomembrane filtration has been reported as an effective means of treating textile effluent water [23].

29.5 Fibers in Filtration Applications

Selection of the right fiber considering their material properties is the prime importance in filter fabric designing. As shown in the textile fiber classifications, many of the fibers belonging to the classifications like the natural fibers such as cotton and wool, man-made regenerated fibers like viscose rayon, and synthetic fibers like polyamides (nylons), polyesters (PET) (aromatic high-performance polyesters), polyolefins (polypropylene, polyethylene), glass (fiberglass), aramids (Nomex, Kevlar), and fluoropolymers (polytetrafluoroethylene [PTFE]) are used for filtration applications. Such textile filters are used in various areas such as metallurgical industries, foundries, cement industries, chalk and lime plants, brick works, ceramic industries, flour mills, medical, pharmaceutical, acoustics and screen printing, workshops, manufacturing halls, control stands to offices, schools, aerospace, automotives, bioclean, disk drives, flat panels, food, hospitals, medical devices, pharmaceuticals, and other electronics. In all these

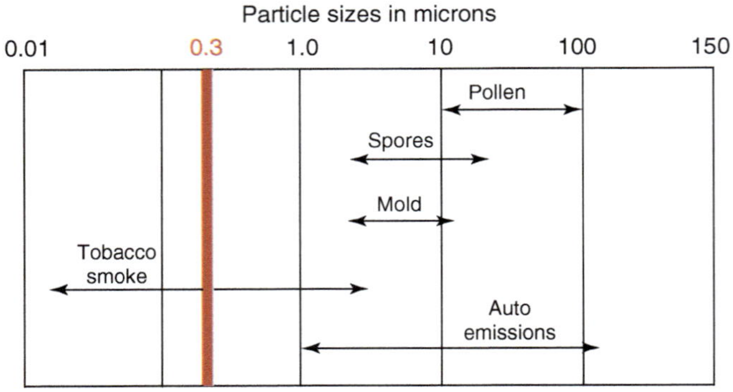

Figure 29.8 Typical class of particles and their physical size.

applications, the contaminants remain mostly in the form of particles whose typical size values are presented in Figure 29.8. It is claimed that the Bosch HEPA Premium Cabin Air Filters provide filtration efficiency of 99.97% at 0.3 μm [24].

Because of the typical nature of textile fibers being fine and flexible along with the other basic thermomechanical and chemical resistance, almost all the textile fibers are suitable candidates for all kinds of filtration in general. However from the historical perspective, before the advent of man-made/synthetic fibers, the natural fibers such as cotton and wool have been used for the development of filter media. Apart from the basic fineness and flexibilities, these two fibers are known for its inherent natural surface roughness and pores due to the polymeric fibril arrangement inside the fiber structures as shown in Figures 29.9 and 29.10 [25, 26].

In addition to the structural nonuniformity, these two fibers are known as the strong hydrophilic fibers, i.e. cotton has moisture regain of 8%, while the wool has 17% moisture regain. Such difference in moisture regain is because of the difference in the polymer molecules in cotton and wool, which are cellulose and protein, respectively. The cellulose molecules contain 6 –OH groups per repeat unit, and the wool protein has about 18 different amino acids in –R as side groups corresponding any one of the groups belonging to acidic, basic, hydroxyl, and hydrophobic type. Of the 18 amino acids, the major amino acids present in

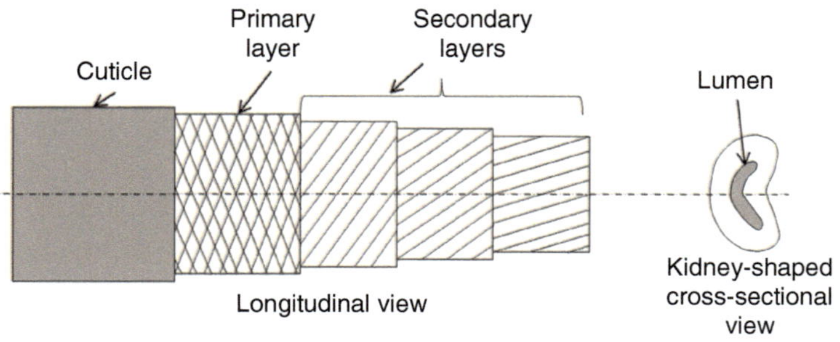

Figure 29.9 Structure of cotton fiber.

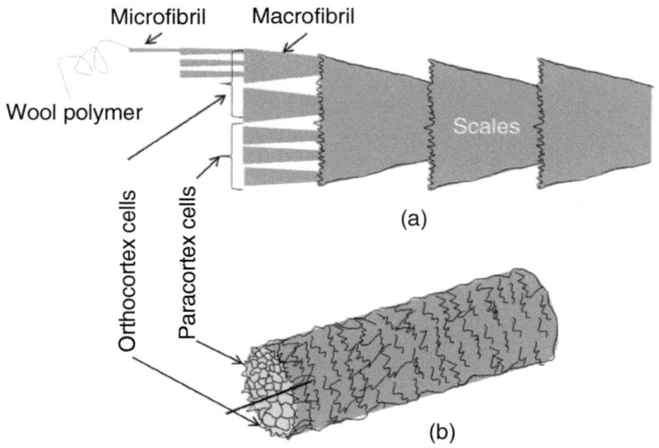

Figure 29.10 Schematic of wool fiber. (a) Scale and fibrils and (b) macroscopic rough surface.

wool fiber are glutamic acid (12%), cystine (11%), serine (10%), glycine (8%), and leucine (7.7%).

The difference is also due to the difference in the degree of crystallinity as in cotton it is 65% while in wool it is only 35%. Because of the structural complexity, wool has low crystallinity and high amorphous region, and hence, wool has high moisture regain. The moisture regain has a practical importance in terms of the diffusion of the water in the structures developed of these fibers. Generally the hydrophilic fibers would have a molecular absorption and transmission, while the hydrophobic fibers would have only surface-level adsorption and transmission as it is regarded as wicking. Additionally, such moisture properties also in general would facilitate biodegradation through the microorganisms. Hence such considerations while selecting the filter media fiber are important based on the kind of substance in the mixture to be separated and the disposal of the filter media.

However, some of the limitations of these fibers in filtration applications were the operating temperature and abrasion resistance. These fibers could be used only up to a temperature of 100 °C and had only average abrasion resistance. Of these two fibers, wool could withstand moisture very well and could be made into a thick felt due to the directional frictional effect.

After the advent of man-made and synthetic fibers, they were preferred the most as a result of improved properties with reference to temperature resistance and chemical and microbial stability. One of the most widely used synthetic fibers is the fiberglass or glass fiber. Some of the main properties that make them to be used in the filtration applications are the high temperature stability >1000 °C and chemical resistance along with the required mechanical properties. Additionally, fiberglass is comparatively cheap due to its raw material, the sand. The major chemical composition of the glass fiber is silica, generally above 50% with other inorganic oxides for the rest. The polymeric form of the silicon dioxide gives a random fine structural arrangement imparting the isotropic characteristics to the glass fiber unlike the other textile fibers that are anisotropic. Such inorganic polymeric fine structure form is responsible for the unique properties of the glass fiber to be used in all the applications and particularly for high temperature (up to 255 °C for continuous operation) and chemical-resistant filtration applications. The glass fibers are lubricated with graphite so that it would not break as a result of rubbing action during handling and cleaning cycle. Graphite is used as lubricant because it can withstand the high operating temperature.

Further invention of the other synthetic fibers in the early twentieth century gave wide options of fibrous materials for developing filter media with techno-economical considerations. Nylon, the first synthetic fiber, has good abrasion resistance and so is more effective in filtering abrasive dusts than the other synthetic fibers invented later. However, nylon is considered to perform poorly under the acidic conditions and also with exposure to light and environment. Polyethylene terephthalate (PET) polyester fiber has superior acid and light resistance in addition to a comparable abrasion resistance. Hence the PET fiber is also used widely in many industrial filtration applications such as foundries, smelters, etc. However, PET is sensitive to alkali and so its use in alkaline environment should be avoided. Polypropylene being chemically more resistant than the other synthetic fibers is also an inexpensive fiber and hence is used in many industrial filtration applications. However, it is limited in its use only to low temperature conditions as its melting point is only 130 °C. The use of polypropylene fiber, in nonwoven form as against the existing roving form as cross-wound package form for textile industrial effluent treatment, has been studied. The study revealed that the nonwoven form resulted in similar particle filtration with lower pressure drop [27].

The synthetic fibers can also be physicochemically modified compared with glass fibers, which makes them more suitable for filtration applications [28]. Because of these multifunctional feasibilities in synthetic fibers, they are mostly preferred. Only on the ground of ecofriendly and sustainable products, natural fibers gain some importance [28]. Moreover, for healthcare applications natural fibers are preferred. For example, Texus Fibre, New Zealand, has created a range of filtration products for personal protective respirators such as face masks and home air purifiers using wool fiber [29]. The regenerated cellulosic fiber, viscose rayon, finds notable applications in filtration for filter paper applications where controlled porosity is required. This fiber has an irregular cross section, by virtue of the fiber formation, and hence it can provide high specific surface area. It is used in food filtration as the fiber is physiologically as well as hygienically safe and is having neutral taste [30]. Similarly, for milk filtration also, the natural fibers are preferred along with the combination of synthetics as a blend, which better meet the requirement of milk filtration than the individuals. The fiber combinations used for the purpose are cotton, viscose, and polyester [31]. As far as the bast fibers like jute and flax are concerned, they are being explored for potential applications. They are already in use for the purpose in geotextiles; however, their main stream use in filtration applications is yet to emerge on the ground of sustainable solutions [27].

Development of high-performance grade textile fibers that exhibit high degree of thermal, mechanical, and chemical properties has also led the way for their use in filtration applications that demand high continuous operating temperatures, which cannot be just met by the aliphatic organic polymer-based fibers, i.e. nylon, polypropylene, PET, etc. Nomex is one such aromatic polyamide fiber

Figure 29.11 Chemical repeat unit structure of (a) Kevlar (poly-paraphenylene terephthalamide) and (b) Nomex (poly-metaphenylene isophthalamide).

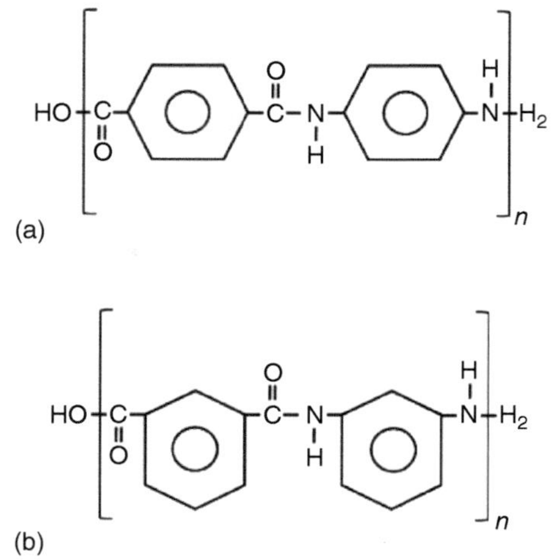

whose chemical name is poly(*m*-phenylene isophthalamide), which is regarded as *meta*-aramid fibers. The Nomex, by trade name, fiber has high temperature stability having limiting oxygen index of 30, T_g of 275 °C, and T_m of 430 °C. Because of these thermal parameters, the Nomex fiber is used for continuous high temperature filtration applications like for filtering dusts from cement coolers, asphalt batch plants, ferroalloy furnaces, and coal dryers. The other well-known aramid is para-aramid fiber, which is known as Kevlar that has high mechanical properties, because of the para links, along with the thermal properties even higher than Nomex. Hence Kevlar can also be used for high temperature filtrations particularly even under certain mechanical strength requirements. The chemical structures of Nomex and Kevlar are shown in Figure 29.11, which are responsible for their superior properties.

Teflon, Ryton, P84, and carbon fibers are some of the other high-performance fibers, which are used for very high temperature filtrations. In addition to temperature stability, Teflon exhibits a very high resistance to acids (except fluorine) and can perform continuously up to 230 °C. Ryton is made from polyphenylene sulfide polymer fiber attached to a polyfluorocarbon scrim. It has good resistance to both acids and alkali and performs continuously at 177 °C. Teflon, Nomex, and Ryton have been used in coal-fired boilers to remove particulate emissions [32].

The use of membrane as a laminate over the fibrous structures has proved to be effective in filtrations. One such effective membrane is the expanded PTFE developed by the Gore-Tex. It has been stated that such structured filtrations used in metal industries, chemical industries, food industries, and coal-fired boilers result in very good emission reduction (99.9%), low pressure drops, increased bag life, and higher air to cloth ratio. Ceramic fibers are also very much found in filtration applications particularly for continuous very high temperature applications of >500 °C like in filters ahead of boiler superheater tube sections in order to remove particles and improve heat transfer in the boiler tubes [33].

Various fibers used in filtration applications and their properties are listed in Table 29.1. Based on these properties, one or more than one materials in combinations could be used for designing the filter bags based on the thermo-mechanical and chemical requirements. In one of the 1992 reports, the cost

Table 29.1 Fibers used in filtration applications and their properties [32–34].

Generic name	Fiber	Density (g/cc)	Maximum temperature		Resistance to					Tensile strength	Relative cost
			Cont.	Surge	Acid	Alkali	Flex abrasion resistance	Oxidizing agent	Hydrolysis		
Natural fiber	Cotton	1.54	82	107	Poor	Excellent	Average	—	—	Good	0.4
Polyolefin	Polypropylene	0.91	95	93	Excellent	Excellent	Good	Poor	Good	Excellent	0.5
	Polyethylene	0.95	85	—	Excellent	Excellent		Poor	Good	Excellent	—
Natural protein	Wool	1.36	93	121	Good	Poor	Average	—	—	Poor	0.8
Polyamide	Nylon	1.14	93	121	Poor to fair	Excellent	Excellent	Poor	Fair	Excellent	0.6
Acrylic	Orlon	1.20	116	127	Very good	Fair	Average	—	—	Excellent	0.7
Polyester	Dacron	1.38	135	163	Good	Fair	Excellent	Fair	Poor	Excellent	0.5
Aromatic polyamide	Nomex	1.14	204	218	Fair	Very good	Very good			Very good	2.0
Fluorocarbon	Teflon	—	232	260	Excellent[a)]	Excellent[a)]	Fair	—	—	Average	6.7
Glass	Fiberglass	2.58	260	260	Good	Poor	Poor to Fair	—	—	Excellent	1.0
Polymer	P84	—	232	260	Good	Fair	Fair	—	—	—	2.5
Polymer	Ryton	—	191	232	Excellent	Excellent	Good	—	—	—	2.5–4.0
Polymer	PBI	1.43	300	—	Very good	Very good	—	—	—	Average	—
Polymer	PBO	1.56	250	—	Excellent	Excellent	—	Good	Good	Excellent	—
Vinyl polymer	PVDC	1.70	85	—	Excellent	Good	—	Excellent	Good	—	—
	PVDF	1.78	100	—	Excellent	Excellent	—	Good	Excellent	—	—
	PTFE	2.10	150+	—	Excellent	Excellent	—	Excellent	Excellent	—	—
	PPS	1.37	150+	—	Excellent	Excellent	—	Fair	Excellent	—	—
	PVC	1.37	80	—	Excellent	Excellent	—	Fair	Excellent	—	—
	PEEK	1.30	150+	—	Good	Good	—	Fair	Excellent	—	—

PBI, polybenzimidazole; PBO, polybenzoxazole; PVDF, polyvinylidene fluoride; PTFE, polytetrafluoroethylene; PPS, poly(phenylene sulfide); PVC, polyvinylchloride.
a) Teflon acid resistance, excellent except poor for fluorine; Teflon alkaline resistance, excellent except poor for trifluoride, chlorine, and molten alkaline metals.

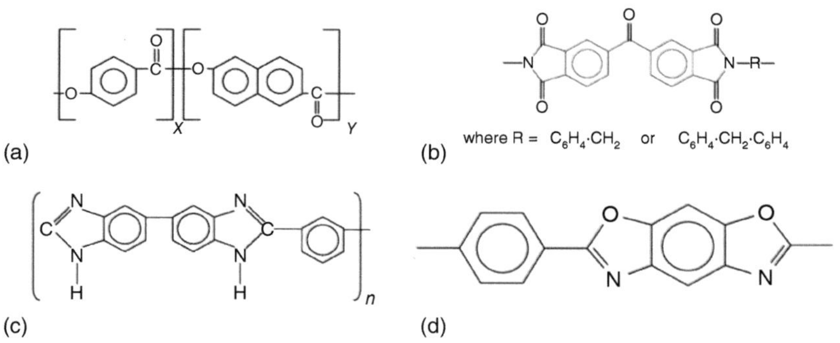

Figure 29.12 Chemical structure of some of the high-performance fibers for filtration applications. (a) Aromatic polyester (Vectron), (b) polyarimids (P84), (c) polybenzimidazole (PBI), and (d) polybenzoxazoles (PBOs, Zylon).

difference between fiberglass and Teflon filter bags could be noticed as much difference as about 220% [34]. The cost of a fiberglass bag 14 ft long and 6 in. in diameter is approximately $35–40, while the same size Teflon bag costs about $115–135. Hence, that the same size filter bag made of Teflon is about 220% higher than the one made of fiberglass. Chemical structure of some of the high-performance fibers is shown in Figure 29.12.

29.6 Factors Governing the Performance of Fibrous Filter Media

29.6.1 Role of Cross Section of the Fiber

Filtration application requires high surface area for a given volume so that the filter media could have large contact with the incoming fluid and have maximum particle contact. This would improve the particle filtration efficiency based on the principles of filtration. In general, natural fibers are having some irregularities over the surface except silk. However, synthetic fibers produced through melt spinning, in general, have round cross section because of the spinneret orifice. Of various mathematic geometries, round/sphere has the least circumference/surface area for a given diameter/volume. When the round and spherical shape is modified immediately, the circumference/surface area would increase according to the degree of cross-sectional deviation from round. These fibers are generally regarded as multilobed structures such as trilobal, octolobal, tape, diabolo (different solid shapes), channeled surface (Coolmax, Gcool), hollow, and bicomponent cross sections. Different types of noncircular cross sections of fibers along with their respective spinneret shapes are shown in Figure 29.13. Figure 29.14 shows the various types of bicomponent fibers produced for especially built-in properties.

Man-made fibers with irregular cross sections are used in filtration applications with some value-added functions [35]. It is reported that the multilobed cross-sectional fiber assists the particle and molecular filtration probably due to

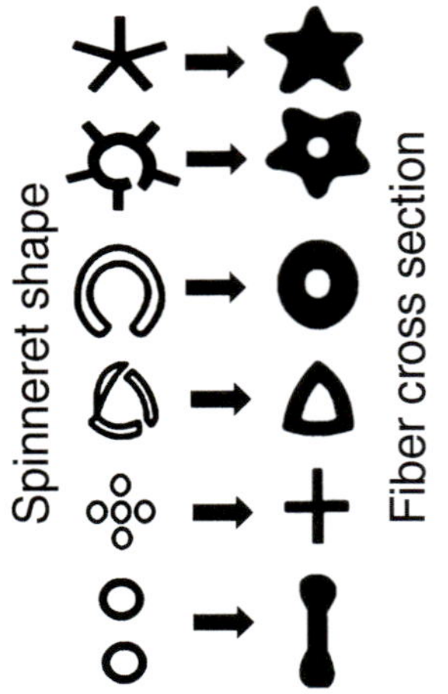

Figure 29.13 Various types of fiber cross sections along with the respective spinneret shapes.

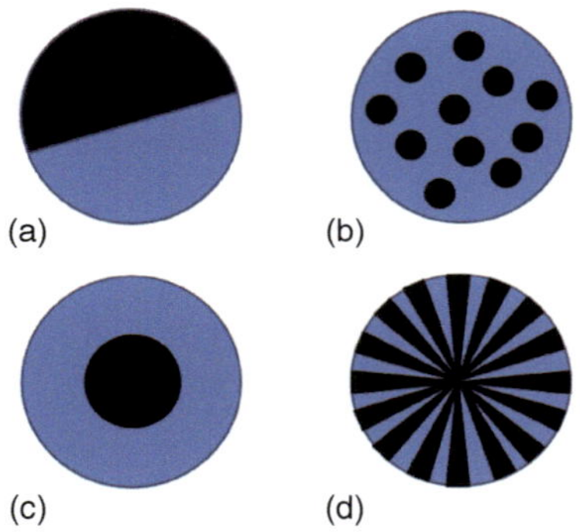

Figure 29.14 Types of bicomponent fiber: (a) side by side, (b) sea island, (c) core–sheath, and (d) splittable.

the increased surface area [36]. Active agents like acid or alkaline components are placed on the longitudinal slots of the multilobed fibers. These reactants react with the molecular and particle contaminants and fix them with the substrate. Such reagents could also be incorporated through some adsorptive particles in the slots of the fibers. In Cornell University, researchers have tried to study the Y fiber combinations on the packing density and found low packing density with relatively large area to volume ratio for triple Y fiber combinations. Such fiber combinations help particle trapping by Brownian motion [37].

Special channeled polyester fibers named as 4DG™, as one of the members of the Capillary Channel Polymer™ (CCP™), have many unique fiber geometries and have 2.3–2.8 times higher surface area compared with same denier round fibers because of the deep groove fibers with eight surface channels as shown in Figure 29.15 [38].

The CCP fibers have been used in tobacco smoke filters as its large surface channels could hold high content of tobacco smoke-modifying agents, such as flavorants [39]. Hence the channeled geometry could be an efficient way of capturing particles, and these kinds of structures are promising in filtration research [38].

Figure 29.15 Cross-sectional image of CCP™ fiber.

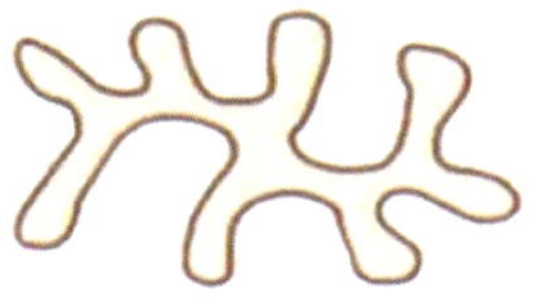

29.6.2 Role of Pore Size

The bulk properties of the filter media would depend on the type of material, the fibers used for its development, and hence the factors/properties of fibers are of importance in this context. Such properties of fibers mainly influence the kind of filtration applications, efficiency, and the durability of the filter media. For example, if the filtration requirement is to remove particles from hot chemicals, the choice of the fiber should be such that it has both chemical resistance and temperature resistance along with the high degree of chemical inertness even under hot conditions. This is an extreme example for which the fiber choice could be a fiberglass; however, for only chemical filtrations, the inert olefin fibers could be considered but not under hot conditions as they are sensitive to heat. Similarly fibers with high degree of mechanical performance such as strength, elongation, creep, and modulus might be required for pressure filtrations under ambient and elevated temperature conditions. Accordingly, the properties of fibers could be grouped and could be regarded as factors governing the choice and the performance of the fibrous filter media. The relevant properties of fibers can be classified as morphological properties, mechanical properties, thermal properties, chemical properties, and resistant properties to various environmental factors such as sunlight, microbes, and reactive particles in the atmosphere. There are other properties of textile fibers that might have an indirect influence on the filter media such as the moisture properties and electrical conductive properties [5, 26].

The morphological properties can be further subdivided into macromorphological properties and fine morphological properties and have notable influence on the filtration performance. Particularly, the macromorphological properties would decide the kind of pore surface profiles that has direct influence on the particle interaction. The other properties such as the mechanical properties could cover range of properties such as the strength, specific strength, modulus, elongation, elasticity, dimensional stability, abrasion resistance, creep, and stress relaxation. The mechanical properties are considered the most as it governs many of the filtration requirements. Thermal properties are mainly important to maintain the structural dimension under the intended working temperature.

Similarly, the other properties like chemical resistance are important, which again depend on the inherent nature of the polymer molecule. In this case the chemical groups involved in the polymer chains are expected not to respond in any one of the possible chemical reaction mechanisms such as the electrophilic–nucleophilic reactions and redox and free radical reactions. Polymer chains without any polar functional groups but with only the alkyl chains like polyethylene and polypropylenes are considered to have good resistance to these kinds of chemical reaction mechanisms, and so they are considered good for chemical filtrations. Apart from this the polymers with cyclic structure, cross-linked structures, and high activation energy are also considered to have good chemical resistance along with the thermal resistance.

Generally, the microbial resistance of fibers made of artificial polymers like polyester (PET), nylon, acrylic, and other high-performance fibers is considered good and better than the natural fibers like cotton, silk, wool, etc. The possible reason could be not as simple as those given for thermal and chemical resistance but could be regarded as complex by virtue of the inherent growing and multiplying tendency of the microbes over the natural substances than those on the man-made substances. The other properties such as the resistance to sunlight and reactive particles in the atmosphere are also indirectly related with the chemical and thermal resistance. The properties like moisture and electrical properties do have a notable influence based on the type of filtration applications. In general, moisture properties impart the filter media to work with the principle of interaction with moisture in the incoming fluid at a molecular level and ensure moisture separation, for example, the removal of moisture from diesel in using cellulosic diesel filters. The electrical properties in turn influenced by moisture govern the kind and the quantity of surface charges, which might affect the particle through similar and opposite charge interaction effect, resulting in enhanced and/or selective particle separation.

29.7 Characterization of Filter Media

Characterization of filter media is a vast topic and could vary based on the particular applications. However, broadly, it can be listed as the dimensions, stiffness, mechanical properties such as tensile, tear strength, burst strength, elongation, and elasticity; density, porosity, pore size and structure – bubble point, mean flow pore size, image analysis, microscopy, X-ray microtomography, liquid extrusion porosimetry, liquid intrusion porosimetry, gas vapor adsorption; pressure drop, filtration efficiency – single pass, multi-pass efficiency, cumulative and fractional efficiency, dust loading capacity, cake release properties, etc. Some of the other properties are the moisture properties, formaldehyde contents, air permeability, fold endurance, water repellency, flammability and resistance to chemicals, redox reactions, and bacterial attack [13, 40, 41].

Standard test methods for these parameters have been established by many global organizations, some of which are the American Association of Textile Chemists and Colorists (AATCC); American National Standards Institute (ANSI); American Society of Heating, Refrigerating and Air-Conditioning Engineers (ASHRAE); ASTM International (ASTM); BSI British Standards (BSI); German Institute for Standardization (DIN); European Disposables and Nonwovens Association (EDANA); US Environmental Protection Agency (EPA); US Food and Drug Administration (FDA); Association of the Nonwoven Fabrics Industry (INDA); International Organization for Standardization (ISO); Japanese Standards Association (JSA); etc. Based on the country and the seller–buyer interest, any one of the methods could be followed. The standard numbers under each of the organizations could be found in the respective web pages and are also summarized by Irwin M. Hutten [3].

29.8 Future Prospects

With the increasing thrust on technical textiles and technology developments, the need for filter media is increasing. With the matching material properties, textile fibers are found to be suitable candidates to prepare the filter media through any one form. Most suitable form in many cases is the nonwoven form as it turns out to be techno-economically viable. Among the varieties of fibers and the choices available, certainly the end-use application requirements have to be considered, and a suitable fiber has to be chosen accordingly. For general applications polypropylene, polyester (PET), and nylon are used; however, when the filters are expected to perform at high temperatures, the high-performance grade fibers may be preferred. Nowadays with the global emphasis on ecofriendly and sustainable products, natural fibers are also being revived for filtration applications. As of now, the global market potential for filtration nonwoven textiles is estimated to be $5.4 B with the market share as shown in Table 29.2 for each category of applications.

Among various filtration applications, automotive industry, water purification systems, and indoor air conditioning systems (heating, ventilation, and air conditioning [HVAC]) are the three major areas of applications. Healthcare industry and manufacturing industry requirements also go beyond 10% each. The other minor areas are the advanced technological sectors, hydrocarbon processing, and food and beverages.

With increasing use of commodity products, the need for filtration media is going to increase with higher rate ever year. Textile fibers along with the existing suitable properties, further feasibilities for modifications and functionalization based on the ever-increasing requirements makes them suitable ideal material for the filtration media, and hence their consumption would also be high.

Table 29.2 Market potential for filtration textiles.

Market application	Market share (%)
Transportation	22
Water filtration	20
HVAC	14
Food and beverage	2
Healthcare	11
Manufacturing	10
Advanced technology	6
Hydrocarbon processing	5

References

1 Hearle, J.W.S. (2001). *High-Performance Fibres*. New York: Woodhead Publishing Ltd.

2 Sundarrajan, S., Tan, K.L., Lim, S.H., and Ramakrishna, S. (2014). Electrospun nanofibers for air filtration applications. *Procedia Engineering* 75: 159–163.

3 Sutherland, K. (2008). *Filters and Filtration Handbook*, 5e. New York: Butterworth-Heinemann, Elsevier.

4 Tanchis, G. (2008). *The Nonwovens – Reference Books of Textile Technologies*. Italian Trade Commission.

5 Morton, W.E. and Hearle, J.W.S. (2008). *Physical Properties of Textile Fibres*. England: The Textile Institute, Woodhead Publishing Limited.

6 Purchas, D. and Sutherland, K. (2002). An introduction to filter media. In: *Handbook of Filter Media* (eds. Derek B Purchas and Ken Sutherland), Chapter 1, 2e, 1. Elsevier Advanced Technology. ISBN: 1856173755.

7 Chen, T., Wu, C., Liu, R. et al. (2011). Effect of hot vapor filtration on the characterization of bio-oil from rice husks with fast pyrolysis in a fluidized-bed reactor. *Bioresource Technology* 102 (10): 6178–6185.

8 Darby, J.L. and Lawler, D.F. (1990). Ripening in depth filtration textiles: effect of particle size on removal and head loss. *Environmental Science and Technology* 24 (7): 1069–1079.

9 Holdich, R.G., Cumming, I.W., and Smith, I.D. (1998). Crossflow microfiltration of oil in water dispersions using surface filtration with imposed fluid rotation. *Journal of Membrane Science* 143 (1–2): 263–274.

10 Sioutopoulos, D.C. and Karabelas, A.J. (2012). Correlation of organic fouling resistances in RO and UF membrane filtration under constant flux and constant pressure. *Journal of Membrane Science* 407–408 (15): 34–46.

11 Raoul, B. (1957). Pressure filters. US Patent 2, 799, 397 A.

12 Matteson, M.J. (1986). *Filtration Textiles: Principles and Practices*, 2e. CRC Press.

13 Hutten, I.M. (2007). Testing of nonwoven filter meida. In: *Handbook of Nonwoven Filter Media* (ed. Irwin M. Hutten), Chapter 6, 245–290. Elsevier Science and Technology.

14 Eichhorn, S., Hearle, J.W.S., Jaffe, M., and Kikutani, T. (2009). *Handbook of Textile Fibre Structure: Fundamentals and Manufactured Polymer Fibres*, vol. 1. Elsevier.

15 UNITEX (1971). Textile Fibres: Classification by Type – Terms and Definitions. Constitution.

16 Eichhorn, S., Hearle, J.W.S., Jaffe, M., and Kikutani, T. (2009). *Natural, Regenerated, Inorganic and Specialist Fibres*, vol. 2. Elsevier.

17 Gupta, V.B. and Kothari, V.K. (2012). *Manufactured Fibre Technology*. Springer Science & Business Media.

18 Wisniowski, M.-T. (2017). General properties of fiber polymers and fibers macro properties – Part V. http://artquill.blogspot.in/2013/08/general-properties-of-fiber-polymers.html (viewed 17 April 2017).

19 Yasuda, T. and Fujiwara, Y. (1970). Method of forming microfibers. US Patent 3, 549, 734 A.
20 Mukhopadhyay, A. and Choudhary, A.K. (2013). Performance of filter media as function of fibre fineness in pulse jet filtration system. *Textiles and Light Industrial Science and Technology (TLIST)* 2 (1): 13–26.
21 Periyasamy, S., Gupta, D., and Gulrajani, M.L. (2007). Nanoscale surface roughening of mulberry silk by monochromatic VUV excimer lamp. 103 (6): 4102–4106.
22 Periyasamy, S., Krishna Prasad, G., Chattopadhyay, S.K. et al. (2017). Micro-roughening of polyamide fabric using protease enzyme for improving adhesion strength of rubber-polyamide composite. 37 (3): 297–306.
23 Naidu, L.D., Saravanan, S., Goel, M. et al. (2016). A novel technique for detoxification of phenol from wastewater: nanoparticle assisted nano filtration (NANF). *Journal of Environmental Health Science and Engineering* 14 (9): 1–12.
24 BOSCH Invented for Life. HEPA premium cabin filter, USA. www.boschautoparts.com. (Accessed 18 April 2017)
25 Koushik, C.V. (1979). *The Helical Structure in the Secondary Wall of the Cotton Fibre*. University of Strathclyde. Department of Fibre Science.
26 Hearle, J.W.S. and Peters, R.H. (2013). *Fibre Structure*. Elsevier.
27 Harshavardhan, G., Kavin, P., and Krishnan, A. (2015). Development of non-woven filter for textile effluent filtration. B.Tech Project Work Thesis. Guided by Dr. G. Thilagavathi, Department of Textile Technology, PSG College of Technology.
28 Skomra, E. (2010). Production and characterization of novel air filtration media. All Dissertations. Paper 683. Clemson University, TigerPrints.
29 Texus Wool Fibre (2017). Natural Fibres That Improve Human Health, 111a Kerwyn Ave, East Tamaki, Auckland, New Zealand. http://lanaco.co.nz/texusfibre/assets/File/Texus_Brochure_August2015.pdf (viewed 18 April 2017).
30 Wimmer, P. and Zacharias, J. (2015). Viscose speciality fibres – bio-based fibres for filtration. *Filtration & Separation* 52 (3): 36–39.
31 Delaval. Efficient milk filtration. DeLaval International AB, 2008, Borosilicate Glass Fiber Filters Applications by Grade, Sterlitech Corporation, USA. www.sterlitech.com. (Accessed 18 April 2017)
32 Belba, V.H., Grubb, W.T., and Chang, R.L. (1992). The potential of pulse-jet baghouse for utility boilers. Part 1: A world-wide survey of users. *Journal of the Air and Waste Management Association* 42 (2): 209–218.
33 McKenna, J.D. and Turner, J.H. (1989). *Fabric Filter-Baghouses I, Theory, Design, and Selection*. Roanoke, VA: ETS.
34 Mohurle, N. and Thakare, N.R. (2013). Analysis on fabric filtration material for pulse jet fabric filter. *International Journal of Emerging Technology and Advanced Engineering* 3 (6): 603–609.
35 Zhang, W. and Yao, L. (2016). Review of the patented electrostatic air filter-materials and designs controlling technology. *BAOJ Nanotech* 2 (1): 9.
36 Rohrbach, R., Bause, D., Unger, P. et al. (2008). Complex shaped fiber for particle and molecular filtration. US Patent 7, 442, 223, B2.

37 Zhu, H. and Hinestroza, J.P. (2009). Collection efficiency for filters with staggered parallel Y and triple Y fibers: a numerical study. *Journal of Engineered Fibers and Fabrics* 4 (1): 16–25.
38 Vaughn, E. and Ramachandran, G. (2002). Fiberglass vs. synthetic air filtration media. *Journal of Engineered Fibers and Fabrics*: 41–53.
39 Phillips, B.M., Wilson, S.A., and Pollock, M.A. (1994). Tobacco smoke filter. US Patent 5, 356, 704.
40 Jena, A. and Gupta, K. *Characterization of Pore Structure of Filtration Media*. New York: Porous Materials, Inc.
41 Pike, R.D., Lassig, J.J., Shipp, P.W. Jr., and Williams, B.J. (2001). Nonwoven filter media. US Patent 6, 169, 045 B1.

30

Fibrous Materials for Thermal Protection

Gouwen Song[1] *and Yun Su*[2,3]

[1] *Department of Apparel, Events and Hospitality Management (AESHM), Iowa State University, Ames, IA 50011, USA*
[2] *College of Fashion and Design, Donghua University, Shanghai 200051, China*
[3] *Key Laboratory of Clothing Design and Technology, Donghua University, Ministry of Education, Shanghai 200051, China*

30.1 Introduction

Thermal hazardous environments are manifold and depend on the type of fire ground, for example, residential, manufacturing, wild land, storage, or mercantile fire [1]. Firefighters while exposed to fires and thermal environments encounter not only with the fatal injuries but also with nonfatal injuries, such as skin burn injuries, strain, bruise, and smoke or gas inhalation [2]. It is estimated that every year over 30 000 firefighters suffer injuries on the job in United States and many of them lose their lives while fighting fires, rescuing people, or responding to hazardous material incidents [2, 3]. Although extensive efforts have been dedicated to improve the health and safety of firefighters, this profession continues to be one of the world's most dangerous industries [4].

According to some reports [5, 6], thermal environmental conditions were usually identified as routine, hazardous, and emergency. An emergency thermal environment generally appears in a flashover condition that the temperature can rise above 600 °C with a heat flux range of 20–209.34 kW/m^2. Firefighter in this thermal environment has several seconds to react [7]. The routine and hazardous thermal environments can be defined as low-level radiative heat exposure, and the corresponding heat flux ranges from 5 to 20 kW/m^2 [5]. Even though the intensity of heat exposure does not cause the thermal degradation of protective material for longtime exposure, the majority of burn accidents occur in these thermal exposure situations [8, 9]. Therefore, in order to minimize or prevent fatal and nonfatal injuries from possible accidents, firefighters are required to wear protective clothing that is characterized by flame retardance and thermal insulation [10]. However, it is reported that thermal protection provided from thermal protective clothing is determined by various factors, such as fabric properties, air gap size, moisture content, and garment design. Additionally, the development of the proper type of fibrous materials is absolutely vital for the improvement of thermal protective performance of clothing against various heat exposures.

Handbook of Fibrous Materials, First Edition. Edited by Jinlian Hu, Bipin Kumar, and Jing Lu.

In this chapter, an overview of the literature on performance requirements of thermal protective clothing is presented. Also, fibrous materials suitable for thermal protection are summarized in detail, including structure, properties, and theories. Assessment methods of thermal protective performance and applications of fibrous materials are discussed. The effect of various factors on thermal protective and comfort performance of clothing is reviewed. Finally, the future trends for thermal protection are given in order to develop advanced thermal protective clothing that can provide better occupational health and safety for firefighters.

30.2 Performance Requirements of Thermal Protective Clothing

According to standard NFPA 1971-97 and EN 469-05, a typical firefighting ensemble is usually composed of three different fabric layers, i.e. an outer shell, a moisture barrier, and a thermal liner (including a layer of apertured, closed cell foam and a layer of facecloth fabric). In order to minimize the flame burn, contact burn, and steam burn or scald, protective clothing should provide effective thermal and moisture protection from flash fires, high-intensity thermal radiation, and hot liquid and steam. The development in firefighting protective clothing changes the protective requirements from the single thermal hazard to multiple thermal hazards, attaching importance on human ergonomics and comfort. Thus, performance requirements of each layer fabric are presented in Figure 30.1.

The outer shell is designed to provide flame resistance, protection against heat radiation, and resistance to water and certain levels of abrasion, since the outer shell is directly exposed to flame, radiation, hot liquid, and hot steam. Thus, blended fabrics are generally selected as the outer shell in order to meet these performance requirements and reduce cost of production. The preferred fiber is constructed of a flame- and heat-resistant material, such as woven aramid and polybenzimidazole ("PBI," a trademark of Celanese Corp.) fibers. Commercially

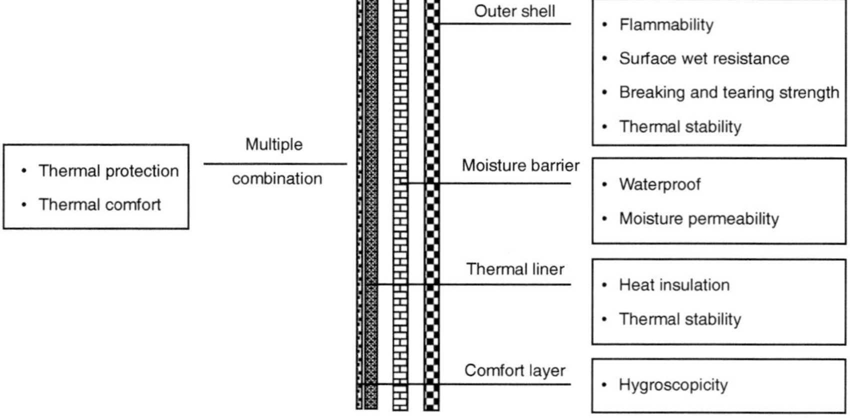

Figure 30.1 Performance requirements of thermal protective clothing.

available aramid materials mainly include Nomex and Kevlar (both are trademarks of E.I. DuPont, Inc.).

The vapor barrier is intended to protect the firefighter from steam and harmful chemicals. It was reported from the etiology of injuries to firefighters that flame burns just caused injury to 20% firefighters, while 65% patients received steam burns or scalds, especially for the poor sealing regions of clothing (such as the neck and hand) [11]. In addition, thermal protective clothing while providing thermal protection for wearer also brings about thermal stress due to the collection of moisture in the thermal liner and friction between the outer shell and the thermal liner layers. The design of moisture barrier also promotes water vapor transport outwardly from the wearer, which reduces wet weight gain from sources of moisture [12]. A moisture barrier commonly consists of a semipermeable membrane bonded or laminated to a fabric substrate, e.g. GORE-TEX (a trademark of W. L. Gore & Associates, Inc.). It was reported that GORE-TEX is a microporous polymeric film of polytetrafluoroethylene (PTFE), with pore volumes reaching 82% and maximum pore sizes of 0.2 μm [13]. Even though PTFE is hydrophobic, the extremely large number of pores and the thinness of the film allow the diffusion of gases [14]. Such moisture barriers are impermeable to liquid moisture but allow water vapor to pass through.

The thermal liner acts as an insulating medium against heat conduction, as well as providing thermal comfort for wearer. In order to reduce conduction heat transfer from thermal environment to human body, a thermal liner is typically made of a batting of aramid fiber quilted to a woven facecloth of spun yarn fiber. This is because the fabric structure can increase the amount of embedding air within thermal liner and reduce clothing weight to weaken the wearer's stress. The aramid fiber preferably is made of a flame- and heat-resistant material, such as Nomex and Kevlar. The preferred yarn fiber is fire-resistant cotton or viscose in order to provide better thermal comfort.

30.3 Fibrous Materials Suitable for Thermal Protection

Fire-retardant fibers can be classified into two types: chemically modified flame-resistant fiber and inherently flame-resistant fiber. The chemically modified flame-resistant fibers, including natural and synthetic fibers, are produced by a textile finishing process depending on the properties of flame retardant. The inherently flame-resistant fiber has special molecular structure to get high decomposition temperature, such as *meta*-aramid and *para*-aramid. The burning behavior of fibers is often determined by thermal transition temperatures and thermodynamic parameters, such as softening and melting temperature, and rate of heat release. The rate of heat release determines burning hazard, indicating the rate of fire spread and severity of burns. Table 30.1 shows the transition temperatures and thermodynamic parameters of the commonly fibrous materials. It can be found that the inherently flame-resistant fibers have higher thermal transition temperature compared with most fibers chemically modified flame retardant, since they have different mechanisms of flame resistance. The limiting oxygen index (LOI) of inherently flame-resistant fibers is more than 29%, meaning that they possess better flammability.

Table 30.1 Thermal transition of the commonly fibrous materials.

Category	Fiber type	Softening temperature (°C)	Melting temperature (°C)	Pyrolysis temperature (°C)	Igniting temperature (°C)	LOI (%)	Heat release rate (kg/J)	Flame-resistant mode
Chemically modified flame-resistant fiber	Cotton	—	—	350	350	18.4	19	Chemical finish
	Wool	—	—	245	600	25	27	
	Viscose	—	—	350	420	18.9	19	Additive introduced
	Polyester	80–90	255	420–447	480	20–21	24	Copolymeric modifications
	Nylon	50	215	431	450	20–21.5	39	
	Acrylic	100	>220	290	>250	18.2	32	
	Modacrylic	<80	>240	273	690	29–30	—	
Inherently flame-resistant fiber	*meta*-Aramid	275	375	410	>500	29–30	30	Acromatic homo- or copolymer
	para-Aramid	340	560	>590	>550	29	—	
	PBI	435	—	580	>600	>41	—	
	Kermel	285	—	411	—	30–32	—	

Source: Horrocks et al. 1988 [15]. Reproduced with permission of Taylor & Francis.

30.3.1 Chemically Modified Flame Retardant Fibers

Chemically modified fire-retardant fibers include natural (e.g. cotton, wool, and viscose) and synthetic fiber (e.g. polyester, acrylic, and nylon). There are various kinds of methods to improve the flame resistance of fibrous materials. Some flame retardant additives, such as halogen, nitrogen, silicon, and phosphorous, are usually incorporated into the spinning dopes during the manufacture of fire-resistant fibers [16]. Also, flame-retardant finishes can be applied on the surface of natural or synthetic substrate fibers. Generally speaking, these fire-resistant mechanisms involve in three aspects of burning behavior: fuel (from thermally degraded or pyrolyzed fibers), heat (from ignition and combustion), and oxygen (from the air), as described in Figure 30.2 [17].

The principles to improve flame resistance of fiber can be recognized as solid/condensed and gas/vapor phase (see Figure 30.3) [18, 19]. These flame-retardant strategies are related to combustion mechanism of fibrous materials, as can be seen from Figure 30.2. The solid/condensed phase strategies are to reduce the absorbed thermal energy from degradation reaction (a) or increase decomposition temperature by inserting inherently flame-resistant fibers (b). However, the most effective flame-retardant methods are to promote char

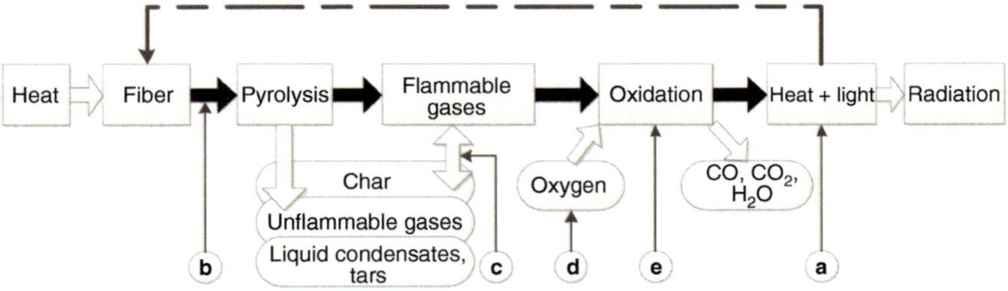

Figure 30.2 Combustion mechanism of fibrous materials.

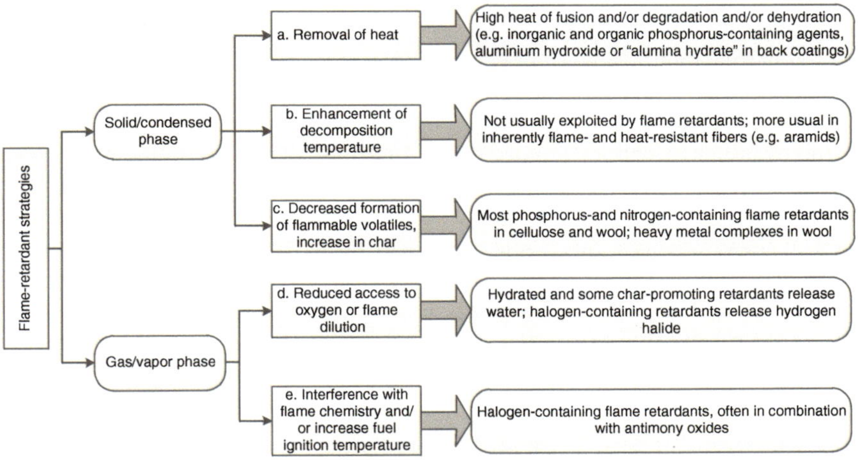

Figure 30.3 Flame-retardant strategies of fibrous materials.

formation by converting the organic fiber structure to a carbonaceous residue or char and hence reduce volatile formation (Figure 30.2c). Therefore, char-forming flame retardants offer both flame and heat resistance to a textile fiber. For example, most phosphorus- and nitrogen-containing retardants, when present in cellulose and wool, promote char formation and reduce volatile formation during heat exposure. In addition, the existence of elements like nitrogen and sulfur is known to enhance synergistically the performance of phosphorus-containing retardants by further increasing char-forming tendencies [19]. With regard to the gas/vapor-phase strategies, the flame-resistant principle is to affect the oxidation reaction of fibrous materials (see Figure 30.2). For one thing, hydrated and some char-promoting retardants can release water, while halogen-containing retardants produce hydrogen halide. These compounds can reduce the access to oxygen or flame dilution by forming protective barrier on the fiber surface (d) [20]. For another, the flammable gases generated from decomposition reaction can react with oxidation, which contributes to decrease the amount of oxygen used in combustion reaction (Figure 30.2e). The LOI of fibrous materials has an increase due to the reduced access to oxygen and increase of the igniting temperature.

All flame-retardant cottons are usually produced by after-treating fabrics chemically, depending on chemical character and cost of flame retardant. A textile finishing process also yields flame-retardant properties having varying degrees of durability to various laundering processes. The most effective flame retardants contain either phosphorus or antimony–bromine-based systems [19], such as Flacavon WP and Flacavon F12/97 (Schill & Seilacher), which can be applied to cotton fibers for improving the flammability. But this generates a perception of unacceptable environmental hazard and may cause wear casualties due to the inhalation of the toxic gases produced when the fibers catch fire.

Among the area of flammability of all the natural fibers, wool fiber has the highest inherent non-flammability and the pyrolyzing temperature. It has a relatively high LOI value of about 25% and a low flame temperature of about 600 °C. But wool fiber is not suitable for thermal protective clothing, since the pyrolyzing temperature is relatively low of around 245 °C. In order to meet performance requirement for thermal protection, zirconium hexafluoride complexes can be used to chemically finish wool fiber, such as Zirpro (IWS), Pyrovatex CP (Ciba) and Aflammit ZR (Thor).

The conventional synthetic fibers may be provided flame resistance during production by either incorporation of a flame retardant additive in the polymer melt or solution prior to extrusion. Also, the method of copolymeric modification can be used to improve fire retardance of synthetic fibers. Synthetic fibers produced in these ways are often said to be inherently flame retardant. However, the absence of polyamides reflects their high softening and melting reactivities and hence poor flame-retardant compatibilities. No char-promoting flame retardants exist for any of the conventional synthetic fibers so that it is not easy to offer high-performance flame and heat resistance.

30.3.2 Inherently Flame Retardant Fibers

The flame-retardant effect in chemically modified flame-retardant fibers can be reduced after several washing. However, the inherently flame-retardant fibers have fundamentally combustion-resistant structures (i.e. all-aromatic polymer) that can better resist heat and flame for long-term usage. According to Table 30.1, most of inherently flame-retardant fibers decompose above 375 °C or so. Their all-aromatic structures are responsible for their low or non-thermoplasticity and high pyrolysis temperatures. In addition, their low flammabilities are dependent on their high char-forming potentials and LOI values.

Since high-performance *meta*-aramid was successfully developed by DuPont in the early 1960, Nomex fiber has been widely used for military, aerospace, and firefighting industry [21]. *meta*-Aramid possesses good tensile and other mechanical properties. The presence of aromatic groups can provide high thermal stability and flame resistance (see Figure 30.4). There has no aging under heat exposure of 220 °C for long time. The thermal transition temperatures of fiber are greatly more than that of the commonly natural and synthetic fibers (see Table 30.1). The char begins to form up to 400 °C, while the fabric can char quickly for high temperature of 900–1500 °C and thus lead to the increase of thickness to provide better thermal protection for wearer.

However, Nomex fiber easily shrinks when exposed to high-intensity heat flux, which can reduce thermal protective performance of protective materials. In order to avoid the drawback, Nomex fiber is usually blended with Kevlar fiber that has better strength and abrasion resistance, such as a blend of 95% Nomex and 5% Kevlar fibers [22]. Kevlar fiber developed by DuPont (Kevlar) and AKZO (Twaron) in the 1970s is one of *para*-aramid. The structure of *para*-aramid is given in Figure 30.5. The main difference between *meta*- and *para*-aramid is the positions of the aromatic groups along the backbone chain. Thus, *para*-aramid fibers comprise extreme symmetry within its polymer chains and have greater interchain links and crystallinity to provide higher tensile strength, rigidity, and moduli. Also, *para*-aramid fibers are high heat- and flame-resistant material. According to Table 30.1, its transition temperatures are larger than that of *meta*-aramid fibers. There exist no flame spread, shrinking, and melting phenomena when exposed to flash fire, which is appropriate for thermal protection.

Development work on aromatic PBI was conducted in 1960s. Later, PBI was used by NASA and the Air Force Materials Lab to make the astronaut's clothing. In 1983, the PBI fibers were manufactured commercially by Celanese Corporation [23, 24]. Figure 30.6 presents the molecular structure of PBI polymer. Table 30.1 shows that the decomposition of PBI in air occurs at

Figure 30.4 Structure of *meta*-aramid.

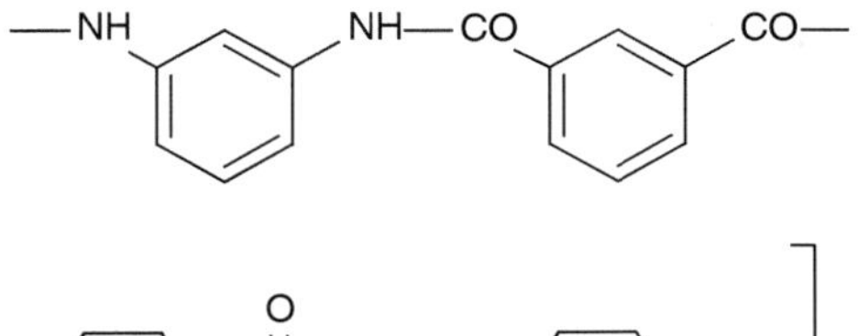

Figure 30.5 Structure of *para*-aramid.

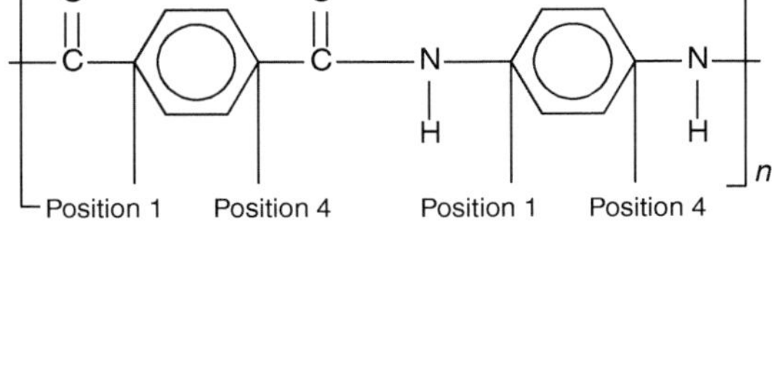

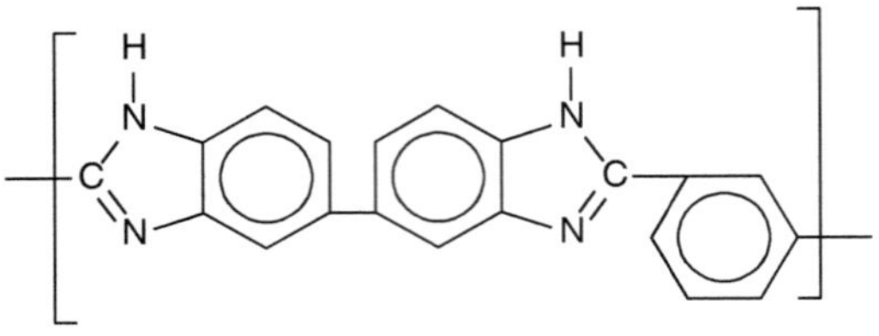

Figure 30.6 Structure of PBI polymer.

approximately 580 °C and the fiber can withstand high temperatures without severe loss of mechanical properties. In an inertial atmosphere, the temperature of weight loss occurs at 600 °C [25]. Long-term temperature aging can lead to rapid loss of mechanical properties owing to oxidative degradation, while PBI can durably maintain its properties in temperatures as low as −196 °C. The LOI value of PBI fiber is more than 41%. In addition, PBI is typically resistant to organic acids, chlorinated solvents, and strong acids. Due to the several attractive properties, such as high glass transition temperature, chemical resistance, and retention of good mechanical properties, PBI has been popularly used in high-performance polymer blends. PBI was generally introduced as an outer shell protective fabric that is typically composed of 40% PBI and 60% *para*-aramid.

Other popular type of *meta*-aramid fiber is poly(aramid-imide) fiber, like Kermel fiber. The molecular structure of poly(aramid-imide) consists of a high proportion of aromatic groups and combined double bonds (see Figure 30.7) that is similar with Nomex. Kermel fiber was developed by Rhone-Poulenc of France in 1971. The thermal properties of Kermel is also approximate to Nomex fiber, providing durable flame resistance even if for washing of 200 times. Meantime, Kermel fiber possesses excellent heat insulation and presents no melt and afterflame for long-time exposure of 250 °C. The Kermel fiber is developed through polycondensation of toluene diisocyanate and trimellitic anhydride [26]. A distinction of Kermel fiber is the circular transverse section that makes fabric soft for providing good comfort. The combustion of Kermel fiber just generates a small amount of smoke and toxic products, which is far less than most natural and synthetic fibers. But the Kermel fiber is commonly combined with cotton or viscose due to the expensive production costs. The most typical blended fabric is 50% Kermel and 50% flame-resistant viscose (Lenzing from Austria), which has excellent flame retardance and comfort.

Also worth mentioning is polybenzoxazole (PBO) from Dow Company in United States and Zylon (Toyobo), which comprises poly(*p*-phenylene-2 and 6-benzobisoxazole), as shown in Figure 30.8. The PBO fiber is a high modulus and high tenacity fiber that are about twice as high as respective values for

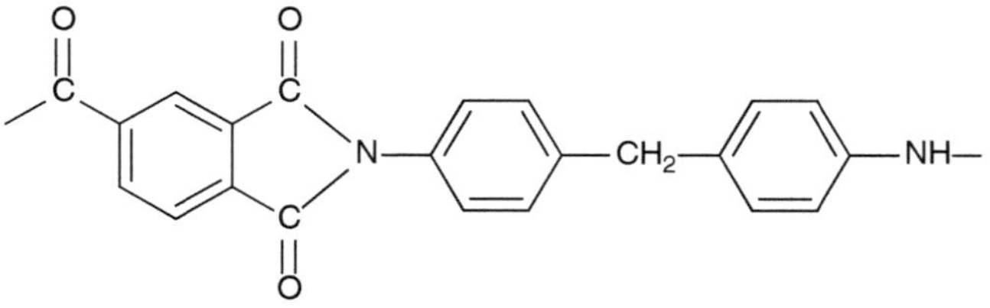

Figure 30.7 Structure of polyamide-imide.

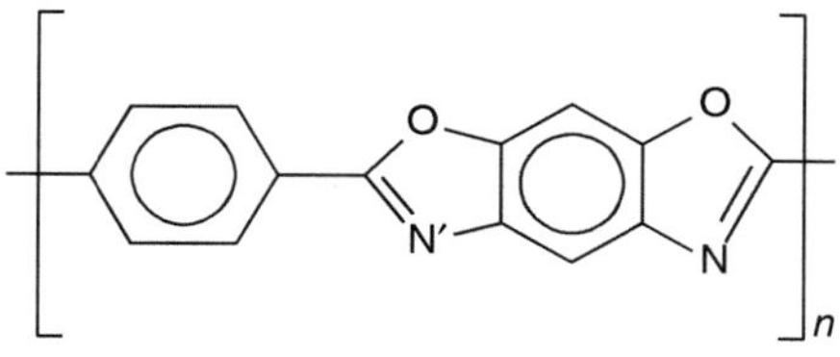

Figure 30.8 Structure of polybenzoxazole fiber.

meta-aramids and *para*-aramids [27]. Its LOI value is around 68%, which is far superior to other flame-retardant fibers. Liu et al. [26] compared the thermal property of three kinds of high-performance Kermel, Kevlar, and PBO. The results demonstrated that PBO fiber has the highest thermal degradation temperature. The initial temperatures of decomposition are, respectively, 556.5 °C (PBO), 464.2 °C (Kevlar), and 411.2 °C (Kermel). The mechanical property retention of PBO when heated is higher for providing a very high thermal stability compared with Kevlar and Kermel.

30.4 Performance Standards and Evaluation Method Development

The protective performance of fabric or clothing can be evaluated based on the laboratory simulation methods. Many related assessment standards and methods have been developed in recent decades. The evaluation of fibrous materials for potential use in thermal protective clothing generally involves two steps. First, the fibrous materials are tested to ensure their ability of flame resistance and thermal stability. Flammability test is carried out to evaluate this performance of protective materials that directly determine if the protective materials are suitable for thermal protection. Second, thermal protective performance of protective clothing should be evaluated for confirming the ability of resisting heat transfer, including conduction, convection, and radiation.

30.4.1 Flammability Test

Flammability is defined as the relative ease of ignition and relative ability to sustain combustion. The burning behavior of fabrics is influenced by a number of factors, such as the nature of the igniting source and time of its impingement, the fabric orientation and point of ignition (e.g. at the edge or face of the fabric or top or bottom), the ambient temperature and relative humidity, and the velocity of the air and fabric structural variables [19]. For flammability evaluation, various standardized test methods have been developed. The commonly evaluation methods include LOI, forty-five degree flammability test, vertical flammability test, and smoke concentration test.

The LOI is widely used to characterize the flammability of fibrous materials. The LOI represents the minimum volume percent of oxygen in a mixture of oxygen and nitrogen that can support flaming combustion of a material at 23 ± 2 °C. LOI is usually measured based on benchtop tester, as shown in Figure 30.9. A small test specimen is supported vertically by specimen holder. A mixture

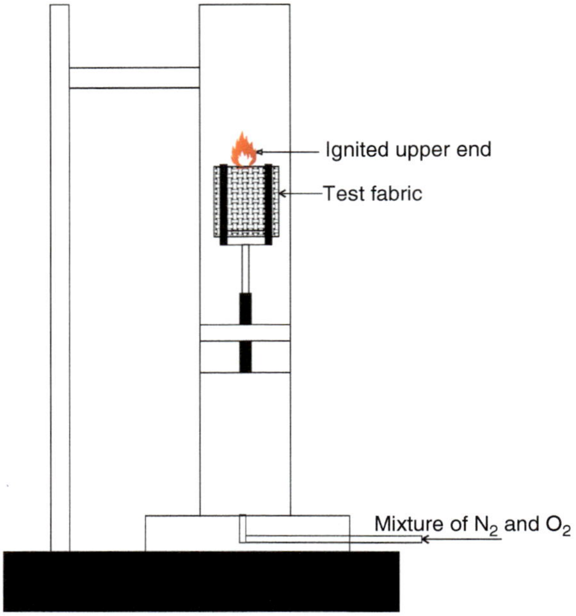

Figure 30.9 Frame design of LOI test apparatus from ASTM D2863.

of oxygen and nitrogen with 23 ± 2 °C is flowed upward through a transparent chimney at a rate 40 ± 2 mm/s. A flame igniter that consists of a tube with an inside diameter of 2 mm is inserted into the chimney to ignite the test specimen. The period for which burning continues and the length of specimen burnt is observed during each burning. By testing a series of specimens in different oxygen concentrations, the LOI is determined by gas measurement and control system. The larger LOI value means that the test specimen is characterized by better flammability and thermal stability. According to the LOI value of fibrous materials, the burning behavior can be divided into two types: fire-retardant fiber (LOI ≧ 26%) and inflammable fiber (LOI < 26%). The LOI value of the commonly fire-resistant fibers is given in Table 30.1.

LOI, while it proves to be a very effective indicator of ease of ignition, has not achieved the status of an official test within the textile field. This is because the sample ignition occurs at the top to give a vertically downward burning geometry, which is considered to be unrepresentative of the ignition geometry in the real situation. Furthermore, the exact LOI value is influenced by fabric structural variables for the same fiber type and is not single-valued for a given fiber type or blend [28]. However, it finds significant use in developing new flame retardants and optimizing levels of application to fibers and textiles.

The flammability of textile fabrics can be evaluated according forty-five degree and vertical flammability test as well (Figure 30.10), specified, respectively, by standard ASTM D1230 and ASTM D3659. According to standard ASTM D1230, a test specimen with 50 by 150 mm is held in the flammability tester at an angle of 45°. A standardized flame is applied to the surface near the lower end for

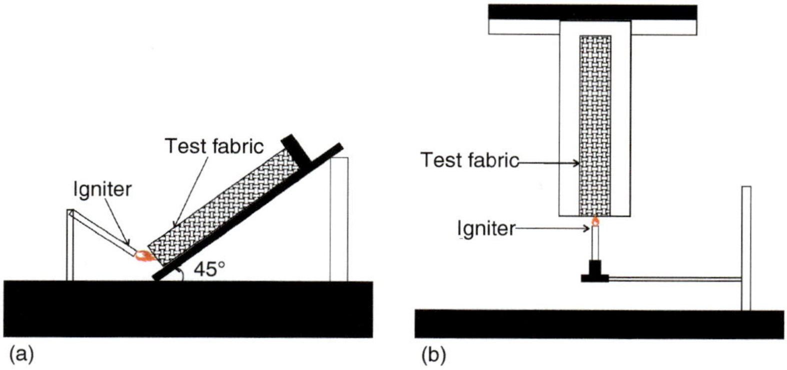

Figure 30.10 (a) Forty-five degree and (b) vertical flammability test apparatuses.

1 ± 0.1 seconds. An automatic electric timer is used to record the required time for the flame to spread up the fabric for a distance of 127 ± 0.2 mm, as an index of assessing the flaming degree of fibrous materials. The subsequent burning behavior of the specimen, such as ignites, chars, melts, or fuses, is observed to judge the ease of ignition and flammability.

Compared with forty-five degree flammability test, vertical flammability test is carried out for assessing the flammability properties of fabrics in a vertical configuration. The bottom edge of test specimen is ignited for three seconds using a small, open flame that can simulate the flaming process of an A-line-type garment on a manikin [20]. The afterflame and afterglow times are measured during the exposure. After the exposure, mass loss and char length of test specimen is calculated and used as an indicator of the material's ability to support combustion. In actual fire scenario, firefighter can be in a flame to perform emergency rescue so that protective clothing can be ignited at any angle. Therefore, the flammability properties of fabrics are extremely important for protecting them from flame exposure. In addition, protective clothing can be used to resist the spread of fire, or firefighter keeps a certain distance with an open fire. In view of the situation, the thermal insulation of protective materials that is the ability to resist convective, conductive, and radiative heat transfer is particularly important. Therefore, thermal protective performance of flame-retardant materials should be further evaluated.

30.4.2 Thermal Protective Performance Evaluation

Thermal protective clothing can resist heat transfer from thermal environment to human body. Thermal protective performance provided by clothing is directly related to worker's safety. Thus, assessment of thermal protective performance has a significant impact on the selection of protective materials. The laboratory simulation methods can be used to evaluate thermal protective performance of fabric or clothing, such as the benchtop tests [29–39] and the full-scaled manikin tests [40, 41], as shown in Table 30.2. The tested specimen is usually exposed to the simulated thermal environmental conditions, such as flame, thermal

Table 30.2 Standards for evaluating thermal protective performance (TPP) of clothing.

Standard	Test apparatus	Heat source	Evaluation index
NFPA 1971-2007 [33]	TPP test device	50/50 radiant/convective heat flux of 84 kW/m^2	TPP
ASTM F2700-2008 [36]	TPP test device	50/50 radiant/convective heat flux of 84 kW/m^2	HTP
ASTM F2703-2008 [38]	TPP test device	50/50 radiant/convective heat flux of 84 kW/m^2	TPE
NFPA 1977 [32]	RPP test device	100% radiant heat flux of 84 kW/m^2	RPP
ASTM F1939-2008 [35]	RPP test device	100% radiant heat flux of 21 and 84 kW/m^2	RHR
ASTM F2702-2008 [37]	RPP test device	100% radiant heat flux of 21 and 84 kW/m^2	RHP
ASTM F2731-2011 [39]	SET test device	100% radiant heat flux of 8.5 ± 0.5 kW/m^2	MET
ASTM F1060-2008 [34]	Hot contact tester	Hot surface with a temperature up to 360 and a pressure of 3 kPa	Skin burn time
ISO 6942-2002 [30]	RPP test device	100% radiant heat flux ranging from 20 to 80 kW/m^2	TF and RHTI
ISO 17492-2003 [31]	TPP test device	50/50 radiant/convective heat flux of 80 kW/m^2	TPI and HTI
ISO 9151-1995 [29]	Flame exposure apparatus	Flash fire of 80 kW/m^2	TF and HTI
ISO 13506-2008 [40]	Instrumented manikin	Flash fire of 84 ± 2.5 kW/m^2	Skin burn degree and burn area
ASTM F1930-2011 [41]	Instrumented manikin	Flash fire of 84 ± 2.5 kW/m^2	Skin burn degree and burn area

radiation, and hot surface. The thermal sensor at the rear of specimen records thermal histories in the process of heat exposure, which can be used to calculate the heat transfer rate through fabric (ISO 6942-2002 and ISO 17492-2003) or the required time to cause skin burn (NFPA 1971, ASTM F1939-2008, and ASTM F1060-2008).

30.4.2.1 Benchtop Testing

In the past decades, most researchers used TPP test device and RPP device to assess thermal protective performance of protective materials that are, respectively, confirmed with standards NFPA 1971 [33] and NFPA 1977 [32] (see Figure 30.11). The Stoll criterion developed by Stoll and Chianta [42] is employed in these test methods that are suitable to characterize performance of materials used in protective clothing in 50/50 radiant/convective heat flux of 84 kW/m^2 and 100% radiant heat flux of 21, 42, and 84 kW/m^2. The TPP and RPP

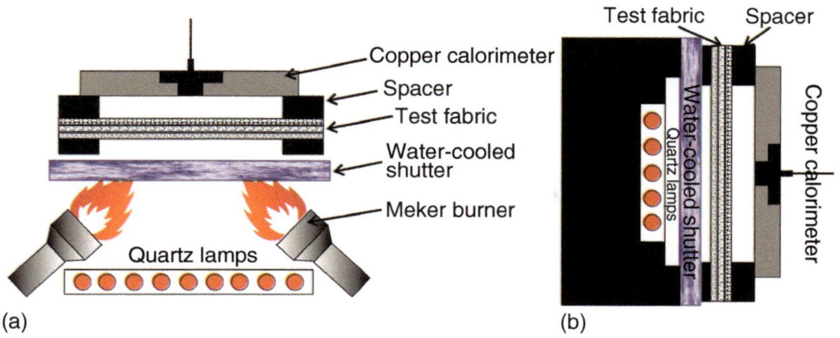

Figure 30.11 Thermal protective performance test apparatus. (a) TPP tester and (b) RPP tester.

values are employed to evaluate thermal and radiant protective performance, which can be obtained in Eq. (30.1):

$$\text{TPP/RPP} = \text{burn time} \times \text{calibrated burner or radiant heat flux value (cal/(cm}^2\text{ s))} \quad (30.1)$$

For meeting the performance requirement of standard NFPA 1971, a TPP rating of at least 35 is required for a fabric assembly used in firefighting protective clothing. It means that firefighter wearing this protective clothing can withstand a flash fire exposure (84 kW/m^2) for 17.5 seconds without causing second-degree burns. The RPP rating of fabric for thermal protection specified by NFPA 1977 is more than 20, which can provide 20 seconds protection time for firefighters under an approximately 42 kW/m^2 radiant exposure.

Although TPP and RPP tests have been widely used for characterize performance of materials, there are still many drawbacks. For instance, they are only suitable to copper disk calorimeter, constant heat flux intensity, and short-time thermal exposure as the measurement accuracy of metal sensors has a decrease tendency with prolonging duration time [10, 43, 44]. Furthermore, the discharge of the stored thermal energy in the fabric system after the exposure is not considered in the conventional TPP and RPP tests. Therefore, some new evaluation indexes have been put forward in order to effectively assess thermal protective performance of the clothing system.

As is described in Table 30.2, TPP, heat transfer performance (HTP), and thermal performance estimate (TPE) are used to evaluate thermal protection for the same heat exposure condition. The differences among the above three indexes lie in calculative methods of the second-degree burn time (see Figure 30.12). These calculative methods are based on Stoll criteria that speculate the relationship between thermal energy and tolerance time to second-degree burns under constant heat exposure level. Regarding to the TPP, it is defined as the relationship between the temperature rising worked out by a conversion equation according to ASTM E457 and second-degree burn time that is given by

$$T = 8.871t^{0.2905} + T_0 \quad (30.2)$$

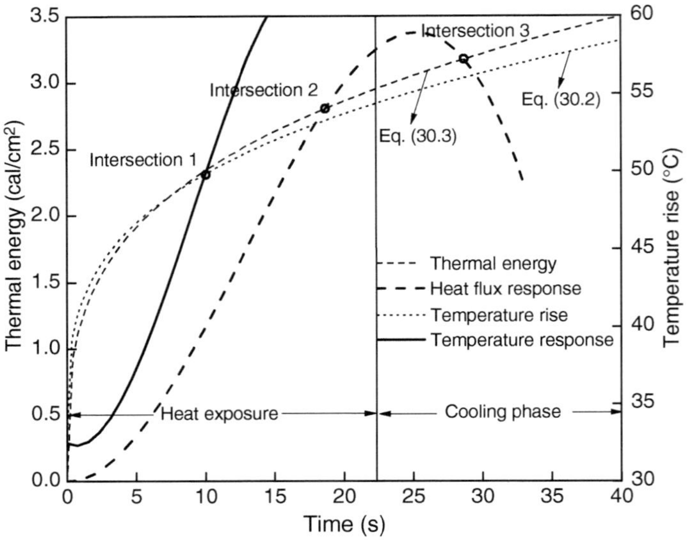

Figure 30.12 Comparison of different evaluation methods based on Stoll criteria.

where T_0 is the initial temperature of sensor. Its purpose is simpler to acquire second-degree burn injuries, but it is merely confined to the TPP sensors because different type sensors have different calculative formulas between the heat flux and the temperature. While the HTP is on the basis of the relationship between thermal energy transferred to human tissue and second-degree burn time, written as

$$q = 1.1991t^{0.2901} \tag{30.3}$$

therefore, the HTP has a larger range of application, which is not only for copper disk calorimeter but also for other sensors, e.g. skin-simulant sensor. But both the above two evaluation indexes are not used to investigate stored energy burns of skin tissues as the exposure duration does not include the cooling process. The TPE can be employed to account for skin burn resulting from the release of stored energy in clothing system during the cooling phase. That is to say, when there is merely an intersection between Eq. (30.2) and heat flux response curve in Figure 30.12, the obtained time is the burn time considering stored thermal energy. Compared with the TPP and the HTP, the TPE can provide the comprehensive skin burn data, including the heat transfer during heat exposure and stored heat transfer during cool down. In addition, a stored thermal energy test apparatus was employed to evaluate the transmitted and stored heat within fabrics under low heat flux conditions [39]. A new index called minimum exposure time (MET) was introduced to evaluate skin burn due to transferred and stored heat.

Other simple methods to assess thermal protection were developed without involving skin burn injuries, such as ISO 6942-2002, ISO 17492-2003, and ISO 9151-1995 (see Table 30.2). Among these standards, heat transmission factor (TF) is the fraction of heat flux transmitted through a specimen exposed to a source of radiant heat, which is numerically equal to the ratio of the transmitted energy to the incident heat flux. In addition, heat transfer index (HTI) is the mean time in seconds to achieve a temperature rise of $12 \pm 0.2\,^{\circ}\mathrm{C}$ (or $24 \pm 0.2\,^{\circ}\mathrm{C}$) when testing by this method with a specified incident heat flux density. Their principles are quite straightforward and suitable for all heat exposures owing to not involving skin heat transfer, but they are only used to rate the performance of different flame-retardant fabrics, not for assessing skin burns.

30.4.2.2 Full-Scaled Manikin Testing

Although these benchtop tests can effectively evaluate the heat transfer behavior of fabrics in various environmental conditions, it is extremely difficult to replace full-scaled manikin tests for evaluating thermal protection of the clothing. This is because manikin test can observe ignition and burning of the garment and heraldry, shrinkage or sagging in all directions after flaming, hole generation, smoke generation, and structural failure of seams. It can also be used to estimate the extent and nature of skin burn that a person would suffer if wearing the test garment under flash fire or radiant heat exposure. Thus, manikin experiments are great important tools to simulate the actual wearing state of clothing and present the heat and moisture transfer behavior in fire environment, protective clothing, and human body.

Over the years, full-scaled manikin test systems have been rapidly established for evaluating thermal protective performance of clothing. Most manikin test systems are employed to simulate a flash fire of $84\,\mathrm{kW/m^2}$ using 8 or 12 propane gas torches, such as PyroMan of North Carolina State University [45], flame manikin Harry Burns of University of Alberta [46], and Thermo-Man of DuPont [47]. These manikins are tested in a standing position in initially quiescent air. Controlled air motion for simulating wind effects or body movement is not currently considered, but it is possible to move the manikin through a stationary flame. These manikins that represent an adult human using a flame-resistant, thermally stable, glass fiber-reinforced vinyl ester resin are equipped with more than 100 thermal energy sensors distributed uniformly within each area on the manikin surface. These sensor types mainly include copper, guarded copper slug, skin-simulant sensor, and embedded sensor, which can record the variation in skin temperature or heat flux over a preset time interval during and after heat exposure. These measurement data can be used to further calculate the time to cause second- or third-degree skin burn in different parts of human body and the overall percentage of burn area based on Henriques burn integral [48], as shown in Figure 30.13.

According to ASTM F1930-2011 and ISO 13506-2008, the test apparatus before dressing the manikin is essential to be calibrated. When the nude manikin is exposed to flames, the average exposure heat flux level is typically equal to $84\,\mathrm{kW/m^2}$ with a standard deviation of 5%. A exposure duration with a range of 3–5 seconds is selected for the exposure of the dressed manikin, while the thermal sensors shall collect data for at least 60 seconds for ensuring the discharge of thermal energy stored in protective clothing.

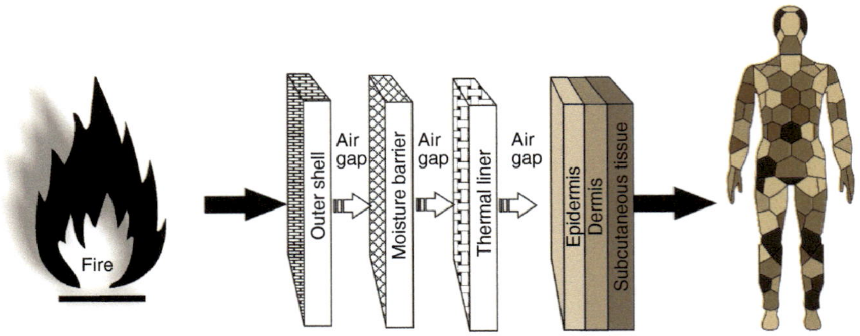

Figure 30.13 Thermal protective performance evaluation based on full-scaled manikin.

Even though the flash fire exposure is intense situation and extremely dangerous, most firefighters characteristically perform their duties at a relatively safe distance from the flame front and only require protection from low to moderate levels of radiant heat. Similar to flame manikin, RadMan test system was developed by North Carolina State University in order to evaluate radiant protective performance in a realistic radiant heat exposure that might be seen in the field for wildland firefighters [49]. As described in Figure 30.14, a radiant heat panel employed in RadMan system can produce a heat flux of 5–21 kW/m^2. The manikin is constructed of a metallic and fiber-layered structure for mimicking human skin. For further considering the effect of human physiology, a water-cooled system is introduced into RadMan skin surface to simulate the heat saturation and blood flow effects. Other than PyroMan, 70 RdF foil sensors are employed in RadMan system. This type of sensor can reduce the measurement errors for long duration exposure. By combining the skin heat transfer model and Henriques burn injury model, second- and third-degree burn times are obtained for evaluating the performance of clothing under a preset radiant heat exposure.

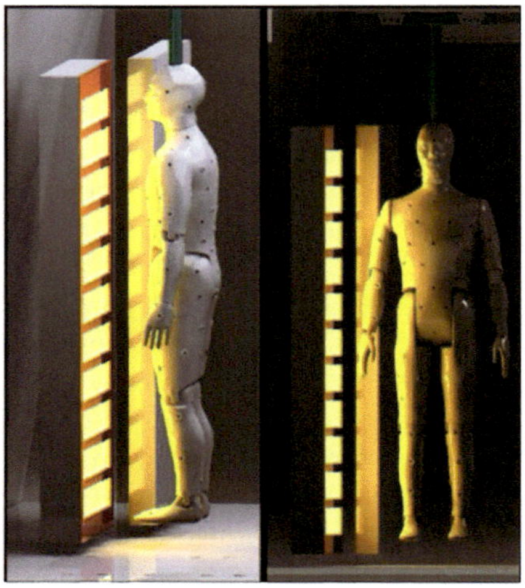

Figure 30.14 Thermal protective performance evaluation based on RadMan. Source: Watson 2014 [49]. Reproduced with permission of NC state university.

30.5 Influencing Factors of Thermal Protective Performance

Thermal protective performance of clothing is affected by fabric basic properties, such as physical properties and thermophysical properties [45]. Due to the specific geometry of human body shape and unique fabric properties, the air gap between the garment and human body is unevenly distributed over the body when the thermal protective clothing is worn on the human body [50]. Previous studies, including experimental investigation and numerical simulation, showed that the thermal performance of protective clothing also greatly depends on the air gap size [51, 52], the air volume [50], and the air gap position entrapped within the multilayer [53]. In addition, some studies indicated that moisture from high-humidity environment and body perspiration has a great influence on thermal protection as well [54]. The main reasons are that moisture can change thermophysical properties, such as thermal conductivity and specific heat capacity, and optical properties [55]. Moreover, moisture absorption/desorption and phase change can absorb or release a large amount of energy to affect heat transfer process in firefighter's clothing system, even resulting in skin steam burn or scald [10].

30.5.1 Effect of Fabric Basic Properties

Fabric physical properties play an important role in thermal protection of clothing, such as fabric structure, weight, thickness, porosity, and density. Many of these parameters are interrelated, and the dependence of protective performance of fabrics on these parameters is complex. These parameters can affect heat transfer rate in protective clothing, including conduction, radiation, and convection [56]. The convective heat transfer is generally ignored, since the porosity of fabric is enough small [57]. Conductive heat transfer in protective clothing mainly depends on thermal conductivity of fabric. Since thermal conductivity of air is so small (0.023 W/(m K) at 300 K) compared with most common insulating materials, the heat transfer rate can reduce if the air content embedded in clothing increases [58]. The radiative heat transfer is mostly determined by fabric's extinction coefficient that depends on fabric thickness and transmissivity [59]. Thus, thickness is the primary factor to determine thermal protective performance in radiant heat exposure and flash fire [60, 61]. A fabric with lower density and high weight can store more steady air to decrease conductive heat transfer [62, 63].

In addition, fabric's heat capacity affects thermal protection of clothing, as it can determine the capacity of storing thermal energy. During the firefighting operation, the thermal energy from the fire ground is transferred to the human skin through firefighting protective clothing, but a large amount of energy is stored in the clothing system [8]. The stored energy can be discharged to the ambient and the human skin in natural or forced ways due to body movement or external pressure after the exposure [64, 65]. It was reported that about 50% of the total energy to cause second-degree skin burn comes from the discharge of the stored thermal energy in the multilayer fabric under flash fire [9]. The stored thermal burn can be reduced by increasing heat capacity or decreasing the discharged thermal energy to skin surface. Also, the optical properties of the fabric have an important influence on the thermal protective performance of firefighters' protective clothing, especially in intense thermal environments where radiation is the key mode of heat transfer in a fabric system with air gaps. Radiative heat transfer in a protective clothing is quite complex, involving absorbing, emitting, scattering, and transmitting [56]. The larger transmissivity and less reflectivity can increase the radiative heat transfer from the fabric system to the ambience and exacerbate skin burn.

30.5.2 Effect of Air Gap Size

The air gap size and volume between the clothing and human body can be obtained by using a portable three-dimensional (3D) body scanner [66]. The 3D images of the nude and clothed manikin can be captured and compared based on post-processing techniques. The results revealed that the air gap below the clothing is unevenly distributed, while the average air gap size in a protect clothing is approximate to 25 mm in real wearing situation [50]. The air gap size is dependent not only on the body shape but also on fabric properties and garment size [67].

Some researchers predicted that there exists an optimum air gap between the clothing and the body, which can provide the best thermal protection. Thus, different air gap sizes were employed to evaluate the thermal protective performance based on benchtop testers, such as TPP tester, RPP tester, and SET tester. It was found that the effect of the air gap size on thermal protection provided by the fabric system changed over the type of the heat source. For the flash fire of an 80 kW/m^2 heat flux, the rising of the air gap size could increase the thermal protective performance of flame-resistant fabrics [51, 52]. When the fabric system was exposed to an 84 kW/m^2 radiant heat flux, the thermal protective performance increased first and then reduced with the air gap size [68]. Furthermore, the effect of the air gap size in low-level thermal radiation (2, 5, 10 kW/m^2) was investigated using a self-designed bench-scale apparatus, showing that the thermal protective performance was greatly influenced by the size of the air gap and the location entrapped in multilayer fabric [53]. It was reported that the optimum air gap size was when the natural convection occurs [69]. Therefore, flow visualization experiments was conducted by Torvi and Dale [69] to investigate the occurrence of natural convection, predicting that the optimum air gap size was about 6–7 mm. The fabric system with air gaps of 9–12 mm in different moisture content seemed to provide maximum thermal protection from flash fire [70]. When the air gap presented different relative humidities, around 12–15 mm was the critical air gap size [71]. Therefore, it is found that the optimum air gap size has no certain value or does not exist, depending on the type of heat source, moisture content within fabric, and relative humidity in air gap. Additionally, most air gaps between the fabric sample and the test sensor were horizontally created in benchtop tests [33, 38]. But the air gap enclosure should be vertical in actual wearing state. There is a significant difference for the occurrence of natural convection under air gap enclosures of different directions due to the location of hot surface and the buoyant force [58]. For example, the natural convection heat transfer in the vertical air gap enclosure should begin when the Rayleigh number is more than 1000, while the convective heat transfer within the air gap occurs when the Ra is more than 1708 [58]. Therefore, the effect of the vertical air gap should be further discussed.

30.5.3 Effect of Moisture Content

It is inevitable to sweat a lot for firefighters in fire ground while they suppress unwanted fires and rescue victims. Also, the firefighting lance is popularly used by firefighters to suppress fire. The presence of moisture from skin sweat or ambience can wet thermal protective clothing. Some studies reported that moisture content and distribution within the fabric system obviously affected thermal protection of fabric under various fire environments. Barker et al. [72] studied the impact of moisture content on the thermal protective performance under radiant heat flux of 6.3 kW/m^2 and found that multilayer fabrics with 15% moisture had the worst thermal protective performance. The study of Mäkinen et al. [73] showed that there was a decrease in thermal protective capacity with the increment of moisture content in radiant heat exposure of 20 kW/m^2. Furthermore, the thermal protective performance of single-layer fabrics that contains 70–80% moisture decreased to 35% under radiant heat exposure of 84 kW/m^2, while moisture had a positive effect on thermal properties in 50/50 radiant/convective heat flux of 84 kW/m^2 [74]. The main reason was that convective heat transfer between environment and clothing could speed up moisture evaporation and took away a section of heat. In contrast, radiant heat transfer had a complicated effect on moisture heat transfer with the variation of radiative intensity [55, 75].

Also, the moisture with different layer fabrics presented different influence tendencies on heat protection [76]. The moisture stored in the outer shell increased the thermal protective performance in low- and high-intensity thermal radiation [77]. But the moisture within the thermal liner, while providing the higher thermal protection in low-level radiation, increased the rate of heat transfer in protective clothing under high-level thermal radiation [76]. It can be attributed to the fact that moisture evaporation rate of outer shell is far larger than that of thermal liner and moisture barrier [77, 78]. Therefore, moisture applied in multilayer fabric system had an complex effect on thermal protection of the clothing, depending on the moisture content and distribution, the type of heat source, and the essential properties of flame-resistant fabric.

30.6 Future Trends

Significant research has been done in the field of thermal protective clothing to provide higher thermal protective, but the development in some smart materials and new technology creates new possible for further improvement of thermal protective clothing. These materials readily interact with human/environmental conditions to produce change in material properties. For example, shape memory materials can change their shapes with an increase of temperature, since the shape memory effect results in an object reverting to a previously held shape when heated [79]. The selection of shape memory materials and its actuation temperature is key problem. Due to the importance of air gap size between different clothing layers, and clothing backside and body for thermal protection, the shape memory materials can be inserted in fabric layers to regulate the air gap size. But the rising of air gap size uncertainly increases thermal protective performance of clothing, as the occurrence of convective heat transfer for larger air gaps can speed up heat transfer from protective clothing to body [59]. Thus, the optimum air gap size should be further studied for different heat exposure conditions, especially for vertical air gap.

The development in intelligent thermal protective clothing has been introduced in past decades. The weaving of wires and sensors into protective clothing can monitor some fire environmental and physiological parameters, such as ambient temperature, skin temperature, heart rate, and heat stress. However, the time to cause skin burn or minimum heat exposure time was not predicted as an alert information for guiding the firefighting operations of firefighter. The monitoring data should be further analyzed for indicating the safety staying time and zone in fire ground conditions based on skin heat transfer model and Henriques burn model. Moreover, some active cooling methods (e.g. evaporative cooling, water-retaining fiber, ice pack, phase change materials, and ventilation cooling vest) should be combined with the intelligent monitoring system. These materials can be embedded in fabric layers to interact with a human body and produce thermoregulated effect by absorbing surplus body heat. But the cooling effect and duration of these methods is the urgent problem to be solved at present.

In addition, the relatively new technology of flame retardant is intumescent treatments. Most flame retardant are normally in the form of thin and low insulation coatings on a lining fabric, but these coatings will instantly form an inert insulative char to protect the wearer when activated by excessive heat or flame [27]. The flame retardant, its application technology, and the product must be environmentally acceptable for reducing inhalation of toxic fumes and gases. Furthermore, thermal hazards that firefighters are usually exposed not only include flash fire and high-intensity thermal radiation but also contact with hot objects, hot liquid splashes, and hot steam [80, 81]. The distinction of heat sources can result in the change of protective highlight. Protection from hot objects is to increase heat insulation of protective materials, while the reduction of hot liquid and steam penetration is a key method for protecting scalds or steam burns. Therefore, further study should develop multifunctional thermal protective clothing to meet multiple functional requirements.

References

1 Ali, A.H. and Mohmmed, R. (2015). A review of the firefighting fabrics for flashover temperature. *International Journal of Engineering Sciences and Research Technology* 4 (3): 247–257.

2 Karter, M.J. and NFP Association (2009). *Patterns of Firefighter Fireground Injuries*. National Fire Protection Association Quincy.

3 Fahy, R.F., LeBlanc, P.R., and Molis, J.L. (2006). *Firefighter Fatalities in the United States-2005*. National Fire Protection Association. Fire Analysis and Research Division.

4 Walton, S.M., Conrad, K.M., Furner, S.E., and Samo, D.G. (2003). Cause, type, and workers' compensation costs of injury to fire fighters. *American Journal of Industrial Medicine* 43 (4): 454–458.

5 Abbott, N.J. and Schulman, S. (1976). Protection from fire: nonflammable clothing — A review. *Fire Technology* 12 (3): 204–218.

6 Foster, J. and Roberts, G. (1995). Measurements of the firefighting environment – summary report. *Fire Engineers Journal* 55: 30–34.

7 Barker, R. (2005). A review of gaps and limitations in test methods for first responder protective clothing and equipment. *Report Presented to National Personal Protection Technology Laboratory*.

8 Guowen, S., Paskaluk, S., Sati, R. et al. (2010). Thermal protective performance of protective clothing used for low radiant heat protection. *Textile Research Journal* 81 (3): 311–323.

9 Song, G., Gholamreza, F., and Cao, W. (2011). Analyzing thermal stored energy and effect on protective performance. *Textile Research Journal* 81 (11): 1124–1138.

10 Barker, R.L. (2005). *A Review of Gaps and Limitations in Test Methods for First Responder Protective Clothing and Equipment*, 1–98. National Institute for Occupational Safety and Health.

11 Kahn, S.A., Patel, J.H., Lentz, C.W., and Bell, D.E. (2012). Firefighter burn injuries: predictable patterns influenced by turnout gear. *Journal of Burn Care and Research* 33 (1): 152–156.

12 Aldridge, D. (1997). Firefighter garment with combination facecloth and moisture barrier. Google Patents.

13 Gohlke, D.J. and Tanner, J.C. (1976). Gore-Tex® waterproof breathable laminates. *Journal of Industrial Textiles* 6 (1): 28–38.

14 Reischl, U. and Stransky, A. (1980). Comparative assessment of Goretex™ and Neoprene™ vapor barriers in a firefighter turn-out coat. *Textile Research Journal* 50 (11): 643–647.

15 Horrocks, A., Tune, M., and Cegielka, L. (1988). The burning behaviour of textiles and its assessment by oxygen-index methods. *Textile Progress* 18 (1–3): 1–186.

16 Tsukada, M., Khan, M.M.R., Tanaka, T., and Morikawa, H. (2011). Thermal characteristics and physical properties of silk fabrics grafted with phosphorous flame retardant agents. *Textile Research Journal* 81 (15): 1541–1548.

17 Horrocks, A. (1983). An introduction to the burning behaviour of cellulosic fibres. *Journal of the Society of Dyers and Colourists* 99 (7–8): 191–197.

18 Banks, M., Ebdon, J.R., and Johnson, M. (1993). Influence of covalently bound phosphorus-containing groups on the flammability of poly(vinyl alcohol), poly(ethylene-*co*-vinyl alcohol) and low-density polyethylene. *Polymer* 34 (21): 4547–4556.

19 Horrocks, A.R. and Price, D. (2001). *Fire Retardant Materials*. Woodhead Publishing.

20 Song, G. (2016). *Thermal Protective Clothing for Firefighters*. Woodhead Publishing.
21 Behnke, W.P., Chapin, R.S., and Fierro, J.F. (1978). Poly(*m*-phenylene isophthalamide) and poly(*p*-phenylene terephthalamide) fibers. Google Patents.
22 Wang, C.C. and Chen, C.C. (2005). Some physical properties of various amine-pretreated Nomex Aramid yarns. *Journal of Applied Polymer Science* 96 (1): 70–76.
23 Jackson, R.H. (1978). PBI fiber and fabric—properties and performance. *Textile Research Journal* 48 (6): 314–319.
24 Coffin, D.R., Serad, G.A., Hicks, H.L., and Montgomery, R.T. (1982). Properties and applications of celanese PBI—polybenzimidazole fiber. *Textile Research Journal* 52 (7): 466–472.
25 Chung, T.S., Glick, M., and Powers, E.J. (1993). Polybenzimidazole and polysulfone blends. *Polymer Engineering and Science* 33 (16): 1042–1048.
26 Liu, X.-Y., Tao, Y.-K., Wang, X.-L., and Yu, W.-D. (2009). The thermal property of high performance fiber. *Journal of Xi'an Polytechnic University* 23 (2): 267–271.
27 Shishoo, R. (2002). Recent developments in materials for use in protective clothing. *International Journal of Clothing Science and Technology* 14 (3/4): 201–215.
28 Hendrix, J., Drake, G., and Reeves, W. (1972). Effects of fabric weight and construction on OI values for cotton cellulose. *Journal of Fire and Flammability*: 338–345.
29 ISO/TC 94 ISO 9151 (1995). *I. Protective Clothing Against Heat and Flame Determination of Heat Transmission on Exposure to Flame*. Geneva, Switzerland: ISO, 2004.
30 ISO/TC 94 ISO 6942 (2002). *I. Protective Clothing-Protection Against Heat and Fire Method of Test: Evaluation of Materials and Material Assemblies When Exposed to a Source of Radiant Heat*. Geneva, Switzerland: ISO, 2004.
31 ISO/TC 94/SC13 ISO 17492 (2003). *I. Clothing for Protection Against Heat and Flame Determination of Heat Transmission on Exposure to Both Flame and Radiant Heat*. Geneva, Switzerland: ISO, 2004.
32 NFP Association (1998). *Standard on Protective Clothing and Equipment for Wildland Fire Fighting*. National Fire Protection Association.
33 NFP Association (2006). *Standard on Protective Ensembles for Structural Fire Fighting and Proximity Fire Fighting*. National Fire Protection Association.
34 ASTM F1060-08 A (2008). *Standard Test Method for Thermal Protective Performance of Materials for Protective Clothing for Hot Surface Contact*. West Conshohocken, PA: ASTM International.
35 ASTM F1939-15 A (2015). *Standard Test Method for Radiant Heat Resistance of Flame Resistant Clothing Materials with Continuous Heating*. West Conshohocken, PA: ASTM International.
36 ASTM F2700-13 A (2013). *Standard Test Method for Unsteady-State Heat Transfer Evaluation of Flame Resistant Materials for Clothing with Continuous Heating*. West Conshohocken, PA: ASTM International.

37 ASTM F2702-15 A (2013). *Standard Test Method for Radiant Heat Performance of Flame Resistant Clothing Materials with Burn Injury Prediction*. West Conshohocken, PA: ASTM International.

38 ASTM F2703-13 A (2013). *Standard Test Method for Unsteady-State Heat Transfer Evaluation of Flame Resistant Materials for Clothing with Burn Injury Prediction*. West Conshohocken, PA: ASTM International.

39 ASTM F2731-11 A (2011). *Standard Test Method for Measuring the Transmitted and Stored Energy of Firefighter Protective Clothing Systems*. West Conshohocken, PA: ASTM International.

40 ISO/TC 94 ISO/DIS 13506 (2004). *I. Protective Clothing Against Heat and Flame Test Method for Complete Garments Prediction of Burn Injury Using an Instrumented Manikin*. Geneva, Switzerland: ISO.

41 ASTM F1930-15 A (2015). *Standard Test Method for Evaluation of Flame Resistant Clothing for Protection Against Fire Simulations Using an Instrumented Manikin*. West Conshohocken, PA: ASTM International.

42 Stoll, A.M. and Chianta, M.A. (1968). A method and rating system for evaluation of thermal protection. DTIC Document.

43 DeMars, K., Henleson, W., and Lies, M. (2000). Thermal measurements for fire fighters' protective clothing. In: *Thermal Measurements: The Foundation of Fire Standards*, ASTM STP1427. West Conshohocken, PA: American Society for Testing and Materials.

44 Grimes, R., Mulligan, J., Hamouda, H. et al. (1996). The design of a surface heat flux transducer for use in fabric thermal protection testing. *ASTM Special Technical Publication* (1237): 607–624.

45 Song, G. (2003). Modeling thermal protection outfits for fire exposures. Doctoral Thesis. North Carolina State University.

46 Sipe, J.E. (2004). *Development of An Instrumented Dynamic Mannequin Test to Rate the Thermal Protection Provided by Protective Clothing*. Worcester Polytechnic Institute.

47 Behnke, W.P., Geshury, A.J., and Barker, R.L. Thermo-Man® and thermo-leg: large scale test methods for evaluating thermal protective performance. In: *Performance of Protective Clothing*, vol. 4, 1992. ASTM International.

48 Henriques, F. Jr. (1947). Studies of thermal injury; the predictability and the significance of thermally induced rate processes leading to irreversible epidermal injury. *Archives of Pathology* 43 (5): 489–502.

49 Watson, K.L. (2014). From radiant protective performance to RadManTM: the role of clothing materials in protecting against radiant heat exposures in wildland forest fires. Master Thesis. North Carolina State University.

50 Song, G. (2007). Clothing air gap layers and thermal protective performance in single layer garment. *Journal of Industrial Textiles* 36 (3): 193–205.

51 Torvi, D.A., Dale, J.D., and Faulkner, B. (1999). Influence of air gaps on bench-top test results of flame resistant fabrics. *Journal of Fire Protection Engineering* 10 (1): 1–12.

52 Sawcyn, C.M.J. (2003). Heat transfer model of horizontal air gaps in bench top testing of thermal protective fabrics. Master Thesis. University of Saskatchewan Saskatoon.

53 Fu, M., Weng, W., and Yuan, H. (2014). Effects of multiple air gaps on the thermal performance of firefighter protective clothing under low-level heat exposure. *Textile Research Journal* 84 (9): 968–978.
54 Su, Y. and Li, J. (2016). Analyzing steam transfer though various flame-retardant fabric assemblies in radiant heat exposure. *Journal of Industrial Textiles* https://doi.org/10.1177/1528083716674907.
55 Fu, M., Yuan, M.Q., and Weng, W.G. (2015). Modeling of heat and moisture transfer within firefighter protective clothing with the moisture absorption of thermal radiation. *International Journal of Thermal Sciences*: 96201–96210.
56 Su, Y., He, J., and Li, J. (2016). An improved model to analyze radiative heat transfer in flame-resistant fabrics exposed to low-level radiation. *Textile Research Journal* https://doi.org/10.1177/0040517516660892.
57 Fan, J.T., Luo, Z.X., and Li, Y. (2001). Heat and moisture transfer with sorption and condensation in porous clothing assemblies and numerical simulation (vol 43, pg 2989, 2000). *International Journal of Heat and Mass Transfer* 44 (5): 1079–1079.
58 Cengel, Y.A. (2003). *Heat Transfer: A Practical Approach*. McGraw-Hill.
59 Torvi, D.A. (1997). Heat transfer in thin fibrous materials under high heat flux conditions. Doctoral Thesis. University of Alberta.
60 Sun, G., Yoo, H., Zhang, X., and Pan, N. (2000). Radiant protective and transport properties of fabrics used by wildland firefighters. *Textile Research Journal* 70 (7): 567–573.
61 Day, M. and Sturgeon, P. (1987). Thermal radiative protection of fire fighters' protective clothing. *Fire Technology* 23 (1): 49–59.
62 Shalev, I. and Barker, R.L. (1984). Protective fabrics: a comparison of laboratory methods for evaluating thermal protective performance in convective/radiant exposures. *Textile Research Journal* 54 (10): 648–654.
63 Song, G., Paskaluk, S., Sati, R., and Crown, E.M. (2010). Thermal protective performance of protective clothing used for low radiant heat protection. *Textile Research Journal* https://doi.org/10.1177/0040517510380108.
64 Su, Y., He, J., and Li, J. (2016). Modeling the transmitted and stored energy in multilayer protective clothing under low-level radiant exposure. *Applied Thermal Engineering*: 931295–931303.
65 Su, Y., He, J., and Li, J. (2016). A model of heat transfer in firefighting protective clothing during compression after radiant heat exposure. *Journal of Industrial Textiles* https://doi.org/10.1177/1528083716644289.
66 Wang, M., Li, X., and Li, J. (2015). A new approach to quantify the thermal shrinkage of fire protective clothing after flash fire exposure. *Textile Research Journal* 86 (6): 580–592.
67 Lu, Y., Song, G., and Li, J. (2014). A novel approach for fit analysis of thermal protective clothing using three-dimensional body scanning. *Applied Ergonomics* 45 (6): 1439–1446.
68 Zhu, F.L. and Zhang, W.Y. (2006). Evaluation of thermal performance of flame-resistant fabrics considering thermal wave influence in human skin model. *Journal of Fire Sciences* 24 (6): 465–485.
69 Torvi, D.A. and Dale, J.D. (1999). Heat transfer in thin fibrous materials under high heat flux. *Fire Technology* 35 (3): 210–231.

70 Lu, Y., Li, J., Li, X., and Song, G. (2013). The effect of air gaps in moist protective clothing on protection from heat and flame. *Journal of Fire Sciences* 31 (2): 99–111.

71 Li, J., Lu, Y., and Li, X. (2012). Effect of relative humidity coupled with air gap on heat transfer of flame-resistant fabrics exposed to flash fires. *Textile Research Journal* 82 (12): 1235–1243.

72 Barker, R.L., Guerth-Schacher, C., Grimes, R.V., and Hamouda, H. (2006). Effects of moisture on the thermal protective performance of firefighter protective clothing in low-level radiant heat exposures. *Textile Research Journal* 76 (1): 27–31.

73 Mäkinen, H., Smolander, J., and Vuorinen, H. (1988). Simulation of the effect of moisture content in underwear and on the skin surface on steam burns of fire fighters. In: *Performance of Protective Clothing: Second Symposium, ASTM STP*, 415–421.

74 Lee, Y.M. and Barker, R.L. (1986). Effect of moisture on the thermal protective performance of heat-resistant fabrics. *Journal of Fire Sciences* 4 (5): 315–331.

75 Nazaré, S. and Madrzykowski, D. (2014). A Review of Test Methods for Determining Protective Capabilities of Fire Fighter Protective Clothing from Steam. NIST TN–1861, Gaithersburg, MD.

76 Keiser, C. and Rossi, R.M. (2008). Temperature analysis for the prediction of steam formation and transfer in multilayer thermal protective clothing at low level thermal radiation. *Textile Research Journal* 78 (11): 1025–1035.

77 Lawson, L.K., Crown, E.M., Ackerman, M.Y., and Dale, J.D. (2004). Moisture effects in heat transfer through clothing systems for wildland firefighters. *International Journal of Occupational Safety and Ergonomics* 10 (3): 227–238.

78 Rossi, R.M. and Zimmerli, T. (1996). *Influence of Humidity on the Radiant, Convective and Contact Heat Transmission Through Protective Clothing Materials*, ASTM Special Technical Publication 1237, 269–280. ASTM International.

79 Otsuka, K. and Wayman, C.M. (1999). *Shape Memory Materials*. Cambridge University Press.

80 Lu, Y., Song, G., Zeng, H., and Zhang, L. (2013). Characterizing factors affecting the hot liquid penetration performance of fabrics for protective clothing. *Textile Research Journal* 84 (2): 174–186.

81 Su, Y. and Li, J. (2016). Development of a test device to characterize thermal protective performance of fabrics against hot steam and thermal radiation. *Measurement Science and Technology* 27 (12): 1–9.

31

Comfort Management of Fibrous Materials

Chengjiao Zhang[1] and Faming Wang[2]

[1] Soochow University, Department of Fashion Design and Engineering, Laboratory for Clothing Physiology and Ergonomics, Suzhou, China
[2] Department of Architecture, School of Architecture and Art, Central South University, Changsha, Hunan, China

31.1 Introduction

Human beings are unique animals, and one of the most distinct differences between human beings and other species is the ability to bioengineer his thermal environment correspondingly to changes in the surrounding environment via adjusting clothing. The clothing, air temperature, radiant temperature, humidity, air movement, and metabolic heat have been regarded as the six fundamental factors that determine human thermal balance [1, 2].

This chapter focuses on human thermoregulation and shows the importance of clothing materials in maintaining the body thermal comfort. The interactions between the human body and the clothing will be discussed with a special emphasis on the materials and factors that could affect the comfort properties of the garment. Besides, the chapter also discusses the assessment techniques for comfort evaluation and the trends of clothing and materials for comfort management.

31.2 Human Thermal Regulation and Heat Transfer Mechanisms

Human beings are homoeothermic and needed to maintain body temperature by thermoregulatory and/or behavioral adjustments to adapt to changing environments. For a person at thermoneutral state, the internal temperature maintains at around 37.0 °C and indicates that a dynamic heat balance is built up between the body and its thermal environment. That is, numerically, the heat produced inside the body must be balanced by heat losses to the environments. The heat input source varies depending on the ambient temperature; the heat source within the body is generally the heat generated by metabolism in normal climatic conditions (see Figure 31.1). However, under extreme conditions, when the ambient temperature is higher than skin temperature, the heat input is combined by heat transfer

Handbook of Fibrous Materials, First Edition. Edited by Jinlian Hu, Bipin Kumar, and Jing Lu.

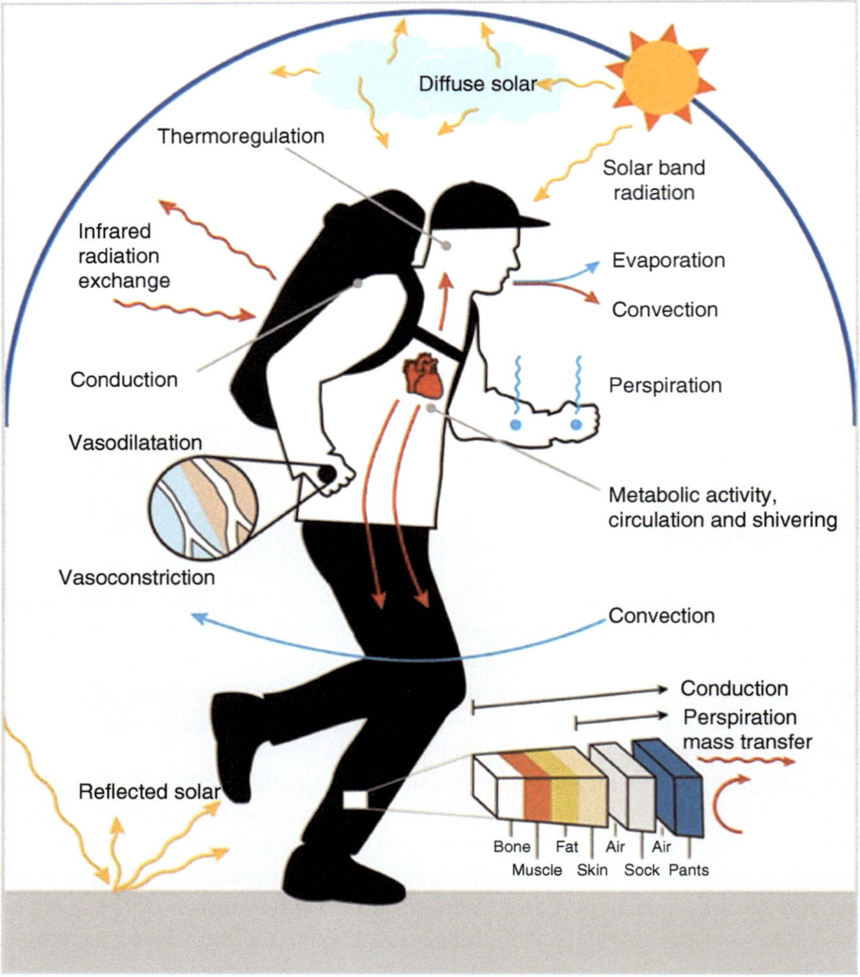

Figure 31.1 Human thermoregulation and heat transfer mechanisms.

from the surrounding environment and heat generated by metabolism within the body [2].

Heat transfers away from the body to the surrounding environment via several ways [3]:

- Dry heat transfer via conduction, convection, and radiation. Conductive heat transfer occurs between solid surfaces contacting with each other; convective heat transfer occurs between a solid surface and its surrounding fluid, e.g. air or water; radiation indicates heat transfer via emission or absorption of electromagnetic waves.
- Sweat evaporation.
- Heat transfer via respiration.

The body temperature would increase if the heat input sources were numerically greater than heat outputs, and it would decrease correspondingly if there was more heat drawn away from the body than the heat input. Ideally, the heat generation and heat loss should be equal. Thus a heat balance equation (in W or W/m^2) for the human body can be expressed based on this principle:

$$M \pm W = S \pm R \pm C \pm K - E \tag{31.1}$$

where M is the metabolic heat production, W is the energy consumed/absorbed by mechanical work, E is the heat transfer via evaporation, R is heat transfer by radiation, C is heat transfer via convection, K is heat transfer via conduction, and S is heat storage.

The heat balance equation for the human body can be expressed in many forms. However, all equations have the same underlying concept and involve three key elements; those for heat generation in the body, heat transfer, and heat storage (see Figure 31.2). The metabolic rate produces energy that enables a human to do mechanical work, and the resting energy is transferred into heat and mainly released into the environment via dry heat loss (i.e. conduction, convection, and radiation), sweat evaporation, and respiration. The heat storage S reflects the thermal debt situation of the body. When the body is in a thermally neutral state, S is zero. S becomes positive when the heat loss is less than the heat production, which causes rise in body temperature. On the other hand, the S will be negative when the heat loss is greater than the heat production, which leads to fall in body temperature.

Goldman [4] described that the heat production is about 1 met (defined as $58.15\ W/m^2$) for an average adult man (who has a body surface area of $1.8\ m^2$) at rest. Usually, about 12% of the resting heat production is drawn away by respiration; another 12% is taken away via evaporation on the skin surface. The remainder (about 76%) of the resting metabolic heat production is transferred from the body by convection and radiation in a comfortable environment; the relative proportion of convective to radiation losses is affected by the ambient air movement. With minimal clothing in still air, about 60% of the resting heat production is lost by radiation, and air movement can facilitate extra convective heat loss.

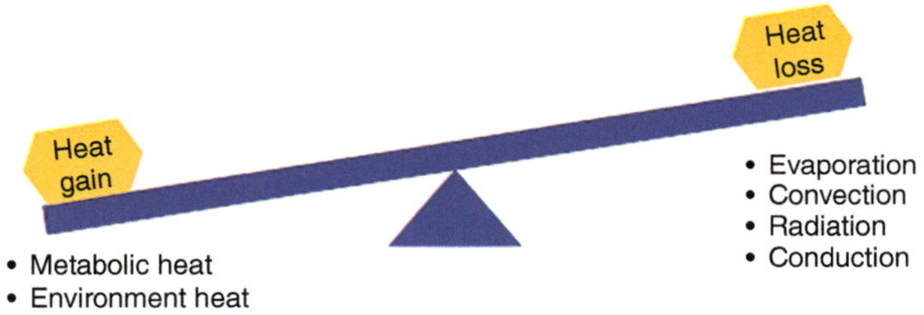

Figure 31.2 A schematic chart on human thermal balance.

31.3 Clothing Comfort

Clothing acts as both a barrier and a transporter between the human body and its surrounding environment. On one hand the clothing can protect the wearer from environmental hazards such as cold and heat. Besides, the clothing could also manage the heat and moisture balance the body needs. The interaction between the human body and clothing system involves four basic elements: physical processes between the clothing and surrounding environment, physiological processes within the body, and neurophysiological and psychological processes [5]. The physical process between the body, clothing, and environment involves heat and moisture transfer, which affects the level of thermal comfort for the human body [6].

Comfort indicates a complex state of mind that is affected by many physical, physiological, and psychological factors. Four different types of comfort may be classified from documented papers: thermal comfort, sensorial comfort, garment fit, and psychological comfort (e.g. esthetics [3]). As defined in ASHRAE 55 [7] and ISO 7730 [8], thermal comfort is indicated as the condition of mind that expresses satisfaction with the thermal environment. Comfort feeling is a psychological status, not directly related to physical environment or physiological state. Thermal comfort is affected by the six basic factors, among the six basic factors. Among the six basic factors, the air temperature, mean radiant temperature, air movement, and humidity represent the four key environmental factors that affect thermal comfort. These four parameters could directly affect the heat transfer between the body and its surrounding environment. The rate of convective heat transfer is a linear function of the temperature gradient between skin temperature and the ambient air temperature. The metabolic heat production is another major parameter defining thermal comfort because the individual metabolic heat production is the essential element of heat balance. The sixth parameter affecting thermal exchange between the human body and its environment is clothing, which acts as a thermal and moisture barrier between the body and the surrounding environment [4].

Many efforts have been made to develop thermal indexes to assess thermal comfort of environments. In the 1920s, researchers on the behalf of the American Society of Heating and Ventilating Engineers (ASHVE) proposed the effective temperature (ET) index based on a comprehensive series of studies [9, 10]. The ET index proposed the concept of standard environment as the comfort index value. ET is still used nowadays but is not generally recommended to use, although it has been found useful for assessing hot environments. Afterward, the resultant temperature [11] and the equivalent temperature [12, 13] were proposed to determine human thermal index.

In 1970, Fanger published the book "Thermal Comfort," and it was regarded as the most significant landmark in thermal comfort research and practice. The book outlines the conditions essential for thermal comfort and also describes methods and principles for thermal environment evaluation and analysis with respect to thermal comfort. The methods that he proposed in the book are now the most influential and widely used throughout the world. In his book, he recognized that it is the combined thermal effect of all (six basic parameters) physical factors that determines human thermal comfort and three criteria should be fulfilled for human to reach the thermal comfort status: (i) the body is in heat balance; (ii) the sweat rate is within comfort limits; (iii) the mean skin temperature is within comfort limits [1].

31.4 Heat and Moisture Transfer Through Textiles

31.4.1 Heat Transfer Through Textiles

Heat transfer indicates that the energy transfers between two mediums because of the temperature gradient, e.g. the heat transfers from the medium with high-temperature to medium with lower temperature. Heat transfer will continue until the two media reach the same temperature, i.e. dynamic heat exchange equilibrium is reached. The amount of the transferred heat depends on temperature gradient and the thermal resistance between the two media. Heat can be transferred in fabrics by conduction through air and fibers, convection of the air within the fabric structure, and radiation from fiber to fiber.

Conductive heat transfer occurs at the interface between two physical contacting objects. The heat flow is depended by the temperature gradient, the larger the temperature gradient, the faster the heat flow between the two surfaces. In clothing system, conductive heat transfers between two contact fabrics or between fabrics and the human skin.

Convective heat transfer is caused by the molecular motion of a fluid or the air. The air density decreases after the heat absorption when it contacts to a high-temperature surface. The difference in air density induced by the temperature gradient forces warmed air to rise, and thus natural convective heat transfer occurs. Moreover, air movement could significantly impact convection and accelerate heat exchange and forced convective heat transfer. In clothing, convective heat flow mainly depends on the openness of the fabric and is not affected by the fabric thickness.

Radiative heat exchange indicates heat transferred by electromagnetic waves between objects with temperature gradient through space. The radiative heat transfer in textile materials could be significantly affected by the material density, construe, and finishing process of the fabrics [6].

The heat transfer in fabrics becomes much more complicated and is normally considered as coupled heat and moisture transfer when the presence of moisture or sweating [14]. Generally, the total heat transfer consists of a dry heat transfer component, moisture, and a liquid transfer component: dry heat loss includes conduction, radiation, and convection; the second form indicates insensible moisture diffusion; and the last form involves the diffusion of liquid water (e.g. sweat).

31.4.2 Moisture Transfer Through Textiles

Moisture exists in both liquid and vapor forms. The transfer of moisture through textiles is a complicated process, which could largely affect thermal comfort property of the fibrous materials. Textile products should exhibit an enough good moisture transfer ability to provide wearers thermal comfort. Water vapor transfer takes place by diffusion and forced convection. Vapor diffusion indicates the process that the water molecules migrate through the textile substance, driven by either the concentration gradient or by forced convection. Water vapor always moves from high vapor concentration regions to low vapor concentration areas. Forced convection also enhances water vapor transmission.

Moisture vapor diffusion is depended by the structure of the yarn and the fabric, as well as the interstices that are formed between fibers within yarns. Generally, fabrics with larger interstices normally allow rapid diffusion of water. In addition, vapor transfer is also affected by moisture adsorption, desorption between fibers and the surrounding air (see Figure 31.3). Adsorption indicates water molecules that are attracted to fabric surface, and it has a linear relationship with the fiber surface area within a fabric, e.g. a larger surface area promises larger amount of water adsorbed. Desorption is the process of molecular diffuses through the material through moisture adsorption and absorption. The heat absorption and release through moisture absorption or desorption further complicates the moisture transfer process within textile materials.

To characterize the vapor diffusion property, numerous methods have been proposed, and several international standards have been developed. The method proposed in the standards could be divided into two categories according to their principles, namely, the gravimetric method [15–17] and the hot plate method [18, 19]. For gravimetric methods, the water vapor transmission ability are obtained by measuring the mass transferred within in a certain time (usually 24 hours) and thus are usually time-consuming. For the hot plate method, the water vapor transmission ability is evaluated by measuring the water vapor resistance of a fabric.

Liquid water transfers through textiles via wetting and wicking. Wetting process indicates the displacement of the fiber–air interface with a fiber–liquid interface. Wicking is the process that liquid transfers into fiber assemblies via capillary effect. During wetting process, the liquid wets the fiber surface before transferring in and through the pores at the amorphous regions of fibers. The spreading of the liquid on the fiber surface depends on the surface energy

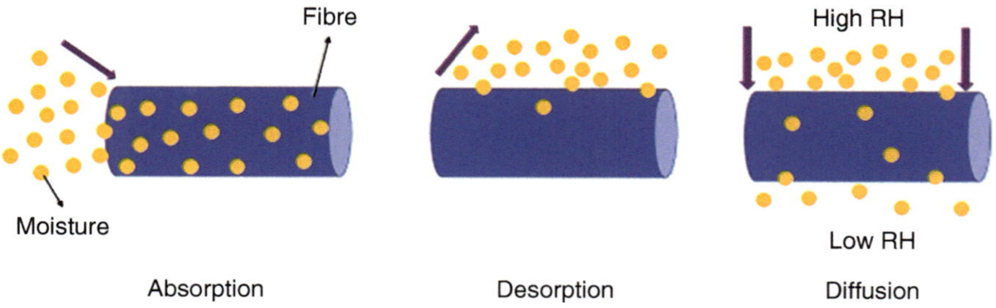

Figure 31.3 Absorption, diffusion, and desorption.

of the material and the liquid. The fabric will have a higher wettability if the surface energy of the fabric could overcome the free surface energy of the liquid. Otherwise, the fabric will have a lower wettability [20–22].

Wicking only occurs under the criteria that the fibers are wetted and there are capillary spaces between them, and it is drawn by capillary forces, and the wicking ability indicates the ability to maintain liquid flow in the capillaries. Depending on the direction of water flow, water flows in the plane of a fabric is always called "transverse wicking," "in-plane wicking," or "horizontal wicking." On the other hand, liquid flow perpendicular to the plane of the fabric is usually referred to as "vertical wicking" [23]. Both wetting and wicking behaviors are mainly determined by the surface tension of the fabric and the liquid. In addition, the wettability and wicking effect are also affected by the hydrophobicity of the fiber material [24]. Fabrics made of fibers with different hydrophobicity allow water molecules to penetrate into different fiber regions and thus affects the moisture regain. Fiber moisture regain affects wicking performance. Fibers with lower moisture regain tend to increase the wicking performance. Polyester fabric with a very low moisture regain can be developed to create a high wicking fabric. Besides, the capillary effect is also significantly affected by cross-sectional shapes of the fiber. Zhang et al. [25] compared the capillary effect of six types of fibers with different cross-sectional shapes and found that the best wicking effect was obtained from the concave cross-sectional shape. Currently, numerous methods have been proposed to characterize the wettability and wicking property of fabrics, and a number of standards are also available. For instance, the AATCC TM 197 [26] standard for testing fabric vertical wicking and AATCC TM 198 [27] for horizontal wicking, DIN 53924 [28] for the wicking height, the AATCC TM 39 [29], and ISO 4920 [30] standards are used to evaluate the fabric wettability.

31.5 Assessment of Materials and Clothing

With the advancement of textile technology, a shift of attention from using natural materials (cotton, fur) to man-made fiber or combinations of both has taken place. In clothing development, it is impossible to test all possible combinations because of the wide spectrum of material combinations at present. It requires a systematic approach in clothing development, and a five-stage evaluation system advocated by Goldman [31] and Umbach [32] provides a solution to the problem and has been widely used in the clothing research community. It is schematically presented in Figure 31.4. Prior to the evaluation process, a detailed analysis is necessary to be conducted on the requirements of the task of the intended clothing user. Then, a selection of fabrics and materials should be carried out, which are tested for a number of physical properties (Stage 1). During Stage 2, a prediction for the overall clothing characteristics can be made using the fabric characteristics, which can be put into the now-readily available thermal models that could analyze the heat exchange for a hypothetical user (e.g. such as the Required Insulation Index IREQ [33]), or actually combine this heat exchange approach with a physiological human thermoregulatory model, which allows a

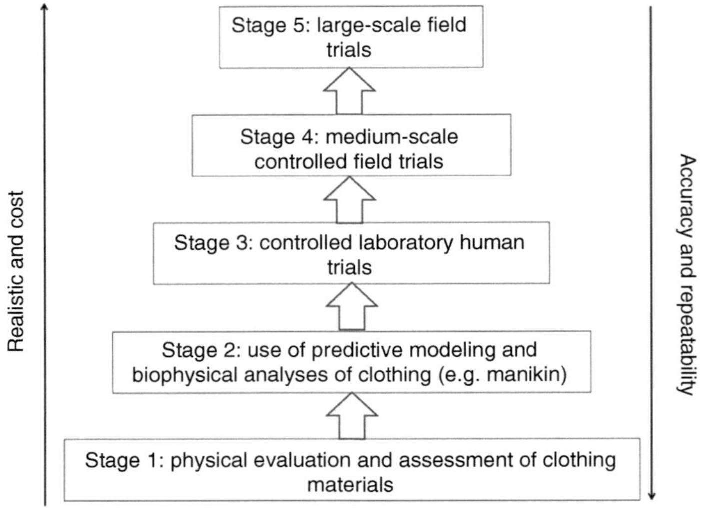

Figure 31.4 Five stages in the development and assessment of clothing systems proposed by Goldman [31] and Umbach [32].

prediction of the body's thermophysiological responses and the associated risks to the wearers' health or comfort (Stage 2) [34, 35].

In Stage 3, clothing wearing test will be conducted under defined conditions in a climatic chamber, and detailed data of the wearer will be collected. The outcome of this testing may help narrow down the material or fabric choices, which can then be evaluated in small- or large-scale field trials. Time and cost required in these two stages will increase over previous stages. Nevertheless, crucial information will be obtained in these two stages.

31.5.1 Material and Fabric Testing

For comparison of different fabrics, several tests on heat and vapor resistance are available, and many are now defined in international standards. A guarded hot plate is extensively used for heat resistance measurement of fabrics, which measures the amount of heat lost through a sample at a certain temperature gradient between the plate and the environment. From this, thermal resistance of the fabric can be calculated as

$$R_{\text{ct}} = \frac{T_{\text{plate}} - T_a}{H_{\text{dry}}} - R_0 \tag{31.2}$$

where R_{ct} is the heat resistance of the fabric sample ((m^2 K)/W), T_{plate} is the mean hot plate surface temperature (°C), T_a is the ambient temperature (°C), H_{dry} is the dry heat loss per square meter of plate area (W/m^2), and R_0·is the heat resistance measured without a sample present ((m^2 K)/W).

Different equipment forms are in use for the fabric thermal insulation test and are presented in ISO 11092 [19], CAN/CGSB 4.2 No. 78.1-2001 [36], ASTM

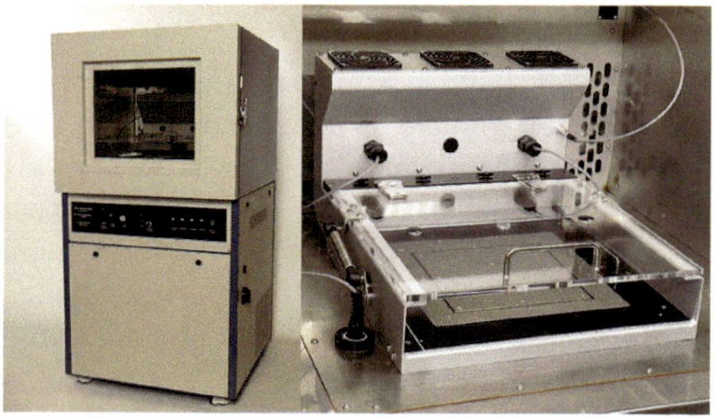

Figure 31.5 The sweating guarded hotplate housed in a climate chamber.

D1518-85 [37], and BS 4745 [38]. However, the equipment styles are underlying the same principle, though the design differs a lot. Some clamp the fabric between a hot and a cold plate, others add an air layer between the sample and the cold plate, or use only a hot plate, while the other side of the sample is exposed to ambient air.

For the guarded hot plate (see Figure 31.5), the fabric's vapor resistance is calculated as

$$R_{\text{et}} = \frac{P_{\text{plate}} - P_a}{H - H_{\text{dry}}} - R_{\text{et0}} \tag{31.3}$$

where R_{et} is the heat resistance of fabric sample ((m^2 Pa)/W), P_{plate} is the mean hot plate surface vapor pressure (Pa), P_a is the ambient vapor pressure (Pa), H is the total heat loss per square meter of plate area (W/m^2), H_{dry} is the dry heat loss per square meter of plate area (W/m^2), and R_{et0} is the vapor resistance measured without a sample present ((m^2 K)/W).

A number of methods for determination of fabric vapor resistance are presented in BS 7209 [39], CAN/CGSB-4.2 No. 49-99 [40], ASTM F1868 [18], ASTM E96 [15], and ISO 11092 [19]. Test methods can vary dramatically in complexity and cost.

Besides the heat and vapor resistances, Woodcock [41] proposed the permeability index (i_m). The moisture permeability index of a material is defined as the ratio of the actual evaporative heat flow capability between the skin and the environment to the sensible heat flow capability as compared with the Lewis Ratio. The permeability index (i_m) is computed as

$$i_m = S \times \frac{R_{\text{ct}}}{R_{\text{et}}} \tag{31.4}$$

where S is the Lewis constant (60.6 Pa/K).

This permeability index varies between zero (totally impermeable) to the unity (air) (though the typical upper limit is due to its definition around 0.5). As mentioned before, in order to avoid moisture accumulation, the highest number is the best here. In addition to parameters aforementioned, the air permeability, waterproofness, and wicking property of the fabric are the characteristics that are usually evaluated.

31.5.2 Full-Scale Clothing Testing

As the step from fabric testing to laboratory human wear trials is quite big, it is necessary to have an intermediate step to evaluate the complete prototype of garments and ensembles on a physical apparatus. The thermal manikins developed during the last decades provide a practical solution to this problem. Thermal manikins are now available in a variety of sizes (baby or adult), shapes (male or female), and technologies (metal, resin shells, water filled fabric shells, etc.), and many of which now can "sweat" by applying uniformly wetted fabric "skin" or sweating output unit.

Moreover, with the technology development, thermal manikins (see Figure 31.6) available nowadays could be regulated regimentally; thus the local insulation can be studied and the whole body insulation. This allows a better analysis of "cold spots" in the clothing design. Most current manikins have 16–34 body segments, and each segment can be individually evaluated. Measurements are typically performed with the manikin standing static and with the manikin performing walking movements. Also measurements without wind (static) and with wind can be performed. In this way, effects of pumping, forced ventilation, and forced convection can be investigated, which is crucial for clothing evaluation [42, 43].

31.5.3 Clothing Thermal Insulation

Clothing thermal insulation is one of the most important parameters to characterize clothing thermal comfort. Thermal insulation of full-scale clothing is normally determined by a thermal manikin. Thermal manikins are manufactured in the human shape (see Figure 31.2), and the testing principle is the same as that of a guarded hot plate. Generally, three calculation methods might be used to

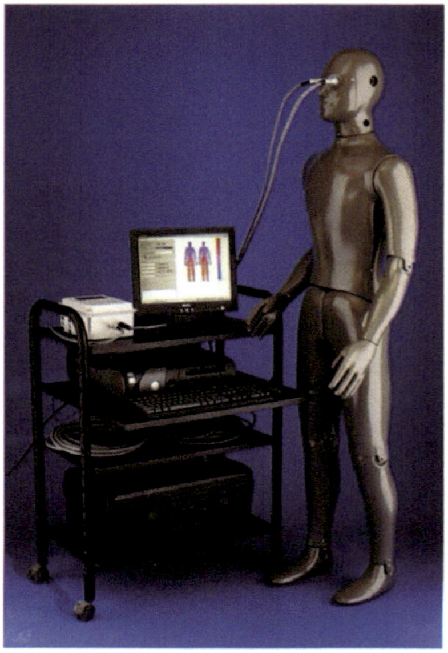

Figure 31.6 The Newton-type sweating thermal manikin (Thermetrics LLC, Seattle).

calculate clothing thermal insulation: the parallel method, the serial method, and the global method [44–46].

In the parallel method, thermal insulation is obtained by utilizing its reciprocal relation relationship with heat transfer coefficient. The heat transfer coefficient is calculated by averaging local heat transfer of each manikin segments. Thus the thermal insulation could be calculated as

$$I_t = \frac{1}{0.155} \sum_{i=1}^{i=n} \frac{a_i}{A} \frac{(T_{\mathrm{si}} - T_a)}{H_{\mathrm{ci}}} \tag{31.5}$$

where I_t is the total thermal insulation of the clothing (clo), n is the manikin's segment number, a_i is the surface area of the segment i (m^2), A is the total surface area of the manikin, T_{si} is the surface temperature of the segment i (°C), T_a is the air temperature (°C), and H_{ci} is the heat flux of the segment i (W/m^2).

Varies from the parallel method, the thermal insulation in the serial method is calculated by the averaged local insulation of each manikin segment, and the calculation is presented:

$$I_t = \frac{\sum a_i \cdot (t_i - t_a)}{\sum (a_i \cdot H_i)} \tag{31.6}$$

where a_i is ratio of the surface area of segment i to the total surface area of the manikin, t_a is the air temperature (°C), t_i is the segmental temperature (°C), and H_i is the heat flux of the segment i (W/m^2).

In addition to the parallel and the serial methods, the global method may also be used to calculate clothing thermal insulation [46]. It is calculated basing on the segmental heat loss, area-weighted surface temperature, and body segmental surface area, and the calculation is given as

$$I_g = \frac{\sum a_i \cdot t_i - t_a}{\sum (a_i \cdot H_i)} \tag{31.7}$$

where I_g is the global thermal insulation of the clothing, a_i is ratio of the surface area of segment i to the total surface area of the manikin, t_a is the air temperature (°C); t_i is the segmental temperature (°C), and H_i is the heat flux of the segment i (W/m^2).

Clothing insulation values calculated by different method may differ from each other depending on the clothing properties [47–49]. Generally speaking, thermal insulation values calculated by these three methods will be the same if clothing insulation distributes uniformly over the manikin body (i.e. homogeneous clothing insulation). However, if the clothing insulation is not evenly distributed over the manikin segments (i.e. inhomogeneous clothing insulation), the global and the serial methods will yield the same insulation value if the segmental heat flux is uniform over the manikin. In addition, the insulation calculated by the serial method is always higher than that calculated by the parallel method. Moreover, the global and parallel methods will result in the same thermal insulation value if the surface temperature is distributed uniformly over the manikin body; thus the insulation value obtained by the serial method will always be higher than that obtained by the global method [48–51].

31.5.4 Clothing Vapor Resistance

Evaporative resistance (or water vapor resistance) is the widely used physical parameter to quantify the moisture transfer property of full-scale clothing. The clothing evaporative resistance of a full-scale clothing ensemble can be calculated by either the heat loss option or the mass loss option [52]:

$$R_{\text{et}} = \frac{(p_s - p_a)A}{H_e - \frac{(T_s - T_a)A}{R_t}} \tag{31.8}$$

$$R_{\text{et}} = \frac{(p_s - p_a)A}{\lambda\left(\frac{dm}{dt}\right)} \tag{31.9}$$

where R_{et} is the total evaporative resistance of clothing ensemble and surface air layer ((kPa m^2)/W), p_s is water vapor pressure at the manikin's sweating surface (kPa), p_a is water vapor pressure in the air flowing over the clothing (kPa), A is area of the manikin's surface that is sweating (m^2), H_e is power required for sweating areas (W), T_s and T_a are fabric "skin" surface temperature and the ambient temperature, respectively (°C), and R_t is the total thermal resistance of the clothing ensemble and surface air layer ((°C m^2)/W). λ is the heat vaporization of water at the measured surface temperature (W); dm/dt is the evaporation rate of moisture leaving the manikin's sweating surface (g/min).

To determine clothing evaporative resistance, the simulation of human profuse sweating on the manikin is required (i.e. simulating a person with a fully wetted skin surface) [119]. In summary, three methods might be applied to achieve the simulation of human sweating [53–55]: pre-wet fabric "skin," manikins equipped with pumps and pipes to regulate water supply through sweat nozzles to the whole manikin body surface (a piece of highly stretchable fabric "skin" is also required for the purpose of distribution of sweat throughout the entire manikin body [see Figure 31.7]) and sweating fabric manikins featured with a water filled body covered by a waterproof but permeable fabric "skin." The pre-wetted fabric "skin" method is effective, but the "skin" tends to be dried out after a certain testing period, which is dependent upon the type of clothing tested, the test condition, and how much moisture the "skin" could hold with no dripping.

The second method (i.e. active sweating method) is the most often used, and this method solved the drying-out issue observed on the pre-wetted "skin" method. Cautions should be made on the adjustment of appropriate sweating rate for testing [56]. The sweating rate must be set to ensure the whole "skin" could maintain fully saturated throughout the entire testing period. Otherwise, partially dried out "skin" will significantly impact the determined clothing evaporative resistance, which tends to be significantly higher because the evaporative heat loss is much smaller [57–59]. Another issue with this method is that the surface temperature of the fabric "skin" is uncontrolled. Due to the sweat evaporation, the fabric "skin" surface temperature is found to be much lower than the controlled manikin surface [58, 59]. This will also largely impact the precision and accuracy of clothing evaporative resistance. Hence, it is highly recommended to future sweating manikins to control the fabric "skin" temperature rather than the manikin surface temperature during the wet tests.

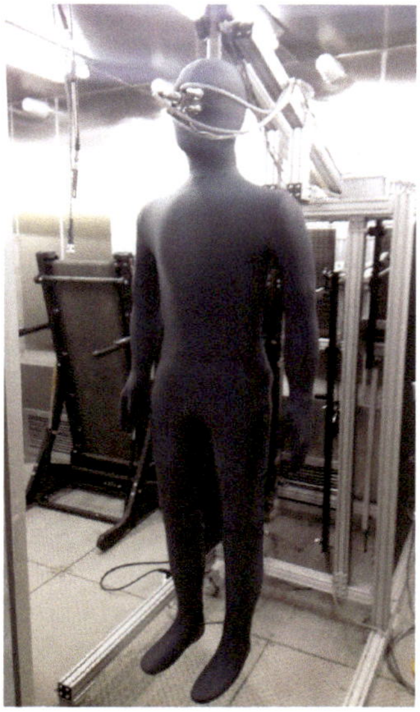

Figure 31.7 A sweating manikin is wearing the saturated knitted fabric "skin."

The third typical type of commercialized sweating manikin is the "Walter" manikin [60]. From a technical viewpoint, the only merit of the "Walter" manikin is that, unlike many other sweating manikins that regulate the manikin surface temperature, "Walter" controls its fabric "skin" surface temperature. This design contributes to enhancing the measurement accuracy on clothing evaporative resistance. Nevertheless, the "Walter" sweating manikin has many limitations. First, this manikin has only one segment, and thus it is unable to measure clothing localized evaporative resistance. Second, the "Walter" manikin can only mimic human vaporous sweating. Hence, it fails to detect the effect of absorption of liquid sweat on clothing heat and mass transfer properties (unless intentionally wetting the tested clothing using additional water). Besides, the maximum perspiration rate of the waterproof and permeable Gore-Tex "skin" is a constant, and it is found that the perspiration rate is rather too low. It is found that the use of Gore-Tex "skin" changes the driving force for water evaporation. Normally, the main driving force for water evaporation is the water vapor pressure gradient between the skin surface and the environment [61]. However, in such test conditions where the environment permits much more sweat/water evaporation than the "skin," clothing evaporative resistance determined by the "Walter" manikin seems incorrect (i.e. the value tends to be much higher than those determined on other types of sweating manikin). Therefore, all these drawbacks greatly limit the application of the "Walter" manikin to determine clothing moisture transfer property of full-scale clothing.

It should be noted that the selection of two calculation options should be made with caution [57]. Otherwise, the reported clothing evaporative resistance data might be incorrect. For those manikins with a function of manikin surface temperature controlling (e.g. the "Newton" sweating manikin) and tests performed in isothermal conditions (i.e. $T_{\text{manikin}} = T_{\text{air}} = T_r$), the mass loss option is suggested.

The heat loss method significantly overestimates the real clothing evaporative resistance. If the manikin cannot record mass loss during the test period, the heat loss option should be corrected before use [62]. This correction model is found to be reliable and easy to use. Also, the fabric "skin" surface temperature should be used when computing the water vapor pressure at the saturated fabric "skin" surface. Wang et al. [63] established a universal equation to calculate the fabric "skin" temperature based on the heat loss observed from the test and the manikin surface temperature. This equation has been widely used by scientists to estimate the fabric "skin" temperature.

Though it is highly recommended to carry out clothing evaporative resistance in an isothermal condition, many published studies reported data from the tests performed in non-isothermal conditions. For tests performed in non-isothermal conditions, how to choose the options to calculate clothing apparent evaporative resistance (i.e. the evaporative resistance determined in non-isothermal conditions) becomes more complicated; further investigations are still required to explore this. It has been found that the condensation could significantly affect the heat exchange between the manikin and the test condition, especially for garments with limited vapor permeability [64, 65]. Thus, the mass loss method might be incorrect when being used to calculate clothing vapor resistance. Lastly, it should be emphasized that the clothing apparent evaporative resistance is only comparable with that determined at exactly the same test environmental condition.

31.5.5 Human Trials

Human wearing trials can provide valuable information that cannot be obtained from laboratory tests with test equipment (see Figure 31.8). The human wearing trials are mainly conducted in and out climatic chambers in laboratory. In climatic chambers, outdoor conditions could be mimicked with high accuracy, as well as if not more important reproducibly. In the controlled climatic chamber, it is possible to compare the performance of a number of garments on the same test subjects and thus get a good comparison of their properties in practical

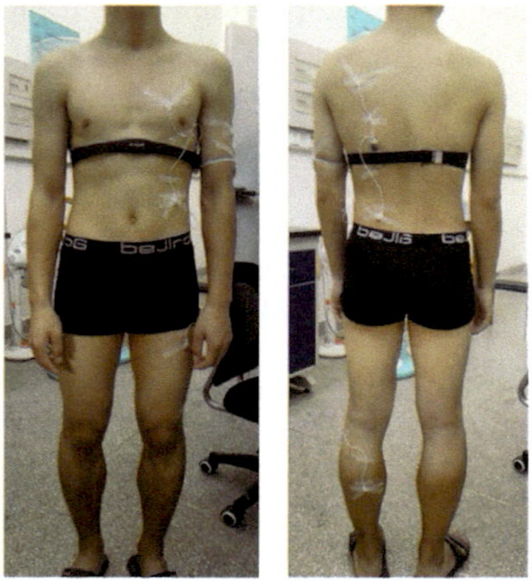

Figure 31.8 Skin temperature measurements on human subjects.

applications. However, the human wearing trials are usually very time-consuming and therefore costly. Also the repeatability is lower compared with measurements carried out by thermal manikins. Besides, the measurement performed by human trials needs to be repeated by several participants in order to make the results and conclusions reliable and statistically comparable.

31.5.6 Fiber and Thermal Comfort

Fiber is the fundamental component of yarns, and usually it is the yarns that made into porous fabrics via weaving or knitting techniques. In addition, nonwoven fabrics can be constructed directly from fibers or filaments without the process of yarns. One key factor for fabrics is thermal resistance, and the trapped still air in the fabric structure contributes a lot to the heat barrier because the thermal resistance of air is eight times higher than that of fibers. According to Farnworth [66], the effective thermal conductivity of the fabric system is partly due to the air volume and partly due to the fiber fractions of the system. Thus using the effective thermal conductivity can be calculated with Eq. (31.10):

$$K_{sys} = (1 - P_\varphi)K_a + P_\varphi K_f \tag{31.10}$$

where P_φ is the packing factor, K_a is the thermal conductivity of the air volume fraction, and K_f is the thermal conductivity of the fibrous component (Table 31.1 shows the typical thermal conductivity of various fibers).

From Eq. (31.10) it could be concluded that the more amount of still air trapped in the fabric structure, the higher the thermal insulation. Fibers affect the fabric thermal insulation via two approaches, namely, preventing air movement and providing a shield to radiative heat transfer. The thermal insulation of a fabric is determined by fiber physical properties, the structure of the fabric, and the

Table 31.1 Thermal conductivity of various fibers.

Textile fibers	Thermal conductivity (W/(m K))
Cotton	0.243
Nylon	0.171
Polyester staple	0.127
Polyester filament	0.157
Flax	0.344
Polypropylene	0.111
Rayon staple fiber	0.237
Wool	0.165
Silk	0.118
Aramid	0.192

Source: Hearle and Morton 2008 [67]. Reproduced with permission of Elsevier.

boundary air layer existing on the fabric surface. When air moves toward the fiber or fabric, its velocity of air adjacent to the fiber or fabric surface area keeps relative still to the fiber or fabric, which forms boundary air layer. Therefore, the existing boundary air layer, along with the existing still air, increases thermal insulation and provides considerable resistance to heat transfer through the fabric [6].

As mentioned above, the thermal insulation of the fabric depends on the physical structure of fiber, including its fineness, crimp, and length. Generally, the fineness of the fiber has a linear relationship with the thermal insulation because of the high surface-to-volume ratio of fine fibers. As a result of the high surface-to-volume ratio of fine fibers, many small spaces for still air between fibers can be formed, which limits the heat transfer through conduction and convection, resulting into a higher thermal insulation.

Microfiber has gained exclusive attention recently and has been used in clothing to improve clothing function because of its larger surface area. Fiber crimp can trap more still air by creating interstices in the yarn structure and thus increases the fabric thermal insulation. For example, the high thermal insulation of wool fibers can be explained by the natural crimp, which maintains a high volume of still air, explaining its traditional applications in cold weather clothing.

Additionally, Bogaty et al. [68] observed that fabric thermal insulation has close relation with the fiber arrangement. Fiber with parallel and perpendicular arrangements to the fabric surface showed improved thermal insulation. Besides, Kong et al. [69] described that fiber configuration, and its porosity can significantly impact the heat and moisture transfer. It was found that fabrics or garments made from hollow fibers could provide higher thermal insulation values because of the large still air volume trapped within the fabric. Because of its unique characteristics, hollow fibers can be used in applications like swimming suits. Hollow fibers can provide buoyancy for the body in the water and prevent human body from being cooled.

Besides the hollow fiber, with the developing of fiber technologies, functional fibers that are found growing importance may include the following:

- *Bicomponent or multicomponent fibers*: which are mainly obtained either via polymer blending or graft polymerization. For example, bicomponent synthetic fibers with low melting temperature sheath.
- Fibers manufactured by materials with inherent coloration and with a high degree of color fastness.
- Hollow fibers with a filled core or a core fiber spun with a sheath for protection or enhancement of the core properties. For example, glycol-based fillings that buffer heat changes [70].
- *Modified fibers*: fibers whose basic properties have been modified after certain physical or chemical process. For example, to increase the melting point of polyester fibers [71].

31.6 Fabric and Thermal Comfort

The heat and moisture transfer in a fabric depends on the yarn structure, the fiber geometry, and fabric construction. Compared with a flat filament, a spun yarn or textured filament yarn can trap more still air and therefore can obtain better moisture and heat barrier, i.e. higher thermal insulation. The heat and moisture transport is also affected by the degree of yarn twist, the higher twist yarns, the more compact, resulting in lower thermal insulation because of less trapped air volume. On the other hand, the compact yarns may facilitate the capillary effect, resulting in improvement in moisture transfer.

It is proven by previous studies that the fabric thickness is the most important parameter in determining thermal insulation [72]. Generally, the thicker the fabric, the more amount of still air trapped in the fabric, providing higher thermal insulation. It is observed in some cases that there is a linear relation between the fabric thermal insulation and the fabric thickness. Compared with a single layer thick fabric, fabrics with multilayer present even higher insulation. For multilayer fabrics, however, the thermal resistance and thickness are not a linear relation [73].

The heat and moisture transfer capability is also affected by the fabric structures because different fabric structures have different levels of porosity, resulting in different volumes of still air entrapped in the fabric [118]. On average, knitted fabrics entrap more still air compared with woven fabrics, although the yarn density of weaving and knitting is a factor as well. Additionally, pile or napped constructions in fabrics can improve fabric thermal insulation, which is because the yarns or fibers perpendicular to the surface provide numerous spaces for entrapped air. These fabric insulation can be optimized when they are used as an inner layer next to the skin.

Furthermore, the fabric thermal insulation performance is also affected by its weight. However, clothing weight may deteriorate the comfort property, so it is important to balance high thermal insulation with low weight to improve comfort, and it is especially true for cold weather clothing. Lightweight fabrics can be obtained by bonding knitted fabric with woven fabric using polyurethane foam. The low density of polyurethane foam provides the sandwiched fabric-foam-fabric structure with high insulation and light weight.

In a study by Frydrych et al. [74], they compared fabrics made from regular cotton and man-made cellulose fibers. It was found that thermal insulation differed noticeably both with different types of fiber and different weaving patterns using same fiber. They also described a difference in the thermal absorption of the fabrics when different finishing processes were applied. A similar study was also performed on different clothing materials using a sweating cylinder [75]. In their study, the effectiveness of different material combinations on the thermal insulation was investigated, and significant different in thermal insulation was determined between different material combinations.

Multilayer fabrics increase the overall thickness of the garment, which contributes to increase in the total insulation of the clothing system. The multilayer clothing system normally comprises three layers: the outer layer, the middle layer, and the inner layer. The inner layer, together with the skin, forms the microclimate. Heat and moisture transport starts directly from this inner layer. The middle layer, usually made from fabrics that could entrap a large amount of air, normally contributes most to the thermal insulation. The main function of the outer layer is to provide sufficient protection to prevent the garment from being injured. The outer layer is normally made of materials with excellent mechanical performance. In addition, functional or chemical finishing can be introduced to the outer layer in order to obtain some desirable properties such as self-cleaning. In addition to the physical properties of fiber, yarn and fabric, the thermal properties can be changed through functional coating and finishing applied to the surface of yarn or fabric.

31.7 Personal Conditioning Clothing for Improving Wear Comfort

31.7.1 Personal Cooling Clothing System

Heat strain has been regarded as one of the underlying causes for heat-related injuries and work accident, and even heat-associated high morbidity and mortality are attributed to the elicited body in extreme environments. When the ambient temperature is higher than that of the body skin in summer, sweat evaporation is the only avenue for human body to dissipate heat to environments [76]. Besides, human body may even receive conductive, convective, and radiative heat from the ambient in such extreme environments [77]. Thus, the body thermal balance would be easily disrupted in such conditions with the elevated body thermophysiological (such as core temperature rise) [78]. When the core temperature reaching 40 °C, cellular damage occurs rapidly, initiating a cascade of events that may lead to organ failure and possible fatality [79].

Recently, there has been a shift in attention from a focus on cooling the entire macro-environment (mainly using air conditioner) to cooling the microenvironment around the human body (using personal cooling strategies [PCSs]) [80]. Generally speaking, PCSs are energy saving, more flexible and user-friendly. Air fan (a type of PCS) is a direct cooling strategy and proved to be effective below some temperature limits [81, 82]. Besides the environmental limits, the less portability of the air fans limiting its use to active people. Regarding this, portable personal cooling clothing (PCC) may serve as an alternative to mitigate body heat strain in hot conditions [83, 84].

External cooling source may be incorporated into clothing to create a cooler the microenvironment between the human body and clothing to promote body heat dissipation. Various types of PCC have been developed during last few decades, and they could generally be divided into five categories: phase change materials (PCMs) for cooling clothing, evaporative cooling clothing (ECC), liquid cooling clothing (LCC), air cooling clothing (ACC), and the hybrid cooling clothing combining two or more cooling techniques [85–87]. The PCMs (such as frozen gels and ice packs) are usually placed closely to the body to absorb body heat during phase change. The cooling effect of PCMs mainly relies on their melting temperature and heat fusion, as well as their mass and the body coverage area. ACC utilizes the circulated natural or cold air to enhance evaporative or/and convective heat loss. The current existing ACC are generally configured in two forms: incorporating arrayed fine hoses with air pores into clothing to direct air flow (natural or cold air) to the body skin, powered by an external air compressor, and

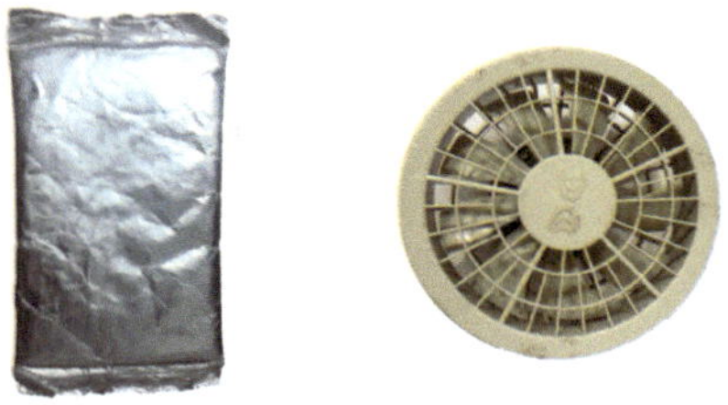

(a) (b)

Figure 31.9 (a) Phase change material pack and (b) the air ventilation fan.

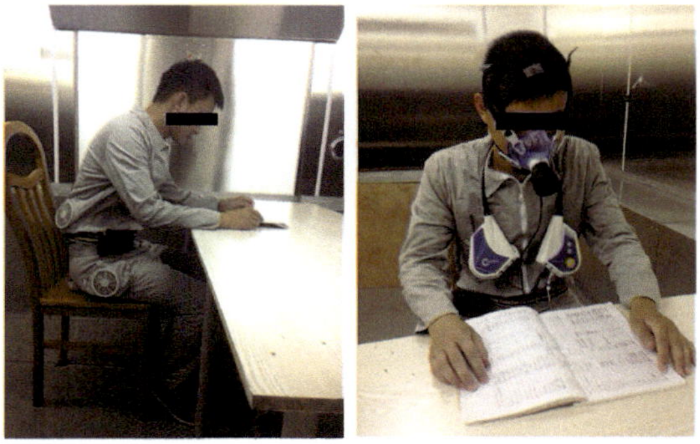

(a) (b)

Figure 31.10 Hybrid personal cooling clothing incorporated with phase change materials and ventilation fans.

embedding air fans into clothing to make efficient air ventilation (natural air), powered by portable batteries (see Figure 31.9). ACC circulating natural air works on the same rational as air fans, both of which utilizes the increased air velocity around the body skin to enhance evaporative heat loss. LCC utilizes the positive temperature gradient between the human skin and the liquid to promote the conductive and convective heat dissipation. It was developed by incorporating a network of fine tubes into clothing in close contact with the body skin, and the circulating cold liquid (such as water) in tubes, pumps from a refrigerating system, removes excessive body heat. Hybrid cooling strategy (only one study was retrieved) combining liquid and air cooling sources has been shown to bring large cooling effects in reducing body physiological heat strain and improving human performance in hot and moderate humid environment [88–91] (see Figure 31.10). ECC utilizes water stored in the clothing to absorb heat from the human body to evaporate, and this type of PCC is normally light (lower than 1 kg), cheap, and convenient to use.

All the aforementioned studies focus on the cooling effects of PCC in specific fields such as firefighting, military, and biological and chemical protection fields where PCC is worn under protective clothing [92–94]. The cooling effect of the PCC worn as the outer layer might be different from that of PCC used as the interlayer, as the heat exchange between PCC and the ambient is more enhanced, and sweat evaporation is also less restricted compared with conditions where PCC is used as an outer layer [95].

31.7.2 Personal Heating Clothing System

Since there is no heating, ventilation, and air-conditioning (HVAC) system in some world spots such as Yangtze River urban areas in China, people there may experience cold discomfort at their ears, cheeks, hands, and feet while sitting in unheated cold environments in winter (below 10.0 °C). One possible way to minimize the body heat loss in cold environments is to wear high insulating traditional cold protective clothing or wear clothing in corporate with auxiliary heating [55]. Normally, traditional cold weather clothing has a multilayer structure and is usually thick and heavy. Besides, traditional cold weather clothing restricts the human body movement, reduces the manual dexterity, and thereby impairs the human work performance [96]. Moreover, the total thermal insulation of multilayer clothing has an upper limit, i.e. around 4.3 clo [97]. Hence, the usage of traditional clothing to improve body thermal comfort has been limited in extremely cold environments or long duration mild cold exposures [96]. Thereby, auxiliary heating sources implanted into clothing provide a promising solution to this problem. Properly designed heated clothing has been proven to be capable of providing the human body with sufficient protection in cold environments without sacrificing freedom of body movement.

The heating clothing currently available nowadays may be divided into four types: electrically heated clothing (EHC, see Figure 31.11); chemically heated clothing (CHC), harnessing heat generated via chemical reaction; phase change heating clothing incorporated with PCMs, and air/fluid flow heated clothing. Heated gloves and socks have been extensively examined, and they have evidently improved the thermal comfort in the hands/feet and manual dexterity in unconventional cryogenic environments [96]. Choi et al. [98] observed the clothing incorporated PCM exhibited higher microclimatic temperatures and improved thermal sensations compared with the non-heated clothing, but it had no effect on the rectal and mean skin temperatures. Coca et al. [99] described that a liquid heated garment (LHG), with 24 °C inletting water, could be useful in stabilizing the rectal temperature and maintaining comfort before and during exercise. Flouris et al. [100] also observed a higher mean body temperature during exercising in LHG compared with the non-heating and pre-heating (in LHG before exercise) scenarios, and the last two scenarios triggered higher frequency of cold-induced vasodilation (CIVD) in extremities. Chan and Burton [101] examined the performance of a CHC system (i.e. a jacket equipped with 30 heating pads and a pair of gloves incorporated with 4 chemical heating pads), and they observed that the CHC could maintain the torso skin temperature, but it

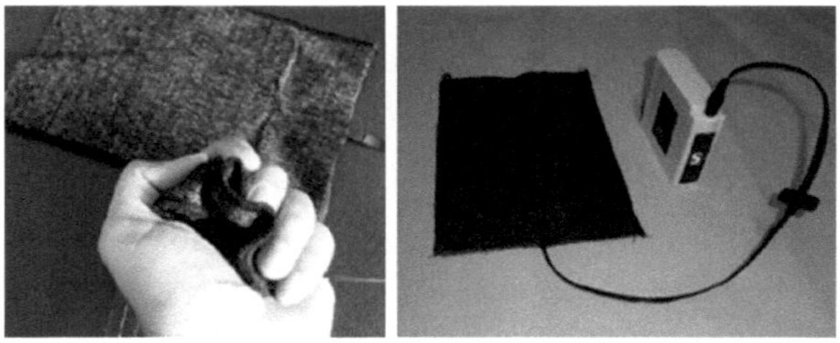

Figure 31.11 Heating fabric pads for engineering electrically heated clothing (EHC).

had no effect on other body parts that were not covered by the heating sachets in cold water. Kirsi et al. [102] investigated the performance of an electrically heated vest in a very cold environment (i.e. −15 °C) while having subjects standing or doing light work. The vest could significantly raise local skin temperatures and improve the thermal sensation of the torso but exhibited limited effect on both core and extremity temperatures. Furthermore, the effectiveness of the heated clothing mainly depends on the heating power and the exposing temperatures. For instance, Rantanen et al. [103] discovered that the electrically heated shirt provided sufficient protection to wearers in normal winter environment in Finland. However, decrease in the skin temperature was observed in wearers in extremely cold environments (i.e. from −14 to −20 °C). Heated clothing with a large heating power was also found to be able to improve both local and overall body thermal comfort (such as promoted finger blood flow and dexterity) when exposing to −25 °C for three hours [104].

31.7.3 Smart Heating Sleeping Bag

Sleeping bags have been regarded as essential protective products for sleeping outdoors, especially under cold environments. On applications, especially for expedition and mountaineering, the maximal thermal insulation with relatively lightweight is of the most important parameters for a sleeping bag [116]. Currently, sleeping bags have different shapes (e.g. mummy, rectangular, and quit shaped) and usually have a three-layer configuration, namely, the outer layer, an inner layer, and the filler [105]. The outer and inner layers are usually made of synthetic materials, and the filler material may be cotton, down, or synthetic fibers.

In order to characterize the performance of the sleeping bag, many models have been developed to estimate the minimal operating temperature of the sleeping bags, e.g. the Goldman's model, the KSU model, the Belding's model, the IREQ model, and the Havenith model [106]. In contrast to the minimal operating temperatures, the EN 13537 [107] standard proposed a model to predict the comfort operating temperature. The EN 13537 [107] listed out four label temperatures: maximum temperature, comfort temperature, limit temperature, and extreme temperature. Among the four above labeled temperatures, the comfort and limit temperatures are considered to be the most important temperatures. The comfort temperature is defined as the lower temperature threshold at which a standard woman could sleep for eight hours with a global thermal equilibrium state and just without feeling cold at a relaxed posture. The limit temperature is the lower temperature threshold at which a standard man could sleep for eight hours with a global thermal equilibrium state and just without feeling at a curved up posture [107]. Unfortunately, the EN 13537 [107] has not been widely adapted by sleeping bag manufacturers. Manufactures tend to label their products by different methods. For example, there are manufacturers labeling the temperatures based on customers' feedbacks, and others may label the temperatures based on the sleeping bag's physical properties such as the thickness, weight, and loftiness. Therefore, the thermal performance of sleeping bags labeled with same temperature ranges may vary greatly [108].

Even though there are quite a few models available for predicting minimal operating temperature, all those models have been questioned for mainly focusing on the global thermal comfort of the wearers. The models appear to have less consideration to the local thermal comfort, especially for the human feet. Compared with other body regions, such human extremities as the feet are more sensitive to cold environments, and their temperatures will decrease rapidly when being exposed to cold due to their larger specific surface area and relatively low local metabolic rate [109]. Compared with the other body parts, the specific surface area of human feet is around 2.5–3.0 times larger than that of the whole body, indicating it will lose heat more rapidly when exposing to cold. The decreased temperatures will lead to thermal discomfort, and further decrease may cause pain feeling in the toe and feet. Moreover, the accumulation of local thermal discomfort at the feet can significantly influence the whole body thermal comfort [110, 111].

Local thermal discomfort at the body extremities such as the feet has been widely reported, albeit that the mean skin temperature of the wearers is still within the thermoneutral range (i.e. 32–34 °C) when using sleeping bags at their predefined comfort temperatures [110, 112, 113, 119]. Huang [114] observed a similar tendency of decrease in toe temperatures on both male and female subjects when comparing the five models (i.e. the EN 13537 [107] model, the Goldman's model, the KSU model, the Belding's model, and the den Hartog's model). Lin et al. [112] reported that the foot temperature underwent a continuously and rapidly drop in sleeping bags under the EN 13537 [107] defined operating temperatures. Within the two hours exposure, the foot temperature dropped to almost 14 °C, at which strong cold sensations occurred even the mean skin temperature was well maintained within the thermal neutral range (i.e. 32–34 °C).

A series of studies have been conducted by the authors [110, 113, 115] to investigate the improvement in local and whole body thermal comfort using smart electrically heated sleeping bags. A smart heating sleeping bag that could response to the changes of the foot temperature was developed (see Figure 31.12). It was found that, compared with the traditional sleeping bag, the newly developed smart heating sleeping bag could maintain the foot temperature within the comfort range throughout the eight-hour trials under the EN 13537 [107] defined comfort temperatures, whereas linearly decreasing in foot

Figure 31.12 A smart electrical heating sleeping bag.

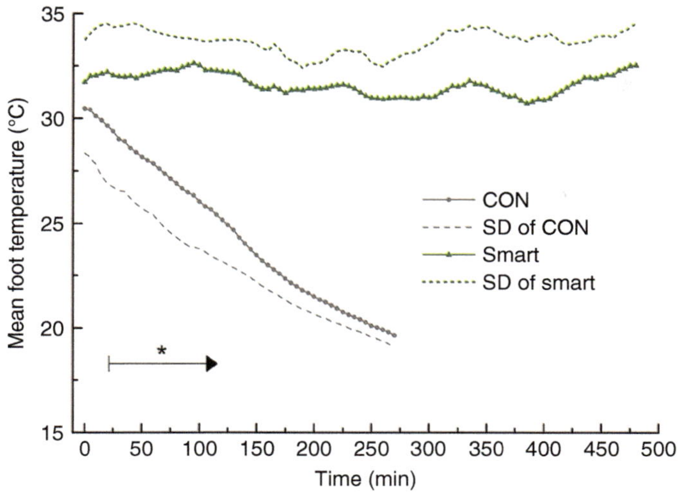

Figure 31.13 Time course changes in the mean foot temperature in CON (i.e. the traditional sleeping bag) and smart (i.e. the smart electrically heated sleeping bag). $^*p < 0.05$.

temperatures were observed in the traditional sleeping bags (see Figure 31.13). Additionally, it was noted that both local and overall thermal comfort were greatly improved in the smart heating sleeping bag.

31.8 Conclusions

Comfort management property plays an important role in fibrous materials, particularly for textiles and apparel. Comfort management includes regulation of body temperature and moisture transfer (including absorption, adsorption, desorption, wetting, wicking, drying, etc.). Due to advancing technology, more advanced intelligent and smart textiles will appear, which may greatly facilitate the improvement of the comfort property of textiles and clothing. In the future, flexible and small size sensors and actuators will be embedded into textiles, and this makes the textile even smarter. Personal cooling and heating clothing will be regulated based on the wearers' real-time physiological and psychological status. On the other hand, more advanced test equipment will be also developed. Measurements of heat and moisture transfer properties of textiles will become quicker, more precise, and more reproducible and thereby contribute to improving the precision of mathematical models.

References

1 Fanger, P.O. (1970). *Thermal Comfort*. Copenhagen: Danish Technical Press.
2 Parsons, K.C. (2002). *Human Thermal Environment*. London: Taylor & Francis.

3 Rossi, R. (2009). Comfort and thermoregulatory requirements in cold weather clothing. In: *Textiles for Cold Weather Apparel* (ed. J.T. Williams), 3–18. New York: CRC Press.

4 Goldman, R.F. (1988). *Biomedical Effects of Clothing on Thermal Control and Strain, Handbook on Clothing, Biomedical Effects of Military Clothing and Equipment Systems*, 1–18. Soesterberg: TNO Institute of Perception.

5 Li, Y. (2001). The science of clothing comfort. *Textile Progress* 31: 1–135.

6 Song, G. (2009). Thermal insulation properties of textiles and clothing. In: *Textiles for Cold Weather Apparel* (ed. J.T. Williams), 19–32. New York: CRC Press.

7 ASHRAE 55 (2013). Thermal Environmental Conditions for Human Occupancy. In: . Atlanta: American Society of Heating Refrigeration and Air Conditioning Engineers, American Society of Heating Refrigeration and Air Conditioning Engineers, Inc.

8 ISO 7730 (1984). *Moderate Thermal Environments-determination of the PMV and PPD Indices and Specification of the Conditions for Thermal Comfort*. Geneva: International Standards Organization.

9 Houghton, F.C. and Yagloglou, C.P. (1923). Determining equal comfort lines. *Transactions of American Society of Heating and Ventilation Engineers* 29: 165–176.

10 Houghton, F.C. and Yagloglou, C.P. (1924). Cooling effect on human beings produced by various air velocities. *Transactions of American Society of Heating and Ventilation Engineers* 30: 193–212.

11 Missenard, A. (1948). A thermique des ambiences: equivalences de passage, equivalences deséjours. *Chaleur et Industrie* 276: 159–172.

12 Dufton, A.F. (1929). The eupatheostat. *Journal of Scientific Instruments* 6: 249–251.

13 Dufton, A.F. (1936). The equivalent temperature of a warmed room. *Journal of the Institution of Heating & Ventilating Engineers* 4: 227–229.

14 Henry, P.S.H. (1939). Diffusion in absorbing media. *Proceedings of the Royal Society of London Series A* 171: 215–241.

15 ASTM E96 (2016). *Standard Test Methods for Water Vapor Transmission of Materials*. West Conshohocken, PA: American Society for Testing and Materials.

16 ISO 15496 (2004). *Textiles – Measurement of Water Vapor Permeability of Textiles for the Purpose of Quality Control*. Geneva: International Organisation for Standardisation.

17 ISO 2528 (1995). *Sheet Materials. Determination of Water Vapour Transmission Rate-Gravimetric (dish) Method*. Geneva: International Organisation for Standardisation.

18 ASTM F1868 (2014). *Standard Test Method for Thermal and Evaporative Resistance of Clothing Materials Using a Sweating Hot Plate*. West Conshohocken, PA: American Society for Testing and Materials.

19 ISO 11092 (2014). *Textiles-Physiological Effects-Determination of Physiological Properties-Measurement of Thermal and Water-vapour Resistance under Steady-state Conditions (Sweating Guarded-hotplate Test)*. Geneva: International Organisation for Standardisation.

20 Karppinen, T., Kassamakov, I., Haeggstrom, E., and Stor-Pellinen, J. (2004). Measuring paper wetting processes with laser transmission. *Measurement Science and Technology* 15: 1223–1229.

21 Rengasamy, R.S. (2006). Wetting phenomena in fibrous materials. In: *Thermal and Moisture Transport in Fibrous Materials* (eds. N. Pan and P. Gibson), 156–187. Cambridge: Woodhead Publishing.

22 Uddin, F. and Lomas, M. (2010). Wettability of easy-care finished cotton. *Fibres and Textiles in Eastern Europe* 18: 56–60.

23 Birrfelder, P., Dorrestijn, M., Roth, C., and Rossi, R.M. (2013). Effect of fiber count and knit structure on intra- and inter-yarn transport of liquid water. *Textile Research Journal* 83: 1477–1488.

24 Zhong, W. (2006). Surface tension, wetting and wicking. In: *Thermal and Moisture Transport in Fibrous Materials* (eds. N. Pan and P. Gibson), 136–155. Cambridge: Woodhead Publishing.

25 Zhang, Y., Wang, H., Zhang, C., and Chen, Y. (2007). Modeling of capillary flow in shaped polymer fiber bundles. *Journal of Materials Science* 42: 8035–8039.

26 AATCC TM 197 (2013). *Vertical Wicking of Textiles*. Research Triangle Park, NC: American Association of Textile Chemists and Colorists.

27 AATCC TM 198 (2013). *Horizontal Wicking of Textiles*. Research Triangle Park, NC: American Association of Textile Chemists and Colorists.

28 DIN 53924 (1997). *Testing of Textiles - Velocity of Soaking Water of Textile Fabrics (Method by Determining the Rising Height)*. Berlin: German Institute for Standardization.

29 AATCC TM 39 (1977). *Evaluation of Wettability*. Research Triangle Park, NC: American Association of Textile Chemists and Colorists.

30 ISO 4920 (2012). *Textile Fabrics -- Determination of Resistance to Surface Wetting (Spray Test)*. Geneva: International Standards Organization.

31 Goldman, R.F. (1974). Clothing design for comfort and work performance in extreme thermal environments. *Transactions of the New York Academy of Sciences* 36: 531–544.

32 Umbach, K.H. (1983). Evaluation of comfort characteristics of clothing by use of laboratory measurements and predictive calculations. *Proceedings of the 1981 International Conference on Protective Clothing Systems*, Stockholm, Sweden.

33 ISO 11079 (2007). *Ergonomics of the Thermal Environment-Determination and Interpretation of Cold Stress When Using Required Clothing Insulation (IREQ) and Local Cooling Effects*. Geneva: International Organisation for Standardisation.

34 Lotens, W.A. (1993). Heat transfer from humans wearing clothing. PhD destination. Netherland: Technische Universiteit Delft.

35 Wissler, E.H. and Havenith, G. (2009). A simple theoretical model of heat and moisture transport in multi-layer garments in cool ambient air. *European Journal of Applied Physiology* 105: 797–808.

36 CAN/CGSB 4.2 No. 78.1-2001 (2001). *Textile Test Methods-Thermal Protective Performance of Material for Clothing*. Quebec: Canadian General Standards Board.

37 ASTM D1518-85 (2003). *Standard Test Method for Thermal Transmittance of Textile*. West Conshohocken, PA: American Society for Testing and Materials.
38 BS 4745 (2005). *Determination of the Thermal Resistance of Textiles*. London: British Standards Institute.
39 BS 7209 (1990). *Specification for Water Vapour Permeable Apparel Fabrics*. London: British Standards Institute.
40 CAN/CGSB-4.2 No. 49-99 (1999). *Textile Test Methods Resistance of Materials to Water Vapour Diffusion*. Quebec: Canadian General Standards Board.
41 Woodcock, A.H. (1962). Moisture transfer in textile systems. *Textile Research Journal* 8: 628–633.
42 Lu, Y., Wang, F., Wan, X. et al. (2015). Clothing resultant thermal insulation determined on a moveable thermal manikin: Part I effects of wind and body movement on total insulation. *International Journal of Biometeorology* 59: 1475–1486.
43 Lu, Y., Wang, F., Wan, X. et al. (2015). Clothing resultant thermal insulation determined on a moveable thermal manikin: Part II effects of wind and body movement on localised insulation. *International Journal of Biometeorology* 59: 1487–1498.
44 ASTM F1291 (2015). *Standard Test Method for Measuring the Thermal Insulation of Clothing Using a Heated Manikin*. West Conshohocken, PA: American Society for Testing and Materials.
45 ISO 15831 (2004). *Clothing-Physiological Effects-Measurement of Thermal Insulation by Means of a Thermal Manikin*. Geneva: International Organisation for Standardisation.
46 ISO 9920 (2007). *Ergonomics of the Thermal Environment. Estimation of Thermal Insulation and Water Vapour Resistance of a Clothing Ensemble*. Geneva: International Organisation for Standardisation.
47 Huang, J. (2008). Calculation of thermal insulation of clothing from mannequin test. *Measurement Techniques* 51: 428–435.
48 Kuklane, K., Gao, C., Wang, F., and Holmér, I. (2012). Parallel and serial methods of calculating thermal insulation in European manikin standards. *International Journal of Occupational Safety and Ergonomics* 18: 171–179.
49 Xu, X., Endrusick, T., Gonzalez, J. et al. (2008). Comparison of parallel and serial methods for determining clothing insulation. *Journal of ASTM International* 5: 1–6.
50 Huang, J. (2012). Theoretical analysis of three methods for calculating thermal insulation of clothing from thermal manikin. *Annals of Occupational Hygiene* 56: 728–735.
51 Oliveira, A.V.M., Gaspar, A.R., and Quintela, D.A. (2008). Measurements of clothing insulation with a thermal manikin operating under the thermal comfort regulation mode: comparative analysis of the calculation methods. *European Journal of Applied Physiology* 104: 679–688.
52 ASTM F2370 (2015). *Standard Test Method for Measuring the Evaporative Resistance of Clothing Using a Sweating Manikin*. West Conshohocken, PA: American Society for Testing and Materials.

53 Wang, F. (2011). Clothing evaporative resistance: its measurements and application in prediction of heat strain. PhD dissertation. Lund, Sweden: Lund University.

54 Wang, F., Gao, C., Kuklane, K., and Holmér, I. (2009). A study on evaporative resistances of two skins designed for thermal manikin Tore under different environmental conditions. *Journal of Fiber Bioengineering and Informatics* 1: 301–305.

55 Wang, F., Gao, C., Kuklane, K., and Holmér, I. (2010). A review of technology of personal heating garments. *International Journal of Occupational Safety and Ergonomics* 16: 387–404.

56 Lu, Y., Wang, F., Peng, H. et al. (2016). Effect of sweating set rate on clothing real evaporative resistance determined on a sweating thermal manikin in a so-called isothermal condition ($T_{manikin}=T_a=T_r$). *International Journal of Biometeorology* 60: 481–488.

57 Wang, F., Gao, C., Kuklane, K., and Holmér, I. (2011). Determination of clothing evaporative resistance on a sweating thermal manikin in an isothermal condition: heat loss method or mass loss method? *Annals of Occupational Hygiene* 55: 775–783.

58 Wang, F., Gao, C., Kuklane, K., and Holmér, I. (2012). A comparison of three different calculation methods for clothing evaporative resistance. *Proceedings of The 5th European Conference on Protective Clothing and NOKOBETEF 10 (ECPC)*, Valencia, Spain.

59 Wang, F., Kuklane, K., Gao, C., and Holmér, I. (2012). Effect of temperature difference between manikin and wet fabric skin surfaces on clothing evaporative resistance: how much error is there? *International Journal of Biometeorology* 56: 177–182.

60 Fan, J., Chen, Y., and Zhang, W. (2001). A perspiring fabric thermal manikin: its development and use. In: *Proceedings of the Fourth International Meeting on Thermal Manikins*, September 2001, pp. 7–12.

61 Lu, Y., Wang, F., and Peng, H. (2016). Effect of two sweating simulation methods on clothing evaporative resistance in a so-called isothermal condition. *International Journal of Biometeorology* 60: 1041–1049.

62 Wang, F., Zhang, C., and Lu, Y. (2015). Correction of the heat loss method for calculating clothing real evaporative resistance. *Journal of Thermal Biology* 52: 45–51.

63 Wang, F., Kuklane, K., Gao, C., and Holmér, I. (2010). Development and validity of a universal empirical equation to predict skin surface temperature on thermal manikins. *Journal of Thermal Biology* 35: 197–203.

64 Havenith, G., Fogarty, A., Bartlett, R. et al. (2008). Male and female upper body sweat distribution during running measured with technical absorbents. *European Journal of Applied Physiology* 104: 245–255.

65 Havenith, G., Richards, M., Wang, X. et al. (2008). Apparent latent heat of evaporation from clothing: attenuation and 'heat pipe' effects. *Journal of Applied Physiology* 104: 142–149.

66 Farnworth, B. (1983). Mechanisms of heat flow through clothing insulation. *Textile Research Journal* 53: 717–725.

67 Hearle, J.W.S. and Morton, W.E. (2008). *Physical Properties of Textile Fibres*, 4e. Cambridge: Woodhead Publishing.

68 Bogaty, H., Hollies, N.R.S., and Harris, M. (1957). Some thermal properties of fabrics: Part I: The effect of fiber arrangement. *Textile Research Journal* 27: 445–449.

69 Kong, L.X., Li, Y., She, F.H. et al. (2001). Effect of fiber geometry and porosity on heat and moisture transfer in textiles. *Proceedings of the Textile Institute 81st World Conference - An Odyssey in Fibers and Spaces*, Melbourne, Australia.

70 Virgo, T.L. and Frost, C.M. (1985). Filled hollow fibres. *Textile Research Journal* 56: 737–743.

71 Zhu, S., Shi, M., Tian, M. et al. (2015). Burning behavior of irradiated PET flame-retardant fabrics impregnated with sensitizer. *Materials Letters* 160: 58–60.

72 Holcombe, B.V. and Hoschke, B.N. (1983). Dry heat transfer characteristics of underwear fabrics. *Textile Research Journal* 53: 368–374.

73 Wilson, C.A., Niven, B.E., and Laing, R.M. (1999). Estimating thermal resistance of the bedding assembly from thickness of materials. *International Journal of Clothing Science and Technology* 11: 262–267.

74 Frydrych, I., Dziworska, G., and Bilska, J. (2002). Comparative analysis of the thermal insulation properties of fabrics made of natural and man-made cellulose fibres. *Fibres & Textiles in Eastern Europe* 10 (4): 40–44.

75 Celcar, D., Meinander, H., and Gersak, J. (2008). A study of the influence of different clothing materials on heat and moisture transmission through clothing materials, evaluate using a sweating cylinder. *International Journal of Clothing Science and Technology* 20: 119–130.

76 Jay, O., Cramer, M.N., Ravanelli, N.M., and Hodder, S.G. (2015). Should electric fans be used during a heat wave? *Applied Ergonomics* 46: 137–143.

77 Cheung, S.S. and McLellan, T.M. (1998). Heat acclimation, aerobic fitness, and hydration effects on tolerance during uncompensable heat stress. *Journal of Applied Physiology* 84: 1731–1739.

78 Kenny, G.P. and Jay, O. (2013). Thermometry, calorimetry, and mean body temperature during heat stress. *Comprehensive Physiology* 3: 1689–1719.

79 Bowler, K. (1981). Heat death and cellular heat injury. *Journal of Thermal Biology* 6: 171–178.

80 Tham, K.W. and Pantelic, J. (2010). Performance evaluation of the coupling of a desktop personalized ventilation air terminal device and desk mounted fans. *Building and Environment* 45: 1941–1950.

81 Zhang, C., Song, W., and Wang, F. (2016). Evaporative cooling techniques to reduce body heat strain in heatwaves. *Proceedings of the 11th International Meeting on Thermal Manikin and Modelling (11i3m)*, Suzhou, China.

82 Ravanelli, N.M., Hodder, S.G., Havenith, G., and Jay, O. (2015). Heart rate and body temperature responses to extreme heat and humidity with and without electric fans. *JAMA* 313: 724–725.

83 Chan, A.P., Song, W., and Yang, Y. (2015). Meta-analysis of the effects of microclimate cooling systems on human performance under thermal stressful environments: potential applications to occupational workers. *Journal of Thermal Biology* 49–50: 16–32.
84 Gao, C., Kuklane, K., Wang, F., and Holmér, I. (2012). Personal cooling with phase change materials to improve thermal comfort from a heat wave perspective. *Indoor Air* 22: 523–530.
85 House, J.R., Lunt, H.C., Taylor, R. et al. (2013). The impact of a phase-change cooling vest on heat strain and the effect of different cooling pack melting temperatures. *European Journal of Applied Physiology* 113: 1223–1231.
86 Kenny, G.P., Schissler, A.R., Stapleton, J. et al. (2011). Ice cooling vest on tolerance for exercise under uncompensable heat stress. *Journal of Occupational and Environmental Hygiene* 8: 484–491.
87 Song, W. and Wang, F. (2016). The hybrid personal cooling system (PCS) could effectively reduce the heat strain while exercising in a hot and moderate humid environment. *Ergonomics* 59: 1009–1018.
88 Barhood, M.J., Davey, S., House, J.R., and Tipton, M.J. (2009). Post-exercise cooling techniques in hot, humid conditions. *European Journal of Applied Physiology* 107: 385–396.
89 Lai, D., Wei, F., Lu, Y., and Wang, F. (2017). Evaluation of a hybrid personal cooling system using a manikin operated in constant temperature mode and thermoregulatory model control mode in warm conditions. *Textile Research Journal* 87: 46–56.
90 Semeniuk, K.M., Dionne, J.P., Makris, A., and Medina, T. (2005). Evaluating the physiological performance of a liquid cooling garment used to control heat stress in hazmat protective ensembles. *Journal of ASTM International* 2: 1–9.
91 Song, W., Wang, F., and Wei, F. (2016). Hybrid cooling clothing to improve thermal comfort of office workers in a hot indoor environment. *Building and Environment* 100: 92–101.
92 Hadid, A., Yanovich, T., Erlich, T. et al. (2008). Effect of a personal ambient ventilation system on physiological strain during heat stress wearing a ballistic vest. *European Journal of Applied Physiology* 104: 311–319.
93 Reffeltrath, P.A. (2006). Heat stress reduction of helicopter crew wearing a ventilated vest. *Aviation Space and Environmental Medicine* 77: 545–550.
94 Vellerand, A.L., Michas, R.D., Frim, J., and Ackles, K.N. (1991). Heat balance of subjects wearing protective clothing with liquid or air-cooled vest. *Aviation Space and Environmental Medicine* 62: 383–391.
95 Teunissen, L.P.J., Wang, L., Chou, S.N. et al. (2014). Evaluation of two cooling systems under a firefighter coverall. *Applied Ergonomics* 45: 1433–1438.
96 Risikko, T. and Anttonen, H. (1997). Use of personal heaters in cold work. *Proceedings of the 1st International Symposium on Problems with Cold Work*, Stockholm, Sweden.
97 Hollies, N.R.S. and Goldman, R.F. (1977). *Clothing Comfort: Interaction of Thermal, Ventilation, Construction and Assessment Factors*. Ann Arbor, MI: Ann Arbor Science Publishers Inc.

98 Choi, K., Chung, H., Lee, B. et al. (2005). Clothing temperature changes of phase change material-treated warm-up in cold and warm environments. *Fibers and Polymers* 6: 343–347.

99 Coca, A., Koscheyev, V.S., Dancisak, M.J., and Leon, G.R. (2002). Optimization of thermal regimes within a liquid cooling/warming garment for body heat balance during exercise. *Medicine and Science in Sports and Exercise* 34: S220.

100 Flouris, A., Westwood, D., Mekjavic, I., and Cheung, S. (2008). Effect of body temperature on cold induced vasodilation. *European Journal of Applied Physiology* 104: 491–499.

101 Chan, C.Y.L. and Burton, D.R. (1982). A low level supplementary heating system for free dives. *Ocean Engineering* 9: 331–346.

102 Kirsi, H., Rissanen, S., Rintamaki, H., and Hzvarinen, V. (2013). Clothing and skin temperatures and heat flow while wearing far infrared heating vest in the cold-a thermal manikin and test subject study. *Proceedings of the 15th International Conference on Environmental Ergonomics (ICEE)*, Queenstown, New Zealand.

103 Rantanen, J., Vuorela, T., Kukkonen, K. et al. (2001). Improving human thermal comfort with smart clothing. *Proceedings of 2001 IEEE International Conference on System, Man and Cybernetics*, New York.

104 Brajkovic, D., Ducharme, M.B., and Frim, J. (1998). Influence of localized auxiliary heating on hand comfort during cold exposure. *Journal of Applied Physiology* 85: 2054–2065.

105 Zuo, J. (2004). Factors affecting the insulation value of sleeping bag systems. PhD dissertation. Manhattan, NY, USA: Kansas State University.

106 Huang, J. (2008). Prediction of air temperature for thermal comfort of people using sleeping bags: a review. *International Journal of Biometeorology* 52: 717–723.

107 EN 13537 (2012). *Requirements for Sleeping Bags*. Brussels: European Committee for Standardization.

108 McCullough, E.A., Huang, J., and Jones, B.W. (2005). Evaluation of EN 13537 and other models for predicting temperature ratings of sleeping bags. *The 11th International Conference on Environmental Ergonomics (ICEE)*, Ystad, Sweden.

109 Kuklane, K. (2009). Footwear for cold weather conditions. In: *Textiles for Cold Weather Apparel* (ed. J.T. Williams), 342–373. New York: CRC Press.

110 Zhang, C., Xu, P., Lai, D. et al. (2016). Electrically heated sleeping bags could improve the human feet thermal comfort in cold outdoor environments. *Industria Textila* 57: 164–173.

111 Song, W.F., Zhang, C.J., Lai, D.D., Wang, F.M., Kuklane, K. (2016). Use of a novel smart heating sleeping bag to improve wearers' local thermal comfort in the feet. *Scientific Report* 6: 19326.

112 Lin, L.Y., Wang, F., Kuklane, K. et al. (2013). A laboratory validation study of comfort and limit temperatures of four sleeping bags defined according to EN13537 (2002). *Applied Ergonomics* 44: 321–326.

113 Zhang, C., Ren, C., Li, Y. et al. (2016). Designing a smart electrical heating sleeping bag to improve wearers' feet thermal comfort while sleeping in a

cold ambient environment. *Textile Research Journal* https://doi.org/10.1177/0040517516651104.

114 Huang, J. (2003). Evaluation of heat loss models for predicting temperature ratings of sleeping bags. PhD dissertation. Manhattan, NY: Kansas State University.

115 Song, W., Zhang, C., Lai, D. et al. (2016). A novel smart heating sleeping bag to improve wearers' local thermal comfort in the feet. *Scientific Reports* 6: 19326.

116 Johnson, R. (2004). Pick the best sleeping bag. *Outdoor Life* 211: 132.

117 Ogulata, T. and Serin, M. (2010). Investigation of porosity and air permeability values of plain knitted fabrics. *Fibres & Textiles in Eastern Europe* 18: 7–10.

118 Taylor, N.A.S., Machado-Moreira, C.A., van den Heuvel, A.M.J. et al. (2009). The roles of hands and feet in temperature regulation in hot and cold environments. *Proceedings of the 13th International Conference on Environmental Ergonomics*, Boston, MA, USA.

119 Wang, F., Kuklane, K., Gao, C., and Holmér, I. (2010). Development and validity of a universal empirical equation to predict skin surface temperature on thermal manikins. *Journal of Thermal Biology* 35: 197–203.

32

Fibers for Radiation Protection

Boris Mahltig

Hochschule Niederrhein, University of Applied Sciences, Faculty of Textile and Clothing Technology, Webschulstraße 31, 41065 Mönchengladbach, Germany

32.1 Introduction

This chapter will present and discuss fibrous materials for radiation protection. For this, fibers for radiation protection are the aim of further discussion. However, at first, some important questions have to be checked before taking a view on different fibers and their properties. These questions are: what is the meaning of the term "radiation protection" and why a radiation protection is important?

Radiation protection can be easily described as protection of human, animals, or materials against destructive influences from electromagnetic radiation. This protection is gained by a decrease of the radiation intensity to lower values or in best case to zero intensity. This is also achieved by a protective material with low transmission properties for a specific type of radiation. The best way to describe is probably a material acting as a shield, and if you are behind this shield, radiation cannot reach you. Now, a second question comes into being: what may be the reason to avoid exposure to radiation? Simple and very general – it can be stated that by electromagnetic radiation a certain amount of energy is transmitted and the energy is transferred to the radiation-exposed material at a certain amount. Such an energy input can damage a living organism or a material to a certain degree. The kind of damage and its intensity are strongly determined by the type of radiation and its strength.

Different types of electromagnetic radiation are roughly summarized in Figure 32.1 and Table 32.1 [1, 2]. The types of radiation are distinguished by their wavelength λ (m), which is related to the frequency ν (s^{-1}) by the following equation: $\nu = c/\lambda$. In this equation, c is the speed of light with $c = 2.998 \times 10^8$ (m/s). The wavelength is also related to the quantum energy of light $E = h\nu$ with the Planck constant $h = 6.626 \times 10^{-34}$ J s [1].

In a simpler picture, the quantum energy can be understood as the energy that is carried by one single light particle called photon. For this, it is often mentioned that light with shorter wavelength contains higher energy. However, in a more exact description, it should be stated that this is the energy for one photon (photon energy) and not the energy of a radiation in total [2]. Of course, also the light

Handbook of Fibrous Materials, First Edition. Edited by Jinlian Hu, Bipin Kumar, and Jing Lu.

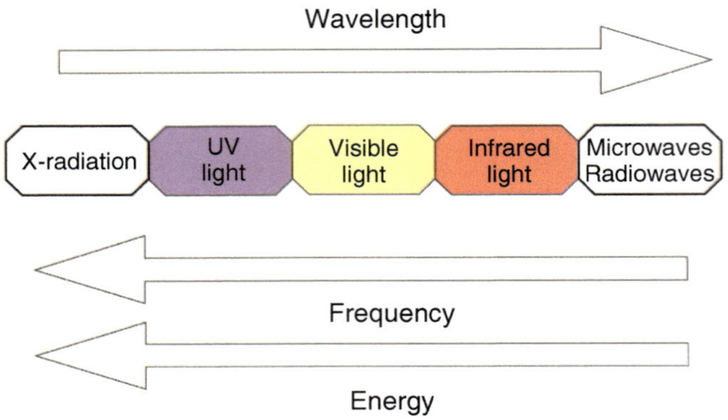

Figure 32.1 Schematic overview on different types of electromagnetic radiation, according to their wavelength, frequency, and energy.

Table 32.1 Overview on different types of radiation, their wavelength, and some potential reasons and applications for radiation protection.

Type of radiation	Range of wavelength (nm)	Reason for protection and potential applications	Protective functional component
Radiowaves, microwaves/electrosmog	$>10^6$	EMI shielding, protection of electronic devices and identity	Electric conductive materials (e.g. elementary metals or carbon in conductive form as graphite, carbon black, or carbon nanotubes [CNTs])
Infrared radiation/IR light	$780–10^6$	Heat management, camouflage	IR reflective materials, e.g. metals or special metal oxides
Visible light/vis	400–780	Optical protection	Dyes or color pigments
Ultraviolet radiation/UV light	10–400	Prevention of sun burn, skin cancer/protection of UV sensitive materials	Organic UV absorbers (e.g. benzotriazoles as Tinuvin); inorganic UV absorbers (e.g. TiO_2, CeO_2, or ZnO)
X-radiation/X-rays	<10	Prevention of radiation damages and cancer	Components containing heavy chemical elements, as lead, $BaTiO_3$, $BaZrO_3$, $BaSO_4$, or Bi_2O_3

Also shown are principles and potential protective components.

intensity influences the energetic impact. This light intensity can be understood as the amplitude of the electromagnetic wave or as the number of photons transporting small units of energy. Many low-energy photons can have in sum the same energy as one high-energy photon.

In summary, electromagnetic radiation can be understood as a spreading wave or a photon containing beam [3]. Besides photons other particles can be part of a beam and form radiation. The probably most prominent ones are α- and β-radiation. α-Radiation contains particles built up from two protons and two neutrons, while β-radiation is formed by electrons. α-, β-, and γ-Radiation are in common understanding often summarized to the general term radioactivity. γ-Radiation is electromagnetic radiation of low wavelength and extremely high photon energy [1].

Another particle radiation is a neutron beam containing neutron particles without any electrical charge. Neutrons have no electrostatic interaction potential to materials due to the absence of charge. For this, their transmission through materials is not hindered by Coulomb interactions to negatively charged electrons or positively charged atomic cores. Compared with other particle radiation, neutrons can transmit very well through most materials. Only if a neutron hits an atomic core its transmission is stopped [4]. Major neutron-absorbing chemical elements are boron and cadmium [5, 6].

An interesting overview on cosmic radiation and possible approaches for protection against them is given by Durante and Cucinotta [7]. In addition to α-, β-, and γ-radiation, cosmic radiation contains significant amounts of protons, muons, neutrons, and several types of heavy ions. It should be clear that this radiation is not just containing many different types of particles. Also these particles are of high energy impact, so a simple only textile based protection is obviously not suitable alone. Only in combination with other protective materials a fiber-based material could result in an suitable material.

After this broad summary according to different types of radiation, the question "why radiation protection is requested" should be picked up again. Now the focus of this question should be on content of these types of radiation, determined by their wavelength as shown in Table 32.1. In Table 32.1 it is clear that each type of radiation has few negative impacts, which makes a radiation protection necessary and in demand (Table 32.1). Roughly the damages caused by radiation can be sort into two categories - acting against human or materials and here especially electrical devices. Radiations with higher wavelength (lower photon energy) like radiowaves or infrared radiation are discussed along with protection and potential applications, e.g. for protection of electrical devices or for heat management systems. The modification of IR reflective properties of textiles is often used to realize a thermal IR camouflage, e.g. for uniforms and corporate wear. In this application a person carrying such a corporate wear is nearly invisible to IR cameras, which is an important issue for military clothes [8].

If this camouflage idea is formally transferred from infrared light to visible light, for visible light as one part of electromagnetic radiation, also the term optical protection can be introduced to the field of radiation protective textiles.

In comparison to these types of radiation, the reason for protection against UV light is much more serious. UV light can cause several serious effects on the skin such as sunburn or skin cancer. Further, the damaging of the eyes even until blindness is possible [9–11]. For this, the main purpose of UV protective fibers or fabrics is to prevent such health issues as skin cancer and sunburns [12].

Another important point is that UV radiation can damage also the fiber material. A UV protective functional fiber can improve the stability of the textile material under the influence of UV light. For example, a higher resistance polyester fiber against UV-B light can be reached after spin doping with a UV protective agent [13].

For radiation with the highest photon energy listed in Table 32.1 (X-radiation), the reason for radiation protection is clear without any doubt. The exposure to X-radiation can cause radiation damages, cell mutation, and cancer and has to be avoided completely [14–16].

Following German law, if X-radiation is used, the operator and the X-radiation have to be in different rooms, separated by appropriate protective material. Only in exceptional cases, e.g. in medicine the in situ observation during an operation, this separation is not possible. However, in these cases personal protection like lead aprons has to be worn by the operator [17–20].

32.2 Structures and Properties of Fibers for Radiation Protection

Before describing and discussing the principles, it is better to discuss some principles of textiles in relation to interaction with radiation. Also, it is necessary to introduce some further fundamental terms to avoid possible misunderstanding in the following subchapters.

Generally, if a textile is exposed to radiation by an incoming beam, the incoming radiation can be roughly separated into three parts related to reflection, absorption, and transmission (Figure 32.2) [12, 21]. First, the transmission T (%) summarizes all radiation that is transmitted through the textile, and the aim of a radiation protective textile is of course to minimize the transmission, in best case to T zero.

Second, the reflection R (%) is related to radiation emitted back. For this, a high reflection will lead to low transmission, so the indication of reflection can be a method to realize radiation protective materials. However, if a dangerous radiation is reflected, it could hit other materials or living persons. Third, the absorption A (%) is the difference between incoming radiation and both reflection and transmission. If the incoming intensity is set to 100, the absorption A (%) can be calculated by A (%) $= 100 - T$ (%) $- A$ (%). The absorption can be understood as radiation uptake or energy uptake by the material, which is exposed to that radiation. For this reason, absorption is the best case to realize a radiation

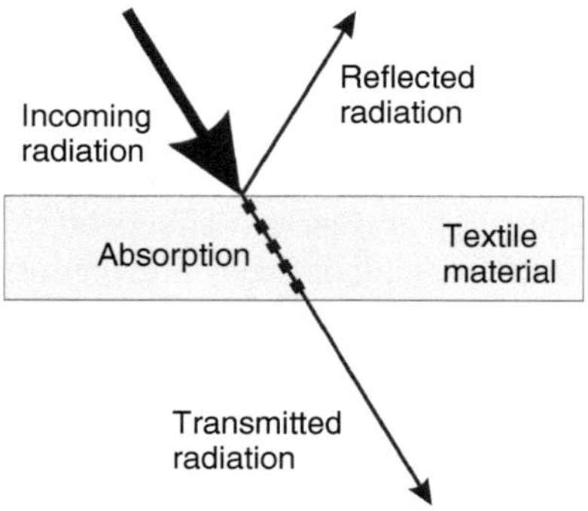

Figure 32.2 Schematic drawing of the radiation transfer through a textile material. The incoming radiation is divided by contact with the textile into reflected, absorbed, and transmitted radiation.

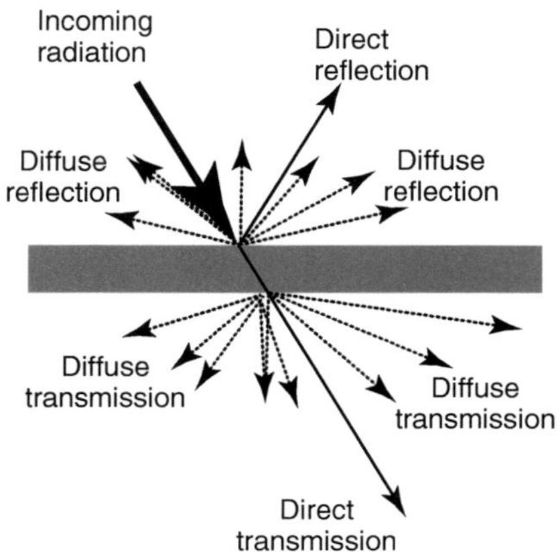

Figure 32.3 Schematic drawing and distinguishing of reflection and transmission (direct and diffuse) of radiation at a textile material.

protection. The absorbed radiation can be transferred to thermal energy, initiate photochemical reactions, or initiate again the emission of radiation in fluorescence or phosphorescence processes [22].

Because photochemical reactions can lead to destruction of materials, in most cases, the transfer of radiation energy to thermal energy is the most wished process for materials with radiation protective properties. This is necessary to gain a long-term stable and active protective material.

A further statement related to transmission and reflection has to be made, which is necessary for measurement arrangements determining the effectivity of a radiation protective material. Both transmission and reflection can be distinguished into a direct and a diffusive part (Figure 32.3) [23, 24].

Direct transmission goes through the textile without changing its direction. This can be described as a direct beam going through the textile material. Besides direct transmission diffuse transmission appears as a result of scattering process that happens during radiation transfer in contact with the textile. This is emitted in all directions, but of course it is similar to direct transmission, which is dangerous. To evaluate the radiation protection quality of a textile material, both direct and diffusive transmission have to be determined and summarized in one number. For this, a measurement arrangement has to be used, which enables the full capture and detection of all diffusive transmitted light. This arrangement can contain a so called integrated sphere, which is collecting the complete diffusive light to a detector [25, 26]. A schematic drawing of such a measurement arrangement is presented in Figure 32.4. Analogously to the transmission also the reflection can be distinguished into a direct reflection and diffusive one. The measurement of the diffuse reflection of visible light is also used to determine the color properties of textiles [27, 28].

The transmission of radiation through a textile is related to its protective function against this type of radiation. For measuring different types of radiation, different types of measurement setup should be used. A rough overview on some examples are given in Figure 32.4 with some schematic drawings in determining the transmission of textile samples for different types of radiation. The general principle for every type of radiation is the same, containing a source

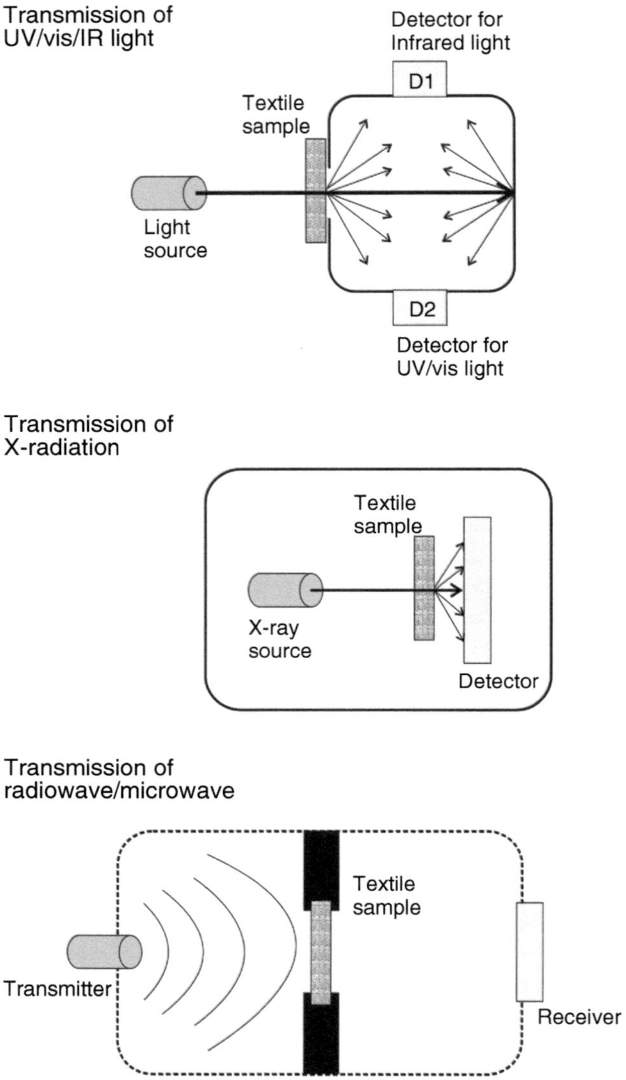

Figure 32.4 Schematic overview on different measurements to determine the transmission of a textile sample for different types of radiation.

for the radiation, a sample holder for the textile sample, and the detector for the radiation. Differences in the arrangements are realized to collect the optimum of transmitted light and for protective reasons.

The first setup is related to the measurement of UV light, visible light, and infrared light. The transmission of those three types of light with the range of wavelengths from 220 to 1400 nm can be determined in one single measurement. An important issue is that the whole transmission (direct and diffusive) is collected in an integrated sphere and recorded by the detectors. The interior of the integrated sphere is equipped with the white pigment barium sulfate, which has also excellent reflective properties for UV and IR light. In the shown arrangement two detectors are used, one for UV and visible light and the other for IR light.

A possible setup for measuring the transmission against X-radiation is presented in Figure 32.4. In this setup the detector for the radiation is place in small distance behind the transmitted textile sample. This small distance is necessary to detect not only the direct but also the diffusive transmitted X-radiation. Another important point due to safety reasons is that the whole measurement setup is placed in a closed box. This protection is necessary to protect the operator from possibly scattered X-radiation.

In the case of measuring the transmission of radiowaves or microwaves, a full covering of the measurement setup is necessary. The reason for this is not only to protect the operator but also to prevent a distortion of the experiment by external radiation sources. Especially to the overall presence of mobile phones, disturbing radiation is nearly everywhere, so this experiment has to be shielded by a suitable coverage. Suitable materials for shielding are polymer materials with embedded carbon black particles as conductive component.

The most common fibers like cotton, polyester, polyamide, or acrylics do not contain any significant radiation protective property [29]. They have only low absorption capacity for any type of electromagnetic radiation. The reflection properties are low as well. Synthetic fibers as polyester or polyamide often contain titanium dioxide as white pigment. By this pigments, besides a high reflectivity for visible light, a certain absorption of UV light is introduced to the fiber. In this case, the fibers carry an additional component – the TiO_2 – containing the protective property [28]. In contrast wool fibers are reported to support UV protection by themselves. Wool is a protein fiber built up by 20 amino acids as natural monomers [30]. From these 20 amino acids, the amino acids tryptophan, tyrosine, and phenylalanine contain aromatic structures. These aromatic structures introduce UV absorption properties to the wool fiber. From those three mentioned amino acids, tryptophan exhibits the best UV absorption properties probable because it contains an aromatic ring system with nitrogen as heteroatom [31, 32]. The presence of aromatic amino acids is probably the reason why for wool high UV protection factor (UPF) values are reported [33]. It should be remarked here that silk mostly is of low UV protective potential, although it is a protein fiber [34, 35]. However, the silk protein is almost built up by the amino acid glycine. Depending on the origin of the silk, the glycine content is in the range of 20–66%. Other amino acids that are as well occurring in high amounts in different silk proteins are alanine, serine, glutamine, and lysine [34]. All those amino acids do not contain any aromatic structure leading to absorption of UV light with wavelength >250 nm. For this, no UV protection property is introduced to pure silk fibers.

Aramid fibers as technical fibers contain a chemical structure with conjugated aromatic systems as repeating unit, or this reason aramid absorbs strongly UV light and also significant amounts of visible light, clearly indicated by the yellow coloration of aramid fibers. However as UV protective fiber, aramid is not useful, due to its high decomposition rates under UV illumination. In contrast, different methods are investigated to stabilize aramid fibers against the influence of UV light [36, 37].

A polymer from natural based resources reported for sun-blocking applications is lignin [38]. Lignin is a biopolymer occurring in higher plants and is present together with cellulose and hemicelluloses in wood [39]. Lignin is evaluated and reported for high-performance broad-spectrum sunscreens [38]. However, similar to aramid lignin exhibits a small stability under long-term exposition to UV light.

In general it should be concluded that if a radiation protection property is wished, a functional additive – carrying this protective property – has to be added, because most common fibers do not contain a radiation protection by themselves. An overview on different types of radiation protective components and additives is given in Table 32.1. In Table 32.1 it is distinguished into different types of radiation requiring different types of protective functional components.

Besides the chemical nature of the protective functional component, its placement in the fiber has a significant influence on its effectivity for radiation protection. A simple schematic overview on different possible arrangements is presented in Figure 32.5. Table 32.2 summarizes some production methods for each arrangement and gives some statements concerning fixation of the function and its durability.

The best radiation protection can be of course gained by a self-functional fiber containing a radiation protective material. In other words, the fiber material is the protective material itself. An example would be a fiber made from pure copper. Due to the high amount of active agent (the copper), the protection is excellent. In other cases, lower effectivity is reached by embedding the same radiation protective material, e.g. as copper pigments, into the fiber during the spinning process. Coating and traditional finishing processes yield to lower amounts of

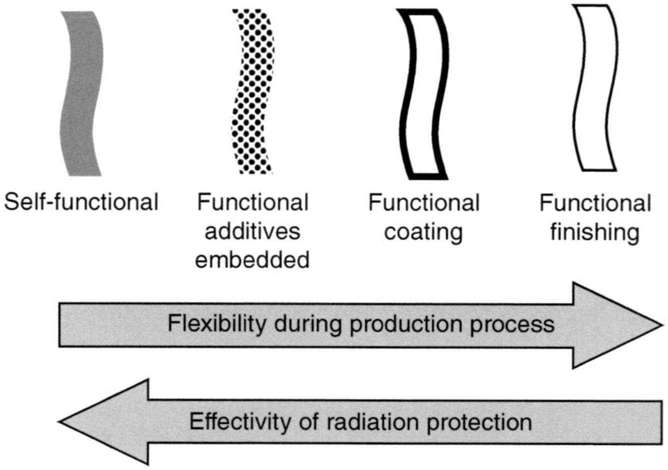

Figure 32.5 View on different positions of a functional component in a fiber. Shown is the relation to effectivity of functionalization – radiation protection – and the flexibility of the production process.

Table 32.2 Comparison of different positions of a functional component in a fiber according to possible production method and place in the textile production chain where the decision for the function is fixed.

Position of function in the fiber	Production method/realization of function	Fixation of function according to textile chain; limits
Self-functional	Selection of self-functional polymer or inorganic material for fiber production	Before fiber production; limited only to few materials
Embedded functional additives	Addition of functional pigments or particles during fiber production by melt spinning/dry or wet spinning	During fiber production; variations possible by combination of different pigments and fiber materials; limited to fiber production process
Functional coating	Coating deposition by different methods as wet chemical method, sol–gel coating, electrodeposition, deposition from gas phase	After fiber production; the mechanical properties of the fibers can be strongly influenced by the coating
Functional finishing	Finishing process analogously to yarn dyeing or fabric dyeing	After fiber or fabric production; the fiber/dye (additive) interaction has to be recognized to gain good fastness properties; compared with other methods smaller amounts of functional additive is applied

applied functional component. The amount of the active components applied is decreased in this order, and by this the effectivity of the protective functionalization is as well determined (Figure 32.5).

However, vice versa the flexibility of the application process is the highest for the finishing application and lowest for the self-functional fiber.

The term flexibility is related to the position in the textile production chain, where the functional additive is added. In other words, if the decision to apply a protective function has to be made in the beginning of the fiber production, the flexibility in the process is low. Vice versa, if the protective function is applied near the end of the textile chain by a dyeing or printing process, the flexibility of the whole application is quite high. An overview in relation to the textile chain is supported in Table 32.2.

The self-functional fiber is produced and dedicated mainly for only one or few types of application. The embedding of functional protective additives inside a synthetic or regenerated fiber enables the production of a fiber containing the advantageous properties of the functional additives and the organic fiber. These fibers contain higher flexibility, due to more possibilities in creation of new material combinations and properties. However, the decision for fiber properties has to be made in the beginning of the fiber production process. The properties are fixed for afterward performed production steps. The highest flexibility is gained by functional finishing usually performed as treatment with an aqueous-based finishing agent at the already prepared yarn or fabric. In best case, a standard conventional polyester or cotton material is supplied and functionalized by a finishing agent, if a certain demand for a certain functional product is requested by the customer. The disadvantage here is that the amount of functional additives applied onto the fiber materials is lower compared with the cases "self-functional" or "embedded additives," so with finishing applications usually lower protection rates are realized. Also by finishing processes usually no inorganic pigments can be applied and fixed permanently onto the fiber surface.

The functional coating is in a certain way similar to the finishing process, because the coating is performed mostly onto already prepared fibers. However, the coating techniques, e.g. vapor deposition or electrochemical deposition, are more challenging and of higher technical effort compared with traditional textile finishing. For this, the coating is related to lower flexibility during production process. However, by coating usually higher amounts of functional additive can be applied to the fiber, so the finally gained effect is advantageous compared with the functional material prepared by finishing.

Radiowaves and microwaves require for effective shielding electric conductive materials [40, 41]. Appropriate materials are conductive metals and their alloys. Metallized fibers containing copper, nickel, silver, or different alloys are often mentioned and are used to produce knitted materials with conductive yarns. Alternatively to metals the coating with conductive polymers can be used [40, 42]. Also useful is carbon in its conductive forms as graphite, carbon black pigments, carbon nanotubes, or graphene and its modifications [43–46]. Carbon fibers with nickel coating are obtained by electroplating method and can be used for electromagnetic interference shielding [47].

Carbon nanotubes can be used as additive in fibers. However, yarns completely prepared from carbon nanotubes are realized [48] and combined in further modification with the carbon type of graphene [49]. Alternatively to carbon fibers, so-called carbonaceous fiber materials can be used to reach effective shielding against microwave radiation in the range of 30–1000 MHz [50]. The preparation starts here from PANOX fiber materials (supplier Carbon Fibers Inc.), which are partly pyrolyzed polyacrylonitrile fibers. On those fibers a further thermal treatment is done, probably to introduce graphite-like structure with electric conductive properties [50].

Protection mechanism against IR radiation is most commonly based on reflective materials [51]. High IR reflection is often gained with pure metals (especially aluminum pigments) [51, 52]. Alternatively special metal oxide pigments are used [53, 54].

Nevertheless, it should be mentioned that especially for near-IR light, organic IR absorbers are developed and described in patent literature [55, 56]. Prominent examples for such organic IR absorbers are metal complexes from dithiene type as depicted in Figure 32.6. The central metal in these complexes can be not only nickel but also other transition metals, e.g. palladium [55]. Other colorless organic IR absorbers are dithiolene metal complexes as presented in Figure 32.7 [56].

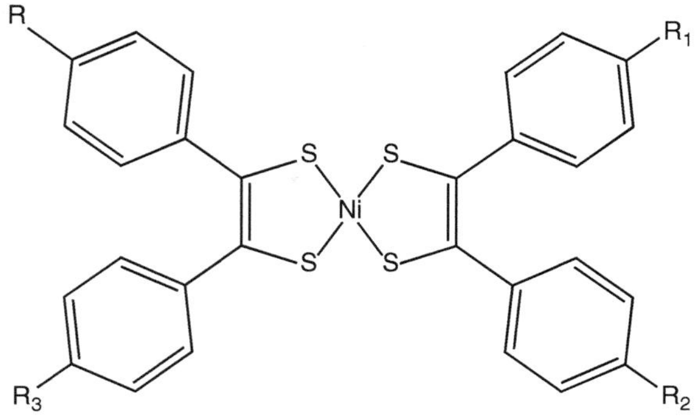

Figure 32.6 Chemical structure of an organic IR absorber – dithiene complex derivative. The central metal atom can be not only nickel but also palladium or platinum. The side groups R to R_3 are ether groups or amine groups [55].

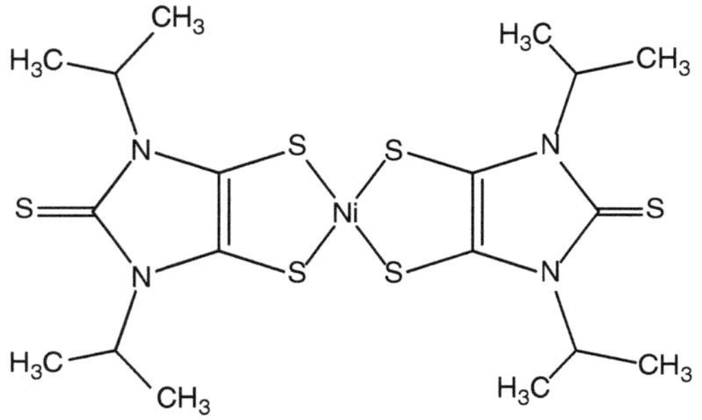

Figure 32.7 Chemical structure of an organic IR absorber – dithiolene metal complex derivative. The central metal atom can be not only nickel but also another metal [56].

Another interesting approach for IR protective materials based on absorption mechanisms are related to transparent conductive oxides (TCO). Such TCO materials are currently developed and modified for textile purposes by the DTNW (Krefeld, Germany) and the CHT (Tübingen, Germany). These institutions developed TCO nanoparticles prepared by a sol–gel process useful as textile finishing agents. By this, comparably high interactions with heat radiation and IR light is introduced to the treated fabric [57]. In further developments, these TCO nanoparticles could be also introduced into synthetic or regenerated fibers by spin doping.

While protection against IR light is realized in most applications by reflection, in most cases the absorption of UV light is used to introduce protection properties. One reason for this is probably that for UV light many excellent UV absorbers are available.

UV absorbers can be roughly distinguished into two groups – organic UV absorbers and inorganic UV absorbers [58]. Inorganic UV absorbers are usually white pigments of semiconductive nature. Prominent examples are titanium dioxide (TiO_2) and zinc oxide (ZnO) but also cerium(IV) oxide (CeO_2) [59]. These pigments can be added into synthetic or regenerated fibers by addition to the spinning mass. An alternative but not commercially used method is the application of TiO_2 or ZnO as sol–gel coating to apply UV protective function [60–63].

A sol–gel finishing agent based on TiO_2 was developed for UV protective purposes by the company CHT (Tübingen, Germany) under the name isys UVP. The stability of this sol–gel system was excellent, so even hydrothermal processes were possible with it without starting gelation [64]. Recently developed TiO_2 materials for UV protective applications are prepared by hydrothermal methods. By this method, manifold different titania nanoparticles can be realized and applied in a sol–gel coating process [65, 66]. A further advantage is the possible combination with additional functional properties as self-cleaning or oil/water separation [66].

Most used UV absorber is the TiO_2, which is often used as white pigment in synthetic fibers (TiO_2; C.I. Pigment White 6) [67]. However, compared with UV protective applications for use as white pigment, only small amounts of titania are used; e.g. a white polyamide fiber is reported to contain only 0.012% titania. In contrast, different UV protective commercially available clothes for outdoor and sportswear contain titanium in the range of 0.22–0.85% [68]. These values are determined from the fibers by ICP-OES method. Further, it was stated that the TiO_2 in these investigated fibers is embedded inside the fibers as additive during the spinning production. For this, the washing stability of those products is almost excellent [68, 69].

Organic UV absorbers can occur in various chemical structures [58]. Some examples are given in Figures 32.8 and 32.9 [70–72]. These UV absorbers contain as basic functional component a chromophore for absorbing UV light as 2-hydroxybenzophenone or 2-(2-hydroxyphenyl)benzotriazole basic structures [70]. Another basic structure is from phenylacrylester derivatives (Figure 32.9) [71, 72].

Two main properties are the main requirements for an organic UV absorber. First, of course they have to absorb UV light over a broad range of wavelength and with high effectivity. Second, the UV absorber should not be decomposed under the exposure of UV light. The uptake of UV light by an organic molecule transfers this molecule to an excited electronic state [73].

Besides the return in the former ground state, this energy uptake can start a photochemical reaction leading to the decomposition of the organic molecule [73]. Organic components used as effective UV absorbers are containing a simple

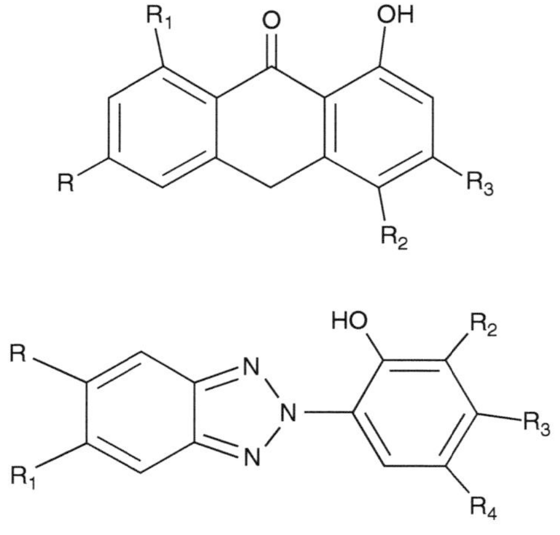

Figure 32.8 Chemical structures of organic UV absorbers with (a) 2-hydroxybenzophenone and (b) 2-(2-hydroxyphenyl)benzotriazole basic structures [70].

Figure 32.9 Chemical structure of an organic UV absorber – type phenylacrylester derivative. Source: Adapted from Rauch et al. 2004 [71] and Böttcher et al. 2006 [72].

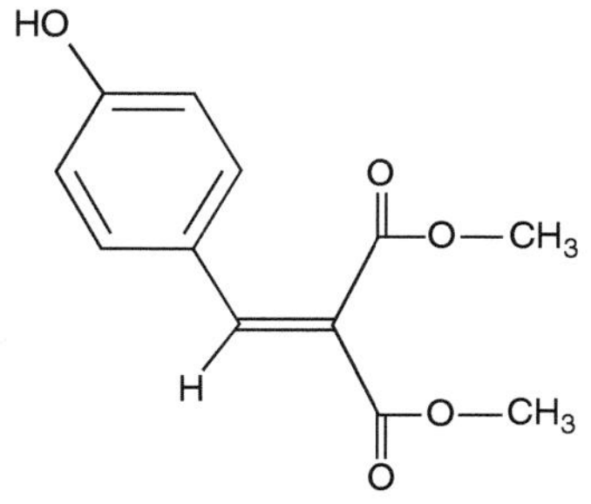

mechanism that converts the uptaken UV light into heat energy; by this a destructive photochemical reaction is avoided [74].

Another important point related to the application of UV absorbers on textile materials is their fixation to the fiber material. This fixation is the main condition for washing stability. To reach an adequate fixation a strong interaction between fiber surface and the UV absorber has to be developed. For this, UV absorbers are developed containing not only the functional chromophore but also functional groups leading to interactions to different fiber surfaces.

Simply spoken, organic UV absorbers can be applied using finishing processes similarly used with traditional dyes. UV absorbers carrying a negative net charge in their molecular structure can be applied like an acid dye onto positively charged fibers as wool or polyamide [69]. The negative charge can be introduced to the UV absorber by a sulfonic group [70]. An example for such a dye is presented in Figure 32.10. This organic UV absorber has the basic structure of oxalic diarylamide. The sulfonic group $-SO_3Na$ is the negatively charged anchor group for application and fixation onto wool or polyamide fiber material. Also an application analogously to a direct dye onto cellulosic fibers is mentioned [70]. This UV absorber is useful not only for applications as clothes but also for other sun protective products as awnings or parasols.

UV absorbers with reactive anchor groups like chlorotriazine or vinylsulfone are dedicated for an application onto cotton or other cellulosic fibers, because of the formation of permanent covalent bonds between this reactive anchor group and the hydroxyl group onto the surface of the cellulosic fiber. An example for such a UV absorber containing both a chlorotriazine group and a vinylsulfone

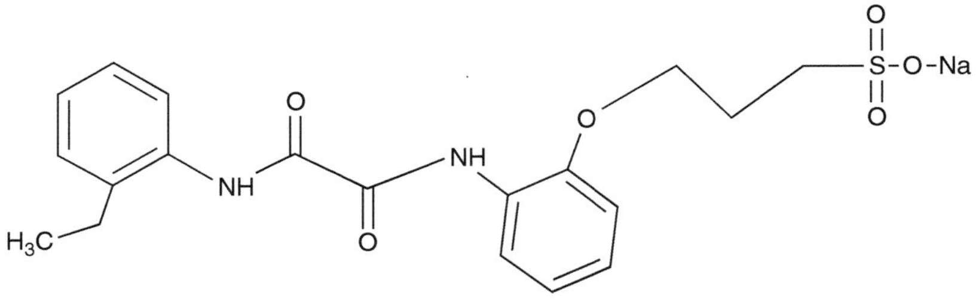

Figure 32.10 Chemical structure of an organic UV absorber with the basic structure of oxalic diacrylamide. The sulfonic group $-SO_3Na$ is the negatively charged anchor group for application and fixation onto wool or polyamide fiber material [70].

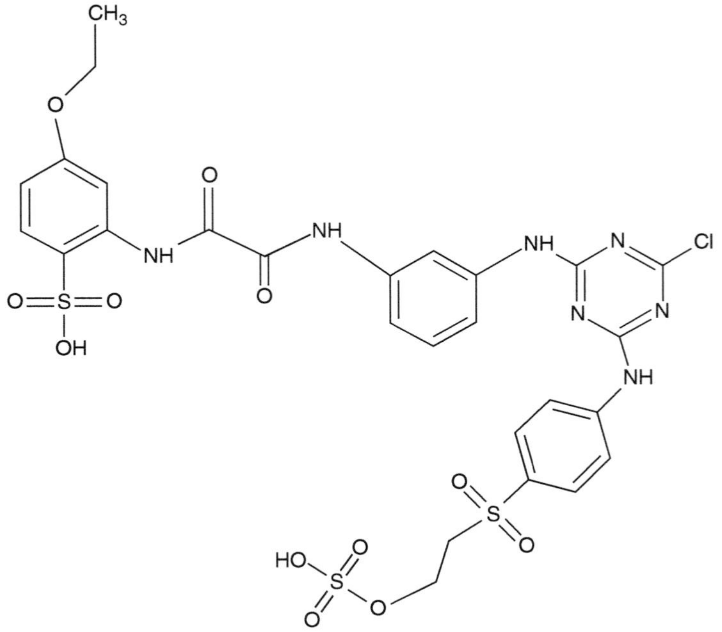

Figure 32.11 Chemical structure of an organic UV absorber with the basic structure of oxalic diacrylamide. This UV absorber contains chlorotriazine group and the vinylsulfone group, two reactive anchors useful especially for the application on cotton or other cellulosic fibers.

group is depicted in Figure 32.11. The basic structure of this UV absorber is of oxalic diacrylamide type [75]. The final vinylsulfone anchor group is formed under the alkaline conditions of the finishing bath, by splitting of a sulfate ion.

For application onto hydrophobic synthetic fibers as polyester, hydrophobic UV absorbers are applied as dispersion using conventional high-temperature (HT) processes.

This application process is similar to the application of conventional disperse dyes in an HT process. HT indicates a process temperature for the dyeing of $T > 100\,°C$, which should be significantly above the glass temperature of the synthetic polyester fiber. Usually process temperatures of 120–130 °C are used to promote the diffusion of the disperse dye into the polymer matrix of the polyester fiber [76]. For UV absorber application an aqueous dispersion based on 2-hydroxybenzophenone derivatives is reported, which is suitable for the application on polyester fabrics (Figure 32.12) [77].

A further mechanism is the transfer of UV light into visible light by fluorescence processes. In this case the energy of the UV light is not converted to heat, as it is the case for common UV absorbers. Instead fluorescence dyes transfer the UV light into visible light. Fluorescence dyes are most commonly used as optical brightener and are useful compounds for UV protective purposes. A prominent structure often found in fluorescence dyes and whitening agents is the stilbene structure. Especially triazinylaminostilbenes are mentioned to be excellent UV protective agents for textile treatment [29]. An example of such a stilbene

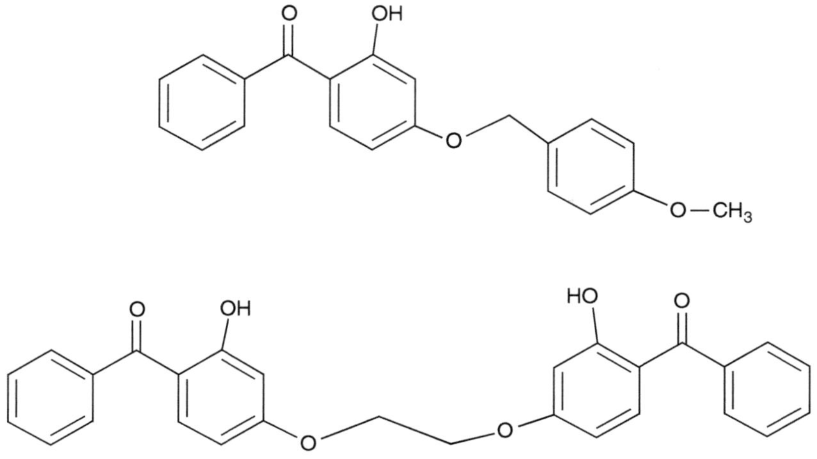

Figure 32.12 Chemical structures of organic UV absorbers. Different examples for UV absorbers of 2-hydroxybenzophenone derivatives for application in aqueous dispersions for the treatment of polyester fibers [77].

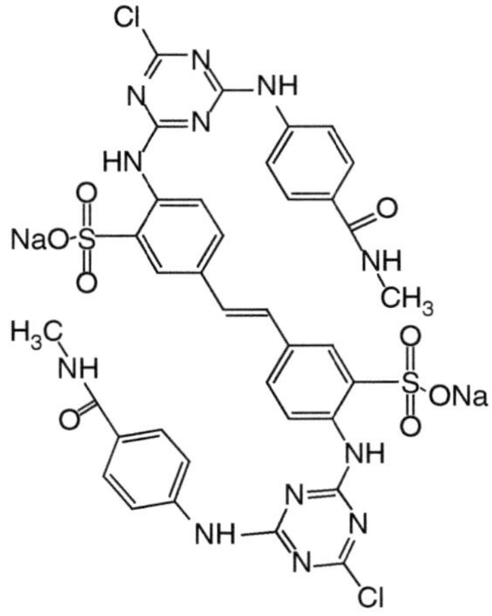

Figure 32.13 Chemical structure of a fluorescence whitening agent (optical brightener) with UV protective properties – stilbene type. Both monochlorotriazine rings are reactive anchors especially for fixation onto cellulosic fibers. The sulfonate groups $-SO_3Na$ support the solubility of this compound in water [78].

derivative is given in Figure 32.13 [78]. The depicted stilbene fluorescence dye contains two monochlorotriazine rings, which are acting as reactive anchor to fix the dye onto cellulosic fibers as cotton or viscose. The sulfonate groups $-SO_3Na$ support the solubility of this fluorescence dye in water [78].

A quite novel development in the field of UV protective materials is the application of graphene materials [79, 80]. Graphene nanoplates can be used as UV absorber onto cotton fibers. Even with amounts of 0.4 wt% applied graphene, significant UV protection can be realized [79]. Alternatively graphene oxide nanosheets can be used to realize UV protective materials based on cotton fibers. The combination with conductive synthetic polymers like polyaniline and this leads not only to the UV protection but also to conductive properties of the modified cotton materials [80].

The absorption capacity of a material for X-ray increases mostly as functions of the atomic weight of the chemical elements building up the material [81, 82]. Textile fibers are usually built up by lightweight atoms as hydrogen, carbon, oxygen, and nitrogen, so common textiles are almost completely transparent for X-ray. Effective X-ray absorbing components are containing heavy metals in elementary form, e.g. lead or in a component like HfO_2, $BaSO_4$, or Bi_2O_3 [83, 84]. For this, a functional component supporting X-ray protective properties to a fiber should contain chemical elements as lead, barium, or bismuth. However the decision which element or compound is the most appropriate has to be done under the aspects of toxicity and costs. Lead, which is traditionally used in lead aprons for X-ray protection purposes, is of high human toxicity and should be therefore not introduced in textile fibers [85]. Chemical element like gold or platinum are may be also effective X-ray absorbers, but their prize is far away from a possible textile application. In comparison, barium sulfate is a suitable compound. Soluble barium compounds like $BaCl_2$ or $Ba(NO_3)_2$ are toxic, because they release the toxic Ba^{2+} ion. However, the solubility of barium sulfate is extremely low, so it can be even used in medical applications [86]. Barium sulfate $BaSO_4$ is a white and nontoxic compound, which is used also as an X-ray contrast agent for medical applications. Its costs are in the range that is acceptable for textile applications. An analogous suitable material is bismuth oxide Bi_2O_3.

Because of the high photon energy of X-radiation and the high penetration depth of this radiation, usually a high amount of X-ray absorptive material is needed to gain a suitable protection against X-radiation. For this, in contrast to the UV protective application, for X-radiation an application of X-ray absorbers in a finishing process is not useful, because by finishing the amount of applied agents is too low. For suitable X-ray protection the functional additive has to be applied in high concentration inside the fiber or with a thick coating on the fabric material. The application of thick coatings with lead or other heavy metal compounds is the state of the art for actually commercialized products, as lead aprons [87–89]. The disadvantage of such thick coatings is that the textile properties are lost and the weight of the material is quite high. Besides the loss of the textile comfort, significant health problems can be the result from carrying high weight protective clothes [90]. Also cracks and defects can occur as results of long-term use and repeating of bending and folding [91–93].

As mentioned above, the durability of the textiles and its radiation protective properties is an important issue for usage. Beside the principal function, the durability is the most important parameter for successful commercialization of new materials. A schematic overview on the aspects of life cycle and durability of textiles for radiation protective purposes is given in Figure 32.14.

After preparation of the textile sample, the protective properties can be determined by transmission measurements with different setups as summarized in Figure 32.14. This measurement proves the principal function of the materials. The question for durability of this function is related to the point that the function be present in the future after a certain duration of usage. On this used sample the transmission measurement can be performed again, and the difference in transmission ΔT before and after usage is a parameter for the durability of the function. If ΔT is zero, best durability is realized.

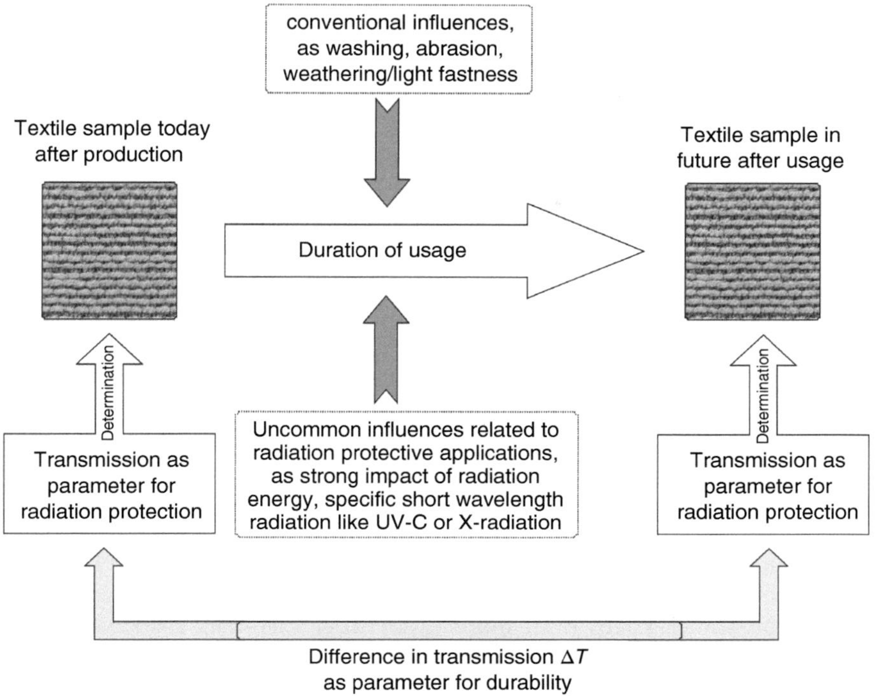

Figure 32.14 Schematic overview on aspects of life cycle and durability of textiles for radiation protective purposes.

For common textiles several conventional influences are taken into account for evaluation of the durability. Mainly these influences are the washing fastness, abrasion stability (crock fastness), and light fastness, which is also sometimes understood as weathering fastness. For durability investigations, usually the textile samples are intensively treated with one of those issues, and afterward the durability is graded with a number. An example could be the testing of the wash fastness of a UV absorber, which is applied in process related to a reactive dye. In this case, the transmission of UV light is tested before and after a repeating washing procedure, which is related to the conditions of usage, e.g. washing several times in a commercial washing machine at 40 °C with standard detergent.

One issue that is especially important for the durability of textile for radiation protective purposes are so-called uncommon influences, which are especially related to the situation of radiation exposition. Especially for applications related to protection against high-energy radiation of UV-C light or X-radiation, this is an important issue. This high-energy radiation with short wavelengths is absorbed by the textile material and can start photochemical reactions, which are able to destruct the textile sample. This important issue has to be tested and investigated especially for protective materials developed for long-term durations.

32.3 Functions, Performance, Evaluations, and Applications

In this section some typical examples for application of radiation protective fibers are summarized. These examples are related to different parts of the electromagnetic spectrum, and besides the principal action of the material also, hints on already successful commercialized materials are given.

32.3.1 Radiowave/Microwave Shielding

The shielding of radiowaves and microwaves is mostly done by electroconductive fibers or yarns [94–96]. Some important suppliers of metallized fibers, yarns, or fabrics are, e.g. Statex GmbH, Marktek Inc., Swicofil, and YSHIELD [97–101]. Often used are fibers from pure metals or synthetic fibers that are metallized – coated with a very thin metal layer [102]. Very prominently reported are silver-coated polyamide fibers, where the silver coating is deposited by electrochemical processes [103]. For electroplating of polyacrylonitrile fibers with silver, the application of primer compounds is useful to improve the durability of the silver coating. Advantageous primer compounds are aminosilanes or mercaptosilanes [104]. Even more exotic fiber materials are silver-coated polybenzoxazole (PBO) filament yarns [105].

The electrochemical deposition of copper or nickel is reported as well. For this, often at first, small amount of other noble is applied to the fiber, probably to promote an afterward copper deposition and especially the fixation of the copper on the fiber surface [106]. An overview on some metallized synthetic fibers and fabrics is given recently [107]. The electrochemical metal deposition is done from solution containing the required metal ions, so the solubility of an appropriate metal salt is a requirement for this deposition technique. Also more noble metals with higher redox potential are deposited by these electrochemical methods more easily. In contrast, metals like aluminum can be only deposited from solution, if strong reductive agents are present, e.g. $LiAlH_4$ [108].

In a US patent Okayasu et al. report on a method described as nonelectrolytically plating of a previously activated polyester textile [109]. The pretreatment is done with compounds of tin and palladium. Afterward metals like nickel, copper, cobalt, chromium, or their alloys are deposited. Twenty years later Onggar and Cherif try to repeat this wet chemical deposition of elementary silver onto polyester textile fabrics [110]. They used a combinatory deposition of silver ions with different amino compounds and a following reduction step with ascorbic acid. This type of deposition process leads normally to the formation of silver nanoparticles on the fiber surface [110]. However, to reach an appropriate conductivity for shielding purposes, the use of nanoparticles is usually not adequate, because of high specific surface area of the nanoparticles.

Another method to realize metallized fiber materials is the vacuum deposition [111–114]. This procedure is especially useful for metals with lower boiling point and less noble metals, which cannot be deposited easily by electrodeposition. Commercially available aluminum-coated fiber materials are usually prepared by

Table 32.3 Overview on products commercialized for EMI shielding by YSHIELD GmbH (Ruhstorf, Germany).

Product	Type	Color appearance	Shielding at 1 GHz (dB)	By EDS detected metals
Evolution	Woven fabric	White	30	Silver, copper
Steel twin	Woven fabric	Light gray	35	Iron, chromium, nickel
Steel gray	Woven fabric	Gray	35	Iron, chromium, nickel
Silver tricot	Knitted fabric	Gray	47	Silver
Silver tulle	Knitted fabric	Gray yellow	50	Silver
HEG03	Mesh	Metallic gray	55	Iron, chromium, nickel
Silver silk	Woven fabric	Metallic silver	60	Silver
HNG80	Woven fabric, netting	Metallic yellow silver	80	Copper, nickel
HNV80	Nonwoven, fleece	Metallic yellow silver	80	Copper, nickel

The shielding rate for radiation with frequency of 1 GHz is supported by the supplier.
Source: Data from YSHIELD GmbH 2016 [101].

this technique. One problem that can occur during vacuum deposition is related to the thermal energy of the evaporated metals. The impact of thermal energy can be transferred to the textile substrate, and by this a destruction of the textile material is possible.

Table 32.3 gives an overview on typical textiles currently offered for electromagnetic shielding [101]. The supplying company – YSHIELD, Germany – reports the shielding effectivity for radiation with 1 GHz frequency, which is in the range of microwave radiation. The shielding effectivity is in the range of 30–80 dB. The color of the fabrics differs from white and gray to a metallic appearance.

Different types of metals are detected on the fiber materials by using energy-dispersive spectroscopy (EDS). Mainly detected are silver, copper, and nickel. Also found are samples containing the metals iron, nickel, and chromium. These three metals are related to the occurrence of metal fibers from stainless steel. These materials are comparable with nonwoven fabrics made from stainless steel and polyester fibers reported by Ozen et al. [115]. From those products, mainly three categories should be distinguished, also shown as scanning electron microscopy (SEM) images in Figure 32.15.

First, there are standard fabrics containing only few metallic fibers. In the SEM images fibers containing chemical elements with higher atomic number are shown with stronger brightness, so the metal-containing fiber can be easily distinguished in these SEM images from standard textile fibers as polyester or cotton. This statement is proven by the EDS measurements supporting a lateral resolution of the element distribution onto the fiber surface (Figure 32.16).

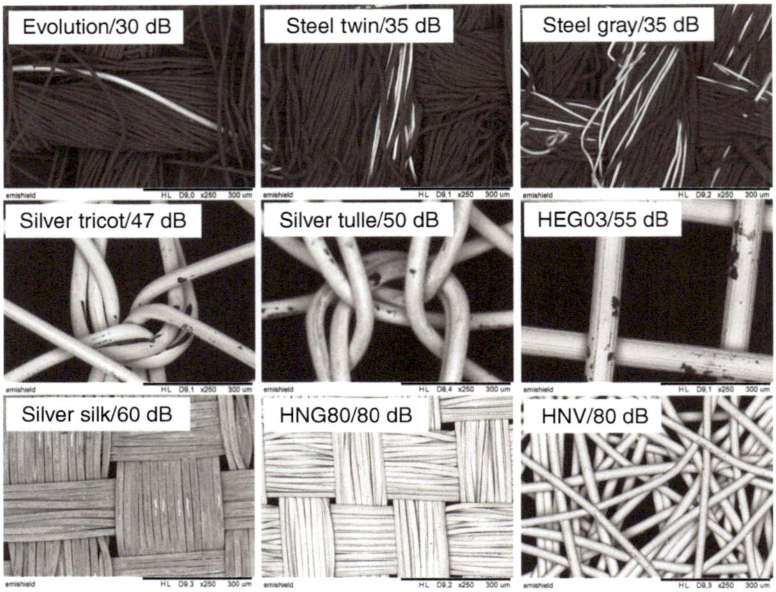

Figure 32.15 Overview on products commercialized for EMI shielding by YSHIELD GmbH (Ruhstorf, Germany); SEM images of different fabric samples recorded with a magnification of 250×. The type of fabric sample and the shielding effect against 1 GHz (supplier information) are directly indicated in each image.

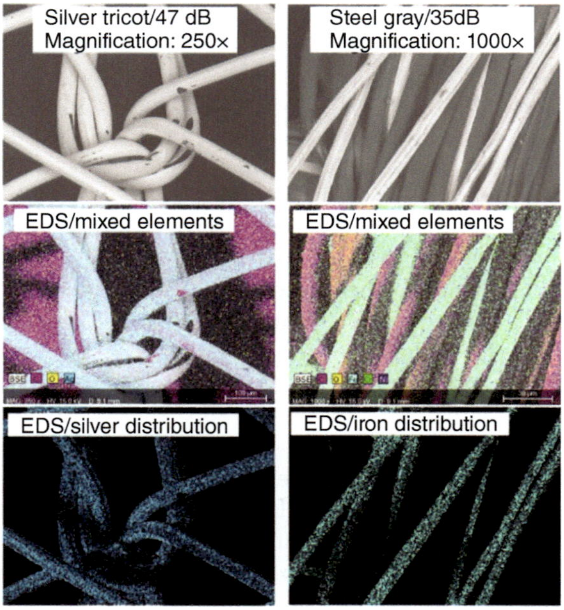

Figure 32.16 Overview on two products commercialized for EMI shielding by YSHIELD GmbH (Ruhstorf, Germany); SEM images and EDS images of different fabrics are shown. The type of fabric sample, the shielding effect against 1 GHz (supplier information), and the magnification for microscopic measurements are directly indicated in each image.

In the first category of shielding materials, shielding rates are in the range of 30–35 dB. These samples have a hand feeling and appearance comparable with common textiles. Their color is white or gray, and dyeability comparable with standard textiles can be expected. Altogether the application as a common textile for clothing and home textiles is possible.

The second category summarizes materials with a shielding rate around 50 dB. These fabrics are built up only from metallized fibers arranged in an open structure. These structures are knitted or an open mesh. Such metallized knitted net fabrics are known and developed in the early 1980s and dedicated especially for the protection against microwave radiation [116]. In an US patent from Ebneth et al., particularly nickel-coated knitted net fabrics are described, and the protective properties against microwave with frequencies from 1 GHz to 36 GHz are tested. A further important field of application is the eye protection [116]. These fabrics have a clearly metallic appearance and hand feeling. They have an open structure, so visible light can transmit in a certain amount. Altogether, these materials are certainly dedicated for application as curtains, window coverages, or technical applications.

The last third category is related to highest shielding rates from 60 to 80 dB. In this category, all fibers are metallized and form a dense structure as woven fabric or nonwoven felt. In summary for these three categories, the effectivity of EMI shielding increases by the number of metallized fibers on a certain area on the fabric, which could be also described as cover factor [117]. An example for a certain product is presented in Figure 32.17 with a bag containing EMI shielding properties commercialized for mobile phones. A mobile phone inside this bag is hidden and cannot reached by electromagnetic radiation. For this, the personal identity is covered. As shown the metal-plated fabric is placed inside the bag. This

Figure 32.17 Bag for mobile phones with EMI shielding properties with metal-plated fabric inside, product of company Electronic Wall (Kinding, Germany).

metal fabric seems to be similar to the silver silk fabric presented in Table 32.2 and Figure 32.17.

32.3.2 IR Shielding/Heat Management

The example of IR shielding is often related to the application in heat management. Those are textile materials like a metallized mesh proposed for the application of a heat protective curtain. If these fabrics are used as curtain for a window, their metallized surface reflects the IR radiation from sun back to outside of the building. By this, parts of the heat radiation are prevented from entering the building, and a heating indoors is avoided. Additionally the open structure of these materials supports still a certain transmittance for visible light, so an additional illumination indoors is not necessary. This fact is important, because illumination realized by artificial lamps is related to heat production and energy consumption. In winter time the situation is vice versa, the temperature inside the building is higher, and an aim would be to keep the heat indoors. Here a curtain with an infrared reflective side oriented to the room is useful.

An SEM images of a knitted fabric dedicated for heat management is presented in Figure 32.18. This fabric contains an open structure, to keep a certain transmittance for visible light. One side of the fabric is coated with aluminum and contains a metallic appearance. The other side is black. The reflection and transmission spectra of such knitted fabrics are presented in Figure 32.19.

Here two types of fabrics are compared. One is aluminum based, and the other one is steel based on the metal side. The transmission is over the investigated spectral range from 220 to 1400 nm wavelength nearly similar in the range of 35–45%. There is no significant difference if the aluminum- or steel-based fabric is measured and if the measurement is done from the metal or the black side of the fabric. A difference is observed by reflection measurement. The knitted fabric

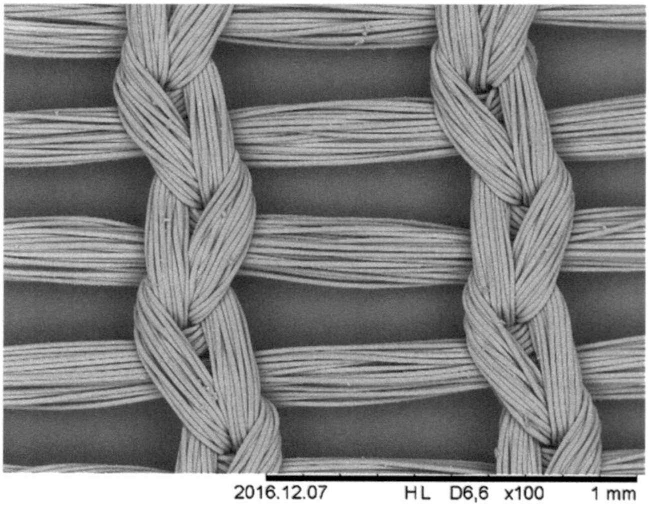

Figure 32.18 SEM image of knitted fabric dedicated for heat management, aluminum-based system.

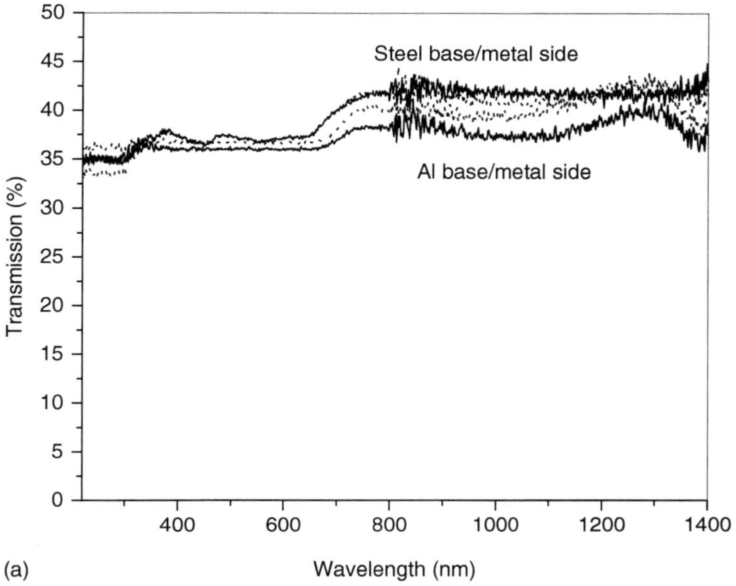

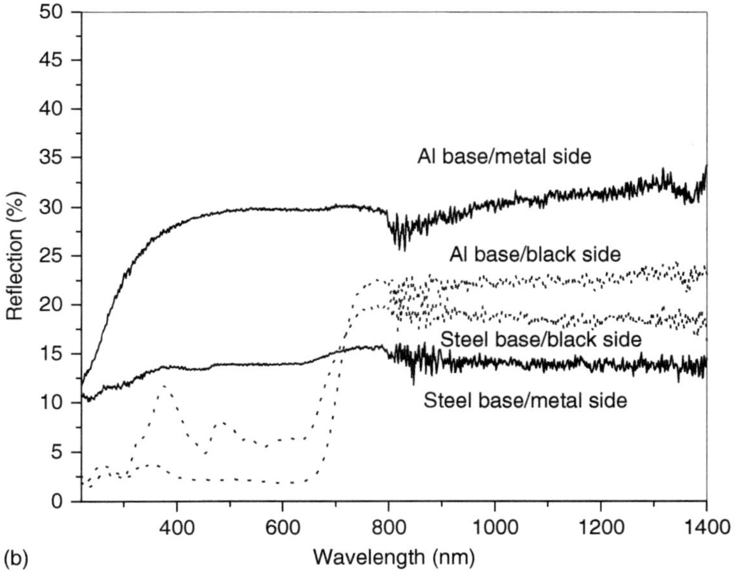

Figure 32.19 Spectra of diffuse transmission (a) and diffuse reflection (b) of fiber mesh from metallized fibers. The spectrum range summarizes UV, visible, and near-IR light. The metallized fiber meshes are measured from metal and from black side. The comparison is done with an aluminum-based mesh and a steel-based mesh.

with aluminum side exhibits the strongest reflection for visible and for infrared light (Figure 32.19). The steel-based system is of lower reflectance for IR light. Further, the reflection from the black side of both materials strongly depends on the type of light. For visible light only low reflection is observed, probably because of the absorption. However infrared light is still reflected in significant ration of around 20%.

32.3.3 UV and IR Protection

For an application supporting both UV and IR protection, a fiber containing a significant amount of TiO_2 can be mentioned. TiO_2 has excellent UV absorption and IR reflection properties, so a fiber with large amounts of this white pigment should be an effective material for this application.

In patent literature it is reported from ultraviolet protective fabrics based on regenerated cellulosic fibers [118]. These cellulosic fibers are prepared by modal or lyocell process and containing TiO_2 as UV absorber. The TiO_2 is applied in amount of 1 wt% and is originally supplied by the company Kronos [118]. Such UV protective cellulosic fibers with TiO_2 are commercialized under the name Tencel Sun by the company Lenzing. Appropriate products are on the market for outdoor apparel since the year 2009 [119].

Such a fiber could be a regenerated cellulosic fiber with incorporated TiO_2 as depicted in Figure 32.20. The SEM image of this fiber is recorded in a measurement mode giving not only the topography but also a material contrast. Areas containing the element titanium are shown brighter compared with the areas containing only cellulose. In fact, the embedded TiO_2 particles can be detected, and it can be also stated that the size of those particles is quite small compared with usually used pigment particles. The use of smaller inorganic particles improves the stability of the cellulosic fibers and enables further production

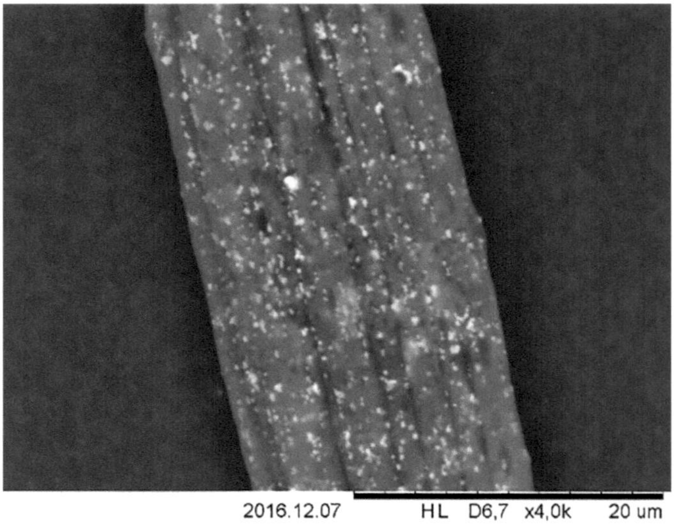

Figure 32.20 SEM image of a viscose fiber with embedded titania, useful for UV protective purposes.

processes as spinning, knitting, or weaving. The elementary content of titanium determined by EDS is around 5% on the fiber surface. This amount of titanium is roughly 10 times more as the amount of TiO_2 used in conventional synthetic fibers as polyester or polyamide. In conventional fiber TiO_2 is used to introduce a certain white color effect to the fiber.

A knitted structure produced form such TiO_2 containing cellulosic fibers contains transmission near to 0% for the complete UV range till 400 nm (Figure 32.21). This zero transmission is reached by the excellent absorption

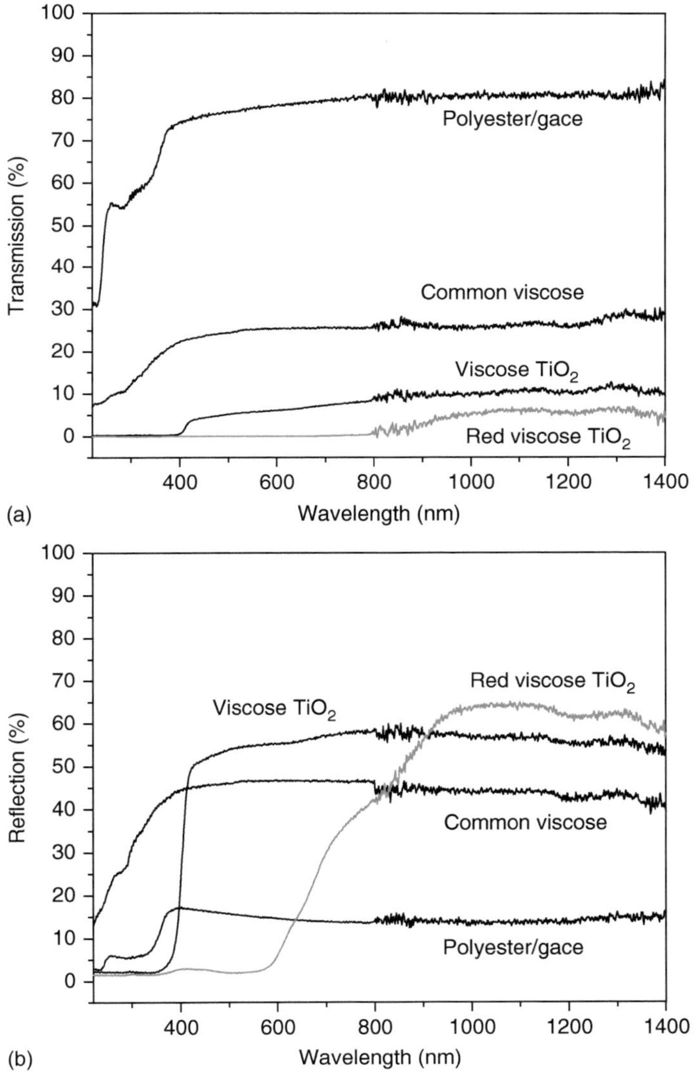

Figure 32.21 Spectra of diffuse transmission (a) and diffuse reflection (b) of different viscose fabrics and an open polyester gace. The spectra range summarizes UV, visible, and near-IR light. The titania containing viscose fibers are compared with a common viscose fabric and a common polyester.

behavior of TiO_2. The transmission of IR light (near-IR light 750–1400 nm is measured) is low with values of smaller 20%. However, the coloration of the textile substrate can have a certain influence. This is shown with spectra from that cellulosic fiber after dyeing with a red dye (Figure 32.21). The reflection of those colored samples depends strongly on the type of radiation. For UV light, the reflection is as well near zero, obviously because all UV light is absorbed by the TiO_2. For the IR light higher reflection of more than 50% are observed, probably because of the good reflectance properties of the embedded TiO_2 particles. For IR light mainly the reflective properties are responsible to gain the IR protective effect. An attractive development could be a combination with an organic absorber for near-IR light applied during a conventional dyeing process.

32.3.4 X-Ray Shielding

Figure 32.22 shows a SEM image of a cellulosic fiber containing $BaSO_4$ as X-ray absorber. The concentration of $BaSO_4$ is 40 wt%. This fiber is produced by the lyocell process at the TITK (Rudolstadt, Germany) [120, 121]. $BaSO_4$ is added to the spinning solution and embedded into the fiber during the wet spinning process. The SEM measurements are done in a measurement modus giving not only the topography information but also a certain material contrast. By this, chemical elements with higher atomic number are shown brighter, so $BaSO_4$ particles near

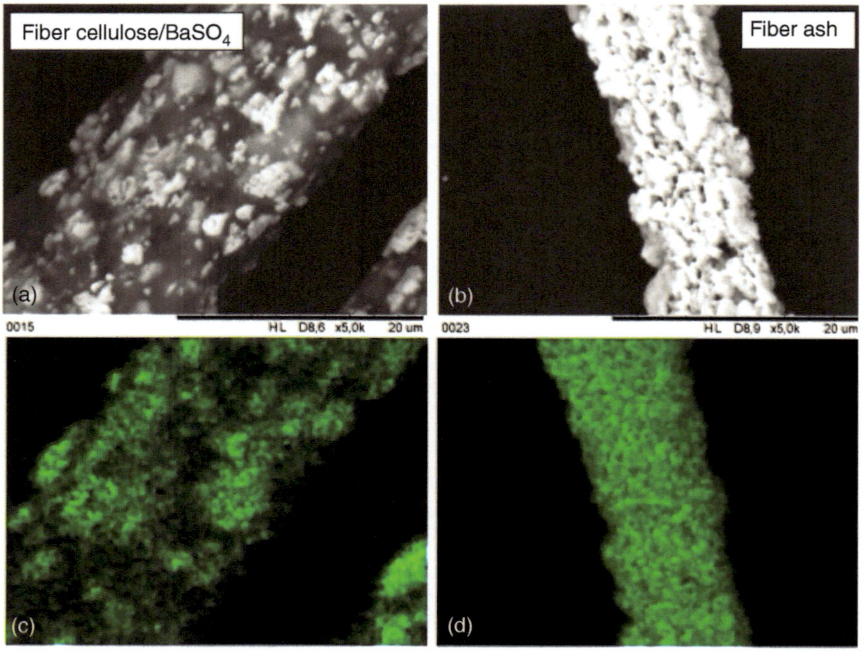

Figure 32.22 Comparison of cellulosic composite fiber with barium sulfate and the ash of this fiber. The images in panels (a, b) are recorded by SEM. The images in panels (c, d) present the distribution of the element barium mapped by EDS measurements.

the fiber surface can be clearly detected as white spots. The related EDS measurements show similar barium distribution on the fiber surface, so the estimation made from SEM investigations is proven.

After ignition of this cellulosic fiber by a flame, a fast burning follows, and a fibrous ash remained as residue. A SEM image of this fiber ash is shown in Figure 32.22. The linear structure of the fiber is still present, and the remaining structure could be well described as a kind of fiber skeleton built up by the non-burning inorganic compound $BaSO_4$. The structure of the ash seems to be denser, and the fiber diameter is slightly decreased. The denser structure is shown in the analogous EDS image presenting the barium distribution in the ash (Figure 32.22).

Such barium sulfate-containing cellulosic fibers are also described as composite fiber containing an organic component, cellulose, and an inorganic component, the barium sulfate as X-ray absorber [121, 122]. Not only barium sulfate but also other inorganic compounds can be embedded inside a cellulosic fiber to realize fiber materials for X-ray protection [84, 121]. An overview on SEM images of different fibers is shown in Figure 32.23. The idea of using barium titanate or barium zirconate instead of the barium sulfate is to replace the sulfur by a heavier element to improve the X-ray absorption properties.

It should be stated here that it is of course quite simple to produce a composite material built up from cellulose and barium sulfate. However, for X-ray

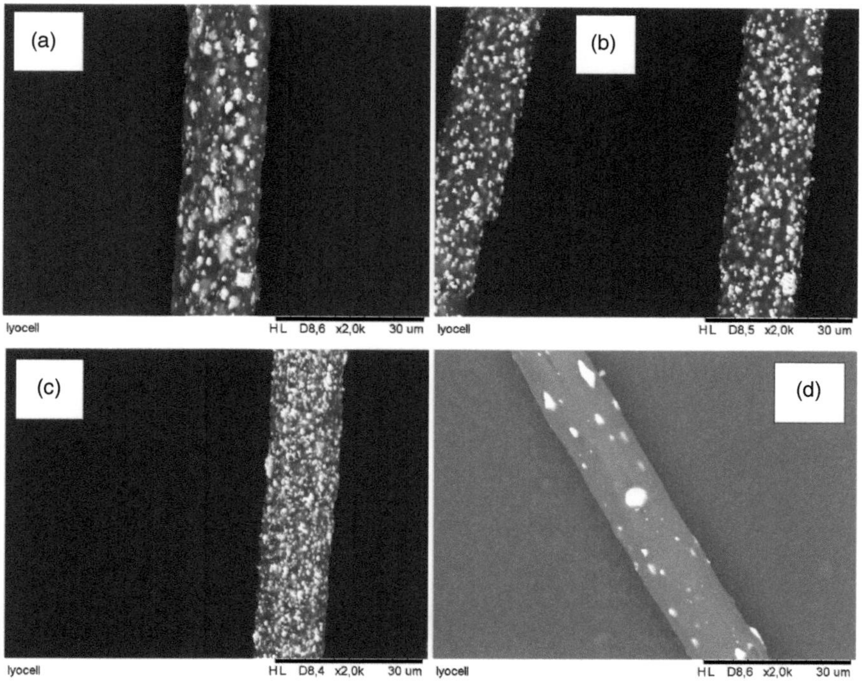

Figure 32.23 SEM images of cellulosic fibers with different embedded inorganic compounds as X-ray absorbers. (a) $BaSO_4$, (b) $BaTiO_2$, (c) $BaZrO_3$, and (d) Bi_2O_3.

protective purposes, fibers with high amounts of inorganic additive are needed, and for those fibers, disadvantageous changes in mechanical stability should be expected. For this, often after the fiber production, the preparation of a nonwoven material is described and investigated for X-ray protective properties [122]. In contrast the preparation of a yarn and later a fabric from those composite fibers is more challenging [84, 121, 123]. During spinning, knitting, or weaving processes, strong mechanical forces are working onto the fiber, and only if the preparation of composite fiber is especially developed and dedicated for textile production that the textile processes can be performed. Also the spinning processes have to be modified to realize an X-ray protective composite yarn [84].

However, the fabric production from such composite yarns is successfully realized, and some product examples are presented in Figures 32.24 and 32.25. It was demonstrated that the production process along the complete textile chain works to realize in the end a clothing product made from X-ray protective fibers [123]. A special idea for application was here to realize a hand protective device as shown in Figure 32.24 or a completely knitted glove for use in mobile X-ray analysis. Mobile analysis of elemental composition is often done by use of X-radiation to determine, for example, the amount of different metals building up an alloy. The operator is usually not exposed to the direct X-ray beam, but backscattering effects and fluorescence can lead to a certain exposition to radiation. For this, a certain protection for the hand and arms is quite suitable.

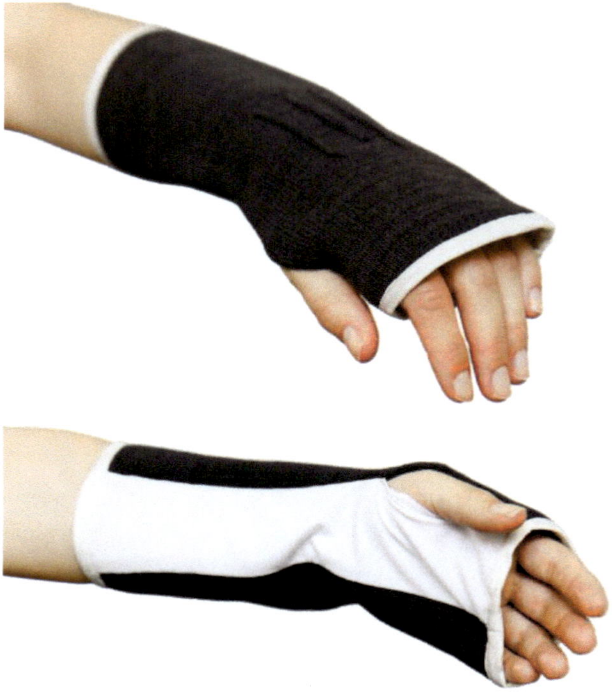

Figure 32.24 Hand protecting device constructed from a cellulose/inorganic composite fiber.

Figure 32.25 Protective jacket constructed from a cellulose/inorganic composite fiber.

The transmission of such textile materials against X-rays is presented in Figure 32.26. The X-ray transmission is shown as function of the photon energy of the X-rays, and it is clearly seen that the transmission is significantly decreased by the developed textile materials. In comparison an ordinary textile has a transmittance for X-ray of near 100%. However it is also obviously that the developed composite textile is not able to support a transmission near to zero. Especially for higher photon energies the transmission is decreases. This absorption behavior is caused by the absorption properties of the used inorganic materials. Also related to this are the marked absorption edges in Figure 32.26,

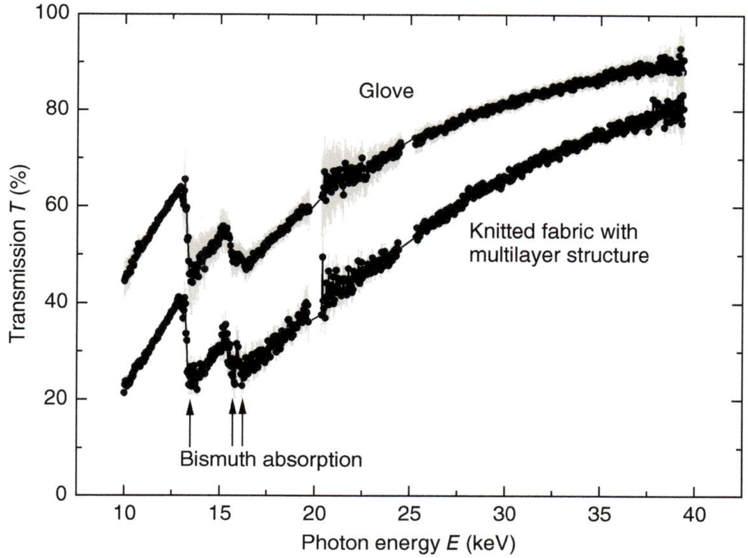

Figure 32.26 Transmission spectra for different knitted structures prepared from X-ray protective fibers.

which are typical for each chemical element. For this, the areas of applications have to be selected carefully, and not only the use as protective materials during material analytics but also the additional use together with traditional lead aprons could be a very perspective field for future applications.

32.4 Future Trends

Radiation protection is an important topic and a significant argument for technical applications and special clothes and of course in marketing. For each type of radiation, a certain reason and demand of radiation protection are raised and significant argument for usage. In general, it can be estimated that the demand for radiation protective fibers and textiles will be increased in the future, especially due to the increasing awareness of the customer to the fact of dangerous risks of radiation. Not only private customers but also the use in official facilities as hospitals, research institutes, police, or military organizations can benefit from radiation protective textile. Besides the protection itself, other applications are important as hiding the electronic identity or the heat camouflage.

A special prospective and attractive field of application will be probably the heat or light management, which is strongly related to radiation protection. The use of light management textile materials is a prospective method to lower the energy consumption for keeping buildings warm. Heating energy is cost saving, so this application is a very promising one.

To reach best protection properties, probable materials will be developed combining different types of functional additives in different arrangement and combinations. Special effective developments could be composite fibers combined with a coating and produced in knitted or woven materials.

Acknowledgments

For fruitful long-term cooperation, many helpful discussions, and experimental help, the presenter would like to thank many colleagues and students. Especially mentioned are Prof. Bendt, Prof. Breckenfelder, Prof. Kyosev, and Prof. Weide (all from Hochschule Niederrhein); Dr. Leisegang (from Saxray/Dresden); Dr. Krieg (from TITK/Rudolstadt); and Prof. Textor (Hochschule Reutlingen). Many thanks are also owed to K. Günther and A. Askani (yarn production); C. Giebing (fabric production); K. Bredlich, F. Bohnert, and N. Brinkert (knitting experiments); and M. Cont (help with the photographs). This work is partly financially supported by IGF project 17783BG/1 of the research association Forschungskuratorium Textil e.V. Reinhardtstraße 12-14, 10117 Berlin, and is supported by the AiF within the scope of the support program of the Industrielle Gemeinschaftsforschung und -entwicklung (IGF) of the Ministry of Economics and Technology due to a decision of the German Bundestag. For the funding of the electron microscope equipment, the author acknowledges very gratefully the program FH Basis of the German federal country North Rhine-Westphalia

NRW. All product and company names mentioned in this chapter may be trademarks of their respective owners, also without labeling.

References

1 Gerthsen, C. and Vogel, H. (1993). *Physik*, 17e. Berlin: Springer-Verlag.

2 Göpel, W. and Ziegler, C. (1994). *Struktur der Materie*. Stuttgart: B.G. Teubner Verlagsgesellschaft.

3 Atkins, P.W. (1993). *Quanten*. Weinheim: VCH-Verlagsgesellschaft.

4 Skoog, D.A. and Leary, J.J. (1996). *Instrumentelle Analytik*. Berlin: Springer-Verlag.

5 Carter, R.S., Palevsky, H., Myers, V.W., and Hughes, D.J. (1953). Thermal neutron absorption cross sections of boron and gold. *Physical Review* 92: 716.

6 Rainwater, J. and Havens, W.W. (1946). Neutron beam spectrometer studies of boron, cadmium, and the energy distribution from paraffin. *Physical Review* 70: 136.

7 Durante, M. and Cucinotta, F.A. (2011). Physical basis of radiation protection in space travel. *Reviews of Modern Physics* 83: 1245–1281.

8 Lottenbach, R. (2016). Innovative fabrics for uniforms, corporate wear and personal protection equipment. *Aachen-Dresden-Denkendorf International Textile Conference*, November 2016.

9 Horneck, G. (1995). Quantification of the biological effectiveness of environmental UV radiation. *Journal of Photochemistry and Photobiology B: Biology* 31: 43–49.

10 Brasch, D.E., Rudolph, J.A., Simon, J.A. et al. (1991). A role for sunlight in skin cancer. *Proceedings of the National Academy of Sciences of the United States of America* 88: 10124–10128.

11 Cerkova, J., Stipek, S., Crkovska, J. et al. (2004). UV rays, the prooxidant/antioxidant imbalance in the cornea and oxidative eye damage. *Physiological Reviews* 53: 1–10.

12 Saravanan, D. (2007). UV protection textile materials. *AUTEX Research Journal* 7: 53–62.

13 Teng, C. and Yu, M. (2003). Preparation and property of poly(ethylene terephthalate) fibers providing ultraviolet radiation protection. *Journal of Applied Polymer Science* 88: 1180–1185.

14 Albertini, R.J. and DeMars, R. (1973). Somatic cell mutation detection and quantification of X-ray induced mutation in cultured, diploid human fibroblasts. *Mutation Research* 18: 199–224.

15 Chambers, C.E., Fetterly, K.A., Holzer, R. et al. (2011). Radiation safety program for the cardiac catheterization laboratory. *Catheterization and Cardiovascular Interventions* 77: 546–556.

16 Chida, K., Kato, M., Kagaya, Y. et al. (2010). Radiation dose and radiation protection for patients and physicians during interventional procedure. *Journal of Radiation Research* 51: 97–105.

17 Fischer, F.M. (2009). Klinischer Vergleich computer-assistierter und konventioneller Implantationsverfahren. PhD thesis. Humboldt-Universität, Berlin.
18 Bohnsack, O. (2005). Reduktion der Strahlenexposition bei CT-Fluoroskopie-gesteuerten Interventionen: Möglichkeiten und Grenzen einer segmentalen Röhrenabschaltung. PhD thesis. Ludwig Maximilians Universität, München.
19 Dresing, K. (2011). Röntgen in Unfallchirurgie und Orthopädie. *Operative Orthopädie und Traumatologie* 1: 70–78.
20 Wucherer, M. and Loose, R. (2005). Berufliche Strahlenexposition. *Radiologie* 45: 291–303.
21 Mahltig, B., Zhang, J., Wu, L. et al. (2017). Effect pigments for textile coating – a review on the broad range of advantageous functionalization. *Journal of Coatings Technology and Research* 14: 35–55.
22 Hesse, M., Meier, H., and Zeeh, B. (2008). *Spectroscopic Methods in Organic Chemistry*. Stuttgart: Georg ThiemeVerlag.
23 Reinert, G., Fuso, F., Hilfiker, R., and Schmidt, E. (1997). UV-protecting properties of textile fabrics and their improvement. *Textile Chemist and Colorist* 29: 36–43.
24 Yu, Y., Hurren, C., Millington, K.R. et al. (2017). Research on the influence of yarn parameters on the ultraviolet protection of yarns. *The Journal of the Textile Institute* 108: 178–188.
25 Gambichler, T., Hatch, K.L., Avermaete, A. et al. (2002). Influence of wetness on the ultraviolet protection factor (UPF) of textiles: in vitro and in vivo measurements. *Photodermatology, Photoimmunology and Photomedicine* 18: 29–35.
26 Neumann, B. (2017). Von UV bis IR. *GIT-Labor-fachzeitschrift* 61: 12–17.
27 Alvarez, J. and Lipp-Symonowicz, B. (2003). Examination of the absorption properties of various fibres in relation to UV radiation. *AUTEX Research Journal* 3: 72–77.
28 Hoffmann, K., Laperre, J., Avermaete, A. et al. (2001). Defined UV protection by apparel textiles. *Archives of Dermatology* 137: 1089–1094.
29 Bajaj, P., Kothari, V.K., and Ghosh, S.B. (2000). Review article – some innovations in UV protective clothing. *Indian Journal of Fibre and Textile Research* 25: 315–329.
30 Asquith, R.S. and Garcia-Dominguez, J.J. (1968). New amino acids in alkali-treated wool. *Coloration Technology* 84: 155–158.
31 Lehninger, A.L., Nelson, D.L., and Cox, M.M. (1994). *Prinzipien der Biochemie*. Heidelberg: Spektrum Akademischer Verlag.
32 Jastorff, B., Störmann, R., and Wölcke, U. (2003). *Struktur-Wirkungs-Denken in der Chemie*. Bremen – Oldenburg: Universitätsverlag Aschenbeck & Isensee.
33 Algaba, I., Riva, A., and Crews, P.C. (2004). Influence of fiber type and fabric porosity on the UPF of summer fabrics. *AATCC Review* 4: 26–31.
34 Mongkholrattanasit, R., Krystufek, J., Wiener, J., and Vikova, M. (2011). Dyeing, fastness, and UV protection properties of silk and wool fabrics dyed

with eucalyptus leaf extract by the exhaustion process. *Fibres & Textiles in Eastern Europe* 19: 94–99.
35 Ebert, G. (1993). *Biopolymere*. Stuttgart: B.G. Teubner Verlagsgesellschaft.
36 Liu, X., Yu, W., and Xu, P. (2008). Improving the photo-stability of high performance aramid fibers by sol–gel treatment. *Fibers and Polymers* 9: 455–460.
37 Milford, J.G.N. (1972). Aromatic polyamide fibers containing ultraviolet light screeners. US Patent 3888821A, filed 2 November 1972 and issued 10 June 1975.
38 Qian, Y., Qiu, X., and Zhu, S. (2015). Lignin: a nature-inspired sun blocker for broad-spectrum sunscreens. *Green Chemistry* 17: 320–324.
39 Jakubke, H.-D. and Jeschkeit, H. (1993). *Concise Encyclopedia Chemistry*. Berlin: Walter de Gruyter.
40 Maity, S., Singha, K., Debnath, P., and Singha, M. (2013). Textiles in electromagnetic radiation protection. *Journal of Safety Engineering* 2: 11–19.
41 Wang, X., Liu, Z., and Zhou, Z. (2014). Virtual metal model for fast computation of shielding effectiveness of blended electromagnetic interference shielding fabric. *International Journal of Applied Electromagnetics and Mechanics* 44: 87–97.
42 Saini, P. and Choudhary, V. (2013). Electrostatic charge dissipation and electromagnetic interference shielding response of polyaniline based conducting fabrics. *Indian Journal of Pure and Applied Physics* 51: 112–117.
43 Azim, S.S., Satheesh, A., Ramu, K.K. et al. (2006). Studies on graphite based conductive paint coatings. *Progress in Organic Coatings* 55: 1–4.
44 Johnson, J.A., Barbato, M.J., Hopkins, S.R., and OMalley, M.J. (2003). Dispersion and film properties of carbon nanofiber pigmented conductive coatings. *Progress in Organic Coatings* 47: 198–206.
45 Singh, B.P., Bharadwaj, P., Choudhary, V., and Mathur, R.B. (2014). Enhanced microwave shielding and mechanical properties of multiwall carbon nanotubes anchored carbon fiber felt reinforced epoxy multiscale composites. *Applied Nanoscience* 4: 421–428.
46 Umrao, S., Gupta, T.K., Kumar, S. et al. (2015). Microwave-assisted synthesis of boron and nitrogen co-doped reduced graphene oxide for the protection of electromagnetic radiation in Ku-band. *ACS Applied Materials & Interfaces* 7: 19831–19842.
47 Wang, R., Yang, H., Wang, J., and Li, F. (2014). The electromagnetic interference shielding of silicone rubber filled with nickel coated carbon fiber. *Polymer Testing* 38: 53–56.
48 Jiang, K., Li, Q., and Fan, S. (2002). Spinning continuous carbon nanotube yarns. *Nature* 419: 801.
49 Foroughi, J., Spinks, G.M., Antiohos, D. et al. (2014). Highly conductive carbon nanotube-graphene hybrid yarn. *Advanced Functional Materials* 24: 5859–5865.
50 McCullough, F.P., Novak, L.R., and Hall, D.M. (1989). Camouflage material. US Patent 5, 312, 678, filed 6 October 1989 and issued 17 May 1994.
51 Gunde, M.K. and Kunaver, M. (2003). Infrared reflection-absorption spectra of metal-effect coatings. *Applied Spectroscopy* 57: 1266–1272.

52 Sutter, C.R., Petelinkar, R.A., and Reeves, R.E. (1998). Infrared reflective visually colored metallic compositions. US Patent 6468647B1, filed 17 February 1998 and issued 22 October 2002.
53 Bendiganavale, A.K. and Malshe, V.C. (2008). Infrared reflective inorganic pigments. *Recent Patents on Chemical Engineering* 1: 67–79.
54 George, G., Vishnu, V.S., and Reddy, M.I.P. (2011). The synthesis characterization and optical properties of silicon praseodymium Y_6MoO_{12} compounds: environmentally benign inorganic pigments with high NIR reflectance. *Dyes and Pigments* 88: 109–115.
55 Kauffman, M.H. (1985). Synthesis of transition metal dithiene complexes. US Patent 4, 593, 113, filed 14 January 1985 and issued 3 June 1986.
56 Lehmann, U. and Heizler, D. (2007). Dithiolene metal complex colorless IR absorbers. US Patent 2010/0021833A1, filed 18 December 2007 and issued 28 January 2010.
57 Straube, T., Mayer-Gall, T., Textor, T., and Gutmann, J.S. (2016). TCO-based nanoparticle coatings for heat-shielding applications in textile architecture. *Aachen-Dresden-Denkendorf International Textile Conference*, November 2016.
58 Hilfiker, R., Kaufmann, W., Reinert, G., and Schmidt, E. (1996). Improving sun protection factors by applying UV-absorbers. *Textile Research Journal* 66: 61–70.
59 Han, Y., Zhao, Y., Xie, C., and Powers, J.M. (2010). Color stability of pigmented maxillofacial silicone elastomer: effects of nano-oxides as opacifiers. *Journal of Color and Appearance in Dentistry* 38: 100–105.
60 Abidi, N., Hequet, E., Tarimala, S., and Dai, L.L. (2007). Cotton fabric surface modification for improved UV radiation protection using sol–gel process. *Journal of Applied Polymer Science* 104: 111–117.
61 Mahltig, B., Böttcher, H., Rauch, K. et al. (2005). Optimized UV protecting coatings by combination of organic and inorganic UV absorbers. *Thin Solid Films* 485: 108–114.
62 Mahltig, B. and Textor, T. (2008). *Nanosols and Textiles*. Singapore: World Scientific.
63 Ghamsari, M.S., Alamdari, S., Han, W., and Park, H.-H. (2017). Impact of nanostructured thin ZnO film in ultraviolet protection. *International Journal of Nanomedicine* 12: 2017–2216.
64 Mahltig, B. and Haufe, H. (2010). Biozidhaltige Nanosole zur Veredlung von weichen und temperaturempfindlichen Materialien. *Farbe & Lack* 116 (3): 27–30.
65 Alebeid, O.K. and Zhao, T. (2014). Anti-ultraviolet treatment by functionalizing cationized cotton with TiO_2 nano-sol and reactive dye. *Textile Research Journal* 85: 449–457.
66 Huang, J.Y., Li, S.H., Ge, M.Z. et al. (2015). Robust superhydrophobic TiO_2 fabrics for UV shielding, self-cleaning and oil–water separation. *Journal of Materials Chemistry A* 3: 2825–2832.
67 Christie, R.M. (2001). *Colour Chemistry*. Cambridge: The Royal Society of Chemistry.

68 Windler, L., Lorenz, C., von Goetz, N. et al. (2012). Release of titanium dioxide from textiles during washing. *Environmental Science and Technology* 46: 8181–8188.
69 Choudhury, A.K.R. (2006). *Textile Preparation and Dyeing*. Enfield: Science Publishers.
70 Reinert, G., Fuso, F., and Hilfiker, R. (1996). Process for increasing the sun protection factor of cellulosic fiber materials. US Patent 5, 914, 444, filed 4 March 1996 and issued 22 June 1999.
71 Rauch, K., Dieckmann, U., Böttcher, H., and Mahltig, B. (2004). UV-schützende transparente Beschichtungen für technische Anwendungen. German Patent DE102004027075A1, filed 2 June 2004 and issued 29 December 2005.
72 Böttcher, H., Mahltig, B., Haufe, H., and Henker, P. (2006). UV-licht absorbierendes material. German Patent DE102006002231A1, filed 17 January 2006 and issued 19 September 2007.
73 Schrader, B. and Rademacher, P. (2009). *Kurzes Lehrbuch der Organischen Chemie*. Berlin: Walter de Gruyter.
74 Küster, B. (1975). UV-Stabilisatoren für aromatische Polyamide. *Lenzinger Berichte* 38: 12–13.
75 Reinert, G., Fuso, F., and Hilfiker, R. (1996). Process for increasing the sun protection factor of cellulosic fibre materials. US Patent 5, 938, 793, filed 31 January 1996 and issued 17 August 1999.
76 Grethe, T., Haase, H., Natarajan, H.S. et al. (2015). Coating process for antimicrobial textile surfaces derived from a polyester dyeing process. *Journal of Coatings Technology and Research* 12: 1133–1141.
77 Heller, J. and Mura, J.-L. (1997). Aqueous dispersions and their use for treating textiles. US Patent 5, 955, 005, filed 23 July 1997 and issued 21 September 1999.
78 Eckhardt, C., Reinehr, D., Metzger, G., and Sauter, H. (1997). Triazine derivatives and their use. US Patent 5, 939, 379, filed 2 June 1997 and issued 17 August 1999.
79 Qu, L., Tian, M., Hu, X. et al. (2014). Functionalization of cotton fabric at low graphene nanoplate content for ultrastrong ultraviolet blocking. *Carbon* 80: 565–574.
80 Tang, X., Tian, M., Qu, L. et al. (2015). Functionalization of cotton fabric with graphene oxide nanosheet and polyaniline for conductive and UV blocking properties. *Synthetic Metals* 202: 82–88.
81 Prince, E., Fuess, H., Hahn, T. et al. (2006). *International Tables of Crystallography*, vol. C. International Union of Crystallography.
82 Hubbell, S.M. and Seltzer, J.H. (1996). *Tables of X-Ray Mass Attenuation Coefficients and Mass Energy-absorption Coefficients from 1 KeV to 20 MeV for Elements Z = 1 to 92 and 48 Additional Substances of Dosimetric Interest*. National Institute of Standards and Technology.
83 Mahltig, B., Haufe, H., Leisegang, T. et al. (2011). Verminderte Röntgentransmission durch beschichtete Textilien. *Melliand Textilberichte* 92: 36–39.

84 Günther, K., Giebing, C., Askani, A. et al. (2015). Cellulose/inorganic-composite fibers for producing textile fabrics of high X-ray absorption properties. *Materials Chemistry and Physics* 167: 125–135.

85 Martinez, T.P. and Cournoyer, M.E. (2001). Lead substitution and elimination study. *Journal of Radioanalytical and Nuclear Chemistry* 249: 397–402.

86 Willmes, A. (1993). *TaschenbuchChemischeSubstanzen*. Thun und Frankfurt am Main: Verlag Harri Deutsch.

87 Azman, N.Z.N., Siddiqui, S.A., and Low, I.M. (2013). Synthesis and characterization of epoxy composites filled with Pb, Bi or W compound for shielding of diagnostic X-rays. *Applied Physics A* 110: 137–144.

88 Lyra, M., Charalambatou, P., Sotiropoulos, M., and Diamantopoulos, S. (2011). Radiation protection of staff in radionuclide therapy. *Radiation Protection Dosimetry* 147: 272–276.

89 Schueler, B.A. (2010). Operator shielding. *Techniques in Vascular and Interventional Radiology* 13: 167–171.

90 Ross, A.M., Segal, J., Borenstein, D. et al. (1997). Prevalence of spinal disc disease among interventional cardiologists. *American Journal of Cardiology* 79: 68–70.

91 Finnerty, M. and Brennan, P.C. (2005). Protective aprons in imaging departments: manufacturer stated lead equivalence values require validation. *European Radiology* 15: 1477–1484.

92 Oyar, O. and Kislalioglu, A. (2012). How protective are the lead aprons we use against ionizing radiation? *Diagnostic and Interventional Radiology* 18: 147–152.

93 Nkubli, B.F., Nzotti, C.C., Nwobi, I.C. et al. (2013). Quality control in radiology units of tertiary healthcare centers in north eastern Nigeria. *Nigerian Journal of Medical Imaging and Radiation Therapy* 2: 26–31.

94 Han, E.G., Kim, E.A., and Oh, K.W. (2001). Electromagnetic interference shielding effectiveness of electroless Cu-plated PET fabrics. *Synthetic Metals* 123: 469–476.

95 Lai, K., Sun, R.J., Chen, M.Y. et al. (2007). Electromagnetic shielding effectiveness of fabrics with metallized polyester filaments. *Textile Research Journal* 77: 242–246.

96 Ziaja, J., Koprowska, J., and Janukiewicz, J. (2008). Using plasma metallisation for manufacture of textile screens against electromagnetic fields. *Fibres & Textiles in Eastern Europe* 16: 64–66.

97 Statex GmbH, Bremen, Germany. www.statex.de. (Accessed 14/01/2020).

98 Marktek Inc. Chesterfield, US. www.marktek-inc.com. (Accessed 14/01/2020).

99 Swicofil AG, Emmenbrücke, Switzerland. www.swicofil.com. (Accessed 14/01/2020).

100 YSHIELD GmbH, Ruhstorf, Germany. www.yshield.com. (Accessed 14/01/2020).

101 YSHIELD GmbH (2016). *Product Catalogue*. Ruhstorf, Germany: YSHIELD GmbH.

102 Marchini, F. (1991). Advanced applications of metallized fibres for electrostatic discharge and radiation shielding. *Journal of Coated Fabrics* 20: 153–166.

103 Bertuleit, K. (1991). Silver coated polyamide: a conductive fabric. *Journal of Coated Fabrics* 20: 211–215.

104 Wang, W., Li, W., Gao, C. et al. (2015). A novel preparation of silver-plated polyacrylonitrile fibers functionalized with antibacterial and electromagnetic shielding properties. *Applied Surface Science* 342: 120–126.

105 Odhiambo, S.A., de Mey, G., Hertleer, C. et al. (2014). Discharge characteristics of poly(3,4-ethylene dioxythiophene):poly(styrenesulfonate) (PEDOT:PSS) textile batteries, comparison of silver coated yarn electrode devices and pure stainless steel filament yarn electrode devices. *Textile Research Journal* 84: 347–354.

106 Orban, R.F. (1985). Continuous process for the sequential coating of polyamide filaments with copper and silver. US Patent 4645574A, filed 2 May 1985 and issued 24 February 1987.

107 Mahltig, B., Darko, D., Günther, K., and Haase, H. (2015). Copper containing coatings for metallized textile fabrics. *Journal of Fashion Technology & Textile Engineering* 3 (1): 1–10.

108 Lee, H.M., Choi, S.-Y., Jung, A., and Ko, S.H. (2013). Highly conductive aluminum textile and paper for flexible and wearable electronics. *Angewandte Chemie International Edition* 52: 7718–7723.

109 Okayasu, S., Shinichi, I., Kaga, U., and Nobatake, A. (1986). Process for producing an electromagnetic radiation-shielding, metallized polyester fiber textile material. US Patent 4, 681, 591, filed 22 May 1986 and issued 21 July 1987.

110 Onggar, T., Hund, H., Hund, R.-D., and Cherif, C. (2014). Silvering of inert PET-textile materials by means of one-bath and two-bath methods and their characteristics. *Fibres & Textiles in Eastern Europe* 22: 108–114.

111 Godley, P. (1948). Deposition of metal on a nonmetallic support. US Patent 2622041A, filed 3 August 1948 and issued 16 December 1952.

112 Manabe, K., Tsutsui, M. Suzuki, T. et al. (1983) Metal-coated fibrous objects. US Patent 4816124A, filed 19 December 1983 and issued 28 March 1989.

113 Yuranova, T., Rincon, A.G., Bozzi, A. et al. (2003). Antibacterial textiles prepared by RF-plasma and vacuum-UV meditated deposition of silver. *Journal of Photochemistry and Photobiology A: Chemistry* 161: 27–34.

114 Amberg, M., Kasdallah, C., Ritter, A., and Hegemann, D. (2012). Influence of residual oils on the adhesion of metal coatings to textiles. *Journal of Adhesion Science and Technology* 24: 123–134.

115 Ozen, M.S., Sancak, E., Beyit, A. et al. (2013). Investigation of electromagnetic shielding properties of needle-punched nonwoven fabrics with stainless steel and polyester fiber. *Textile Research Journal* 83: 849–858.

116 Ebneth, H., Fitzky, H.G., Wolf, G.D., and Giesecke, H. (1982). Use of metallized knitted net fabrics for protection against microwave radiation. US Patent 4, 572, 960, filed 1 November 1982 and issued 25 February 1986.

117 Dubrovski, P.D. and Golob, D. (2009). Effects of woven fabric construction and color on ultraviolet protection. *Textile Research Journal* 79: 351–359.

118 Bisjak, C., Gürtler, A., Dobson, P. et al. (2010). Ultraviolet protective fabrics based on man-made cellulosic fibers. US Patent 0156462A1, filed 3 May 2010 and issued 12 March 2012.

119 Ozgen, B. (2012). New biodegradable fibres, yarn properties and their applications in textiles: a review. *IndustriaTextila* 63: 3–7.

120 Wendler, F., Kosan, B., Krieg, M., and Meister, F. (2009). Possibilities for the physical modification of cellulose shapes using ionic liquids. *Macromolecular Symposia* 280: 112–122.

121 Mahltig, B., Günther, K., Giebing, C. et al. (2014). Inorganic/organic composite fibers for fabrics with X-ray protective properties. *Technical Textiles* 57: 56–58.

122 Qu, L., Tian, M., Zhang, X. et al. (2015). Barium sulfate/regenerated cellulose composite fiber with X-ray radiation resistance. *Journal of Industrial Textiles* 45: 352–367.

123 Mahltig, B., Günther, K., Askani, A. et al. (2016). Clothing form organic/inorganic composite fibers for reduction of X-ray exposition. *Technical Textiles* 59: 20–22.

33

Fibrous Materials for Antimicrobial Applications

Yue Deng, Yang Si, and Gang Sun

University of California Davis, Fiber and Polymer Science, Agricultural & Environmental Chemistry, 251 Everson Hall, Davis, CA 95616, USA

33.1 Introduction

33.1.1 Basic Principles

There are trillions of microorganisms that live in and on a human body [1]. Although some of them help to keep the body healthy, some of them can harm the body. Such kinds of microorganisms include pathogenic virus, gems, bacteria, and fungi. Textiles, which most of them are made of organic materials and have porous structures that hold moisture, provide the essential nutrients for microbes to live and create potentials for infections that can threat the health of human beings. Since textiles are so widely used in daily life, developing antimicrobial fibers becomes an important and valuable direction of research.

An "antimicrobial" is defined to be an agent that can either kill certain types of microbes or slow down their growth [2]. Antimicrobial agents work under different mechanisms and principles. When it comes to textiles, the more conventional way to make fibers antimicrobial is to treat them with some antimicrobial agents, such as *N*-halamines [3, 4], quaternary ammonium compounds (QACs) [5, 6], and photoactive chemicals [7–9]. In some cases, the natural materials themselves, which the fibers are made of, can inhibit the growth of microorganisms. Chitosans are an example of such natural polymers [10, 11]. This is another way to produce antimicrobial fibers. Antimicrobial fibers and their products, as mentioned before, have a variety of applications. For example, socks that claim to be antimicrobial can reduce the reproduction of fungi fed by human sweat and solve the unpleasing feet odor problem. Mildew-resistant shower curtains are another example of antimicrobial textiles as they have resistance against the growth of a type of fungus. Besides antifungals, textiles can also be combined with antibacterial agents and have a wide market in clinics, such as sterilized bandages.

33.1.2 Brief History

The earliest publication that can be found in the Web of Science Core Collection is published in 1972 by N. S. Plotkina, N. V. Vitulska, and V. A. Khokhl, while the

Handbook of Fibrous Materials, First Edition. Edited by Jinlian Hu, Bipin Kumar, and Jing Lu.

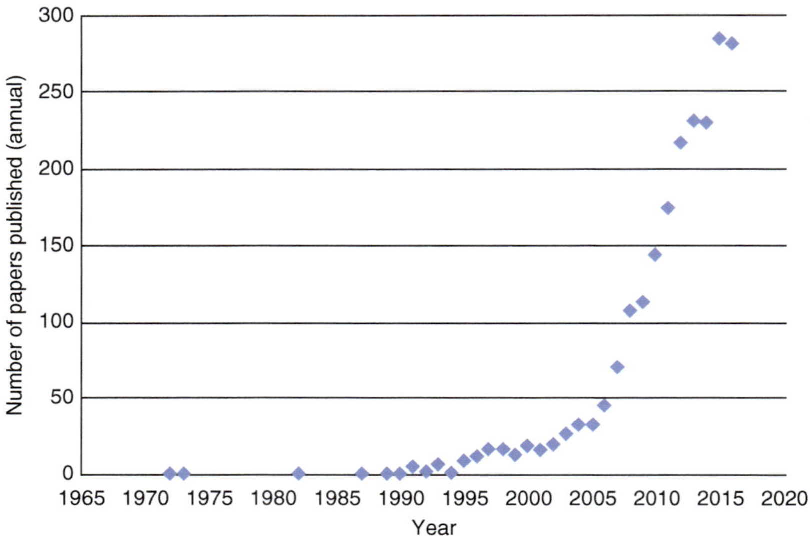

Figure 33.1 Number of publications each year that cover the topic of antimicrobial fibers.

application of antimicrobial textiles began years earlier. Military needs seem to be the main driving force for development of technology back in nineteen and first half of twenty centuries. Demand on microbial fibers was raised in World War II due to the surprisingly short duration of the military tents [2]. It was discovered that the reason that caused those tents to rot so fast is the rapid growth of a filamentous fungus (*Trichoderma reesei*) with the aid of the warm and wet weather in South Pacific.

However, antimicrobial fiber is not a very popular research topic compared with other topics in the biochemical related field. In 2015, one of its peak years, 285 journal papers were published. On the other hand, compared with only a handful publications in the 1980s, this application area is growing rapidly. Researches that were published around 2004 largely stimulated researchers to enter and devote into this field (Figure 33.1). Some of the inspiring studies including the rechargeable *N*-halamine textile [3, 4], applications of silver and other metal oxide nanoparticles on textiles [12–14], textile modification with quaternary ammonium salts [5, 6], and natural microbial agents [15–17]. This chapter will summarize some important discoveries and achievements in fiber modifications with antimicrobial agents. This chapter will also give a brief review on the process of incorporation of these major antimicrobial agents into fabrics and the advantages and disadvantages of each type of modified fibers.

33.2 Quaternary Ammonium Compound Modified Fabrics

QACs are one of the conventional agents that were applied on textiles as a surface treatment to grant the textiles antimicrobial property. Study on QACs started about a hundred years ago, in the early twentieth century [18]. Its antimicrobial property began to draw scientists' attention in the 1960s [19, 20]. QACs were once widely used in communities for disinfections for its low toxicity and good bacteria-eliminating ability. However, later on, people discovered that overuse of QACs seemed to promote severe drug resistance among bacteria [21, 22]. This section will cover the bacterial killing mechanism of QACs to discover the reason why QACs would promote drug resistance based on the mechanism. This section will also discuss the environmental impacts of overusing QACs in late twentieth century.

33.2.1 Antimicrobial Mechanism of QACs

The ammonium group in a QAC brings a positive charge. It reacts with the negatively charged cell membranes and damaged their structures. This leads to a change in permeability of the membranes, causing the cells to lose vital liquids. In the end, the metabolism of the cells is severely disrupted, and the microbes get killed [6, 23].

QACs were once believed to be one of the best antimicrobial agents, because it is effective (at the right concentration), colorless, odorless, and aerobically biodegradable [24]. However, its biodegradability was challenged later on for there were evidences showing high concentration of QACs in natural water bodies [25, 26]. Besides the enrichment problem of QACs in lakes and rivers via sewage water, its bacteria-killing spectrum is also relatively narrow. QACs work better on Gram-positive bacteria than Gram-negative bacteria [27, 28]. It can barely deactivate *Pseudomonas* and cannot kill *Mycobacterium* [29].

In addition, its effectiveness is largely affected by the environment. QAC is a good surfactant and is easily absorbed onto all kinds of surfaces. In hospital practices, QACs can be absorbed onto surfaces of needles, gazes, cotton balls, and glass vessels [30, 31]. Those kinds of equipment can all reduce QAC solution's effective concentration. For example, while using QAC solution to sterilize cotton balls, 1 g of cotton ball can absorb approximately 1.58 g of QAC solution. The QAC solution would work great for sterilizing the cotton fibers. However, the concentration of QAC in the solution stored in the cotton ball would significantly drop to a level much lower than the effective concentration because large amount of QAC is absorbed onto the cotton fibers. Therefore, the cotton ball cannot be used to sterilize wound. The effectiveness of QACs can also be reduced by organic substances like proteins and serums. Since its antimicrobial mechanism is based on the positive charge on ammonium group, any compounds that bring negative charge will antagonize the disinfection ability of QACs. Such negatively charged substances include soap, milk, and washing powder. Any metal ions that can cause ammonium group to participate out will also decrease the activity of QACs [24].

33.2.2 Environmental Impact

The presence of QACs in natural water bodies is mainly influenced by its biodegradability and absorption of QACs onto variety of surfaces to pass through [22, 26].

As mentioned in Section 33.2.1, QACs are considered to be aerobically biodegradable. However, their half-lives are largely affected by factors such as their chemical structure, the oxygen abundance in the environment, and acclimatization of the microorganism that can degrade QACs [32, 33]. Generally, as the length of the alkyl chain increases, so does the time required for the QACs to degrade. On the other hand, longer alkyl chain is one of the popular modifications of QACs to increase its biocidal effectiveness. Therefore, many QACs found in natural water body are QACs with long alkyl chain, which require a considerably long time to be removed.

In addition to the accumulation of QACs in the nature, exposure of high concentration QACs also promote drug resistance in some bacteria. To survive in high concentration of QACs, some bacteria gain resistance to QACs. Main QAC resistance mechanisms include reducing expression of porins [34–36], mobilizing genetic units [37, 38], and elevating expression of efflux pumps [39–41]. These mechanisms could also help such bacteria to survive under some antibiotics. Therefore, the presence of QACs in the environment not only does generate the problem of high-level QAC intakes but also promote drug resistance.

33.2.3 Conclusions

QACs, which are once the most commonly used antimicrobial agents, are now facing many challenges. Its biocidal performance is largely influenced by the environment, which limits its applications. Efforts have been put on this topic to increase the stability of QACs in complex environment and to increase its effectiveness toward more bacteria species. Many new compounds have been synthesized, which have functional groups that can improve either the biocidal effect of QACs or the safety of QACs. For example, the United States Food and Drug Administration has approved that organosilicon quaternary ammonium salt is safe to be used on products that will get contact with the human body directly.

In general, QACs still have some advantages that make them still an important area to study in the field of antimicrobial textiles. They are effective to Gram-positive bacteria and can be strongly bounded onto surface of fabrics. As more novel researches on QACs get published and applied, QACs will be a kind of truly safe and broad-spectrum antimicrobial agent.

33.3 *N*-Halamine Modified Fabrics

The study of the antibacterial nature of *N*-halamine started in the 1980s [42–45]. This organic compound shows high efficiency in water disinfection and was soon applied to produce biocidal polymers in the late 1990s [4, 46–48]. It is still considered as one of the conventional antimicrobial agents for fiber modification, because nanotechnology entered the stage even later. Among all those conventional antimicrobial agents, such as QACs, chitosan, and polyhexamethylene biguanide, the rechargeable characteristic of *N*-halamine makes it stand out and still have many studies going on till today. *N*-Halamine can be bound onto polymers, thin films, and nanofibers and has applications in water disinfection, food packaging, and another antibacterial practices [49–51].

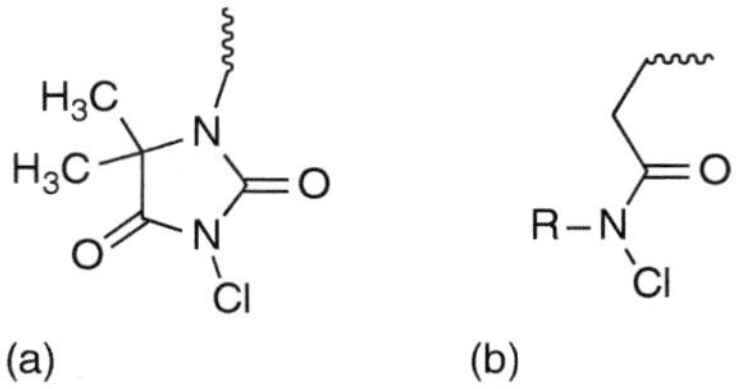

Figure 33.2 Some examples of possible structures of *N*-halamine compounds. (a) Example of a cyclic *N*-halamine and (b) example of an acyclic *N*-halamine.

33.3.1 Basic Principles

N-Halamine compounds have at least one nitrogen–halogen bond (shown in Figure 33.2). In most applications, this halogen is chloride. *N*-Halamine compounds kill microbes based on its oxidizing power provided by this N—X covalent bond. When it gets contact with the microbe, the N—X bond will be reduced to N–H and will release a free halogen ion (usually a Cl^+ ion as N–Cl is the mostly commonly used functional group in *N*-halamines). *N*-Halamine compounds also react with water to release halogen ions and show powerful water disinfection ability. This mechanism is shown in Figures 33.3 and 33.4. As illustrated, this oxidization reaction is also reversible. While putting the reduced functional groups in an environment with halogenating agent in abundance, they can regain the halogen atoms, and the antimicrobial ability can be restored. Dilute chlorine bleach is one of the widely used agents to help *N*-halamine compounds to be recharged with free Cl anions [4, 49, 52]. Since its bacterial killing mechanism is oxidizing, *N*-halamine will not promote microbial resistance.

33.3.2 Modification Method

The methods used for textile modification with *N*-halamine can be categorized into the following ones: cross-linking, grafting, coating, and reactive *N*-halamine treating.

33.3.2.1 Cross-linking

To add *N*-halamine onto surface of a textile via cross-linking, the reaction needs to start with a precursor of *N*-halamine. 1,3-Dimethylol-5,5-dimethylhydantoin (DMDMH) is a precursor of *N*-halamine that itself can also act as a cross-linker in basic environment. It was first successfully applied onto cotton cellulose by Y. Sun and G. Sun [47]. Y.Y. Sun and G. Sun then applied it onto other fabrics and the characteristics of the modified fabrics [3]. It was discovered that DMDMH

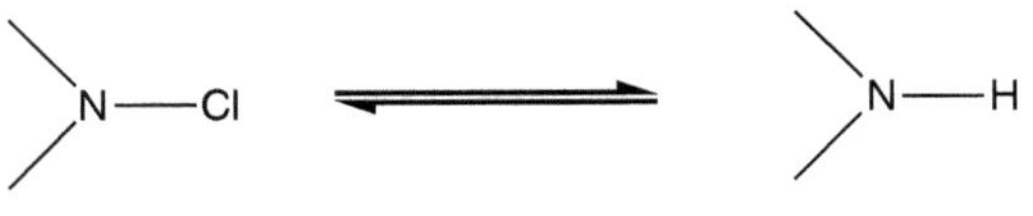

Figure 33.3 Reduce reaction of *N*-halamine (halogenated with chlorine), which is reversible.

Figure 33.4 Reaction of *N*-halamine (halogenated with chlorine) in water. This reaction is reversible.

modified fabrics gain a better durability [52]. However, it was also discovered that DMDMH-treated textiles would slowly release free formaldehyde, which raises safety concerns [51]. Therefore, researchers went on for more safety and reliable *N*-halamine precursors and found out several alternatives.

m-Aminophenylhydantoin (*m*-APH), a synthesized reagent, is one of the solutions to the safety problem of DMDMH. Unlike DMDMH, *m*-APH requires a polycarboxylic acid as a cross-linker to be bound onto the surface of a fabric. 1,2,3,4-Butanetetracarboxylic acid (BTCA) is a most commonly used polycarboxylic acid that gives the best wrinkle recovery with *m*-APH modified cotton in comparison with citric acid, tricarballylic acid, and other popular organic acids in industrial practice [53, 54]. BTCA can also act as the cross-linker for other *N*-halamine precursors to be added onto cellulose, such as 3-(2,3-dihydroxypropyl)-5,5-dimethylimidazolidine-2,4-dione [51], 3-(2,3-dihydroxypropyl)-7,7,9,9-tetramethyl-1,3,8-triazaspiro[4.5]decane-2,4-dione [55], 2-amino-2-methyl-1-propanol [14, 56], 2,2,6,6-tetramethyl piperidinol [56], 1-glycidyl-*s*-triazine-2,4,6-trione, and 1-(2,3-dihydroxypropyl)-*s*-triazine-2,4,6-trione [57].

33.3.2.2 Grafting

For *N*-halamines with vinyl groups, they can be grated onto fabrics through free radical polymerization reaction. For this reaction to happen, strong oxidative initiators, such as potassium persulfate, azobisisobutyronitrile, and ammonium cerium (IV) nitrate, are required.

N-Halamines containing vinyl groups can be further categorized into cyclic *N*-halamines and acyclic *N*-halamine. A typical cyclic *N*-halamine is 3-allyl-5,5-dimethylhydantoin, which has been successfully grafted onto many manmade and natural fabrics, including cotton, nylon, and polyester [4]. Acyclic *N*-halamines can be graded onto cellulose under similar mechanism [58–60]. However, due to its acyclic structure, its wash resistance and storage of chlorine are inferior to cyclic *N*-halamines.

33.3.2.3 Coating

Polymeric *N*-halamines can be applied onto many surfaces including textile via coating [61]. Chemical bonds form between the substrate and the coated polymeric layers, granting the composite reasonable durability. With the help of surfactants, water dispersible *N*-halamine polymers can be formed and be applied onto surfaces via emulsion polymerization. For a particular *N*-halamine monomer, 3-(4′-vinylbenzyl)-5,5-dimethylhydantoin, it can even be applied without using any organic solvent through admicellar polymerization [52, 62].

33.3.2.4 Reactive Reagent Treating

Reactive *N*-halamines refer to those *N*-halamine precursors that can chemically react with fibers to form covalent bonds. Therefore, textiles treated with reactive *N*-halamine have the best durability among all the modification methods. Cyclic *N*-halamines usually have good reactivities. Cyclic *N*-halamines such as 3-glycidyl-5,5-dialkylhydantoin and 1-glycidyl-*s*-triazine-2,4,6-trione are frequently used in reactive *N*-halamine treatments [63–65].

33.3.2.5 Electrospinning

All these processing methods listed in Sections 33.3.2.1 to 33.3.2.4 are embedding a layer or multiple layers of *N*-halamine onto the surface of textiles. However, there is another way to directly incorporate *N*-halamine into fabrics, which is via electrospinning. Electrospinning uses electric force to drive charged polymer solutions to form threads and spins threads to form fabrics. It produces fabrics with high specific surface area and high porosity and is a good method to generate nano- to microfibrous membranes [66–68]. Using water-soluble polymeric *N*-halamines (which were briefly introduced in Section 33.3.2.3), the obtained copolymer solution can be spun into fabrics that have direct incorporation of *N*-halamines. The large surface area makes this kind of fabrics has excellent bacterial killing effect right upon contacting. However, there is also a downside associates with the fast disinfection. *N*-Halamine nanofibrous membranes release chlorine too fast that they cannot stay effective for a long period of time and require frequently recharges.

33.3.3 Conclusions and Future Trend

N-Halamine is a powerful and special antimicrobial agent for fiber modification for the pathogen-killing functional group and can be recharged. Once the attached halogen is consumed, *N*-halamine compound can regain it by putting in an environment with sufficient amount of free halogen. Its bacterial killing mechanism is based on oxidation of the cell structure. Therefore, *N*-halamine works effectively for both Gram-positive and Gram-negative bacteria. It is a truly broad-spectrum antimicrobial agent that merely promotes microbial resistance. Furthermore, it has been discovered and confirmed that *N*-halamine compounds can also store the bacterial killing power until the substances they are in touch with oxidize the *N*-halamine compounds.

However, some conditions can cause *N*-halamine modified textiles to release the stored halogen in a fairly short period of time. Recent research focuses on increased stability of *N*-halamine, especially its durability after washing and under ultraviolet (UV) light. In addition to the effort put on researching its durability and stability, nanotechnology [68] is another direction that provides further space for the development of *N*-halamine. Nanostructures have greater specific area. Therefore, it can enhance the performance and efficiency of *N*-halamine in antimicrobial practices.

33.4 Metals and Metal Oxide Modified Fabrics

The applications of metals and metal oxides in antimicrobial textiles largely relied on the development of nanotechnology. Nanotechnology grants people ways to produce extremely small particles (to $10^{-7}{\sim}10^{-9}$ m), which creates large specific surface areas and largely increases the reactivity of those materials [69, 70]. More importantly, some materials would behave differently in nanoscale [71]. For example, some metal oxides, such as copper oxide [72, 73] and zinc oxide [72–75], show antimicrobial property when they are in the forms of nanoparticles.

Most nanoparticles gain effective biocidal ability by the increased reactivity and contacting area. Their mechanisms of killing microorganisms are based on their own characteristics and are different from each other. This section will briefly talk about several commonly used metals and metal oxides for antimicrobial textile modification.

33.4.1 Silver

Silver has a long history of being used to sterilize water and food. Phoenicians used silver to keep their water clean and milk fresh. Chinese used silver to make food containers that can be used to test the toxicity of food. Today, silver is even applied in more products that people get contact with every day: toilet seats, refrigerator doors, drinking vessels, respirators, and water filters [76–78]. Production of some of those applications requires fibers treated with nanosized silver particles. Silver nanoparticles that can be used in fiber modification need to be inert and charge neutral [76].

There are several ways to produce nanosized silver particles. This section will introduce one of the most common methods. To make silver atoms, a silver salt is reduced in aqueous form with reducing agents. This method involves two agents: the silver salt is treated with a strong reducing agent and then a weak reducing agent. These two reducing agent controls the size of the silver nanoparticles and their dispersity [76, 79, 80].

33.4.2 Oligodynamic Effects of Metals

The antimicrobial nature of silver is called oligodynamic effect, which was discovered and named by Karl Wilhelm von Nägeli in 1893 [81]. Oligodynamic is a biocidal effect that is observed on many heavy metals. This means, not only silver but also other metals such as copper, zinc, gold, tin, and mercury have this biocidal effect [77, 82, 83]. Due to their toxicity toward human beings, besides silver, copper (II) oxide and zinc oxide are the two other metal oxides that are commonly used.

The mechanism of oligodynamic remains unknown. There are two major theories that try to explain this phenomenon. One theory suggests that silver nanoparticles release silver ions in a rate sufficient enough that can disrupt the DNA republication of microbes and so inhibit their growth. Another proposed mechanism believes that silver nanoparticles and ion can react with the cell walls and destroy microbes this way [71, 84–86].

33.4.3 Titanium Dioxide

Unlike copper (II) oxide, titanium dioxide kills bacteria through its photoactive power [72, 73], which will be discussed in Section 33.5. In general, metal oxides like titanium dioxide that has photocatalytic behavior can be made into nanoparticles and used in fiber modification [87–89]. Nanosizing helps them to achieve a much faster reaction rate and a much greater contacting area. Therefore, titanium dioxide and similar metal oxides can kill bacteria effectively.

33.4.4 Conclusions and Future Trend

Nowadays, metals (particularly heavy metals) and metal oxides are a kind of important antimicrobial agent that is widely applied in textile modification. In order to make them work more effectively, in most cases, metals and metal oxides are made into nanoparticles while being used in fiber treatment. Metal oxides like titanium dioxide kill bacteria through photocatalysis, and heavy metals achieve similar result through oligodynamic effect. Since the mechanism of oligodynamic effect is yet not fully understood, a theoretical conclusion stating that metals will not promote resistance cannot be drawn. In general, through the observations in practice, silver nanoparticles are still the most effective and safe broad-spectrum antibacterial agent. However, just as other nanoparticles, nanoparticles of metals and metal oxides such as silver can penetrate human body and might cause damages inside [89–92]. Debate still goes on whether it is truly safe to use silver nanoparticles so widely in commercial products, yet silver nanoparticles are one of the most popular agents for industrial production.

33.5 Photoactive Chemical Modified Fabrics

As it was continuously brought up during this section, microbe infections have caused millions of death. Modified textiles and their products are a good way to control diseases caused by infections and are widely used in environments that require straight hygiene. Antimicrobial fibers usually gain their disinfection power through treatments with antimicrobial agents. A quick review: so far, three kinds of antimicrobial agents have been discussed in Sections 33.2, 33.3 and 33.4, which are QACs, *N*-halamine compounds, and metal (or metal oxide) nanoparticles. This section will cover one more category of agents that have biocidal effect: photoactive chemicals.

33.5.1 LAAAs and Photoinduced Biocidal Effect

LAAA refers to light-activated antimicrobial agent. LAAAs gain their antimicrobial power with the energy of light, so they are all photoactive. Photoactive chemicals can be organic or inorganic. Inorganic compounds, such as titanium dioxide (TiO_2, which was briefly mentioned in Section 33.4.3), and tin dioxide (SnO_2), are photocatalysts [93, 94]. Photocatalysts can speed up photoreaction under presence of light (both visible and invisible). For example, TiO_2 shows extraordinary photochemical activity under UV light.

In general, scientists believe that LAAAs kill cells of pathogenic microorganisms by the reactive oxygen species (ROS) they generate under illumination of light [95]. The mechanism of photoinduced disinfection of inorganic photoactive chemicals is still not yet fully understood. However, it is commonly believed that the ROS generated will react with the cell membranes and break them apart. The deconstruction of cell membranes will lead to loss of vital substances from the inside of the cells and also allow more ROS to get into the cells and attack on more important functional structures. The damages on vital cell structures cause important cell functions and activities to stop and eventually kill the bacteria [96, 97]. Organic LAAAs are mostly strong oxidizing agents. Just like inorganic LAAAs, they produce huge amount of ROS upon shining on light. These ROS then oxidize microbes' cells and inactivate them.

33.5.2 Textile Modification with TiO_2

The photocatalytic nature of TiO_2 was first discovered and published by Prof. Fujishima in 1972 [98]. Many researches have been done on studying and applying its photochemical activity. When it comes to textile modification, TiO_2 has been successfully bound to several kinds of fibers [99–103]. Just like other metal oxides, in order to achieve a better performance, TiO_2 is turned into nanoparticles while being added onto the fibers.

The conventional method was pad-dry-cure [104]. However, since TiO_2 is inorganic yet fiber polymers are organic, the bonds formed between the inorganic and organic are quite weak. This leads to poor durability after washing. To improve the washing resistance of modified textile with TiO_2, lots of studies have been done on inventing a better modification process.

One of the newly discovered methods to immobilize TiO_2 nanoparticles onto textile surfaces is through sol–gel process [103]. Starting with a TiO_2 precursor, at the end of the sol–gel process, a gel, which is a three-dimensional porous structure, will form on the surface of the textile. Drying the gel to eliminate the liquid stored in the pores will leave nanosized TiO_2 particles on the surface that are more strongly bound compared with the conventional pad-dry-cure method.

Pretreating the fabrics with a cross-linker is another way to largely increase the durability of fabrics treated with TiO_2 nanoparticles. Among all possible cross-linkers, the ones with carboxylic acid groups are tested to be the best, because of the covalent bonds and hydrogen bonds that can be formed between TiO_2 nanoparticles and carboxylic acid groups [104–107].

Due to its relatively large bandgap, TiO_2 can only work under UV light. However, UV light does not always present in a sufficient amount. For example, glass can absorbed considerable amount of UV light, so the level of UV radiation is relatively low indoor. Therefore, efforts have also been put in to make TiO_2 nanoparticles work under visible light. Doping is a solution to this problem as the impurity prevents the light-generated active electrons and positive holes from joining together to form stable TiO_2 nanoparticles again [88].

33.5.3 Organic Photoactive Reagents

Most oxidative organic chemicals are more or less having biocidal effect. This section will introduce two major kinds: chemicals that generate ROS through type I photochemical reaction and chemicals that generate oxygen and ROS though type II photochemical reaction.

Anthraquinone and benzophenone derivatives are two representative species that gain biocidal ability through type I photochemical reaction under UV light [108–113]. The ROS production rate of a derivative depends on the structure of the anthraquinone or benzophenone derivative and the available functional groups on the substrate. In general, derivatives with electron-withdrawing groups produce ROS at a higher efficiency. Substrates with good hydrogen-donating ability also help the derivatives to have a better ROS yield. Some benzophenone derivatives are also precursors of perfumes and dyes [112, 113]. Therefore, in addition to produce antimicrobial textiles, they can be used to add color to the processed textiles to make the finished products even more marketable.

Porphyrins and phthalocyanines, on the other hand, generate oxygen and ROS under visible light (with wavelength in the range of 400–800 nm depending on the structure of the particular derivate) via type II photochemical reaction [114–118]. The porphyrin and phthalocyanine derivatives can be excited by visible light at the desired wavelength and then react with oxygen present in the surroundings to generate ground state oxygen and ROS.

ROS has relatively short half-lives. Therefore, even though they are effective in terms of killing Gram-positive bacteria, they show very weak power in inactivating Gram-negative bacteria [115, 116]. The thicker outer cell membrane structure of Gram bacteria prevents ROS from getting into the cell and destroying the vital cell organelles.

33.5.4 Conclusions and Future Trend

This section discussed both inorganic and organic photoactive chemicals that can be used in antimicrobial fiber modification. Since the modified fibers gain their biocidal ability through illumination of light, in general, most photoactive antimicrobial textiles are recyclable and rechargeable. This characteristic of photoactive antimicrobial textiles also raises concerns on their durability.

Semiconductor agents such as TiO_2 have shown poor results after washing, because of the weak bonds formed between the inorganic doping materials and the organic substrates. However, stability is also a problem to organic agents, even though they can form stronger covalent bonds with the textile surface. As photosensitive species, illumination of light will alter the structure of the critical functional groups and eventually cause these agents to lose their bacterial-inactivating ability [115, 119].

Besides stability toward wearing, washing, and light, safety is another issue that is driving scientists and researchers to invest more time and effort on photoactive antimicrobial agents. For inorganic chemicals, nanotechnology is widely applied to increase the efficiency of biocidal performance. However, there is an ongoing debate if nanoparticles are safe to be in contact with the human body directly [120, 121]. There are similar concerns on ROS, which are the major bacterial killing species in LAAA mechanism. It is worried that the bacterial inactivation mechanism is not selective in that those nanoparticles or ROS would not only destroy pathogens but also may attack human cells when they get contact with the cells. There should be more researches studying the actual effects of using photoactive chemical modified textiles on human bodies.

In conclusion, photoactive sensitizers are effective broad-spectrum antimicrobials. Depending on the particular chemical, it can get excited and gain the biocidal effect at a wavelength in the range from visible to UV. Since the bacteria are inactivated by oxidization, photoactive chemicals that can be used in textile modification generally do not promote resistance.

33.6 Natural Antimicrobial Polymers

Sections 33.2, 33.3 and 33.4 cover major antimicrobial agents that can be used to treat fibers to give them disinfection abilities. However, not only synthesized chemicals can inhibit the growth of microorganisms, but also nature-generated chemicals have excellent biocidal effect for animals to use as well. Ancient Egyptians, Ancient Indians, and Chinese started to use herbs to treat diseases and prevent infections thousands of years ago [122–124].

Nature materials tend to have good biocompatibility and low toxicity. Because of these unique properties compared with other synthesized polymers, nature textiles have big advantages in medical practice. They are frequently made into hygiene products and implantable products. This section will focus on natural antimicrobial polymers and discuss their structures, applications, and drawbacks.

33.6.1 Hyaluronic Acid

Hyaluronic acid (HA), which is also known as hyaluronan, is a conjugate base of hyaluronate. HA is a unique glycosaminoglycan for its nonsulfated structure [125, 126]. Figure 33.5 is an illustration of the skeletal formula for HA. Unlike many other glycosaminoglycans, HA is produced in the plasma membrane rather than in the Golgi apparatus. HA is a naturally produced large molecule in which its molecular weight can reach millions. HA is one of the crucial components of the extracellular matrix and can be found in many parts of a body, such as neural and epithelial tissues. In fact, HA was first discovered and isolated from cows' eyes [127].

Due to its excellent biocompatibility, HA is widely used to aid wound healing and repairing. Since HA is vital for cell immigration, it is also seen as an important indicator of tumor development. HA fabrics would be an ideal material for tissue engineering since it is very common in living organs and tissues [128, 129]. However, the fast turnover speed of HA makes it harder to use HA fabrics as a stable constructive material. This situation can be improved by cross-linking

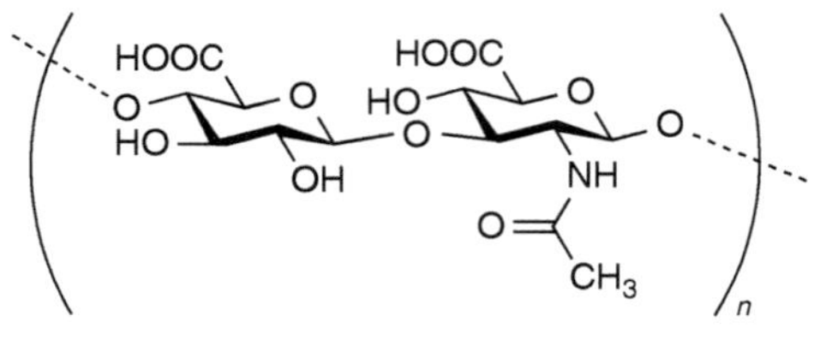

Figure 33.5 Skeletal formula of hyaluronic acid. In each monomer unit, on the left structure is D-glucuronic acid, and on the right is *N*-acetyl glucosamine. They are linked together by alternating β-(1→4) and β-(1→3) glycosidic bonds.

HA fibers and modified HA fibers with other polymers or hydrogels [127–129]. Some successfully commercial examples include Perlane (R), Hylaform (R), and Restylane (R).

33.6.2 Gelatin

Gelatin is a derivative of collagen, which is an abundant protein in mammals that can be found in the extracellular space in many connective tissues [130]. Gelatin can be derived under both acidic and alkaline conditions. Acidic hydrolysis gives gelatin type A, which is positively charged. Basic hydrolysis gives gelatin type B, which is negatively charged [131]. Under specific conditions, the final gelatin product can be transparent. Due to its abundance, gelatin is also very cost-efficient. It also has good biocompatibility and solubility in water. Research shows gelatin can promote cell growth as well [131, 132]. Therefore, gelatin is widely used in medical and cosmetic products. It is also commonly used as gelling agent in food production since gelatin is nontoxic and safe for human intake. Just as HA, gelatin and modified gelatin are one of the most popular materials for tissue engineering and drug delivery research [131, 133].

33.6.3 Alginate

Alginate, which is also called alginic acid or algain, is a natural polysaccharide. (1→4)-Linked β-D-mannuronate and (1→4)-linked α-L-guluronate linked together covalently to form Alginate [134]. Figure 33.6 shows the structure of (1→4)-linked β-D-mannuronate, and Figure 33.7 shows the structure of (1→4)-linked α-L-guluronate. Figure 33.8 is a diagram showing the alginate polymer. Alginates are mostly extracted from brown seaweeds. Alginate refined from different species of brown seaweeds would have small derivatives in their structures [135]. However, those minor differences between alginates do not alter the properties of alginate in general.

Alginate has extraordinary ability of absorbing water [134]. Therefore, it is commonly used in dehydrating process. Alginate is also a gelling agent and is applied in medication for wound healing [134, 136–138]. It can form a gel covering the wound to prevent further infection and keep moisture at a level that is perfect for the damaged tissue to regenerate itself.

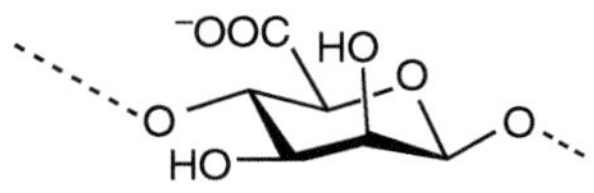

Figure 33.6 Chemical structure of (1→4)-linked β-D-mannuronate.

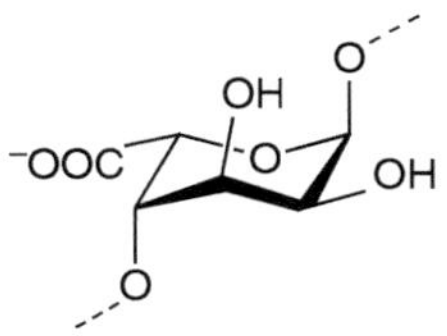

Figure 33.7 Chemical structure of (1→4)-linked α-L-guluronate.

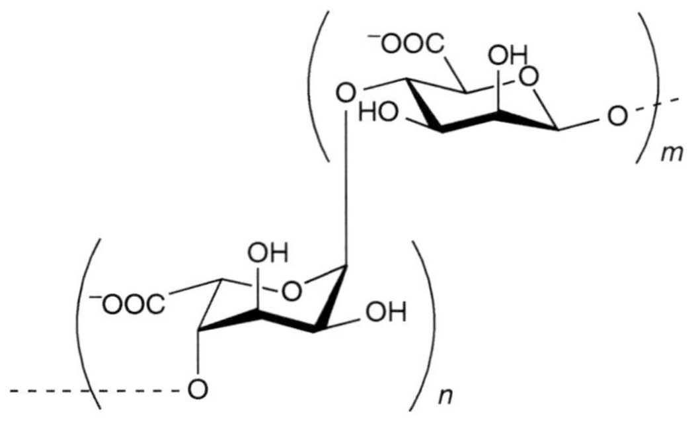

Figure 33.8 Chemical structure of alginate.

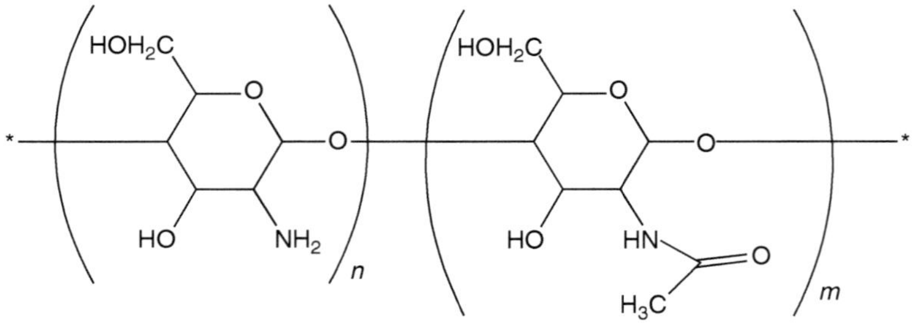

Figure 33.9 Chemical structure of chitosan, which is a copolymer consisting of β-(1→4)-linked 2-acetamido-2-deoxy-D-glucopyranose (shown on the left) and 2-amino-2-deoxy-D-glucopyranose (shown on the right).

33.6.4 Chitosan

Chitosan is a linear nature copolymer that is composed of β-(1→4)-linked 2-acetamido-2-deoxy-D-glucopyranose and 2-amino-2-deoxy-D-glucopyranose, as shown in Figure 33.9 [139]. Chitosan is a derivative of chitin, which is an abundant component in exoskeletons of crustaceans and insects [140]. In industry, chitin is produced from the seafood waste, where the shells of shrimps, crabs, and other marine exoskeletons are processed to get refined chitin. Deacetylation of chitin gives chitosan. This change in the structure of this natural copolymer also makes it more soluble in water, which makes the further processing and modification easier [139].

Similar to other natural polymers, chitosan has good biocompatibility and bioadhesivity. It is also believed to be nontoxic and biodegradable. Combined with its biocidal ability, chitosan has applications in variety of fields, such as medication, food additives, cosmetics, and bioengineering. In addition, due to its structure and available functional groups, chitosan can be processed to take different forms, including hydrogels, fibers, aerogels, thin films, powders, micro- or nanoparticles, and composite [141–144].

The antimicrobial effect of chitosan is highly related to its primary amine groups, which is formed during the deacetylation. Therefore, the acetylation process can affect the antimicrobial ability of chitosan. Its bacterial killing mechanism is based on electrostatic. The primary amine group is electron donating and easily picks up a positive charge [140]. Positively charged amines interact with cell membranes of microbes, which have negative charges, and damage the structures of cells, which can eventually disrupt the DNA replication and other vital functions of the microbe cells and deactivate them.

33.6.5 Conclusions and Future Trend

Natural antimicrobial polymers, just like many other inorganic or organic synthesized antimicrobial agents, show effective biocidal power. Many natural antimicrobials are broad-spectrum bacterial killing agents that work for both Gram-positive and Gram-negative bacteria and fungi. Moreover, until now, it has not yet been reported that any natural microbial has promoted microbial resistance.

Compared with other synthesized antimicrobial agents, the natural agents have a huge advantage that they are biodegradable and have good biocompatibility. This unique characteristic of natural antimicrobial polymers grants them applications in biomaterial and tissue engineering field. Materials like alginate and HA have been proved that they help wounds to heal. Polymers such as chitosan and gelatin have wide usage in drug delivery.

As natural polymers, they present in abundance in living creatures. Therefore, many researchers are studying the method to effectively extract and purify them from plant or animal tissues. Since they can be made into fibers themselves, modifying those natural antimicrobial textiles with other antimicrobial agents is another direction that has great potential and has drawn many attentions.

33.7 Conclusions

Pathogens, a treat to public health and the safety of many other human activities, need to be properly controlled especially in an era when the global population is greater than ever before. Antimicrobial textiles are a good way to deactivate pathogenic bacteria and fungi, because they can be easily made into applications, such as clothes, filters, covering materials, and gaze, which protect people from infections in daily lives.

There are many kinds of antimicrobial agents that can be used in fiber modification to grant fibers biocidal effects. This chapter mainly discussed five kinds of them: quaternary ammonium salts (QACs), *N*-halamine compounds, metals, photoactive chemicals, and natural polymers. Table 33.1 is a summary comparing the biocidal performances of some agents.

Each of these antimicrobial agents has their own advantages. QACs can also soften the fabrics to make them more comfortable to wear. *N*-Halamine compounds can be recharged. Silver nanoparticles have the best bacterial killing effectiveness of all kinds. Photoactive chemicals gain energy from sunlight (300–700 nm), which make the products greener. Natural antimicrobials have applications in tissue engineering and wound healing for their good biocompatibility.

In general, there is not a single antimicrobial agent that is better than everything else in all the aspects. Therefore, an important trend of this industry is

Table 33.1 Comparison among some common antimicrobial agents.

Agents	Effectiveness[a)] against Gram-positive bacteria	Effectiveness[a)] against Gram-negative bacteria	Effectiveness[a)] against Fungi	Microbial resistance	Toxicity[b)]
Quaternary ammonium compounds	+	−	−	Yes	3
N-Halamine	+	+	+	Not reported	1
Silver	+++	+++	+++	Not reported	1
Copper	+	+	+	Yes	2
Chitosan	+	+	+	Not reported	1, b

a) For the biocidal effectiveness evaluation: + means the chemical shows the ability to deactivate the particular kind of microorganisms; − means the chemical does not show the ability to deactivate the particular kind of microorganisms; +++ means the chemical shows outstanding microbial effect at extremely low concentration.
b) For toxicity ranking: 1 means the chemical is lightly toxic or nontoxic, safe for human during regular uses; 2 means the chemical could be toxic if it is used in large amount; 3 means the chemical is toxic and can cause damage to human body and the environment; b means the chemical is biodegradable and biocompatible.

to combine those agents together, along with other novel technology, such as nanotechnology, to further develop antimicrobial fibers with better performance. In the meanwhile, innovation of safer materials that cause less damage to both the environment and the users is another hot direction that much work has been done.

References

1 Sender, R., Fuchs, S., and Milo, R. (2016). Revised estimates for the number of human and bacteria cells in the body. *PLoS Biology* 14 (8): 1–14.
2 Hardin, I.R. and Kim, Y. (2016). Nanotechnology for antimicrobial textiles. In: *Antimicrobial Textiles* (ed. G. Sun), pp. 87–97. Wiley.
3 Sun, Y.Y. and Sun, G. (2001). Novel regenerable N-halamine polymeric biocides. III. Grafting hydantoin-containing monomers onto synthetic fabrics. *Journal of Applied Polymer Science* 81 (6): 1517–1525.
4 Sun, Y. and Sun, G. (2002). Durable and regenerable antimicrobial textile materials prepared by a continuous grafting process. *Journal of Applied Polymer Science* 84 (8): 1592–1599.
5 Son, Y.A. and Sun, G. (2003). Durable antimicrobial nylon 66 fabrics: ionic interactions with quaternary ammonium salts. *Journal of Applied Polymer Science* 90 (8): 2194–2199.
6 Yao, C., Li, X., Neoh, K.G. et al. (2008). Surface modification and antibacterial activity of electrospun polyurethane fibrous membranes with quaternary ammonium moieties. *Journal of Membrane Science* 320 (1–2): 259–267.

7 Mercs, L. and Albrecht, M. (2010). Beyond catalysis: N-heterocyclic carbene complexes as components for medicinal, luminescent, and functional materials applications. *Chemical Society Reviews* 39 (6): 1903–1912.
8 Evans, P. and Sheel, D.W. (2007). Photoactive and antibacterial TiO_2 thin films on stainless steel. *Surface and Coatings Technology* 201 (22–23): 9319–9324.
9 Pant, H.R., Pandeya, D.R., Nam, K.T. et al. (2011). Photocatalytic and antibacterial properties of a TiO_2/nylon-6 electrospun nanocomposite mat containing silver nanoparticles. *Journal of Hazardous Materials* 189 (1–2): 465–471.
10 Rabea, E.I., Badawy, M.E.T., Stevens, C.V. et al. (2003). Chitosan as antimicrobial agent: applications and mode of action. *Biomacromolecules* 4 (6): 1457–1465.
11 Pillai, C.K.S., Paul, W., and Sharma, C.P. (2009). Chitin and chitosan polymers: chemistry, solubility and fiber formation. *Progress in Polymer Science* 34 (7): 641–678.
12 Sharma, V.K., Yngard, R.A., and Lin, Y. (2009). Silver nanoparticles: green synthesis and their antimicrobial activities. *Advances in Colloid and Interface Science* 145 (1–2): 83–96.
13 Melaiye, A., Sun, Z.H., Hindi, K. et al. (2005). Silver(I)-imidazole cyclophane gem-diol complexes encapsulated by electrospun tecophilic nanofibers: formation of nanosilver particles and antimicrobial activity. *Journal of the American Chemical Society* 127 (7): 2285–2291.
14 Duran, N., Marcato, P.D., De Souza, G.I.H. et al. (2007). Antibacterial effect of silver nanoparticles produced by fungal process on textile fabrics and their effluent treatment. *Journal of Biomedical Nanotechnology* 3 (2): 203–208.
15 Cleveland, J., Montville, T.J., Nes, I.F., and Chikindas, M.L. (2001). Bacteriocins: safe, natural antimicrobials for food preservation. *International Journal of Food Microbiology* 71 (1): 1–20.
16 Saleem, M., Nazir, M., Ali, M.S. et al. (2010). Antimicrobial natural products: an update on future antibiotic drug candidates. *Natural Product Reports* 27 (2): 238–254.
17 Singh, R., Jain, A., Panwar, S. et al. (2005). Antimicrobial activity of some natural dyes. *Dyes and Pigments* 66 (2): 99–102.
18 Sisson, W.A. and Saner, W.R. (1939). An X-ray diffraction study of the swelling action of several quaternary ammonium hydroxides on cellulose fibers. *Journal of Physical Chemistry* 43 (6): 687–699.
19 Reck, R.A. and Harwood, H.J. (1953). Antimicrobial activity of quaternary ammonium chlorides – derived from commercial fatty acids. *Industrial and Engineering Chemistry* 45 (5): 1022–1026.
20 Weiner, N.D. and Zografi, G. (1965). Interfacial properties of antimicrobial long-chain quaternary ammonium salts. I. Soluble films at air-water interface. *Journal of Pharmaceutical Sciences* 54 (3): 436–442.
21 Hegstad, K., Langsrud, S., Lunestad, B.T. et al. (2010). Does the wide use of quaternary ammonium compounds enhance the selection and spread

of antimicrobial resistance and thus threaten our health? *Microbial Drug Resistance* 16 (2): 91–104.
22 Buffet-Bataillon, S., Tattevin, P., Bonnaure-Mallet, M., and Jolivet-Gougeon, A. (2012). Emergence of resistance to antibacterial agents: the role of quaternary ammonium compounds–a critical review. *International Journal of Antimicrobial Agents* 39 (5): 381–389.
23 Lu, G., Wu, D., and Fu, R. (2007). Studies on the synthesis and antibacterial activities of polymeric quaternary ammonium salts from dimethylaminoethyl methacrylate. *Reactive and Functional Polymers* 67 (4): 355–366.
24 Chen, C.Z.S., Beck-Tan, N.C., Dhurjati, P. et al. (2000). Quaternary ammonium functionalized poly(propylene imine) dendrimers as effective antimicrobials: structure-activity studies. *Biomacrobolecules* 1 (3): 473–480.
25 Tezel, U. and Pavlostathis, S.G. (2015). Quaternary ammonium disinfectants: microbial adaptation, degradation and ecology. *Current Opinion in Biotechnology* 33: 296–304.
26 Zhang, C., Cui, F., Zeng, G.-M. et al. (2015). Quaternary ammonium compounds (QACs): a review on occurrence, fate and toxicity in the environment. *Science of the Total Environment* 518–519: 352–362.
27 Heir, E., Sundheim, G., and Holck, A.L. (1999). Identification and characterization of quaternary ammonium compound resistant staphylococci from the food industry. *International Journal of Food Microbiology* 48 (3): 211–219.
28 Ioannou, C.J., Hanlon, G.W., and Denyer, S.P. (2007). Action of disinfectant quaternary ammonium compounds against *Staphylococcus aureus*. *Antimicrobial Agents and Chemotherapy* 51 (1): 296–306.
29 Harrison, J.J., Turner, R.J., Joo, D.A. et al. (2008). Copper and quaternary ammonium cations exert synergistic bactericidal and antibiofilm activity against *Pseudomonas aeruginosa*. *Antimicrobial Agents and Chemotherapy* 52 (8): 2870–2881.
30 Antonucci, J.M., Zeiger, D.N., Tang, K. et al. (2011). Synthesis and characterization of dimethacrylates containing quaternary ammonium functionalities for dental applications. *Dental Materials* 28 (2): 219–228.
31 Tseng, C.-C., Pan, Z.-M., and Chang, C.-H. (2016). Application of a quaternary ammonium agent on surgical face masks before use for pre-decontamination of nosocomial infection-related bioaerosols. *Aerosol Science and Technology* 50 (3): 199–210.
32 Games, L.M., King, J.E., and Larson, R.J. (1982). Fate and distribution of a quaternary ammonium surfactant, octadecyltrimethyammonium chloride (OTAC) in wasterwater-treatment. *Environmental Science and Technology* 16 (8): 483–488.
33 Tezel, U., Tandukar, M., Martinez, R.J. et al. (2012). Aerobic biotransformation of *n*-tetradecylbenzyldimethylammonium chloride by an enriched *Pseudomonas* spp. Community. *Environmental Science and Technology* 46 (16): 8714–8722.
34 Tabata, A., Nagamune, H., Maeda, T. et al. (2003). Correlation between resistance of *Pseudomonas aeruginosa* to quaternary ammonium compounds and expression of outer membrane protein OprR. *Antimicrobial Agents and Chemotherapy* 47 (7): 2093–2099.

35 Bore, E., Hebraud, M., Chafsey, I. et al. (2007). Adapted tolerance to benzalkonium chloride in *Escherichia coli* K-12 studied by transcriptome and proteome analyses. *Microbiology-SGM* 153: 935–946.

36 Karatzas, K.A.G., Randall, L.P., Webber, M. et al. (2008). Phenotypic and proteomic characterization of multiply antibiotic-resistant variants of *Salmonella entetica* serovar typhimurium selected following exposure to disinfectants. *Applied and Environmental Microbiology* 74 (5): 1508–1516.

37 Bjorland, J., Steinum, T., Sunde, M. et al. (2003). Novel plasmid-borne gene qacJ mediates resistance to quaternary ammonium compounds in equine *Staphylococcus aureus, Staphylococcus simulans,* and *Staphylococcus intermedius. Antimicrobial Agents and Chemotherapy* 47 (10): 3036–3052.

38 Gillings, M.R., Duan, X.J., Hardwick, S.A. et al. (2008). Gene cassettes encoding resistance to quaternary ammonium compounds: a role in the origin of clinical class 1 integrons? *ISME Journal* 3 (2): 209–215.

39 Littlejohn, T.G., Paulsen, I.T., Gillespie, M.T. et al. (1992). Substrate-specificity and energetics of antiseptic and disinfectant resistance in *Staphylococus aureus. FEMS Microbiology Letters* 95 (2–3): 259–266.

40 McMurry, L.M., Oethinger, M., and Levy, S.B. (1998). Overexpression of marA, soxS, or acrAB produces resistance to triclosan in laboratory and clinical strains of *Escherichia coli. FEMS Microbiology Letters* 166 (2): 305–309.

41 Pages, J.-M., Amaral, L., and Fanning, S. (2011). An original deal for new molecule: reversal of efflux pump activity, a rational strategy to combat gram-negative resistant bacteria. *Current Medicinal Chemistry* 18 (19): 2969–2980.

42 Worley, S.D., Williams, D.E., and Barnela, S.B. (1987). The stabilities of new N-halamine water disinfectants. *Water Research* 21 (8): 983–988.

43 Barnela, S.B., Worley, S.D., and Williams, D.E. (1987). Syntheses and antibacterial activity of new N-halamine compounds. *Journal of Pharmaceutical Sciences* 76 (3): 245–247.

44 Worley, S.D. and Williams, D.E. (1988). Disinfecting water with a mixture of free chlorine and an organic N-halamine. *Journal American Water Works Association* 80 (1): 69–71.

45 Tsao, T., Worley, S.D., and Williams, D.E. (1990). New N-halamine water disinfectants. *Abstracts of Papers of the American Chemical Society 199, 28-BIOT*.

46 Eknoian, M.W., Worley, S.D., Bickert, J., and Williams, J.F. (1999). Novel antimicrobial N-halamine polymer coatings generated by emulsion polymerization. *Polymer* 40: 1367–1371.

47 Sun, Y. and Sun, G. (2001). Novel regenerable N-halamine polymeric biocides. II. Grafting hydantoin-containing monomers onto cotton cellulose. *Journal of Applied Polymer Science* 81: 617–624.

48 Lin, J., Winkelmann, C., Worley, S.D. et al. (2002). Biocidal polyester. *Journal of Applied Polymer Science* 85: 177–182.

49 Si, Y., Cossu, A., Nitin, N. et al. (2017). Mechanically robust and transparent N-halamine grafted PVA-co-PE films with renewable antimicrobial activity. *Macromolecular Bioscience* 17: 1600304.

50 Kenawy, E.-R., Worley, S.D., and Broughton, R. (2007). The chemistry and applications of antimicrobial polymers: a state-of-the-art review. *Biomacromolecules* 8 (5): 1359–1384.

51 Ren, X.H., Kocer, H.B., Worley, S.D. et al. (2009). Rechargeable biocidal cellulose: synthesis and application of 3-(2,3-dihydroxypropyl)-5,5-dimethylimidazolidine-2,4-dione. *Carbohydrate Polymers* 75 (4): 683–687.

52 Sun, Y. and Sun, G. (2001). Durable and refreshable polymeric N-halamine biocides containing 3-(4′-vinylbenzyl)-5,5-dimethylhydantoin. *Journal of Polymer Science Part A: Polymer Chemistry* 39: 3348–3355.

53 Liang, J., Wu, R., Wang, J.W. et al. (2007). N-halamine biocidal coatings. *Journal of Industrial Microbiology and Biotechnology* 34 (2): 157–163.

54 Lee, J., Broughton, R.M., Akdag, A. et al. (2007). Antimicrobial fibers created via polycarboxylic acid durable press finishing. *Textile Research Journal* 77 (8): 604–611.

55 Ren, X.H., Akdag, A., Zhu, C. et al. (2009). Electrospun polyacrylonitrile nanofibrous biomaterials. *Journal of Biomedical Materials Research Part A* 91A: 385–390.

56 Cerkez, I., Kocer, H.B., Worley, S. et al. (2012). Multifunctional cotton fabric: antimicrobial and durable press. *Journal of Applied Polymer Science* 124: 4230–4238.

57 Li, R., Dou, J., Jiang, Q. et al. (2014). Preparation and antimicrobial activity of beta-cyclodextrin derivative copolymers/cellulose acetate nanofibers. *Chemical Engineering Journal* 248: 264–272.

58 Li, R., Sun, M., Jiang, Z. et al. (2014). N-Halamine-bonded cotton fabric with antimicrobial and easy-care properties. *Fibers and Polymers* 15: 234–240.

59 Liu, S. and Sun, G. (2006). Durable and regenerable biocidal polymers: acyclic N-halamine cotton cellulose. *Industrial and Engineering Chemistry Research* 45: 6477–6482.

60 Liu, S. and Sun, G. (2008). New refreshable N-halamine polymeric biocides: N-chlorination of acyclic amide grafted cellulose. *Industrial and Engineering Chemistry Research* 48: 613–618.

61 Luo, J. and Sun, Y. (2008). Acyclic N-halamine coated Kevlar fabric materials: preparation and biocidal functions. *Industrial and Engineering Chemistry Research* 47: 5291–5297.

62 Ren, X., Kou, L., Kocer, H.B. et al. (2008). Antimicrobial coating of an N-halamine biocidal monomer on cotton fibers via admicellar polymerization. *Colloids and Surfaces A: Physicochemical and Engineering Aspects* 317: 711–716.

63 Jiang, J., Chen, Y., Ren, X. et al. (2007). Fabric treated with antimicrobial N-halamine epoxides. *Industrial and Engineering Chemistry Research* 46: 6425–6429.

64 Ma, K., Liu, Y., Xie, Z. et al. (2013). Synthesis of novel N-halamine epoxide based on cyanuric acid and its application for antimicrobial finishing. *Industrial and Engineering Chemistry Research* 52 (22): 7413–7418.

65 Ma, K., Jiang, Z., Li, L. et al. (2014). N-halamine modified polyester fabrics: preparation and biocidal functions. *Fibers and Polymers* 15: 2340–2344.

66 Fong, H., Chun, L., and Reneker, D. (1999). Beaded nanofibers formed during electrospinning. *Polymer* 40: 4585–4592.

67 Greiner, A. and Wendorff, J.H. (2007). Electrospinning: a fascinating method for the preparation of ultrathin fibers. *Angewandte Chemie International Edition* 46: 5670–5703.

68 Si, Y., Li, J., Zhao, C. et al. (2017). Biocidal and rechargeable N-halamine nanofibrous membranes for highly efficient water disinfection. *ACS Biomaterials Science & Engineering* 3 (5): 854–862.

69 MacNaught, A.D. and Wilkinson, A.R. (eds.) (1997). *Compendium of Chemical Terminology: IUPAC Recommendations*, 2e. Blackwell Science.

70 Alemán, J., Chadwick, A.V., He, J. et al. (2007). Definitions of terms relating to the structure and processing of sols, gels, networks, and inorganic-organic hybrid materials (IUPAC Recommendations 2007). *Pure and Applied Chemistry* 79 (10): 1801.

71 Kim, J.S., Kuk, E., Yu, K.N. et al. (2007). Antimicrobial effects of silver nanoparticles. *Nanomedicine: Nanotechnology, Biology and Medicine* 3: 95–101.

72 Heinlaan, M., Ivask, A., Blinova, I. et al. (2008). Toxicity of nanosized and bulk ZnO, CuO and TiO_2 to bacteria *Vibrio fischeri* and crustaceans *Daphnia magna* and *Thamnocephalusplatyurus*. *Chemosphere* 71: 1308–1316.

73 Aruoja, V., Dubourguier, H.-C., Kasemets, K., and Kahru, A. (2009). Toxicity of nanoparticles of CuO, ZnO and TiO_2 to microalgae *Pseudokirchneriella subcapitata*. *Science of the Total Environment* 407: 1461–1468.

74 Wang, R., Xin, J.H., Tao, X.M., and Daoud, W.A. (2004). ZnO nanorods grown on cotton fabrics at low temperature. *Chemical Physics Letters* 398: 250–255.

75 Yadav, A., Prasad, V., Kathe, A.A. et al. (2006). Functional finishing in cotton fabrics using zinc oxide nanoparticles. *Bulletin of Materials Science* 29: 641–645.

76 Marambio-Jones, C. and Hoek, E.M.V. (2010). A review of the antibacterial action of silver nanoparticles. *Journal of Proteome Research* 5: 916–924.

77 Tortora, G.J., Funke, B.R., and Case, C.L. (2004). *Microbiology An Introduction*, 10e. San Francisco, CA: Pearson Benjamin Cummings.

78 Cowan, M.M., Abshire, K.Z., Houk, S.L., and Evans, S.M. (2003). Antimicrobial efficacy of a silver-zeolite matrix coating on stainless steel. *Journal of Industrial Microbiology and Biotechnology* 30: 102–106.

79 Pillai, Z.S. and Kamat, P.V. (2004). What factors control the size and shape of silver nanoparticles in the citrate ion reduction method? *The Journal of Physical Chemistry B* 110: 16248–16253.

80 Gulrajani, M.L., Gupta, D., Periyasamy, S., and Muthu, S.G. (2008). Preparation and application of silver nanoparticles on silk for imparting antimicrobial properties. *Journal of Applied Polymer Science* 108: 614–623.

81 Nägeli, K.W. (1893). Übero ligodynamische Erscheinungen in lebenden Zellen. *Neue Denkschriften der allgemeinen Schweizerischen Gesellschaft für die gesamte Naturwissenschaft*, XXXIII (1).

82 Bauman, R.W. (2012). *Microbiology with Diseases by Body System*, 3e. Benjamin Cummings.

83 Cowan, M.K. (2012). *Microbiology: A Systems Approach*, 3e. Pearson Education, Inc.
84 Choi, O., Deng, K.K., Kim, N.-J. et al. (2008). The inhibitory effects of silver nanoparticles, silver ions, and silver chloride colloids on microbial growth. *Water Research* 42: 3066–3074.
85 Smetana, A.B., Klabunde, K.J., Marchin, G.R., and Sorensen, C.M. (2008). Biocidal activity of nanocrystalline silver powders and partcles. *Langmuir* 24: 7457–7464.
86 Hwang, E.T., Lee, J.H., Chae, Y.J. et al. (2008). Analysis of the toxic mode of action of silver nanoparticles using stress-specific bioluminescent bacteria. *Small* 4: 746–750.
87 Armelao, L., Barreca, D., Bottaro, G. et al. (2007). Photocatalytic and antibacterial activity of TiO_2 and Au/TiO_2 nanosystems. *Nanotechnology* 18 (37): 1–7.
88 Li, Q., Xie, R.C., LI, Y.W. et al. (2007). Enhanced visible-light-induced photocatalytic disinfection of *E-coli* by carbon-sensitized nitrogen-doped titanium oxide. *Environmental Science and Technology* 41 (14): 5050–5056.
89 Nowack, B. and Bucheli, T.D. (2007). Occurrence, behavior and effects of nanoparticles in the environment. *Environmental Science and Technology* 45: 1177–1183.
90 Wijnhoven, S.W.P., Peijnenburg, W.J.G.M., Herberts, C.A. et al. (2009). Decreasing uncertainties in assessing environmental exposure, risk, and ecological implications of nanomaterials. *Environmental Science and Technology* 43: 6458–6462.
91 Nowack, B. (2010). Nanosilver revisited downstream. *Science* 330: 1054–1055.
92 Nowack, B., Krug, H.F., and Height, M. (2011). 120 years of nanosilver history: implications for policy makers. *Environmental Pollution* 150: 5–22.
93 Foster, H.A., Ditta, I.B., Varghese, S., and Steele, A. (2011). Photocatalytic disinfection using titanium dioxide: spectrum and mechanism of antimicrobial activity. *Applied Microbiology and Biotechnology* 90 (6): 1847–1868.
94 Gil-Tomas, J., Dekker, L., Narband, N. et al. (2011). Lethal photosensitisation of bacteria using a tin chlorin e6-glutathione-gold nanoparticle conjugate. *Journal of Materials Chemistry* 21 (12): 4189–4196.
95 Omar, G.S., Wilson, M., and Nair, S.P. (2008). Lethal photosensitization of wound-associated microbes using indocyanine green and near-infrared light. *BMC Microbiology* 8: Article Number 111.
96 Tung, W.S. and Daoud, W.A. (2011). Self-cleaning fibers via nanotechnology: a virtual reality. *Journal of Materials Chemistry* 21 (22): 7858–7869.
97 Lu, Z.X., Zhou, L., Zhang, Z.L. et al. (2003). Cell damage induced by photocatalysis of TiO_2 thin films. *Langmuir* 19 (21): 8765–8768.
98 Fujishima, A. and Honda, K. (1972). Electrochemical photolysis of water at a semiconductor electrode. *Nature* 238 (5358): 37–38.
99 Daoud, W.A., Xin, J.H., and Zhang, Y.H. (2005). Surface functionalization of cellulose fibers with titanium dioxide nanoparticles and their combined bactericidal activities. *Surface Science* 559 (1–3): 69–75.

100 Qi, K.H., Xin, J.H., Daoud, W.A., and Mak, C.L. (2007). Functionalizing polyester fiber with a self-cleaning property using anatase TiO_2 and low-temperature plasma treatment. *International Journal of Applied Ceramic Technology* 4 (6): 554–563.

101 Tung, W.S. and Daoud, W.A. (2008). Photocatalytic formulations for protein fibers: experimental analysis of the effect of preparation on compatibility and photocatalytic activities. *Journal of Colloid and Interface Science* 326 (1): 283–288.

102 Pakdel, E., Daoud, W.A., and Wang, X.G. (2013). Self-cleaning and superhydrophilic wool by TiO_2/SiO_2 nanocomposite. *Applied Surface Science* 275: 397–402.

103 Mahltig, B., Haufe, H., and Bottcher, H. (2005). Functionalisation of textiles by inorganic sol–gel coatings. *Journal of Materials Chemistry* 15 (41): 4385–4398.

104 Qi, K.H., Daoud, W.A., Xin, J.H. et al. (2006). Self-cleaning cotton. *Journal of Materials Chemistry* 16 (47): 4567–4574.

105 Bozzi, A., Yuranova, T., and Kiwi, J. (2005). Self-cleaning of wool-polyamide and polyester textiles by TiO_2-rutile modification under daylight irradiation at ambient temperature. *Journal of Photochemistry and Photobiology A: Chemistry* 172 (1): 27–34.

106 Kiwi, J. and Pulgarin, C. (2010). Innovative self-cleaning and bactericide textiles. *Catalysis Today* 151 (1–2): 2–7.

107 Yuranova, T., Mosteo, R., Bandara, J. et al. (2006). Self-cleaning cotton textiles surfaces modified by photoactive SiO_2/TiO_2 coating. *Journal of Molecular Catalysis A: Chemistry* 244 (1–2): 160–167.

108 Hong, K.H., Liu, N., and Sun, G. (2009). UV-induced graft polymerization of acrylamide on cellulose by using immobilized benzophenone as a photo-initiator. *European Polymer Journal* 45 (8): 2443–2449.

109 Hong, K.H. and Sun, G. (2007). Photocatalytic functional cotton fabrics containing benzophenone chromophoric groups. *Journal of Applied Polymer Science* 106 (4): 2661–2667.

110 Hong, K.H. and Sun, G. (2007). Preparation and properties of benzophenone chromophoric group branched polymer for self-decontamination. *Polymer Engineering and Science* 47 (11): 1750–1755.

111 Hong, K.H. and Sun, G. (2008). Poly(styrene-*co*-vinylbenzophenone) as photoactive antimicrobial and selfdecontaminating materials. *Journal of Applied Polymer Science* 109 (5): 3173–3179.

112 Zhuo, J. and Sun, G. (2013). Antimicrobial functions on cellulose materials introduced by anthraquinone vat dyes. *ACS Applied Materials & Interfaces* 5 (21): 10830–10835.

113 Zhuo, J. and Sun, G. (2014). Light-induced surface graft polymerizations initiated by an anthraquinone dye on cotton fibers. *Carbohydrate Polymers* 112: 158–164.

114 Malik, Z., Hanania, J., and Nitzan, Y. (1990). Bactericidal effects of photoactivated porphyrins – an alternative approach to antimicrobial drugs. *Journal of Photochemistry and Photobiology B: Biology* 5 (3–4): 281–293.

115 Banfi, S., Caruso, E., Buccafurni, L. et al. (2006). Antibacterial activity of tetraaryl-porphyrin photosensitizers: an in vitro study on Gram negative and Gram positive bacteria. *Journal of Photochemistry and Photobiology B: Biology* 85 (1): 28–38.
116 Lambrechts, S.A.G., Aalders, M.C.G., and Van, M.J. (2005). Mechanistic study of the photodynamic inactivation of candida albicans by a cationic porphyrin. *Antimicrobial Agents and Chemotherapy* 49 (5): 2026–2034.
117 Mantareva, V., Kussovski, V., Angelov, I. et al. (2011). Non-aggregated Ga(III)-phthalocyanines in the photodynamic inactivation of planktonic and biofilm cultures of pathogenic microorganisms. *Photochemical & Photobiological Sciences* 10 (1): 91–102.
118 Griffiths, M.A., Wren, B.W., and Wilson, M. (1997). Killing of methicillin-resistant *Staphylococcus aureus* in vitro using aluminium disulphonated phthalocyanine, a light-activated antimicrobial agent. *Journal of Antimicrobial Chemotherapy* 40 (6): 873–876.
119 Segalla, A., Borsarelli, C.D., Braslavsky, S.E. et al. (2002). Photophysical, photochemical and antibacterial photosensitizing properties of a novel octacationic Zn(II)-phthalocyanine. *Photochemistry & Photobiological Sciences* 1 (9): 641–648.
120 Geppert, M., Hohnholt, M., Gaetjen, L. et al. (2009). Accumulation of iron oxide nanoparticles by cultured brain astrocytes. *Jounral of Biomedical Nanotechnology* 5 (3): 285–293.
121 Murr, L.E. (2009). Nanoparticulate materials in antiquity: the good, the bad and the ugly. *Materials Characterization* 60 (4): 261–270.
122 Colegate, S.M. and Molyneux, R.J. (2007). *Bioactive Natural Products: Detection, Isolation, and Structural Determination*. CRC Press.
123 Aboelsoud, N.H. (2010). Herbal medicine in ancient Egypt. *Journal of Medicinal Plant Research* 4 (2): 82–96.
124 Sun, H., Zhang, A., and Wang, X. (2012). Potential role of metabolomic approaches for Chinese medicine syndromes and herbal medicine. *Phytotherapy Research* 26 (10): 1466–1471.
125 Tan, H.P., Ramirez, C.M., Miljkovic, N. et al. (2009). Thermosensitive injectable hyaluronic acid hydrogel for adipose tissue engineering. *Biomaterials* 30 (36): 6844–6853.
126 Schante, C.E., Zuber, G., Herlin, C., and Vandamme, T.F. (2011). Chemical modifications of hyaluronic acid for the synthesis of derivatives for a broad range of biomedical applications. *Carbohydrate Polymers* 85 (3): 469–489.
127 Fakhari, A. and Berkland, C. (2013). Applications and emerging trends of hyaluronic acid in tissue engineering, as a dermal filler and in osteoarthritis treatment. *Acta Biomaterialia* 9 (7): 7081–7092.
128 Burdick, J.A. and Prestwich, G.D. (2011). Hyaluronic acid hydrogels for biomedical applications. *Advanced Materials* 23 (12): H41–H56.
129 Leach, J.B., Bivens, K.A., Patrick, C.W., and Schmidt, C.E. (2003). Photocrosslinked hyaluronic acid hydrogels: natural, biodegradable tissue engineering scaffolds. *Biotechnology and Bioengineering* 82 (5): 578–589.

130 Djagny, K.B., Wang, Z., and Xu, S.Y. (2001). Gelatin: a valuable protein for food and pharmaceutical industries: review. *Critical Reviews in Food Science and Nutrition* 41 (6): 481–492.

131 Siqueira, N.M., Paiva, B., Camassola, M. et al. (2015). Gelatin and galactomannan-based scaffolds: characterization and potential for tissue engineering applications. *Carbohydrate Polymers* 133: 8–18.

132 Li, M.Y., Guo, Y., Wei, Y. et al. (2006). Electrospinning polyaniline-contained gelatin nanofibers for tissue engineering applications. *Biomaterials* 27 (13): 2705–2715.

133 Young, S., Wong, M., Tabata, Y., and Mikos, A.G. (2005). Gelatin as a delivery vehicle for the controlled release of bioactive molecules. *Journal of Controlled Release* 109 (1–3): 256–274.

134 Lee, K.Y. and Mooney, D.J. (2012). Alginate: properties and biomedical applications. *Progress in Polymer Science* 37 (1): 106–126.

135 Wong, T.Y., Preston, L.A., and Schiller, N.L. (2000). Alginate lyase: review of major sources and enzyme characteristics, structure-function analysis, biological roles, and applications. *Annual Review of Microbiology* 54: 289–340.

136 Sun, J. and Tan, H. (2013). Alginate-based biomaterials for regenerative medicine applications. *Materials* 6 (4): 1285–1309.

137 Balakrishnan, B., Mohanty, M., Umashankar, P.R., and Jayakrishnan, A. (2005). Evaluation of an in situ forming hydrogel wound dressing based on oxidized alginate and gelatin. *Biomaterials* 26 (32): 6335–6342.

138 Choi, Y.S., Hong, S.R., Lee, Y.M. et al. (1999). Study on gelatin-containing artificial skin: I. Preparation and characteristics of novel gelatin-alginate sponge. *Biomaterials* 20 (5): 409–417.

139 Dash, M., Chiellini, F., Ottenbrite, R.M., and Chiellini, E. (2011). Chitosan-A versatile semi-synthetic polymer in biomedical applications. *Progress in Polymer Science* 36 (8): 981–1014.

140 Berger, J., Reist, M., Mayer, J.M. et al. (2004). Structure and interactions in covalently and ionically crosslinked chitosan hydrogels for biomedical applications. *European Journal of Pharmaceutics and Biopharmaceutics* 57 (1): 19–34.

141 Agnihotri, S.A., Mallikarjuna, N.N., and Aminabhavi, T.M. (2004). Recent advances on chitosan-based micro- and nanoparticles in drug delivery. *Journal of Controlled Release* 100 (1): 5–28.

142 Suh, J.K.F. and Matthew, H.W.T. (2000). Application of chitosan-based polysaccharide biomaterials in cartilage tissue engineering: a review. *Biomaterials* 21 (24): 2589–2598.

143 Renault, F., Sancey, B., Badot, P.M., and Crini, G. (2009). Chitosan for coagulation/flocculation processes - an eco-friendly approach. *European Polymer Journal* 45 (5): 1337–1348.

144 Lim, S.H. and Hudson, S.M. (2003). Review of chitosan and its derivatives as antimicrobial agents and their uses as textile chemicals. *Journal of Macromolecular Science Polymer Reviews* C43 (2): 223–269.

34

Fibers for Auxetic Applications

Hong Hu and Adeel Zulifqar

The Hong Kong Polytechnic University, Institute of Textiles and Clothing, Hung Hom, Kowloon, Hong Kong, China

34.1 Introduction

Poisson's ratio (PR) is a measure of the Poisson effect, the phenomenon in which a material tends to expand in directions perpendicular to the direction of compression. Conversely, if the material is stretched rather than compressed, it usually tends to contract in the directions transverse to the direction of stretching. It is a common observation when a rubber band is stretched, it becomes noticeably thinner. The PR is denoted by ν and named after Siméon Poisson. It is the signed ratio of transverse strain to axial strain or transversal expansion to axial compression and calculated by using Eq. (34.1):

$$\nu = -\frac{\varepsilon_y}{\varepsilon_x} \tag{34.1}$$

where ε_x is the axial strain and ε_y is the transverse strain.

The PR is one of the fundamental properties of any engineering materials and represents important mechanical characteristics for many applications including in engineering systems, which incorporate textile fibers as structural elements. Mostly the traditional engineering materials (TEMs) have positive PR. In certain rare cases, a material will actually shrink in the transverse direction when compressed (or expand when stretched), which yields a negative value of the PR as shown in Figure 34.1. This occurs due to their specific core structure and the mode of their deformation when the sample is uniaxially loaded. This behavior is termed as auxeticity, and such materials are known as auxetic materials.

The term auxetic was derived from the Greek word auxetikos, which means "that which tends to increase" by K. Evans of the University of Exeter [1]. The known naturally and manmade auxetic materials include metals, silicates, zeolites, laminates, gels, composites foams, fibers, polymers, etc. These materials are incorporated as core materials to produce sandwich panels, drug release systems, and energy absorbance textiles. These materials can also be used for vibration damping and to reduce creep buckling failure. It is claimed that auxetic materials have enhanced mechanical properties like shear modulus, energy

Handbook of Fibrous Materials, First Edition. Edited by Jinlian Hu, Bipin Kumar, and Jing Lu.

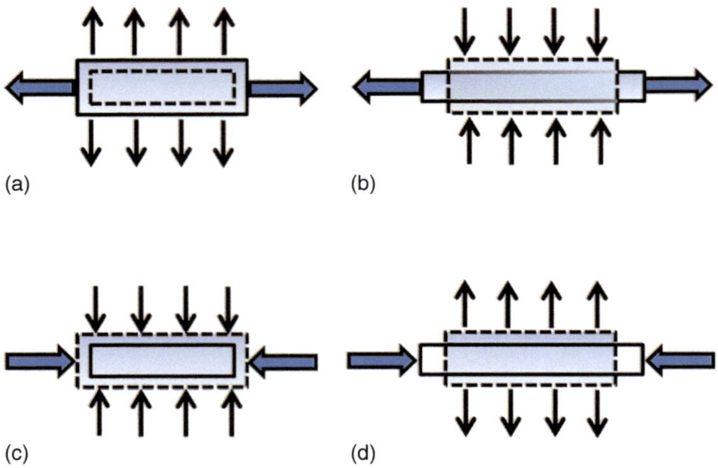

Figure 34.1 Deformation behavior of materials when stretched: (a) auxetic and (b) conventional; deformation behavior of materials when compressed: (c) auxetic and (d) conventional.

absorbance, vibration damping [1–4], sound absorption [5], indentation resistance [6], sync-elastic behavior, and better formability [7]. In recent past years the auxetic fibers have been developed and investigated by many researchers including, polymeric auxetic fibers, auxetic filaments, and moisture-sensitive auxetic fibers.

34.2 Auxetic Structures and Geometries

Auxetic materials have been known for over a hundred years but have only gained attention in recent decades. The PR is an elastic constant and is independent of the material scale. Thus auxetic materials can be single molecules or a particular structure of macroscopic to microscopic level [8–13]. Auxetic materials can be produced by modifying the macrostructure of the material. These materials possess hinge-like features, which can change shape when a force is applied. For example, when a tensile force is applied, the hinge-like structures extends, causing the material to expand in lateral direction. While on application of a compressive force, the hinge-like structure folds even further, and the material contracts in lateral direction. At the macroscale, the auxetic behavior can be exemplified with an elastic cord having an inelastic string wound around it. When a pulling force is applied, the inelastic string gets straight, while the elastic cord stretches and winds helically around it, and the volume of the structure is increased. Auxetic materials usually have open cell structure and have a relatively low density. It is the open cell structure that can be modified to give the desired properties.

Auxetic materials can be made based on two different approaches: firstly, the top-down approach, which involves the manipulation of everyday polymers to give the desired structure and properties, and secondly, the bottom-up approach,

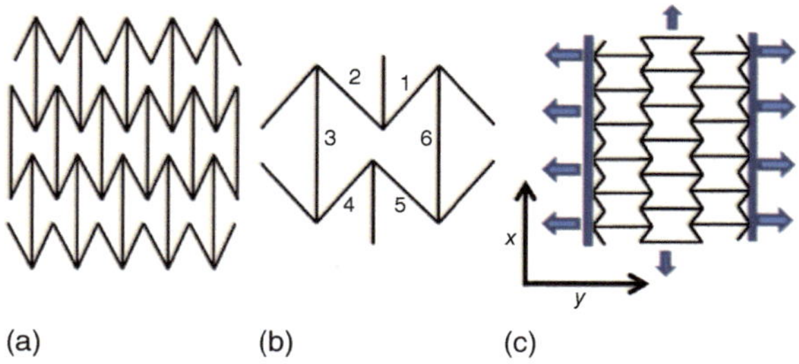

Figure 34.2 Re-entrant hexagon auxetic geometry (a) re-entrant hexagon structure, (b) unit cell, and (c) the structure under stretch.

in which the material is built molecule by molecule on a very small scale. In both techniques the main objective is to create a repeating pattern of unit cells, which possess the hinge-like features. Rod Lakes created the first ever polymer-based auxetic material in 1987 by employing the top-down approach. He applied heat and pressure to ordinary polyurethane foam, which consisted of a honeycomb of hexagonal cells, which resulted in auxetic hexagon [14].

34.2.1 Auxetic Geometries

In past few years, various geometrical structures capable of inducing auxetic behavior into the material have been developed and tested to realize their mechanical properties. These structures include re-entrant model, rotating model, nodule and fibril model, and chiral model. These auxetic structures can be used to develop various fibrous-based auxetic materials. These geometrical structures are explained briefly below.

34.2.1.1 Re-entrant Geometries

Auxetic structures based on re-entrant geometries achieve their auxetic behavior due to the deformation of diagonals along with the rotation around the hinge joints when stretched. Figure 34.2 illustrates one such example of an existing auxetic geometry named as re-entrant hexagonal geometry. The minimal repeating unit or unit cell is shown in Figure 34.2b. When the fabric is subjected to extension as shown in a direction, the fabric also expands in a direction that is perpendicular to the extension direction as shown in Figure 34.2c resulting in the negative PR effect.

34.2.1.2 Rotating Rigid Geometry

Auxetic materials based on rotational rigid geometry are based on deformation theory of geometrical configurations linked at some point of the unit. When such materials are axially stretched in a certain direction, the angles between the edges of adjacent triangles or squares or rectangles may change, and units rotate expanding in both directions. One such geometry is illustrated in Figure 34.3a; the architecture is a rotating rectangles.

The four rigid rectangles are connected at their vertices by hinges, in such a way that the empty spaces between the rectangles form rhombi. The minimal repeating unit or unit cell is highlighted in black color. It is assumed that the rectangle units are rigid and do not change their shape under loading and can

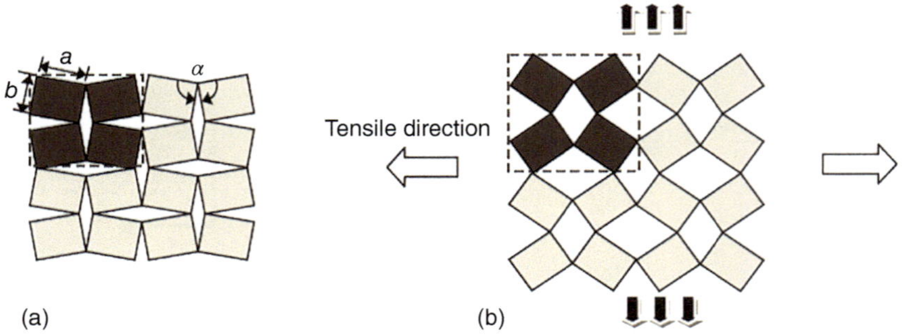

Figure 34.3 Rotating rectangles geometry (a) free state and (b) stretched state. Source: Hu et al. 2011 [15]. Reproduced with permission of SAGE.

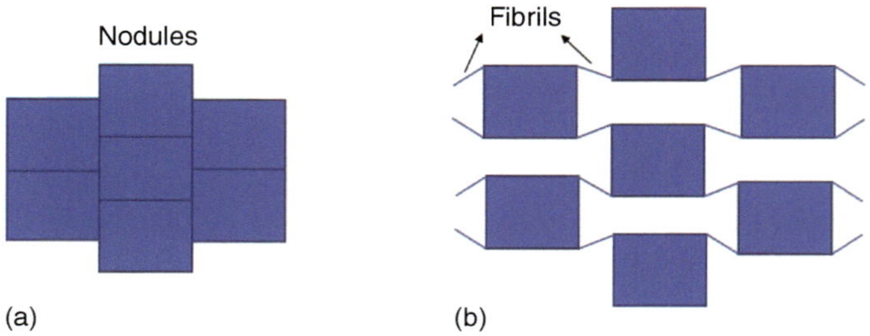

Figure 34.4 Schematic of nodule–fibril model representing microstructure of typical auxetic microporous polymer. (a) The polymer at rest and (b) the polymer at the tensile load. Source: Adapted from Liu and Hu 2010 [17].

rotate freely under loading. When the structure is subjected to an extension in a direction, the structure also expands in a direction that is perpendicular to the extension direction due to the free rotation of the rectangle units as shown in Figure 34.3b resulting in the negative PR effect. The auxetic effect depends on the strain and dimensions of the rectangles [16].

34.2.1.3 Nodule and Fibril-Based Structure

Auxetic materials based on nodule and fibrils achieve auxetic behavior due to the transition of nodes connected by fibrils under tension. One such example is illustrated in Figure 34.4.

In the stress-free state, the nodules overlap with the fibrils wound around them as shown in Figure 34.4a. When the material is stretched in the direction of the fibrils, they get straight, and the nodules rotate and snap into a rigid grid-like arrangement as shown in Figure 34.4b [17].

34.2.1.4 Chiral-Based Auxetic Geometries

Chiral structures are another kind of structures that have been developed for auxetic honeycombs. As shown in Figure 34.5, in this kind of structures, basic chiral units (highlighted in bold) are firstly formed by connecting straight ligaments (ribs) to central nodes, which may be circles or rectangles or other geometrical

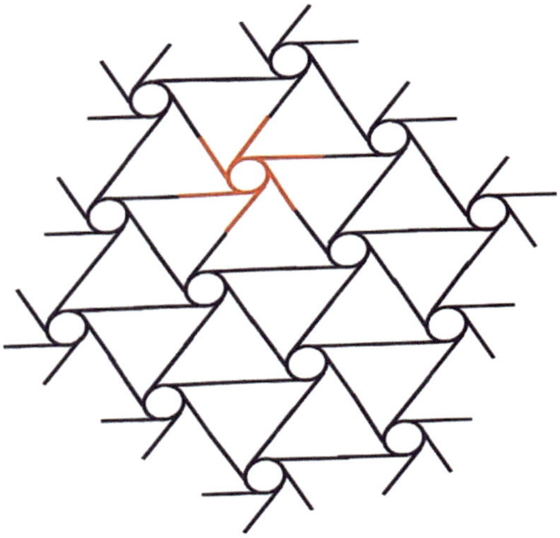

Figure 34.5 Chiral honeycombs formed with the same chiral units. Source: Adapted from Liu and Hu 2010 [17].

forms. The whole chiral structures are then formed by joining the chiral units. The auxetic effects are achieved through wrapping or unwrapping of the ligaments around the nodes in response to an applied force [17].

34.2.1.5 Foldable Geometries

The principle of using foldable structures to create an auxetic effect is that a folded structure can be unfolded when stretched in one direction, increasing the dimensions in lateral direction, thus yielding auxetic effect. One such geometry is illustrated in Figure 34.6. As shown in Figure 34.6a, the geometrical structure used to generate the auxetic effect is a three-dimensional structure formed with parallelogram planes of the same shape and size. These parallelogram planes are connected side to side in a zigzag form.

When stretched in the horizontal or vertical direction, each parallelogram can change its inclined position related to the surface plane of the structure, which

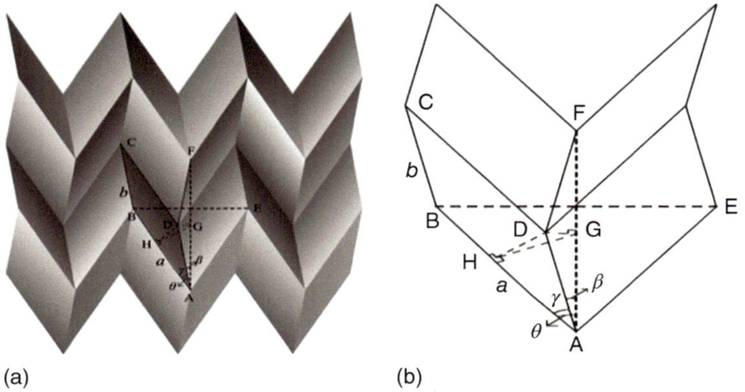

Figure 34.6 Geometrical structure: (a) three-dimensional structure and (b) unit cell. Source: Liu et al. 2009 [18]. Reproduced with permission of SAGE.

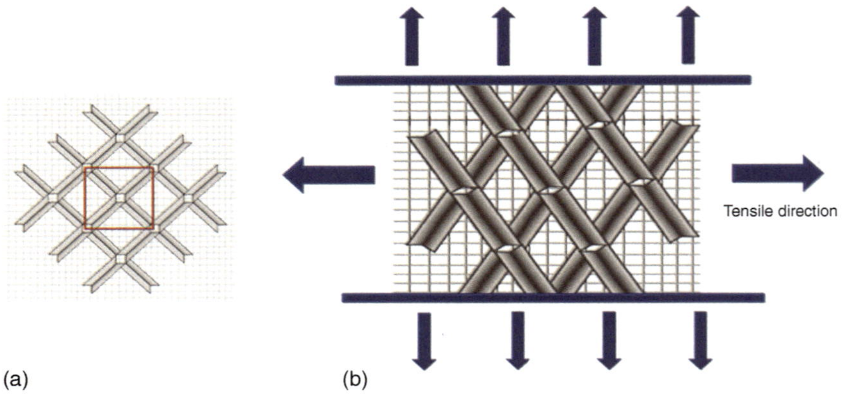

Figure 34.7 Foldable geometries: folded stripes in the oblique fashion in the free state (a) and stretched state (b). Source: Zulifqar et al. 2017 [19]. Reproduced with permission of SAGE.

results in an opening of the whole structure by increasing its dimensions in both the horizontal and vertical directions. Consequently, the auxetic effect can be achieved without changing the shape and size of the parallelogram planes. As the auxetic effect mainly comes from the position change of the parallelogram planes, it is assumed that their shape and size remain unchanged during extension. This means that a, b, and γ are kept constant.

Figures 34.7 and 34.8 illustrate two more foldable geometries that can be used to create the auxetic effect. The architecture of the geometry in Figure 34.7 is formed by placing re-entrant folded stripes in oblique fashion, which means that the stripes intersect each other in diagonal fashion as shown in Figure 34.7a. The empty spaces formed between these stripes are flat and may form a parallelogram shape. The architecture of the geometry in Figure 34.8 is re-entrant in-phase parallel folded stripes in zigzag fashion. This geometry consists of folded stripes placed in zigzag manner running along warp direction or weft direction as shown in Figure 34.8a. The minimal repeating unit or unit cell for both geometries is highlighted by red color box. When the structure is subjected to an extension in a direction, the structure also expands in a direction that is perpendicular to the extension direction due to the flattening of the re-entrant folded oblique stripes, and the transverse dimensions of the structure may remain unchanged or increase resulting in the zero or negative PR effect as shown in Figures 34.7b and 34.8b [19].

34.3 Auxetic Polymeric Fibers and Materials

34.3.1 Liquid Crystalline Polymers and Auxetic Monofilaments

Liquid crystalline polymer (LCP) exhibiting auxetic behavior was also developed. The arrangement of laterally attached rods in a main chain of LCP is shown in

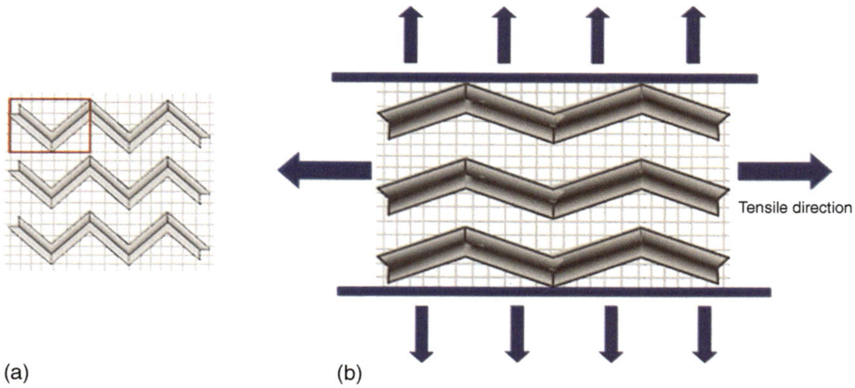

Figure 34.8 Foldable geometries: folded stripes in parallel in-phase zigzag fashion along the weft in the free state (a) and stretched state (b). Source: Zulifqar et al. 2017 [19]. Reproduced with permission of SAGE.

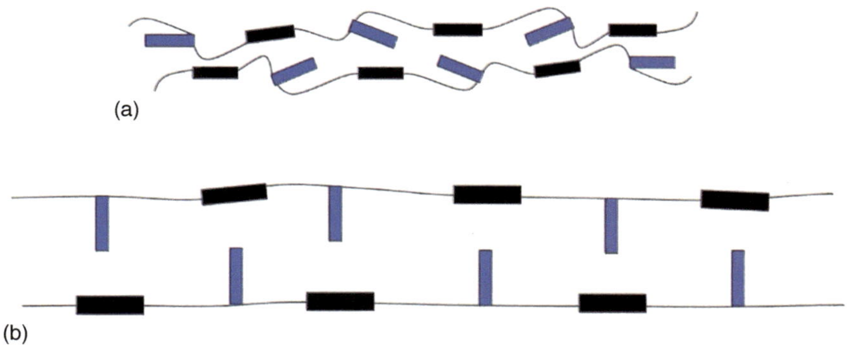

Figure 34.9 Liquid Crystalline Polymer (LCP), arrangement of main chain (a) without deformation; (b) due to stretch along horizontal direction. Source: Adapted from Liu et al. 1998 [20].

Figure 34.9. The nematic field leads to orientation of the laterally attached rods parallel to the polymer chain axis. Under tensile stress as shown in Figure 34.9, full extension of the polymer main chain forces the laterally attached rods normal to the chain axis leading to an expansion in the direction normal to the chain axis and hence to auxetic behavior. It is important to consider that the laterally attached rods should be sufficiently long, in order to increase the interchain distance and ultimately induce the auxetic behavior [20]. Polymeric monofilaments displaying auxetic behavior were also produced. The produced filaments had a microstructure of interconnected surface-melted powder particles. The structure and deformation mechanisms at the microscale, rather than at the molecular level (as compared with conventional filaments extruded from a fully molten polymer), are responsible for enhanced mechanical properties, including the auxetic effect [21].

34.3.2 Auxetic Polymeric Fibers

The auxetic polymeric fibrous material is an area that has a great potential for development. Up until today, auxetic fibers developed include ultra-high molecular weight polyethylene (UHMWPE), polyethylene (PE), polyester, nylon, and polypropylene (PP) fibers. Among these fibers only PP fibers are produced on large scale, and all others were produced on laboratory scale. Auxetic fibers have a wide range of potential applications including fiber-reinforced composites, drug delivery systems, and systems for personal protection. In this section, the auxetic fibers produced and characterized to date are discussed briefly in terms of processing conditions and manufacturing methods.

34.3.3 Auxetic Polypropylene Fibers

Alderson et al. were the first to successfully produce auxetic fibers. They developed a kind of auxetic PP fiber by employing a novel thermal processing technique, based on modified conventional melt spinning technique. This enabled a continuous fabrication process of auxetic PP fibers with diameters of less than 1 mm, and a large value of PR ($\nu = -0.6$) was obtained when measured by using video extensometer. The conventional melt spinning method with some novel modifications is adopted to produce auxetic PP fibers. The set of processing conditions are described including temperature, screw speed, and takeoff speed. It was observed that at a temperature of 159 °C with a screw speed of 1.05 rad/s and takeoff speed of 0.03 m/s, the produced fibers exhibited auxetic behavior. It is important to note that the temperature should be constant throughout the length of the extruder and there should be no drawing. Figure 34.10 shows that the fibers obtained via this processing route have in-phase length and width data, which confirms the auxetic behavior of produced fibers [22].

34.3.3.1 Large-Scale Production of Auxetic PP Fibers

Previously, the auxetic fibers were produced only on laboratory scale. Recently, Kim Alderson et al. produced auxetic PP fibers on large scale, by using a big scale extruder machine as shown in Figure 34.11. The extruder used for this process

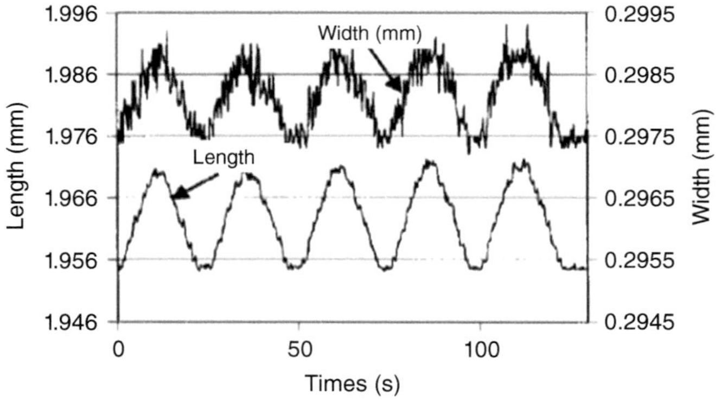

Figure 34.10 Width–length data for polypropylene fibers. Source: Alderson et al. 2002 [22]. Reproduced with permission of Taylor & Francis.

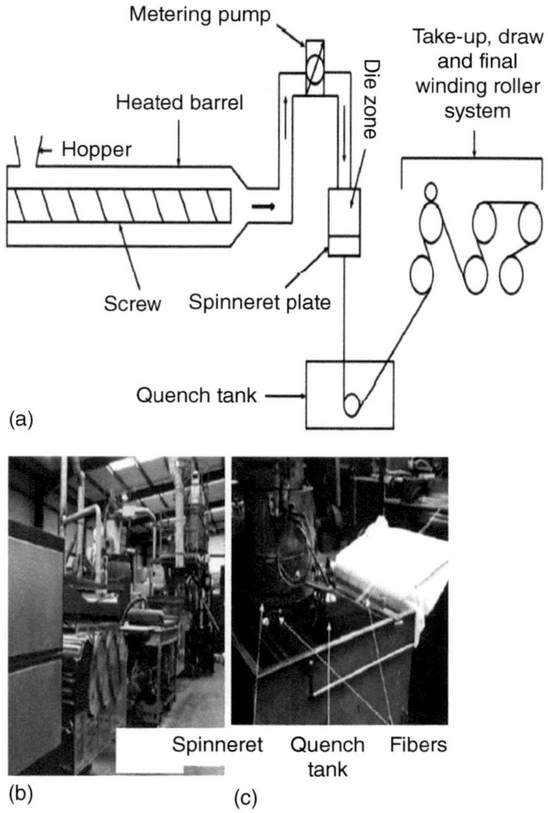

Figure 34.11 (a) Schematic of a general melt extruder, (b) the Davis-Standard thermatic extruder used in the large-scale extrusion, and (c) spinneret and quench tank of the Davis-Standard thermatic extruder. Source: Alderson et al. 2016 [23]. Reproduced with permission of John Wiley & Sons.

was Davis-Standard thermatic extruder by Shakespeare Monofilament UK Ltd., as shown in Figure 34.11b. The auxetic PP fibers produced in this process had good reproducibility, and it was reported that fibers exhibited auxetic behavior over large strain range as compared with the fibers produced in laboratory. The processing conditions to produce auxetic fibers were barrel temperature 200 °C, a screw speed of 12.5 rpm, and a take-up speed of either 1.5 or 3.5 rpm. It is important to mention that the extruded fibers also require ambient air quenching. It is important to consider that unlike laboratory scale production, the barrel temperature of 200 °C was identified in this work to produce auxetic fibers on large scale by using industrial-size extruders. The authors aimed to focus on the thermal characteristics of the polymer in the large-scale extruder and the resulting fiber microstructure, as future work. It is reported that the physical state of the polymer in the larger barrel during extrusion at 200 °C will be determined and will be related to the corresponding state in the smaller barrel laboratory extruder at the 159 °C conditions established for auxetic fibers and also the additional 180–190 °C conditions established for auxetic PP films. It will be necessary to establish the effects of dwell time and thermal equilibration on the nature of the polymer in situ (e.g. homogeneous melt or a partially melted, heterogeneous state). A postproduction assessment of fiber morphology and

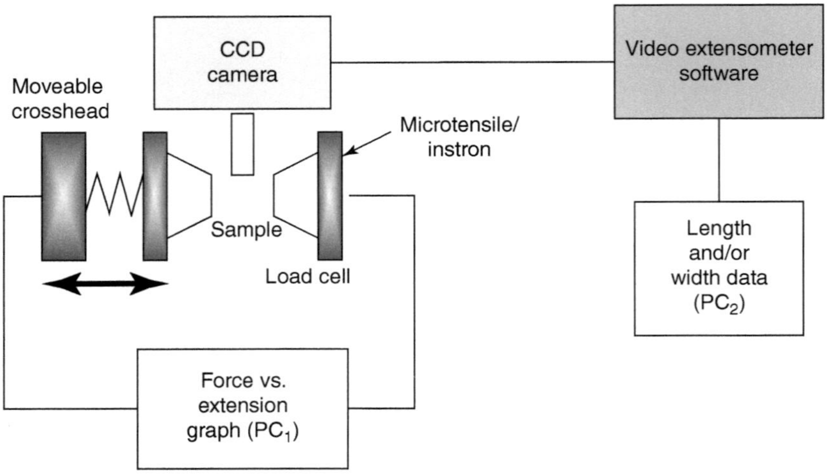

Figure 34.12 Schematic of the video extensometer setup. Source: Alderson et al. 2016 [23]. Reproduced with permission of John Wiley & Sons.

crystallinity will also be undertaken to determine the microstructural features and size scale responsible for the auxetic effect [23].

34.3.3.2 Testing of Auxetic Behavior of PP Fibers

Instron 4300 mechanical testing machine with a 100 N load cell, at a crosshead speed of 2 mm/min, was used to test the auxetic behavior of fibers. The PR was measured using a Messphysik ME46 video extensometer. This consists of a computer software package, which simultaneously measures length and width data from changes in the contrast between markers attached to the fibers along their lengths and the edges of the fibers, which gave the width measurement. The schematic setup of the equipment is shown in Figure 34.12. The software splits the length defined by the fiducial markers into 10 segments, the widths of which are then tracked throughout the test. Occasionally, individual segments can be tracked incorrectly due to features on the fiber surface, lighting issues, and such like. This is evident through the recorded segment width being significantly different to the actual fiber diameter (i.e. adjacent segment widths) or displaying markedly different trends to other segments. In such cases, the identified segment data were removed from the subsequent analysis [23].

Figure 34.13a shows the raw data for conventional PP fiber. This clearly shows as the length increases (bold line), the width decreases. This is conventional behavior and is found in all sections of the fiber. To get the PR from these data, an average true transverse strain vs. true axial strain plot is constructed, and the slope of this gives PR as shown in Figure 34.13b. Figure 34.14a shows the same analysis for auxetic fiber. It can be seen clearly that as the fiber is pulled, it increases in diameter, giving the negative PR by analysis as above as −0.82 as shown in Figure 34.14b. The auxetic effect is then shown for the first time in a fiber produced on a large-scale extruder and persists over the full 5% axial strain range covered in the test. This is an increase in the strain range for auxetic behavior

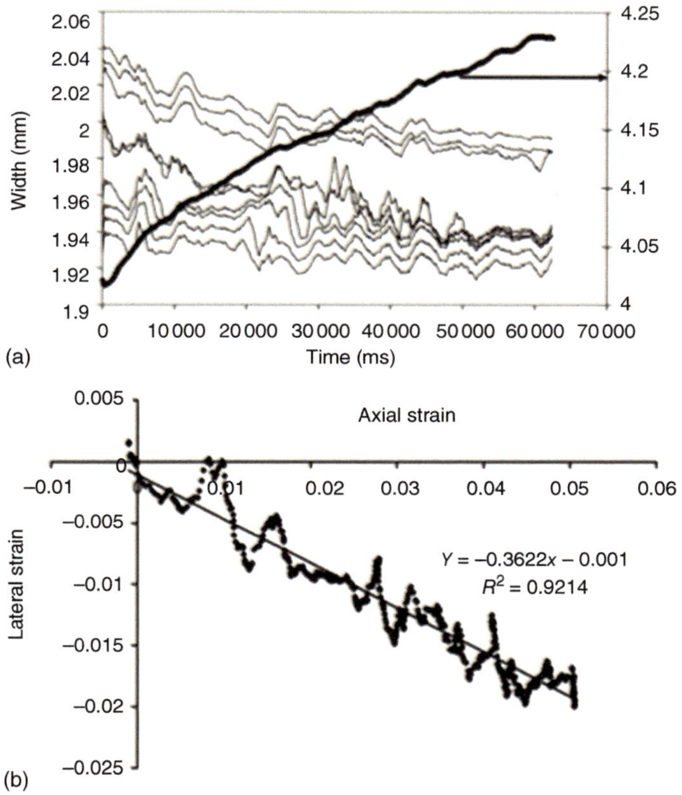

Figure 34.13 (a) Raw width and length data, showing conventional behavior; (b) plot of lateral strain against axial strain, showing Poisson's ratio, +0.36. Source: Alderson et al. 2016 [23]. Reproduced with permission of John Wiley & Sons.

found in melt extruded auxetic fibers, which persists typically up to 1–2% strain in fibers produced previously on a lab-scale extruder [23].

34.3.4 Auxetic Polyethylene

Alderson and Evans produced a microporous form of UHMWPE, capable of exhibiting large negative PR as low as −1.2 depending on the degree of anisotropy in the material. The manufacturing process consisted of three stages including compaction, sintering, and extrusion of UHMWPE fine powder. The function of the compaction is to induce structural integrity to the polymer. The set of compaction conditions including compaction pressure and temperature were described in order to produce auxetic polymer. The investigation revealed that the compaction pressure and temperature must be kept at 0.04 GPa and 110–125 °C, respectively, while holding at these conditions for 10–20 minutes. It was observed that the modulus of the resulting exudate is reduced if the temperature and compaction pressure are lower than the mentioned limits and if higher than the mentioned limits, the polymer particle are deformed. The

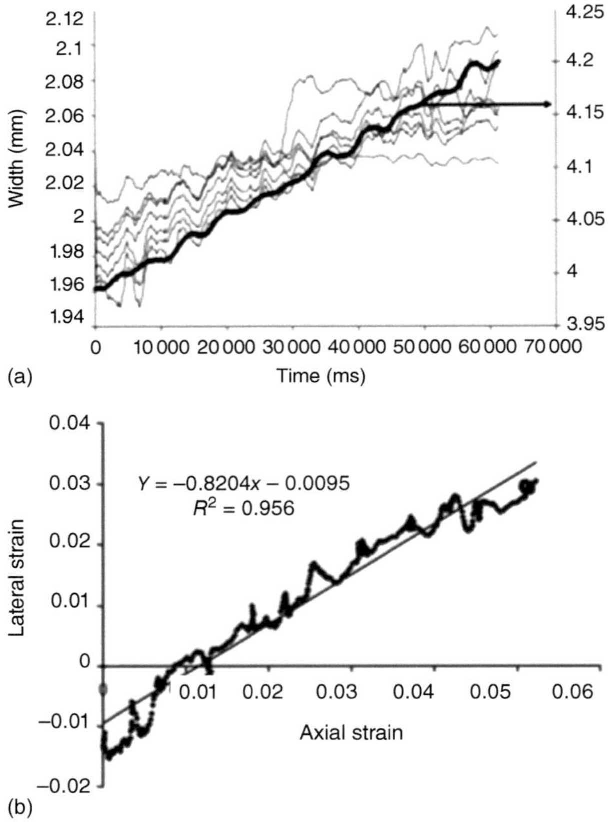

Figure 34.14 (a) Raw width and length data, showing auxetic behavior; (b) plot of lateral strain against axial strain showing Poisson's ratio, −0.82. Source: Alderson et al. 2016 [23]. Reproduced with permission of John Wiley & Sons.

compacted polymer is then reheated up to 160 °C and held at this temperature for 20 minutes. The polymer is then extruded with a speed of 500 mm/min. The geometry of die must be a cone die with a cone semi angle of 30° and a small die capillary length. The die entry diameter of 15 mm and exit diameter of 7–7.5 mm must also be used in order to achieve auxetic extrudate with high structural integrity. The output of these processing conditions is found to be an extrudate as shown in Figure 34.15, with strain-dependent negative PR [24–27].

34.3.5 Auxetic Polyester Fibers

A novel method to produce auxetic polyester fibers has been described. The polyester granules were grinded by using an in-house cryogenic grinding, and particles of less than 150 μm were collected to carry out the extrusion. The extrusion was carried out at higher temperature (230 °C) profile and gradually decreased the temperature to 210 °C until the viscosity of the powder bulk was too high to allow free flow through the die zone. An important consideration for the production of auxetic fibers is maintaining the minimum draw ratio and

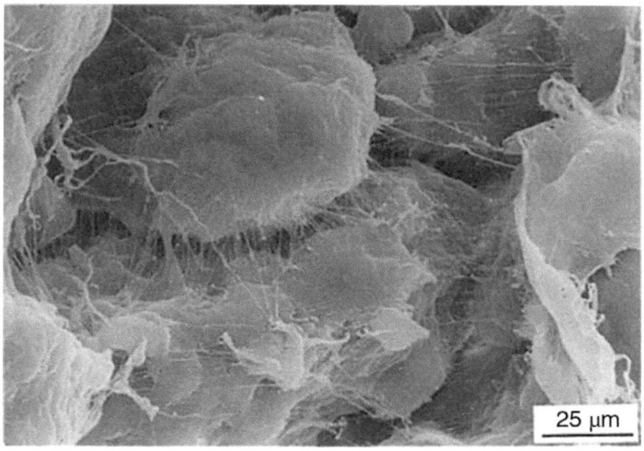

Figure 34.15 Scanning electron microscopy (SEM) images of auxetic microporous UHMWPE. Source: Alderson and Evans 1992 [24]. Reproduced with permission of Elsevier.

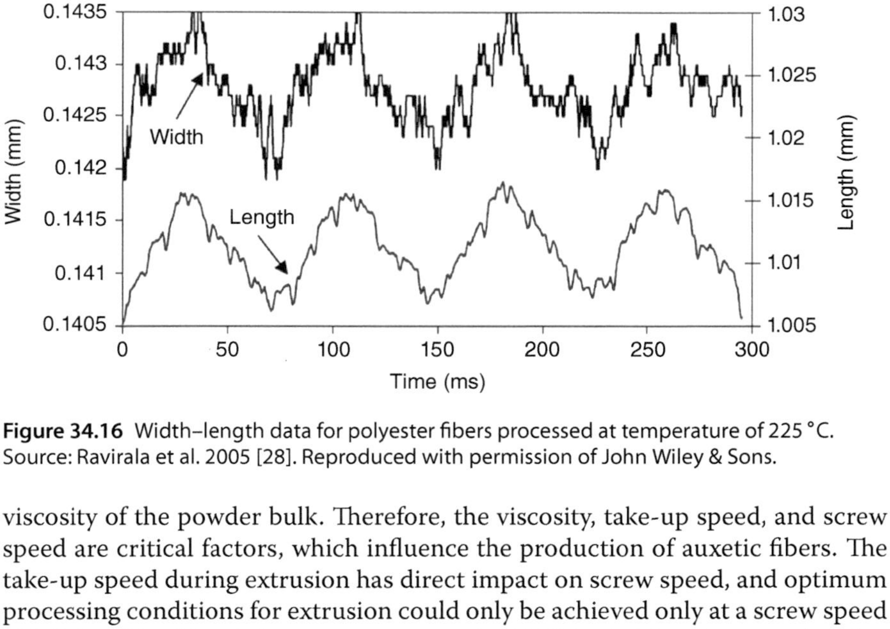

Figure 34.16 Width–length data for polyester fibers processed at temperature of 225 °C. Source: Ravirala et al. 2005 [28]. Reproduced with permission of John Wiley & Sons.

viscosity of the powder bulk. Therefore, the viscosity, take-up speed, and screw speed are critical factors, which influence the production of auxetic fibers. The take-up speed during extrusion has direct impact on screw speed, and optimum processing conditions for extrusion could only be achieved only at a screw speed of 0.525 rad/s with minimum take-up speed of 0.075 m/s. The fibers produced at 225 °C with screw speed 0.525 rad/s and take-up speed 0.075 m/s were found to exhibit auxetic behavior. They have an in-phase length–width data as shown in Figure 34.16. It can be observed that the width increased as the length increases in response to the applied force along the length of the fiber. Similarly, the width decreased as the length decreased on removal of the tensile load. Thus the fiber was confirmed to be auxetic [28, 29].

34.3.6 Auxetic Polyamide Fibers

The nylon powder was successfully used to produce auxetic polyamide fiber. One such development used the nylon powder supplied by nylon colors with 43 μm as average size of powdered particles. First of all, the nylon powder was oven

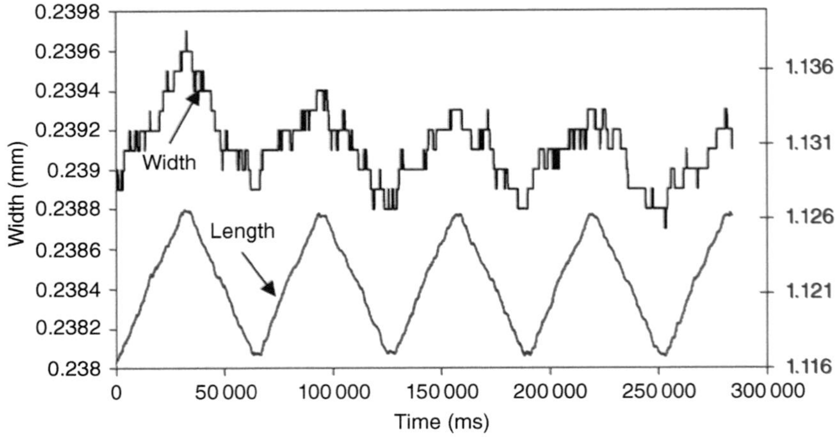

Figure 34.17 Width–length data for polyamide fibers. Source: Ravirala et al. 2005 [28]. Reproduced with permission of John Wiley & Sons.

dried for two days in low vacuum at 80 °C to avoid hydrolysis during extrusion. The powder was then extruded through the melt extruder maintaining constant temperature at 195 °C with screw speed 10 rpm (1.05 rad/s) and take-up speed 2 m/min (0.03 m/s). Figure 34.17 shows the length and width data of the obtained fibers at these processing conditions. It was reported that the length of the fiber increased when the load is applied with increase in width also. When the load was removed, the fiber length decrease combined with the decrease in the width as well. Since the fiber has shown these features in all the continuous cycles of load application and removal, this particular fiber was found to be auxetic [28].

34.3.7 Moisture-Sensitive Auxetic Fibers

The moisture-sensitive auxetic fiber has been invented, which responds not only to external force but also to moisture by using moisture-activated shrinking filament. The invented fiber is a combination of two components, one component is a moisture-sensitive shrinking filament with relatively high modulus of elasticity such as modified cellulosic fibers, e.g. cotton or rayon. The other component is an elastic material of lower modulus of elasticity, for example, siloxane. The fiber is straight in its dry state with no tensile load. When the fiber is in wet state, the moisture-sensitive shrinking component shrinks, and a pulling force is applied along to the elastic component causing it to deform and form helices, and pores are created. When the load (tensile or due to shrinking in wet state) is applied along the length of the auxetic fiber, the wrapped component gets straight, and the diameter of helices formed by the elastic components is increased especially in areas where pores are created. The elastic component of auxetic fibers undergoes opposite displacement in y-direction as compared with the displacement of the elastic component of the adjacent auxetic fiber. These opposite displacements cause the thickness of material to be increased leading to auxetic behavior. The important consideration in development of such yarn is that there must be enough difference between the modulus of elasticity of the materials for two components. Further, the auxetic moisture sensitive yarn can only be produced by arranging two components at different handedness. Helical auxetic yarns made of moisture-sensitive auxetic materials were suggested for functional garment [30]. More advanced work is required in order to increase the range of the materials that can be finished in auxetic fiber form together with the development of predictive models to understand the deformation mechanisms leading to auxetic behavior in fibers.

34.4 Properties and Applications of Auxetic Fibers

The unusual property of auxetic materials is that they are relatively resistant to denting. When an auxetic material is impacted, the compression caused by the impact results in the material compressing toward the point of the impact, thus becoming much denser and resisting the force. When a composite material fractures, the fibers are pulled out of the matrix. This results in the reinforcing effect being lost as the bond between the fiber and the matrix fails. It has been suggested that an auxetic fiber within a composite would resist fiber pullout. When the fiber is pulled, it will expand and effectively lock into the matrix rather than contracting and pulling out easily as a conventional fiber would do as shown in Figure 34.18 [31]. Therefore, the fracture toughness of a composite can be enhanced by using auxetic fiber reinforcements. Auxetic materials are also more resistant to fracture; they expand laterally as a force is applied, and this restricts the growth of any potential cracks in the material.

With the invention of auxetic PP fibers [22], the concept of using embedded auxetic fibers in a softened epoxy resin was employed to produce a single fiber composite [32]. The fiber pullout resistance and energy absorption were tested. In comparative tests with specimens containing positive PR fibers, the auxetic

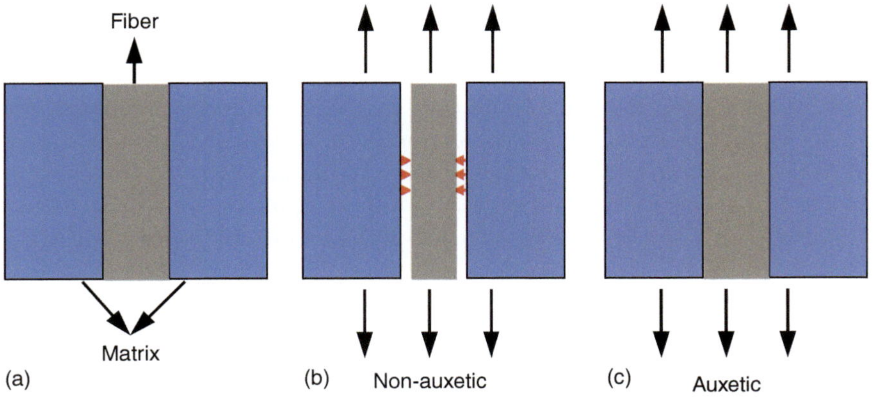

Figure 34.18 Fiber reinforced composites. (a) Fiber Matrix interface; (b) contraction in conventional fibers during stretching; (c) expansion in auxetic fibers during stretching. Source: Wang et al. 2016 [1]. Reproduced with permission of Elsevier.

fiber-locking mechanism is shown to enable the specimen to carry more than twice the maximum load; and in terms of energy absorption, the auxetic fiber is up to three times more difficult to extract than the equivalent positive PR fiber. It has been reported that with auxetic fibers, pullout is resisted because the fibers expand perpendicular to the pullout force. This could help to resist potential failure mechanisms in the composite, such as crack growth. In another study it was reported that in the case of auxetic fiber, the negative PR will induce radial expansion and the tip of the de-bonded fiber during pulling out, to close up, and the stress concentration at that region will be reduced. The radial expansion of the de-bonded fiber increases the contact between the unbonded fiber and matrix leading to an enhanced frictional contact, and more force is required to pull the fiber out [33, 34].

Another potential application for auxetic fibers is in drug delivery systems. The microporous auxetic fibers can be employed to fabricate a wound healing bandage. The drug agent can be entrapped into the microspores of the fiber, and during wound swelling the bandage will receive a stress, and the microspores of auxetic fibers will be opened and release the drug to the infected region. After wound healing the stress on the bandage will be reduced, and drug release will be stopped [35]. The auxetic composites may also find application in the aircraft.

The aircrafts undertake very high shear force when flying. The materials with higher shear strength can optimize the performance of the aircraft. The auxetic composites can have higher shear modulus, and another property worthy to mention, among others, is they have better formability, enabling them easier to make the shape needed. Therefore, the auxetic composites can be good candidate instead of the conventional composites to be used on the aircraft in the future [1]. Nevertheless, other mechanical properties like strength and very high modulus are essential in order to replace conventional fibers with auxetic fibers in these and other applications.

34.5 Conclusions

This chapter mainly discussed the fibers for auxetic applications. The auxetic fibers are very good materials to be used in fiber-reinforced composites; the advantages they possess make them very useful in designing of smart composites. For example, enhanced vibration damping, higher shear modulus, higher energy absorbance, and fracture toughness make them an ideal candidate to be used to fabricate composites for high-profile application areas including aerospace. Although this new breed of composites is not yet mature enough, it is believed that they will be developed and good enough to be used in aerospace engineering instead of conventional composites. However, the disadvantage just like with the auxetic materials is that they are too porous and are not stiff enough to bear the load. Therefore, the auxetic materials are suitable for certain specific applications. Moreover, there are only a few auxetic fibers developed including UHMWPE, PE, polyester, nylon, and PP fibers. Among these fibers only PP fibers are produced on large scale, and all others were produced on laboratory scale.

Therefore, continuous processes to produce auxetic fibers on large scale must be developed. Another important feature that auxetic fibers must have is that they should have the possibility of blending with other materials and spinning them into yarns. In this way the spectrum of fibers for auxetic applications could be broadened. More advanced work is required in order to increase the range of the materials that can be finished in auxetic fiber form together with the development of predictive models to understand the deformation mechanisms leading to auxetic behavior in fibers.

References

1 Wang, Z., Zulifqar, A., and Hu, H. (2016). Auxetic composites in aerospace engineering. In: *Advanced Composite Materials for Aerospace Engineering: Processing, Properties and Applications* (eds, S. Rana and R. Fangueiro), p. 213. Woodhead Publishing.

2 Steffens, F., Rana, S., and Fangueiro, R. (2016). Development of novel auxetic textile structures using high performance fibres. *Materials and Design* 106: 81–89.

3 Jiang, L., Gu, B., and Hu, H. (2016). Auxetic composite made with multilayer orthogonal structural reinforcement. *Composite Structures* 135: 23–29.

4 Bezazi, A. and Scarpa, F. (2007). Mechanical behaviour of conventional and negative Poisson's ratio thermoplastic polyurethane foams under compressive cyclic loading. *International Journal of Fatigue* 29: 922–930.

5 Scarpa, F. and Smith, F. (2004). Passive and MR fluid-coated auxetic PU foam–mechanical, acoustic, and electromagnetic properties. *Journal of Intelligent Material Systems and Structures* 15: 973–979.

6 Chan, N. and Evans, K. (1998). Indentation resilience of conventional and auxetic foams. *Journal of Cellular Plastics* 34: 231–260.

7 Wang, Z. and Hu, H. (2016). Tensile and forming properties of auxetic warp-knitted spacer fabrics. *Textile Research Journal* https://doi.org/10.1177/0040517516660889.

8 Alderson, K., Pickles, A., Neale, P., and Evans, K. (1994). Auxetic polyethylene: the effect of a negative Poisson's ratio on hardness. *Acta Metallurgica et Materialia* 42: 2261–2266.

9 Baughman, R.H., Shacklette, J.M., Zakhidov, A.A., and Stafström, S. (1998). Negative Poisson's ratios as a common feature of cubic metals. *Nature* 392: 362–365.

10 Caddock, B. and Evans, K. (1989). Microporous materials with negative Poisson's ratios. I. Microstructure and mechanical properties. *Journal of Physics D: Applied Physics* 22: 1877.

11 Evans, K., Nkansah, M., and Hutchinson, I. (1994). Auxetic foams: modelling negative Poisson's ratios. *Acta Metallurgica et Materialia* 42: 1289–1294.

12 Lim, T.-C. (2009). Out-of-plane modulus of semi-auxetic laminates. *European Journal of Mechanics - A/Solids* 28: 752–756.

13 Yeganeh-Haeri, A., Weidner, D.J., and Parise, J.B. (1992). Elasticity of a-cristobalite: a silicon dioxide with a negative Poisson's ratio. *Science* 257: 650–652.
14 Lakes, R. (1987). Foam structures with a negative Poisson's ratio. *Science* 235: 1038–1040.
15 Hu, H., Wang, Z., and Liu, S. (2011). Development of auxetic fabrics using flat knitting technology. *Textile Research Journal* 81: 1493–1502.
16 Grima, J.N. and Evans, K.E. (2006). Auxetic behavior from rotating triangles. *Journal of Materials Science* 41: 3193–3196.
17 Liu, Y. and Hu, H. (2010). A review on auxetic structures and polymeric materials. *Scientific Research and Essays* 5: 1052–1063.
18 Liu, Y., Hu, H., Lam, J.K., and Liu, S. (2010). Negative Poisson's ratio weft-knitted fabrics. *Textile Research Journal* 80 (9): 856–863.
19 Zulifqar, A., Hua, T., and Hu, H. (2017). Development of uni-stretch woven fabrics with zero and negative Poisson's ratio. *Textile Research Journal* https://doi.org/10.1177/0040517517715095.
20 Liu, P., He, C., and Griffin, A. (1998). Liquid crystalline polymers as potential auxetic materials: Influence of transverse rods on the polymer mesophase. In: *Abstracts of Papers of the American Chemical Society*, U108.
21 Alderson, A. and Alderson, K. (2005). Expanding materials and applications: exploiting auxetic textiles. *Technical Textiles International* 777: 29–34.
22 Alderson, K., Alderson, A., Smart, G. et al. (2002). Auxetic polypropylene fibres: Part 1-Manufacture and characterisation. *Plastics, Rubber and Composites* 31: 344–349.
23 Alderson, K., Nazaré, S., and Alderson, A. (2016). Large-scale extrusion of auxetic polypropylene fibre. *Physica Status Solidi B* 253: 1279–1287.
24 Alderson, K. and Evans, K. (1992). The fabrication of microporous polyethylene having a negative Poisson's ratio. *Polymer* 33: 4435–4438.
25 Pickles, A., Webber, R., Alderson, K. et al. (1995). The effect of the processing parameters on the fabrication of auxetic polyethylene. *Journal of Materials Science* 30: 4059–4068.
26 Alderson, K., Kettle, A., Neale, P. et al. (1995). The effect of the processing parameters on the fabrication of auxetic polyethylene. *Journal of Materials Science* 30: 4069–4075.
27 Neale, P., Pickles, A., Alderson, K., and Evans, K. (1995). The effect of the processing parameters on the fabrication of auxetic polyethylene. *Journal of Materials Science* 30: 4087–4094.
28 Ravirala, N., Alderson, A., Alderson, K., and Davies, P. (2005). Expanding the range of auxetic polymeric products using a novel melt-spinning route. *Physica Status Solidi B* 242: 653–664.
29 Ravirala, N., Alderson, K.L., Davies, P.J. et al. (2006). Negative Poisson's ratio polyester fibers. *Textile Research Journal* 76: 540–546.
30 Lee, W., Lee, S., Koh, C., and Heo, J. (2010). Moisture sensitive auxetic material. Google Patents.
31 Evans, K.E. and Alderson, A. (2000). Auxetic materials: functional materials and structures from lateral thinking! *Advanced Materials* 12: 617–628.

32 Alderson, K., Simkins, V., Coenen, V. et al. (2005). How to make auxetic fibre reinforced composites. *Physica Status Solidi B* 242: 509–518.
33 Donoghue, J., Alderson, K., and Evans, K. (2009). The fracture toughness of composite laminates with a negative Poisson's ratio. *Physica Status Solidi B* 246: 2011–2017.
34 Simkins, V., Alderson, A., Davies, P., and Alderson, K. (2005). Single fibre pullout tests on auxetic polymeric fibres. *Journal of Materials Science* 40: 4355–4364.
35 Alderson, A., Rasburn, J., Ameer-Beg, S. et al. (2000). An auxetic filter: a tuneable filter displaying enhanced size selectivity or defouling properties. *Industrial and Engineering Chemistry Research* 39: 654–665.

Index

Handbook of Fibrous Materials, First Edition. Edited by Jinlian Hu, Bipin Kumar, and Jing Lu.

d

g

m

n

t